D1229627

Power Electronics Handbook
Components, circuits and applications

F. F. Mazda, DFH, MPhil, DMS, MBIM, CEng, FIEE

Butterworths
London Boston Singapore Sydney Toronto Wellington

PART OF REED INTERNATIONAL P.L.C.

First published 1990

British Library Cataloguing in Publication Data

Mazda, F. F. (Fraidon Framroz) *1939–*
Power electronics handbook.
1. Semiconductor circuits
I. Title
621.38152

ISBN 0-408-03004-6

Library of Congress Cataloging-in-Publication Data

Mazda, F. F.
Power electronics handbook/F. F. Mazda.
p. cm.
Includes bibliographical references.
ISBN 0-408-03004-6
1. Power electronics. I. Title.
TK7881.15.M39 1990
621.31'7—dc20 89-71166
CIP

Composition by Genesis Typesetting, Borough Green, Sevenoaks, Kent
Printed in Great Britain at the University Press, Cambridge

Preface

Fifty years have passed since transistor action was first demonstrated and thirty years have elapsed since the invention of the thyristor, recognised as the workhorse of power electronics. Power electronics is therefore not a new discipline, having moved from the realms of a glamorous invention in search of an application, to that in which it controls vital operations within industry and the home.

Writing a book in an established field is not easy. Other authors have been there before and the newcomer must ensure that he adds to the published material, rather than produces yet another book to gather dust on the shelves of technical bookshops. With this in mind I set out to achieve the following: (i) Collect within a single volume all the material relating to power components, circuit design, and applications; information which would be required by the engineer working in the field. (ii) Provide material in a practical form, with theoretical information presented as formulae, rather than clutter the book with pages and pages of mathematics showing how these formulae have been derived. For the student who needs this extra mathematical treatment reference is made to other books in the bibliography. How far I have been successful in these two aims the reader will judge!

The first chapter briefly describes the processes used in the manufacture of power semiconductors, and the construction and characteristics of the power semiconductors currently available. Chapters 2 to 5 cover devices used in conjunction with power semiconductors. Chapter 2 introduces the methods and components for removing heat generated within the power semiconductor, a vital requirement for effective operation. Chapter 3 describes the low power devices used to control the operation of power semiconductors. Chapter 4 introduces the techniques and components necessary to minimise radio frequency interference generated by power electronic circuits: with the tightening of world legislation, this is clearly an important consideration. In Chapter 5 the protection components and circuits required to protect power components from malfunctions, such as overcurrent or overvoltage, are described.

Chapters 6 to 13 provide detailed information on the arrangements and design of the various types of power semiconductor circuits. Because there are a large number of types of such circuits, this whole field is first surveyed in Chapter 6, to give the student a unified picture.

Chapter 7 describes the most basic type of power circuit, that used for simple static switching. This is extended in Chapter 8 to controlling the value of the a.c. line voltage. Chapter 9 extends the voltage control concept to include rectification, so that the a.c. is controlled to give a variable d.c. voltage.

Although similar in concept to a.c. line control and rectification, the next chapter describes a completely different application, the use of power electronics to vary the frequency of an a.c. supply without first going through d.c.

Chapters 11, 12 and 13 are related, since in these the power semiconductor switches operate from a d.c. supply and need to be forced commutated. Chapter 11 classifies the various commutation systems which exist, and Chapters 12 and 13 describe two prime application areas, those of voltage control and of frequency changing.

Finally Chapter 13 describes some of the most common applications of the power semiconductor circuits introduced in Chapters 6 to 13. The area of electrical machine control is covered in some depth, including their operation, since it represents one of the most important and widespread uses of power semiconductors today.

I started my career as a power electronics engineer at just about the time that the first thyristors were becoming commercially available. From then on I have continued to learn daily, as new developments have been made and new material published. I am grateful to the many authors who have enriched the technical press with their writings and so made this possible. To them, and to the many other power electronic engineers, who have worked to extend our knowledge in this valuable area, this book is gratefully dedicated.

F. F. Mazda
Bishop's Stortford
March 1990

Contents

Part 1

Components

Chapter 1

Power semiconductor devices

1.1 Introduction

1.1.1 Historical development

The field of power electronics is not new. The post- and pre-war periods, from about 1930 to 1950, saw extensive application of power electronics, based primarily on the mercury arc rectifier and the gas-filled tube.

It was in December 1939 that William Shockley first noted the principle of a semiconductor which could be used for controlling electrical power. However, it was not until 23 December 1947, the official date for the invention of the transistor, that a simple point contact transistor was demonstrated by William Shockley, John Bardeen and Walter Brattain, to the executives of Bell Laboratories. It was at this point that semiconductor electronic technology was born.

Two other dates are important when tracking the development of power semiconductors. The integrated circuit was invented by Jack Kelby of Texas Instruments in 1958. Integrated circuits, especially microprocessors, are now used extensively to control power semiconductor devices. Finally, the thyristor, the workhorse of the power semiconductor field, was announced by General Electric in 1957. It was originally called the silicon-controlled rectifier (SCR), to differentiate it from the common silicon rectifier, and it was some time later that the name was changed to thyristor.

1.1.2 Applications

Power semiconductors are used in wide-ranging applications. The following gives only a representative sample.

(i) Industrial applications consist primarily of two areas, motor control and power supplies. The motors which are controlled vary from the very large, as used in steel mills, to the relatively smaller ones, such as in machine tools. Power supplies too come in many shapes and sizes, such as for battery charging, induction heating, electroplating and welding.

(ii) Consumer applications cover many different areas in the home, such as audio amplifiers; heat controls; light dimmers; motor control for food mixers and hand power tools; and security systems.

(iii) Transportation applications, the largest being motor drives for areas such as electric vehicles, locomotives, and fork-lift trucks. Equally important are non-motor drive applications, such as traffic signal control, vehicle electronic ignition and vehicle voltage regulation.
(iv) Aerospace and defence applications include VLF transmitters; power supplies for space, and aircraft; and switching using solid state relays and contactors.

1.1.3 Power semiconductor operation

This chapter describes the construction and characteristics of several types of power semiconductor devices. These can generally be operated in different modes, due either to their construction or the application in which they are used. Four operating modes are considered in this chapter:

(i) One way of differentiating devices is whether they are capable of being controlled, regarding their turn-on point. The power rectifier cannot be so controlled, since it will conduct as soon as the voltage at its anode is more positive than that at its cathode. All other power devices described in this chapter, such as the transistor, thyristor, gate turn-off switch and triac, can be turned on (and sometimes off) by a control signal on an auxiliary input.
(ii) Some power devices can also be operated in a linear or a switching mode. The transistor is the only component described here which is capable of linear operation, so it is the obvious choice for this application. Losses, caused by the product of current through the device and voltage drop across it, are much higher when in the linear mode. Switching devices can handle greater power, since their dissipation is lower. Their power gain is also generally higher, so that they need less drive current to control their operation.
(iii) The third operating mode is the type of signal required to control the power semiconductor devices. Generally, this consists of an electrical signal, although in a large class of devices optical energy is used.
(iv) Finally, the voltage and current capability of power devices can be considerably increased by operating several of them in series or in parallel mode, so that the total voltage and current are shared across several devices.

1.1.4 Device characteristics

Many of the power semiconductor devices described here have special characteristics. There are, however, also many similarities, such as:

(i) The voltage drop across the device when it is carrying current.
(ii) The capability of the device to handle current. Both the steady state current and the peak or overload current-carrying ability are important, since overload capabilities often determine the need for protection.
(iii) The capability of the device to block voltage, both in the direction reverse to that in which it normally conducts, and in the conduction direction, when it has not yet been turned on.

(iv) The maximum rate of rise of current which the device can withstand, without being destroyed due to localised heating, and the maximum rate of rise of voltage in the forward direction which it can withstand, without turning on prematurely.
(v) The switching speed of the device, which influences its switching losses and the maximum frequency at which it can be operated.
(vi) The rate at which a conducting device can recover its blocking capability. This again influences its maximum operating frequency.
(vii) The maximum power dissipation which the device can withstand. This characteristic is often linked to the maximum junction temperature at which the device can operate and its thermal transfer characteristic, that is, its ability to transfer heat to a heatsink.
(viii) The power gain of the device, which is the ratio of the controlled power to the power needed in the control terminal. The higher this gain, the lower the power dissipation in the control electronics of the power semiconductor.

1.2 Fabrication process

The manufacturing processes for power semiconductors closely resemble those used for other semiconductor devices. These consist of the following, which are described further in this section:

(i) Preparation of a wafer of very pure silicon, which forms the base to support the power semiconductor.
(ii) Oxide growth over selected areas of the semiconductor surface, to protect the layers below it from contamination and to form a mask for subsequent processing steps.
(iii) Growth of an epitaxial layer onto the silicon wafer, which forms a controlled layer into which the various parts of the semiconductor device can be formed.
(iv) Photolithography, used to control the areas where the *p* and *n* components are formed.
(v) Diffusion, which is the most common method used to form the *p* and *n* components of the semiconductor device.
(vi) Ion implantation, which is able to produce *p* and *n* areas to a high precision.
(vii) Metal formation, which is used to interconnect the various *p* and *n* parts of the semiconductor devices together, and to provide a base for connection of the silicon to the package of the device.

1.2.1 Crystal preparation

In order to obtain high-quality semiconductor components it is important to start with a semiconductor material which has a very low level of defects in its crystal structure. Several techniques exist for growing bulk semiconductor crystals, some of these being illustrated in Figure 1.1.

In the zone-levelling method a crucible, made from silica, holds the impure silicon, and it is slowly moved along a quartz tube, containing an inert gas. Zoned heating coils are placed along the length of the tube,

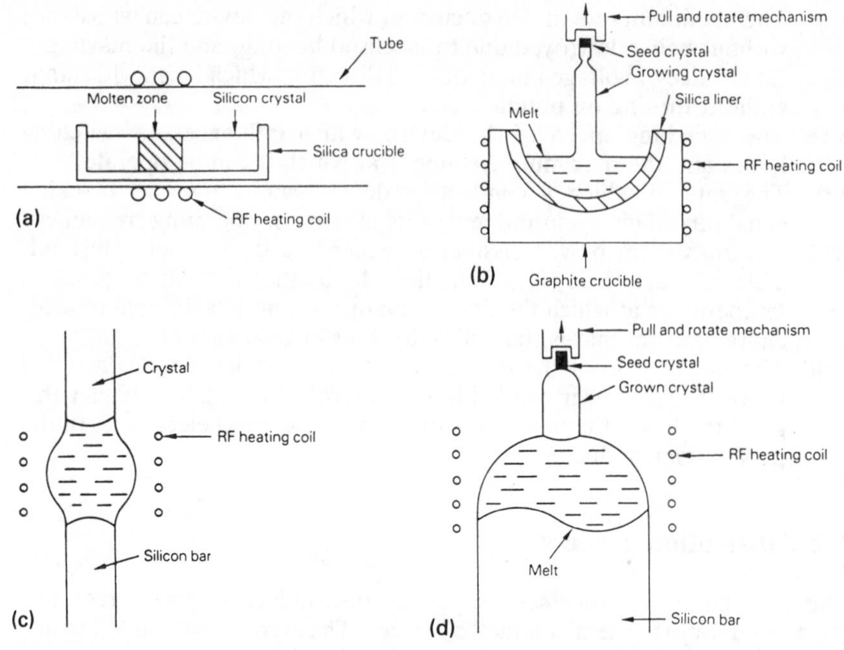

Figure 1.1 Semiconductor bulk crystal purification and growth systems: (a) zone levelling; (b) Czochralski growing; (c) floating zone refining; (d) pedestal pulling

which cause an area in the silicon to melt. This molten area is moved down the silicon, to one end of the crystal, as the silica crucible moves along the tube. Impurities in the silicon are carried by the molten area to the end region, so that after several passes this end can be cut off, leaving a relatively pure section of silicon behind.

Although the zone-levelling method is simple and cheap to operate, it gives a relatively high level of residual impurities and poor crystal structure, since the silicon cools in contact with the silica crucible. The Czochralski growth method overcomes these problems by avoiding contact between the crystal and its crucible. In this system a tiny seed of silicon crystal is lowered into a bath of molten silicon, contained in a graphite crucible. The seed is rotated and slowly withdrawn from the molten bath. This causes the silicon from the melt to settle on the seed crystal and to cool, resulting in the formation of a bar of pure silicon.

The floating zone-refining method uses an r.f. heating coil, which is slowly moved along the length of a silicon bar. This results in a molten layer of silicon moving along the bar, carrying impurities with it, which can eventually be cut off from the end of the silicon bar. The method is therefore similar to the zone-levelling technique of Figure 1.1(a), except that the risk of contamination is considerably reduced, since the silicon is not in contact with a crucible.

The pedestal-pulling technique is similar in principle to the Czochralski method except that the melt is formed in the surface of the impure silicon bar itself, so that risk of contamination from the use of a crucible is once again avoided. Since in both the floating-zone and pedestal-pulling methods the molten silicon is held in place by surface tension, the size of the melt, and therefore of the silicon bars (and wafers) produced by these methods, cannot be very large. These techniques are used for making relatively low-power components.

During preparation of pure silicon, impurities of *p* or *n* type can be added to the melt, to give the final silicon ingots the required resistivity. The ingots are then cut into slices, using a diamond-impregnated saw. The saw cuts are made along the <111> or <100> planes of the crystal. These cuts usually damage the crystal lattice near the silicon surface, resulting in poor resistivity and minority carrier lifetime. The damaged area, which is about 20 μm deep, is removed by etching in a mixture of hydrofluoric and nitric acids, and the surface is then polished to give a strain-free, highly flat region.

1.2.2 Oxide growth

Silicon oxide, also called silica or SiO_2, may be grown and removed several times from the surface of the silicon slice during manufacture of the semiconductor device. The oxide layer is used for diffusion masking, for sealing and passivating the silicon surface, and for insulating the metal interconnections from the silicon. Although the oxide layer may be deposited onto the slice, as done for the epitaxy layer described in the next section, it is more usual to grow it using dry or wet oxygen or steam.

Figure 1.2(a) shows a typical arrangement of the apparatus used for oxidation (and diffusion). The silicon slices are stacked upright in a quartz boat and inserted into a quartz tube. The tube is heated to between 1000°C and 1200°C by zoned heaters, so that the boat is located in an area having a uniform temperature along a length of the tube. Nitrogen, dry oxygen, wet oxygen (which is oxygen bubbled into water at 95°C) or steam can be passed over the slices to grow the oxide layer. A thickness of about 1 μm takes about 4 h to grow and consumes between 0.4 μm and 0.5 μm of the silicon. The colour of the silicon surface changes with the thickness of the oxide layer, due to the shift in the wavelength of the reflected light. This effect is used as an indication of the layer thickness.

1.2.3 Epitaxy growth

Epitaxy means growing a single-crystal silicon structure on the original slice, such that the new structure is essentially a molecular extension of the original silicon. Epitaxy layers can be closely controlled regarding size and resistivity, to about ±10%. This compares favourably with the ±30% resistivity control obtained when pulling from the silicon melt.

Epitaxy apparatus is similar to the oxide growth arrangement shown in Figure 1.2(a). However, r.f. heating coils are normally used and the silicon slices are placed in a graphite boat, which may be coated with quartz to prevent the graphite contaminating the silicon. The bubbler usually contains silicon tetrachloride ($SiCl_4$) to which may be added a controlled amount of an impurity, such as PCl_3. Hydrogen gas is bubbled through this mixture before entering the quartz epitaxy tube.

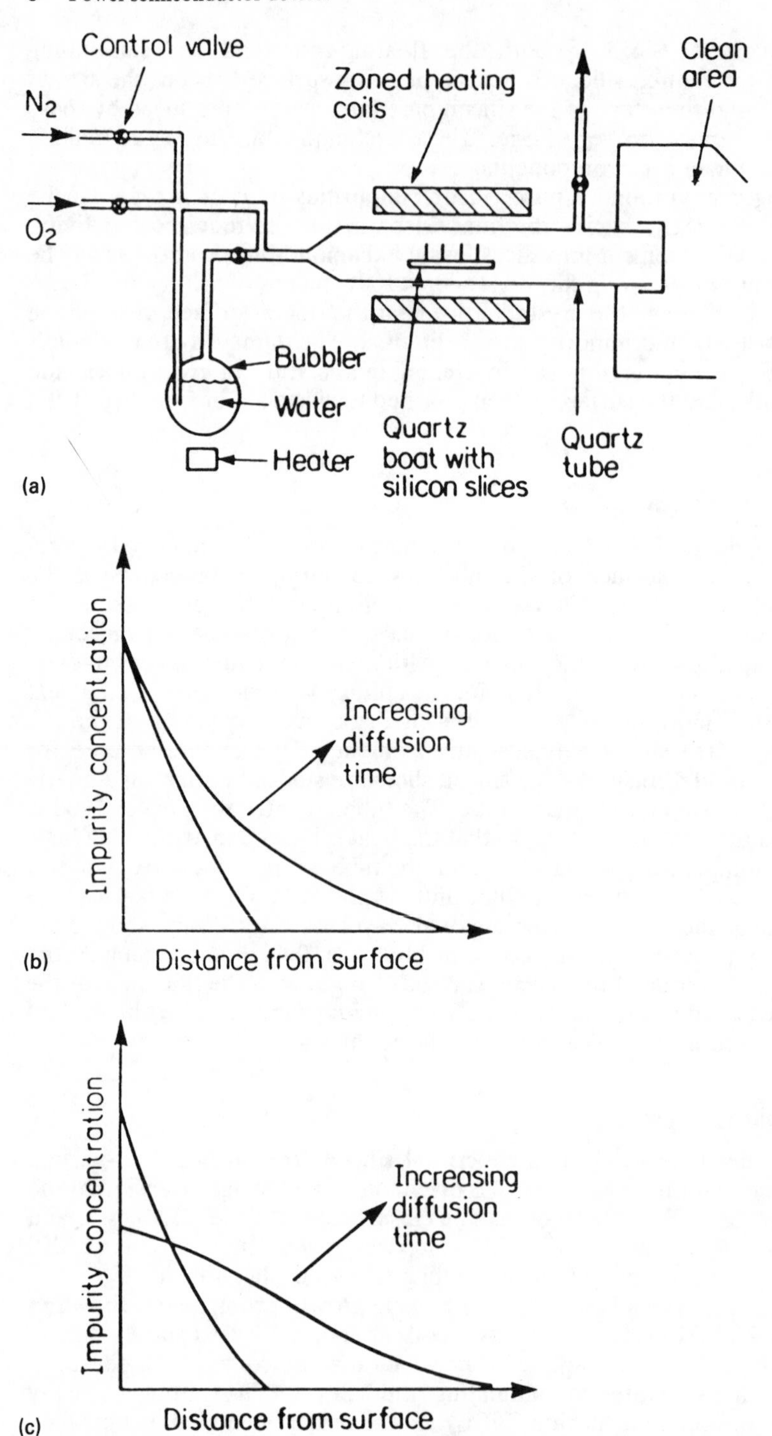

Figure 1.2 Semiconductor diffusion: (a) open-tube arrangement; (b) impurity concentration using error function; (c) Gaussian impurity concentration

Initially the slices are heated to about 1200°C and pure hydrogen and hydrochloric acid vapour are passed over them to etch away any oxide or impurities which may exist on the surface of the silicon. HCl vapour is then turned off and hydrogen bubbled through the $SiCl_4$, the vapour passing over the silicon slices. When this reaches the hot silicon it dissociates and silicon atoms are deposited on the slice, where they rapidly establish themselves as part of the original crystal structure. It is essential to saturate the tube with $SiCl_4$ vapour, to ensure a uniform layer thickness over the whole slice.

The epitaxy layer may be doped with *p*- or *n*-type impurities by introducing these, in the required concentration, into the vapour stream.

1.2.4 Photolithography

The main use of photolithography, in the manufacture of semiconductor components, is to selectively remove areas of oxide or metal from the surface of the silicon slice. To do this the surface is first covered with a thin uniform layer of liquid photoresist. This is best obtained by holding the silicon slice in a vacuum chuck and placing a fixed amount of photoresist onto its centre. The slice is then rotated at very high speeds, to distribute the resist, the excess flying off due to centrifugal forces. The amount left on the slice is clearly a function of the oxide properties and the viscosity of the resist. The slice is then heated for a few minutes in an oven to harden the resist.

The mask, containing the pattern of the area which is to be selectively removed, is then placed in contact with the surface. This may consist of a glass or chrome mask, or a projection of the image onto the surface of the silicon. Once in place, the surface is exposed to ultraviolet light. If negative resist material is used the exposed areas become hardened by the light. The slice is now covered with developer, which dissolves the unexposed resist areas. The slice is then placed in a bath of etchant, such as hydrofluoric acid, which dissolves the exposed areas of the surface, but not the silicon. For applications which require most of the covering surface material to be removed it is more convenient to use a positive resist material, such that the areas exposed to ultraviolet light are dissolved in the developer.

1.2.5 Diffusion

In epitaxy a large area is doped by a closely controlled amount of impurity. Diffusion, on the other hand, enables selective areas to be doped. These areas represent those which are not covered by an oxide layer, so that the photolithographic stage is normally followed by diffusion.

The diffusion furnace resembles the arrangement shown in Figure 1.2. The bubbler contains the impurity and nitrogen is passed through it, on its way to the boat containing the silicon slices. The furnace temperature is kept close to the melting point of silicon, i.e. 1200°C, and at this value the silicon atoms are highly mobile. Impurity atoms readily move through the silicon lattice by substitution, going from a region of high concentration to one of lower density.

Diffusion can be carried out by one of two techniques, as shown in Figure 1.2. In the error function or one-step process the concentration of impurities is kept fixed throughout the diffusion period, giving the curves shown. In Gaussian or two-step diffusion a fixed amount of impurity is present, so that as this moves deeper into the silicon bulk the surface concentration decreases. This gives a flatter dopant distribution of higher resistivity. Generally, diffusion takes place in a slightly oxidising atmosphere. This results in the formation of a glassy layer of impurity on the silicon surface, as well as a slight penetration into the silicon bulk.

The silicon slice can be removed to a second furnace and heated in an inert atmosphere, when the dopants diffuse out of the glassy layer to give an error function distribution. The glassy layer now not only forms a diffusion source but also protects the silicon surface from evaporation, and acts as a getter for impurities from the silicon bulk. For Gaussian diffusion the glassy layer is etched off using hydrofluoric acid, prior to the slice being heated in an inert atmosphere. The critical dopants just below the surface now diffuse into the silicon bulk.

Apart from the open-tube arrangement, shown in Figure 1.2(a), it is possible to diffuse slices by putting them in a sealed quartz container with doped silicon powder and then heating the combination in a quartz furnace. The advantage of this method is that many slices can be diffused simultaneously, giving a larger throughput. However, the quartz container must be broken to remove the slices after diffusion, so the process can prove expensive.

The most commonly used *p*- and *n*-type impurities are boron and phosphorus respectively. Both reach maximum solubility at about 1200°C and have a high diffusion constant. Arsenic, on the other hand, is an *n*-type impurity, which diffuses very slowly. It is used for making the buried layer in transistors, since this must not diffuse appreciably during subsequent high-temperature processes.

Gold is often introduced as an impurity into semiconductor devices. This is a lifetime killer and enables fast-switching components to be built. The gold atom is much smaller than a silicon atom. This means that it does not move through the silicon lattice by substitution, as other impurities do, but very rapidly in between the silicon atoms, that is, intrinsically.

1.2.6 Ion implantation

Ion implantation is a technique for precisely determining the concentration and location of impurities within a silicon slice. In this process selected ions of the required impurity are accelerated into the slice of silicon, where they plough their way through the crystal structure to the required depth. Because of this the material is distorted after ion implantation and it usually needs to go through an annealing stage, in which the atoms are allowed to drift into their places within the lattice.

There are several basic requirements which must be met in any implantation system:

(i) The impurity concentration must be uniform over a given slice, and the process must be accurately reproducible over repeated slices.
(ii) The system must have a high throughput.

(iii) The purity of the dopant must be accurately controlled. Most ion sources produce a range of dopants in addition to the one required. The impurities must be completely removed from the ion stream before it reaches the silicon slice.
(iv) The energy imparted to the ions by the accelerating voltage must be high, to enable them to penetrate the maximum distance likely to be required.

Figure 1.3 shows a simplified arrangement of an ion-implantation system. The ion source produces an abundance of the required dopant.

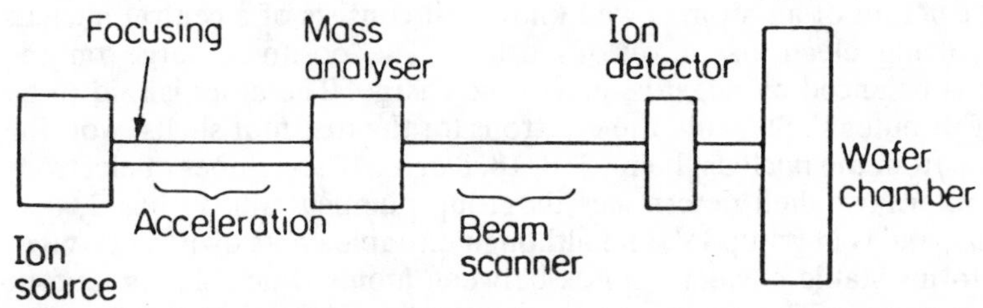

Figure 1.3 Principal parts of an ion-implantation equipment

The source output should remain constant over a long time, to allow reproducible devices to be made without the need for constant adjustment. A focusing system, which is usually electrostatic, is used to focus the ion beam onto the silicon slice. The ion accelerator gives the ions the necessary penetration energy by applying a high voltage across the ions. A mass analyser is used in conjunction with the focusing system to separate out the impurities from the ion beam.

The beam scanner scans the ion beam over the silicon surface, resulting in uniform dopant concentration. In addition,the silicon slices can be moved past the beam for more uniformity. The doping concentration is monitored by measuring the current in the ion detector. This is quite easily done since each ion carries one positive charge unit. The dose imparted to the silicon, measured in ions/cm^2, is equal to the product of the current and the exposure time, divided by the wafer area. The wafer chamber holds the silicon samples. It must be quickly accessible and large enough to hold a useful batch at each operation.

1.2.7 Metal formation

Metal is deposited onto the surface of the semiconductor slice using one of two main methods. In the first system the semiconductor is held face down in the top half of a bell jar, operating in vacuum. The material to be deposited is located at the bottom of the jar and heated until it vaporises and settles as a thin layer onto the semiconductor surface.

The second deposition technique is called sputtering. A bell jar is again used, but now it is filled with an inert gas, such as argon. The semiconductor, again face down in the top half of the jar, is connected to the positive terminal of a high-voltage source and the material to be deposited is placed at the bottom of the jar and connected to the negative terminal. Under the effect of this voltage the argon is ionised and positive ions bombard the cathode, causing it to sputter and emit material, which settles on the semiconductor surface, i.e. the anode.

1.3 Power rectifier principles

1.3.1 Physics of rectification

It is not intended in this brief introduction to delve into atomic physics. Instead some of the basic concepts of semiconductors are introduced.

The structure of an atom is well known. It consists of a central nucleus and revolving electrons in various orbits. The positive charge on the nucleus is balanced by negative electronic charge. The atom is said to be stable if its outer shell is full, the electrons for the first four shells, working outwards from the nucleus, being 2, 8, 18, 8 or 32. The number of electrons in the outermost shell determines its group number. Silicon has 2, 8, 4 electrons, and is in group IV, and although unstable on its own, a crystal of silicon forms stable covalent bonds between atoms. Each shares its four electrons with neighbouring atoms and so has eight orbital electrons. However, if energy is given to the material, say in the form of heat, an electron can break away from its valency bond and cause conduction in the material. This is called intrinsic conduction.

If an impurity of group V (donor) or group III (acceptor) is added to silicon then there will be an abundance or shortage of electrons respectively, causing increased conduction at a given temperature. This is called extrinsic conduction. A shortage of electrons results in a *p*-type doping, where holes are the majority carriers and electrons the minority ones, whereas extra electrons results in an *n*-type material, in which the carrier function is reversed.

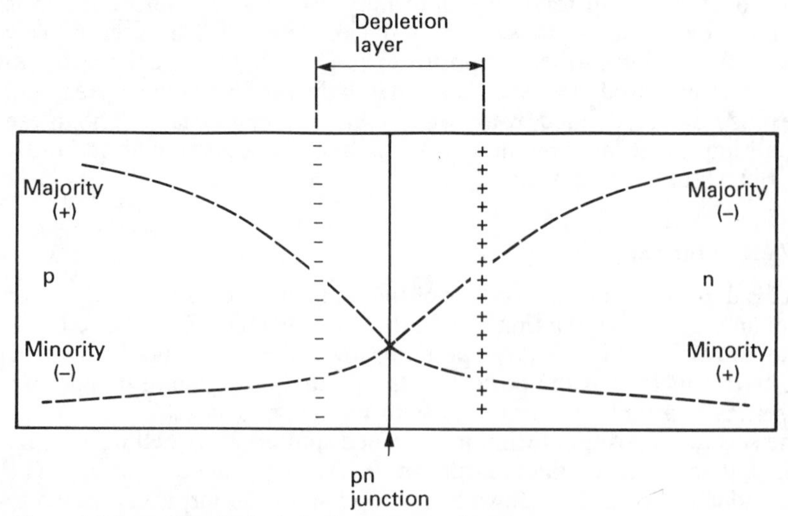

Figure 1.4 Formation of a depletion layer in a *p–n* junction

Holes and electrons are constantly being created in a semiconductor. They disappear due to recombination, a process which can be increased by the presence of specially created trapping centres in the crystal. These hold a carrier until an opposite polarity charge arrives for recombination.

A power rectifier is a two-layer device, similar in principle to a diode, consisting of a *p* and an *n* layer formed within the same semiconductor material. Figure 1.4 shows a simplified arrangement of the two layers. The *p* layer has an abundance of holes (holes are the majority carrier) and the *n* layer has electrons as the majority carriers. Electrons and holes form the minority carriers in the *p* and *n* layers, respectively, as shown in Figure 1.4.

As the junction between the layers is approached, the concentration of *p* and *n* decreases to match that in the other layer. Therefore there is a gradient in the material, which results in the diffusion of holes and electrons across the junction. These cause recombination to occur in the opposite layer, so that a negative charge barrier is formed in the *p* layer, close to the junction, and a positive barrier is formed in the *n* layer, as shown in Figure 1.4. These result in a potential barrier which hinders further diffusions. Because of this barrier only a few holes and electrons, with high kinetic energy, can cross or remain in the junction region. This region therefore has only a few majority carriers and it is known as the depletion region.

Since the depletion layer has only a few charge carriers it is in effect an insulator. The device therefore resembles a capacitor in having two conducting regions separated by an insulator. The width of the layer, and hence the capacitance, is proportional to the applied reverse voltage across the *p–n* junction. The capacitive effect influences the switching performance of the rectifier, as will be seen in the next section.

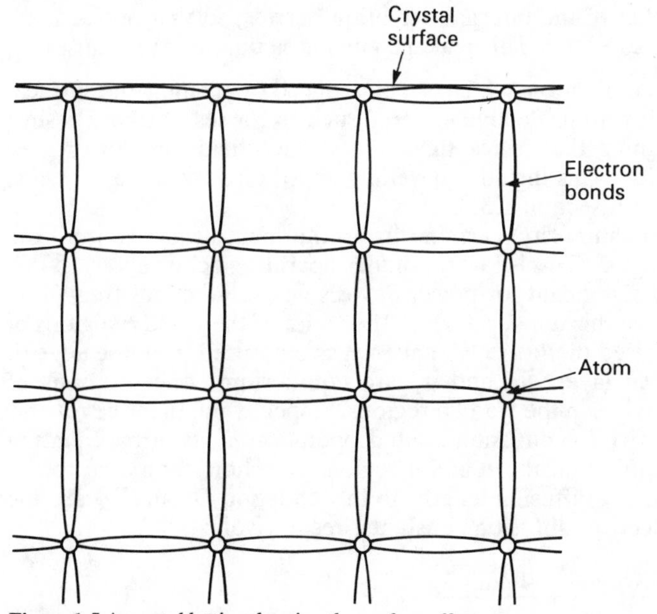

Figure 1.5 A crystal lattice showing the surface effect

The surface of the semiconductor, even if it is completely pure, forms a break in the overall regularity of the crystal structure. This is shown in Figure 1.5. In the whole lattice each silicon atom is bound to its neighbouring atom by two electron bonds, so that each atom is connected to four other atoms. On the surface layer, however, the atom is bound to three other atoms only, so that two holes are unfilled, resulting in a pure semiconductor having a positive surface charge. In practice these surface atoms attract impurity electrons from the atmosphere, so that the surface of a semiconductor is intrinsically impure, resulting in problems in the surface region. It is therefore important, in any semiconductor, to thoroughly clean its surface and then to protect junctions by stabilisers such as oxides and nitrides

The voltage induced in the depletion region is determined by the intrinsic carrier concentration, which for silicon is 1.4×10^{10} per cubic centimetre. This gives a junction voltage of about 560 mV.

1.3.2 Forward and reverse bias

Connecting a battery across the junction in the forward direction, that is, with its positive terminal connected to the *p* layer, will decrease the depletion region, acting with the potential barrier, and cause a current to flow as given by

$$I_F = I_R \left[\exp\left(qV_j/kT\right) - 1\right] \qquad (1.1)$$

where V_j is the voltage across the junction, I_R is the reverse junction current, q is the electron charge, T the absolute temperature, and k is Boltzmann's constant.

When the voltage across the *p–n* junction is reversed, the battery potential helps that of the internal depletion barrier, so that only a small minority current can flow. This leakage current is due to three causes:

(i) Surface contamination, which can be reduced by cleaning the surface; by coating it with protective material such as glasses or silicon resins; or by designing the device such that surface fields are lower than internal fields. This includes bevelling and the use of surface plates, as described in section 1.3.3.

(ii) Diffusion of minority carriers from the neutral areas into the depletion region. This is low for diodes operating below about 100°C, but is more significant for power devices whose junctions frequently operate above this temperature. If the width of the *p* and *n* regions of a reverse-biased diode are W_p and W_n, as measured from the edge of the depletion layer; $\overline{n}_p$ and $\overline{n}_n$ are equilibrium concentrations of minority carriers in the *p* and *n* regions, respectively; then the reverse current density for diffusion-limited operation is given by equation (1.2), assuming that the *p* and *n* regions are short compared to the minority carrier diffusion length. In this equation D_p and D_n are the hole and electron diffusion constants, respectively:

$$I_{ld} = \frac{q\, D_n\, \overline{n}_p}{W_p} + \frac{q\, D_p\, \overline{n}_n}{W_n} \qquad (1.2)$$

(iii) Charge generation within the depletion layer. This is generally the cause of most of the leakage current, and its value is given by equation (1.3), where I_{lg} is the space charge generated leakage current in A per square centimetre; q is the charge on an electron; n_e is the intrinsic electron concentration; W_d is the width of the depletion layer; and τ_s is the space charge generation lifetime:

$$I_{lg} \approx \frac{q n_e W_d}{2\tau_s} \tag{1.3}$$

The current therefore varies directly as the volume of the depletion layer and inversely as the space charge generation lifetime. High-voltage junctions have wide depletion layers, so it is important to have long lifetimes during processing. The charge generation is proportional to n_e so it doubles for approximately every 11°C rise in temperature.

1.3.3 Modified structures

In order to reduce the surface currents, and improve the breakdown characteristics of rectifiers, several techniques can be used. Figure 1.6 shows a method in which the surface, at which the junction between the p and n materials is formed, is bevelled. Both positive and negative bevelling can be used. In positive bevel the part of the crystal which is reduced in volume has a lower concentration of impurities. For negative bevel the converse is true, that is, the part of the crystal which is reduced in volume has the higher concentration of impurities.

Figure 1.6 shows the effect of bevelling on the depletion layer as it approaches the surface. As seen, the layer is wider at the surface than at the centre for positive bevel, giving a lower surface field and therefore

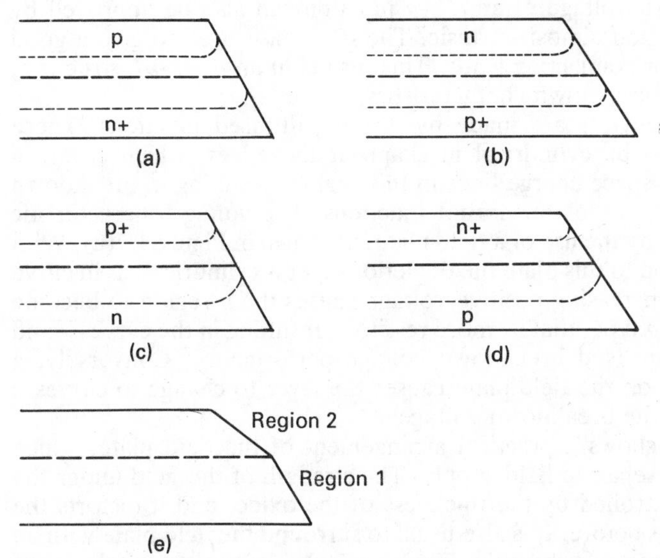

Figure 1.6 Bevelling the semiconductor edge: (a) and (b) positive bevel; (c) and (d) negative bevel; (e) double bevel

leading to less leakage current and a higher breakdown voltage. Since bevelling reduces the metal contact area it is sometimes necessary to use double bevelling, which achieves the same effect on breakdown voltage, but with less shallow bevel angles, and therefore leaves a greater amount of surface contact metal area.

In a *p–n* structure, as the applied reverse field is increased, a point is reached when mobile carriers attain thermal drift velocities, which for silicon are 10^7 cm/s for electrons and 6.5×10^6 cm/s for holes. As the field is further increased beyond this point the velocities of the carriers exceed their thermal drift velocities, and they become 'hot' carriers. These collide with atoms and give enough energy to electrons in valence bands to cause them to move to the conduction band, resulting in hole–electron pair generation. Each new pair is involved in ionisation of further hole–electron pairs. When this process attains infinite rate, avalanche breakdown occurs.

If α_p and α_n are the ionisation coefficients for holes and electrons, then breakdown occurs when equation (1.4) is satisfied. For the case of α_p approximately equal to α_n, and both equal to an average value of α_a, this equation simplifies to equation (1.5):

$$\int_0^{W_{bd}} \alpha_p \exp\left[\int_0^{y} (\alpha_n - \alpha_p)\,dy\right] dx = 1 \qquad (1.4)$$

$$\int_0^{W_{bd}} \alpha_a \, dx = 1 \qquad (1.5)$$

The breakdown voltage of an n^+–p junction can also be improved by making the *p* region almost intrinsic. The p^+ area is used to give a good ohmic contact for connecting leads. This results in an n^+–i–p^+ structure, which has good breakdown characteristics.

Many power devices are made by forming diffused junctions. These junctions tend to be cylindrical in shape at the edges, which causes a distortion of the space charge lines in this region, resulting in breakdown voltages lower than that for abrupt junctions. The voltage characteristic can be improved by the use of a field plate, as shown in Figure 1.7(a). With no voltage applied to this plate the depletion layer is cylindrical, as in curve a. Applying an increasing positive voltage causes the *p* region to become successively less *p* type, that is, more resistive, resulting in the curves b and c. This gives improved breakdown voltage performance. Conversely, a negative voltage on the field plate causes the layer to change to curves d and e, reducing the breakdown voltage.

Figure 1.7(b) shows a practical arrangement of the field plate, which does not need a separate field supply. The strength of the field under the field plate is controlled by the thickness of the oxide, and it distorts the depletion layer as before. It is also usual to surround the field plate with an equipotential ring, as shown in Figure 1.7(c). This prevents the slow drifting breakdown, which field plate structures are prone to exhibit.

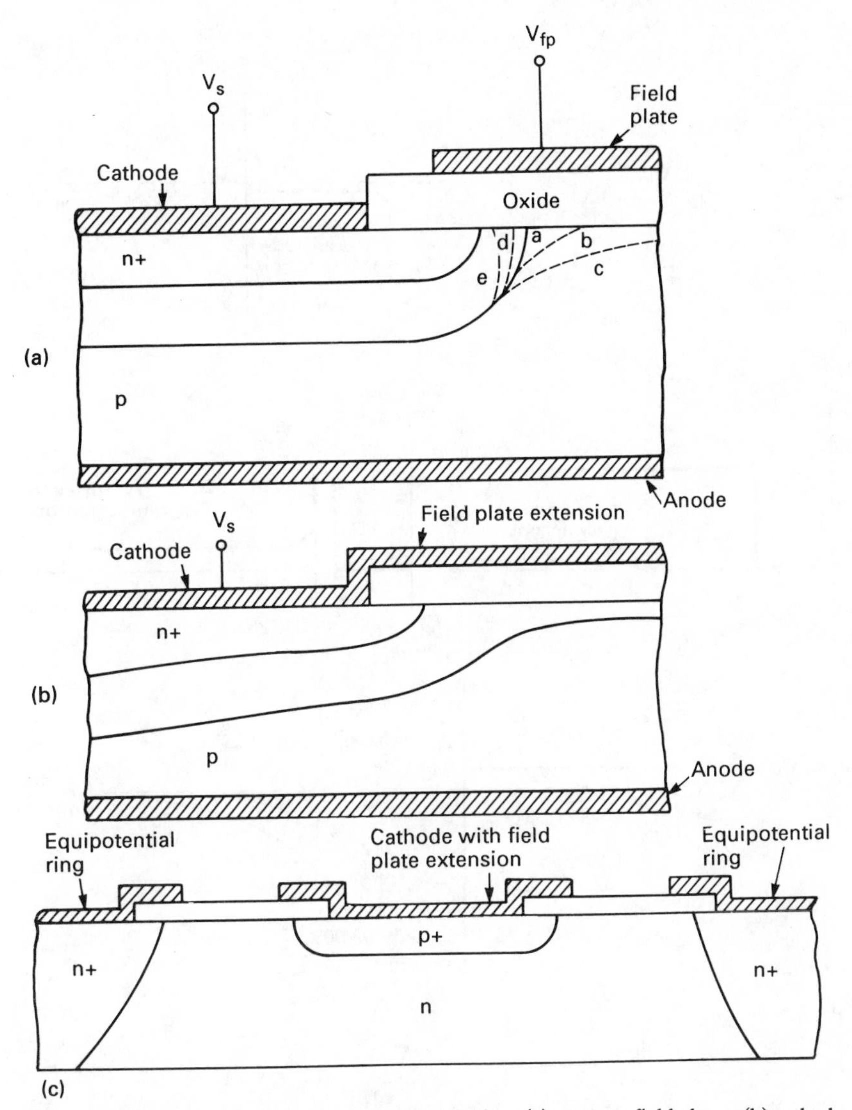

Figure 1.7 Control of depletion layer surface junction: (a) separate field plates; (b) cathode extension plate; (c) addition of equipotential ring

1.4 Power rectifier operation

1.4.1 The diode curve

The symbol for a diode is shown in Figure 1.8(a) and its d.c. characteristic in Figure 1.8(b). When the voltage across the diode is increased in the forward direction the current through the device rises rapidly, once the internal potential barrier, caused by the depletion layer, has been overcome. This curve is temperature sensitive and data sheets normally

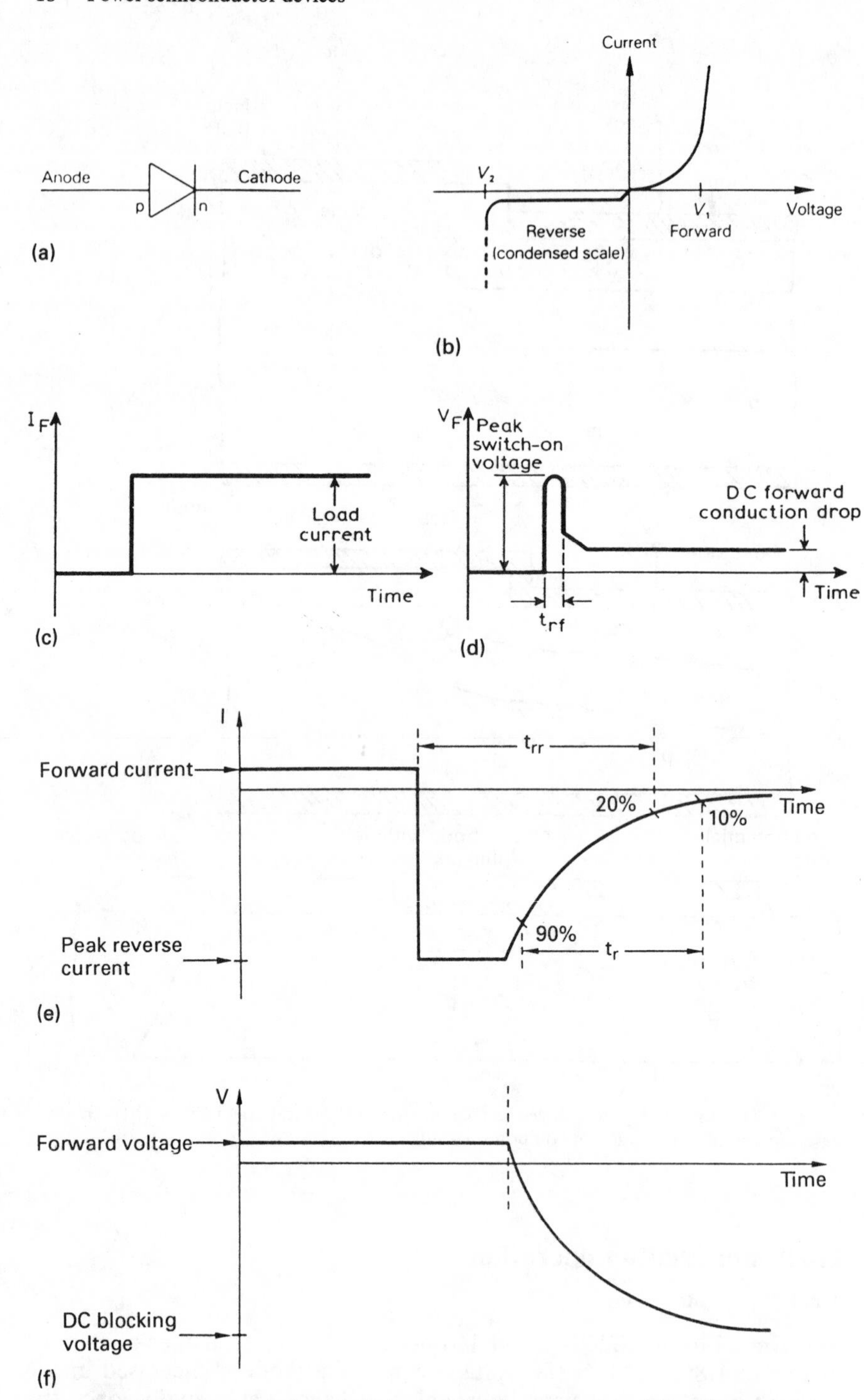

Figure 1.8 Diode characteristics: (a) diode symbol; (b) static characteristic; (c) turn-on characteristic – current curve; (d) turn-on characteristic – voltage curve; (e) turn-off characteristic – current curve; (f) turn-off characteristic – voltage curve

give a typical and a maximum curve, usually at a junction temperature of 125°C.

In the reverse direction the diode blocks and a much lower current flows, equal to the leakage current of the diode. This current increases slowly with applied voltage, until at a high enough reverse voltage, called the avalanche voltage of the device, it breaks down due to avalanche conduction, as described in section 1.3.3. The current through the device now increases very rapidly and since the voltage across the diode is still equal to its breakdown value (V_2), which can be several thousand times the value of the forward voltage (V_1), the diode will dissipate a large amount of power, and could be destroyed

The reverse characteristic curve of the diode is affected much more by temperature than its forward curve. The leakage current increases with temperature, resulting in a lower breakdown voltage.

1.4.2 Rectifier ratings

Data sheets normally specify semiconductor devices by two sets of parameters, ratings and characteristics. The ratings define the maximum values at which the component can be operated without being damaged, and the characteristics indicate its performance under specified conditions.

The following are some of the ratings of a diode:

(i) The maximum reverse voltage. This is usually specified in three ways. The first is the peak working voltage, which defines the normal operating voltage of the device at which it could work indefinitely without any damage. The second is the peak repetitive voltage. This is higher than the peak working voltage, but the diode is capable of withstanding this voltage for a limited period only, this period being specified in the data sheets. The third reverse voltage rating is the peak non-repetitive voltage. This is the voltage which is permitted to occur only infrequently during the life of the device, since it causes the highest power dissipation, and therefore strain, to the silicon die.

(ii) The maximum current rating. This is also specified in three ways, which correspond to the three voltage parameters. For all these three ratings the assumption is made that the junction temperature of the device does not exceed its rated value. The first is the peak working current, which is the maximum current that the rectifier can carry, provided its junction temperature rating is not exceeded. The second is the peak repetitive current, which is the current the rectifier can carry for short periods. The third is the peak non-repetitive current, which the rectifier can carry for a short time and only infrequently during its lifetime. This third current is usually used to blow protection fuses associated with the rectifier. The working current of a rectifier may be specified as a direct current or as an average value.

(iii) The peak power dissipation which the device can withstand.Power dissipation is linked to the device junction temperature and its cooling characteristics.

(iv) The maximum junction temperature at which the rectifier can be operated. Also specified is the maximum storage temperature of the device, and this is given as the highest and lowest temperatures, since the semiconductor can also be damaged by prolonged exposure to sub-zero temperatures.

1.4.3 Rectifier characteristics

The characteristics of a semiconductor component can often be divided into static and dynamic. The following are some of the static characteristics:

(i) Forward voltage drop. This varies with forward current, as shown in Figure 1.8(b), so it is normally specified at a given current or by a graph. The average forward voltage drop is usually specified as an average over a full cycle, at a stated frequency, average forward current and case temperature.
(ii) Reverse leakage current. This is specified at a defined reverse voltage and temperature, or as an average value over one cycle at a given temperature.
(iii) The forward power dissipation. This is the product of the forward voltage drop and current and is given as a curve.
(iv) The reverse power dissipation. This is equal to the reverse leakage current times the reverse voltage, at a defined point in the device curve.
(v) The junction to case thermal resistance. This characteristic, specified in units of degrees Centigrade per watt, indicates the difference in temperature between junction and case, when the rectifier is dissipating power. It is required for designing rectifier cooling systems, as described in Chapter 2.

The dynamic characteristics of a rectifier relate to its switching periods, both on and off. When a rectifier is switched into conduction a finite time is required for the minority carriers to flow across the junction and to prime it to carry the full load current. This can result in an initial peak voltage across the device, as seen in Figure 1.8(d). The load current rises rapidly, being limited by the circuit inductances, but the rectifier is in its fully-on mode after a delay time of t_{rf}. This time is very short, but it does represent a period of large power dissipation, which for very high-frequency operation can start to become important. Data sheets normally specify this forward delay time

When a diode is switched from the forward-conducting to the reverse-biased mode two changes occur. First, the minority charge is swept away as reverse current, in the external supply. This current can be large, often being limited only by the external circuit impedances. It could be described as the surge current through the device junction capacitance, although this is not strictly true. In addition to this external current flow, holes and electrons disappear in the vicinity of the reverse-biased region, due to recombination. The current decays to zero, as the reverse voltage-blocking capability of the diode increases,as shown in Figures 1.8(e) and 1.8(f).

The reverse recovery time of the rectifier is several orders of magnitude greater than its forward turn-on time, so the power dissipation during switch-off is much greater. Clearly the reverse recovery time t_{rr} must be kept as short as possible, in order to limit the device dissipation. However, the decay of current, over time t_r, must not be too abrupt as this will give rise to large voltage spikes in any associated inductive circuits. The ratio of t_r/t_{rr} should be as large as possible and devices which have a relatively high value of this ratio are said to be 'soft'.

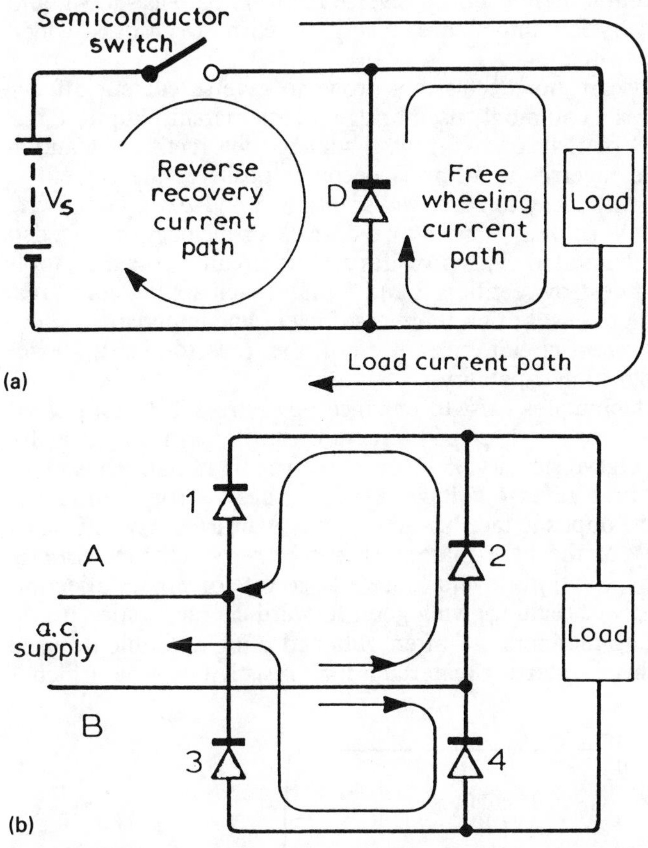

Figure 1.9 Examples of circuit reverse recovery current: (a) chopper circuit; (b) bridge rectifier circuit

In circuit applications the reverse recovery current, which for power devices can be high due to the larger silicon area involved, is often overlooked, with disastrous effects. For instance, consider the chopper circuit shown in Figure 1.9(a).The semiconductor switch is opened and closed at a relatively high frequency and varying duty cycle, in order to control the magnitude of the mean load voltage and current. Suppose the load is inductive. This requires a diode D, often called a free-wheeling diode, to be connected across it, to prevent excessive surges in the switch when it opens.

Ignoring the reverse recovery current in the diode, the system would operate as follows. With the switch closed D is non-conducting and load current is supplied via the d.c. source V_s. When the switch opens the load current free-wheels in D and, provided the loss in this path is not large, the current will be substantially unchanged when the switch is again closed.If the reverse recovery current through D is not ignored then, with the load current free-wheeling through it, the operation is basically as before. However, when the semiconductor switch closes this current transfers to the supply. In addition there is now a reverse recovery path as shown, and since this is of a very low impedance, a surge of current can pass, which could destroy the switch.

Chopper circuits are not exclusively prone to reverse current effects. Figure 1.9(b) shows a commonly used bridge rectifier circuit. Suppose line A is positive to B so that load current is supplied via rectifiers 1 and 4. When the voltage reverses and line B becomes positive, the current in rectifiers 1 and 4 will decay to zero, whilst that in 3 and 2 will increase to support the load. As soon as 1 and 4 turn off a reverse current path exists through them, as illustrated. This provides a short circuit across the supply lines, which could destroy rectifiers 2 and 3. In all such applications great care must be taken to ensure that there is sufficient line impedance, which would limit this short-circuit current until the reverse-biased device recovers its full blocking capability.

Several design techniques exist to enhance power rectifier characteristics. A p^+–p–n–n^+ (or n^+–n–p–p^+) structure can be used in which the base consists of high-resistivity $n-$ (or $p-$) type material, chosen to withstand the required reverse voltage. One face has a p (or n) diffusion made into it and the opposite face has an n^+ (or p^+) diffusion layer. Due to the high resistivity of the base material a good reverse characteristic is obtained, and the n^+ and p^+ layers cause a large rate of carrier injection into the high resistivity region, giving good forward characteristics.

Fast commutating rectifiers are often obtained using a double-diffused construction. In this the starting material is high-resistivity n type, which is

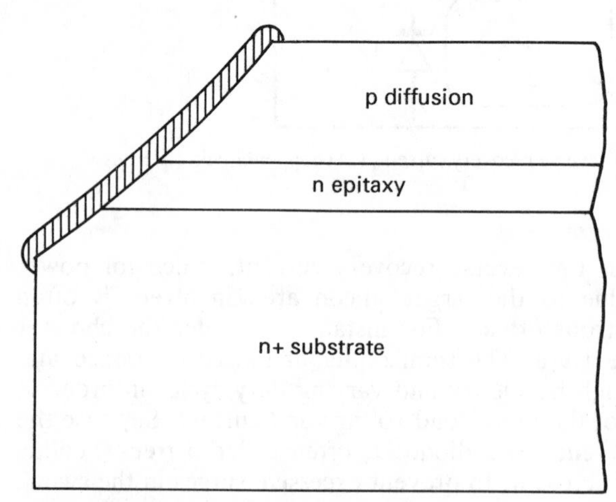

Figure 1.10 An epitaxial diode construction

chosen to withstand the required reverse voltage. Into this material n^+ and p^+ diffusions are made, on opposite sides, for the cathode and anode connections.

Even faster rectifiers are obtained using an epitaxial construction, as shown in Figure 1.10. The base material is now n^+. Into this is grown an n epitaxy layer of high resistivity, which is designed to provide the required reverse voltage blocking capability. A p^+ diffusion or ion implantation is then made, to form the anode layer. This technique is capable of providing devices in which the reverse recovery time t_{rr} is less than 100 ns for a component having a rating of over 1000 V.

1.5 Bipolar transistors

1.5.1 Principle of operation

The bipolar transistor is a three-layer device, which can be made up of *p–n–p* or *n–p–n* layers, as shown in Figures 1.11(a) and 1.11(b). The base region is narrow and lightly doped, unlike the emitter and collector layers, which are heavily doped and comparatively wide. With the biasing arrangements shown the emitter-base region is forward biased so that majority carriers flow across this junction. These carriers are electrons in the case of the *n–p–n* transistor and holes for *p–n–p*.

Since the base is lightly doped only a few holes (for *n–p–n*) or electrons (for *p–n–p*) are available to cross into the emitter. The collector-base region is reverse biased resulting in a small minority carrier flow, which is holes from collector to base and electrons from base to collector, for an *n–p–n* transistor, and vice versa for a *p–n–p* transistor.

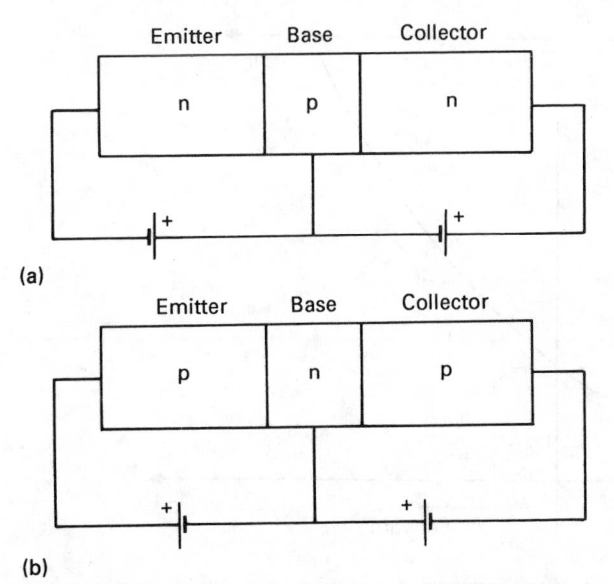

Figure 1.11 Bipolar transistor characteristic: (a) biasing arrangement, *n–p–n* transistor; (b) biasing arrangement, *p–n–p* transistor;

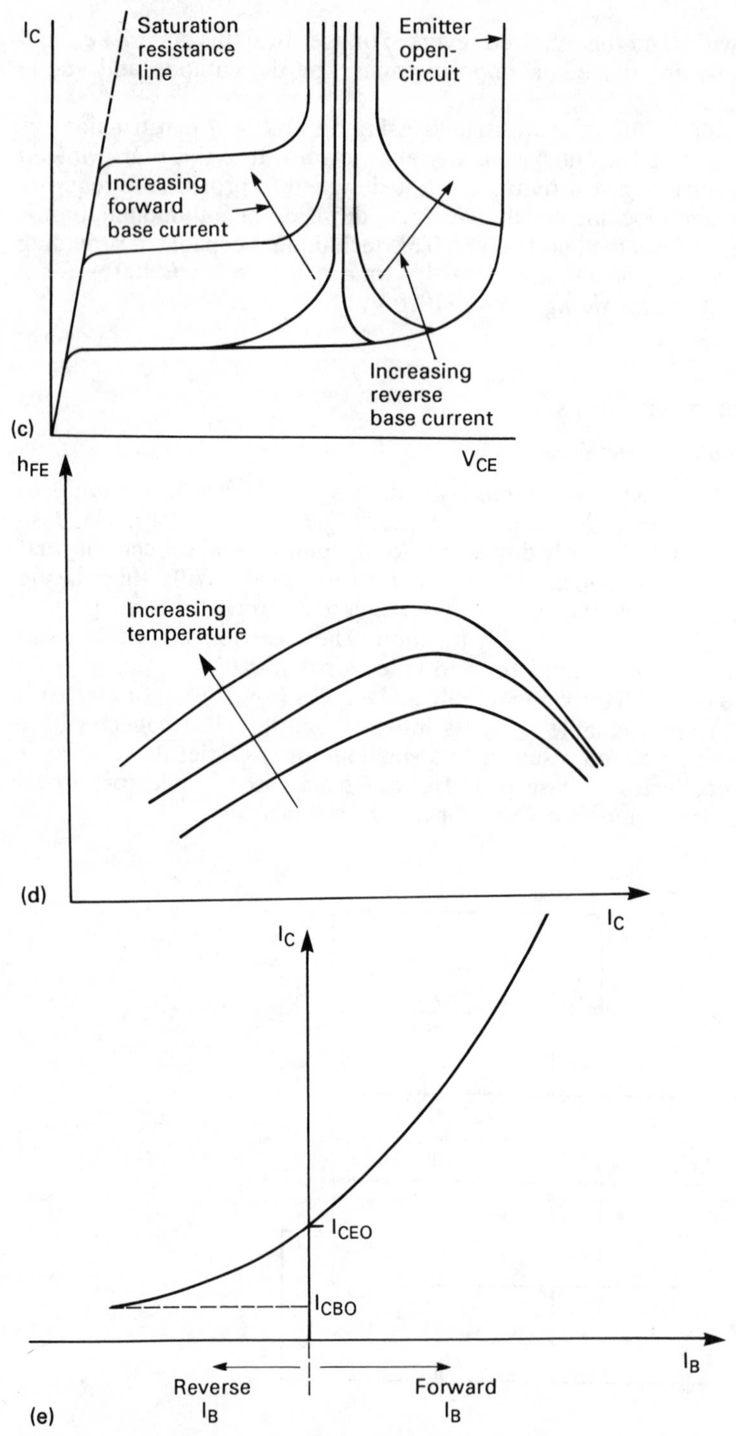

Figure 1.11 continued (c) static characteristic; (d) gain-collector current characteristic; (e) transfer characteristic

The collector bias is much greater than that of the base, and since the base region is narrow, the majority carriers which cross the emitter-base region are diffused across the base and into the edge of the collector depletion layer. These carriers then reach the collector, resulting in collector current flow. A small amount of the majority carriers, crossing from the emitter to the base, combine within the base and this, along with the small number of minority carriers crossing from the base to the emitter, results in a low value of base current.

Transistors can be operated in common base, common emitter or common collector modes. Common emitter is the most popular configuration, common collector being used in emitter follower applications. For common emitter the base current I_B is the input and the collector current I_C is the output. The ratio between these two (I_C/I_B) is called the common emitter forward current transfer ratio h_{FE}. In the transistor a low base-emitter voltage and small base current change results in a large collector current across the high reverse-biased collector-base region. There is therefore power gain within the transistor.

Leakage current is caused within the reverse-biased collector-base junction due to minority carriers. For common emitter this current is multiplied by the gain of the device and flows in the emitter, when the base is open circuit. This leakage current is further increased with temperature.

1.5.2 Characteristics

The main ratings of a power transistor are related to current and voltage. Since the transistor has three terminals, base, collector and emitter, the ratings are associated with each terminal. The maximum collector current is described as the maximum continuous current and the peak current. Since the peak current is only of the order of twice as large as the maximum continuous value, the transistor does not have a high surge rating. The maximum current is usually limited by the gain of the device and by the bonding wires.

There are also several voltage ratings for a transistor, depending on which junction is being considered and the state of the third terminal. For example V_{CBO} is the breakdown voltage between the collector and base junctions with the emitter open circuit; V_{CES} is the breakdown voltage between the collector and emitter with the base short-circuited to the emitter; V_{CEO} is the collector emitter breakdown voltage with the base open circuit.

The output characteristic of a bipolar transistor is shown in Figure 1.11(c). For very low values of collector-emitter voltage the collector does not gather up electrons which pass through the base region. This process becomes more efficient as V_{CE} increases, giving a rapid rise in collector current. After the knee of the curve has passed, most of the carriers generated at the base-emitter junction are gathered up by the collector, so that increasing V_{CE} does not appreciably affect I_C. This is known as the saturation region of the characteristic. The larger the value of base current, the higher the collector current at which saturation occurs. Therefore a bipolar transistor is essentially a current-operated device. At very large values of V_{CE} the carriers attain sufficient energy to cause avalanche breakdown.

There are two gains which are important in bipolar transistors, the large signal or d.c. gain (h_{FE}) and the small signal or a.c. gain (h_{fe}), given by

$$h_{FE} = \frac{I_c}{I_B} \tag{1.6}$$

$$h_{fe} = \left.\frac{\Delta I_c}{\Delta I_B}\right|_{I_c} \tag{1.7}$$

For switching applications the d.c. gain is the most important, and is given by the ratio of the d.c. values of the collector and base currents. It falls off slowly at low values of collector current, as shown in Figure 1.11(d), and also increases slightly with V_{CE}, which accounts for the slope in the output characteristic of the transistor. The d.c. current gain falls off rapidly at high values of collector current, as the saturation resistance line is approached. The gain also increases with temperature.

For linear applications, and for low-frequency and audio amplifiers, the a.c. gain is of more relevance. It is the ratio of the change in collector current for a small change in the base current, and its value is determined by the magnitude of the steady state collector current.

The bipolar transistor is said to be in saturation when increasing base current gives no further increase in collector current. The voltage drop across the external terminals of the device is now known as the saturation voltage. Data sheets usually give the collector to emitter saturation voltage ($V_{CE(SAT)}$) and the base to emitter voltage ($V_{BE(SAT)}$), and specify the conditions, such as collector current, base current and temperature, under which they were measured. At saturation both collector-emitter and base-emitter junctions are forward biased and the junction voltages oppose. The net junction voltage appearing across the collector-emitter forms one part of the saturation voltage, the drop in the saturation resistance (R_{CE}) forming the other. Data sheets sometimes specify the value of R_{CE} instead of $V_{CE(SAT)}$.

Transistor sustaining voltage is defined as the minimum collector-emitter breakdown voltage. It can be specified with the base open circuit, short

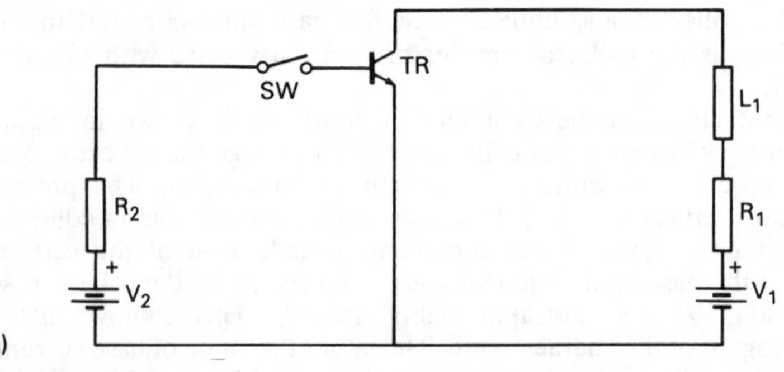

Figure 1.12 Bipolar transistor test circuits and waveforms: (a) test circuit to measure $V_{CEO(SUS)}$;

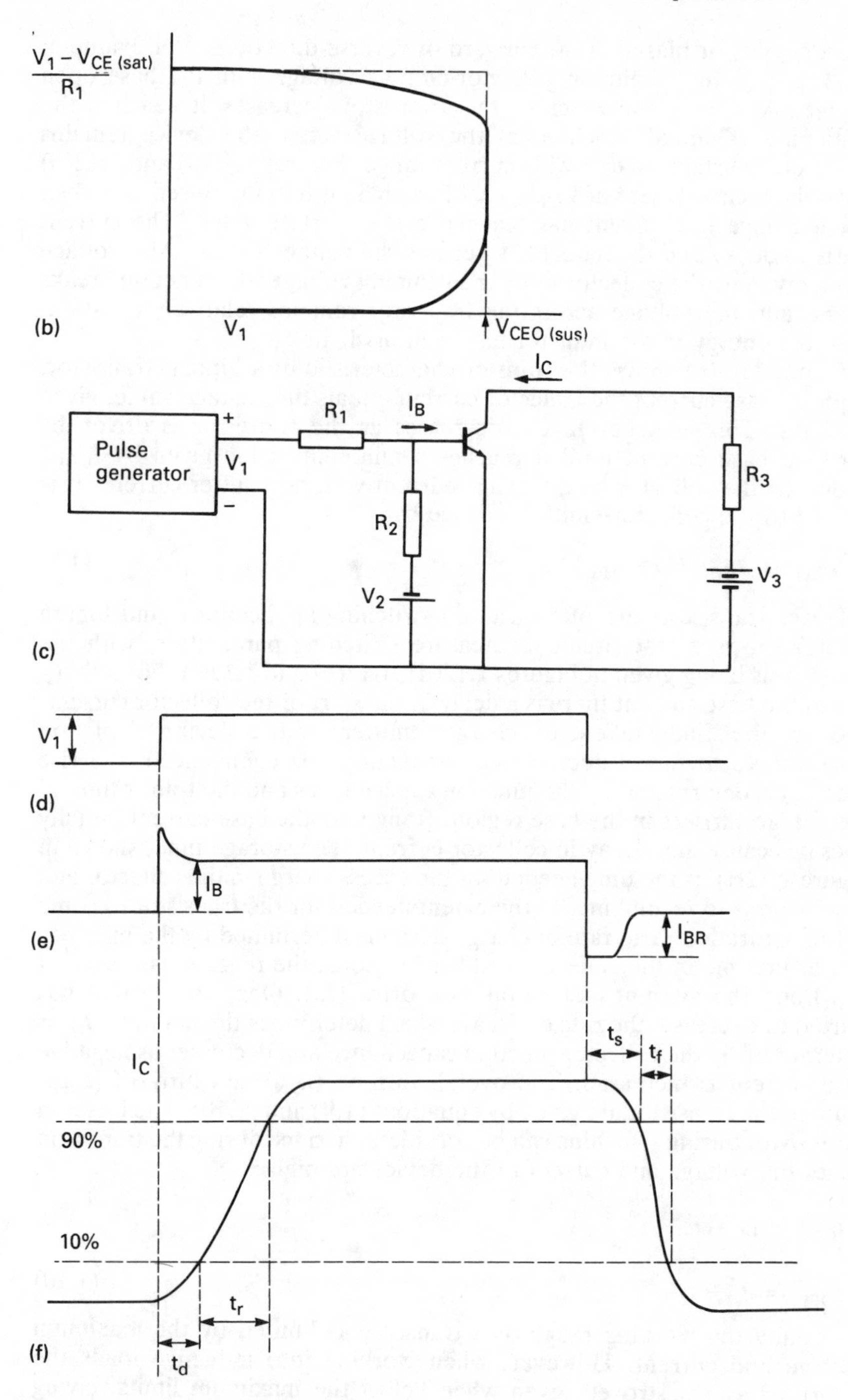

Figure 1.12 continued (b) waveforms for circuit (a); (c) test circuit to measure switching parameters; (d) base voltage waveform for circuit (c); (e) base current waveform for circuit (c); (f) collector current waveform for circuit (c)

short circuit, or biased in the forward or reverse directions. For example, $V_{CEO(SUS)}$ is the minimum collector-emitter voltage with the base open circuit. As the voltage across the transistor increases it reaches the sustaining value, at which point the voltage across the device remains relatively constant, over a wide current range. Figures 1.12(a) and 1.12(b) show the measurement of $V_{CEO(SUS)}$. Switch S_w is initially closed, and then opened once the current has reached a steady state value. The current starts to decay and the inductor L_1 causes the voltage to rise. At a voltage level, given by the collector-emitter sustaining voltage, the junction breaks down, and the voltage across the transistor remains relatively constant, until the energy in the inductor has been dissipated.

Figure 1.11(e) shows the transfer characteristic of a bipolar transistor. With no base current the collector current equals the leakage value, given by I_{CEO}. This leakage current decreases as the transistor is driven by negative base current until it reaches a minimum value given by I_{CBO}, known as the collector-base leakage current with no emitter current. It is related to the collector-emitter leakage by

$$I_{CEO} = I_{CBO}(1 + h_{FE}) \tag{1.8}$$

Power transistors are often used in switching applications, and Figure 1.12(c) shows a test circuit to measure switching parameters, with the waveforms being given in Figures 1.12(d), 1.12(e) and 1.12(f). For a sharp rise in the base current there is a delay in the start of the collector current, due to the time taken to charge emitter and collector depletion capacitances to new values. Once conduction has commenced, the rise time (t_r) is determined by the junction capacitances and the transit time of the charge carriers in the base region. Removing the base current initially does not cause any decay in collector current. The storage time, shown in Figure 1.12(f), is the time needed for the excess charge in the collector and base regions to recombine, to the extent needed for the transistor to come out of saturation. The rate of charge decay is determined by the minority carrier lifetime in the collector and base regions, the reverse base current (I_{BR}) and the amount of turn-on base drive (I_B). Once the current has started to decrease, the rate of decay, which determines the fall time (t_f), is determined by the collector junction capacitance and decreases as negative base current is increased. The overall turn-on (t_{ON}) and turn-off (t_{OFF}) times of the transistor are given by equations (1.9) and (1.10). The losses in a transistor during switching can be considerable since during the transition stages the voltage and current of the device are high:

$$t_{ON} = t_d + t_r \tag{1.9}$$

$$t_{OFF} = t_s + t_f \tag{1.10}$$

Usually the working range of a transistor is limited by the maximum voltage and current. However, when working into inductive loads the device can be destroyed, even when below the maximum limits, giving second breakdown. This is due to hot spots caused by current concentration, resulting in local thermal runaway. Current concentrations occur due to causes such as unstable lateral temperature distribution, base region potential drops, uneven base widths, and uneven mounting of the silicon chip onto the heat sink.

Figure 1.13(a) shows the second breakdown characteristic of a bipolar power transistor. For a given base current the collector current (I_C) will

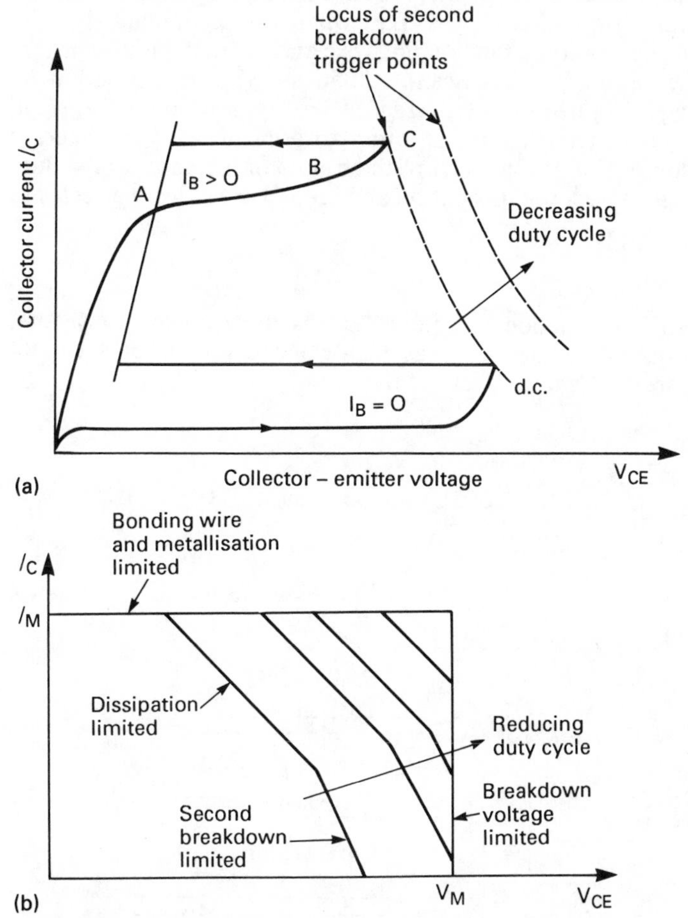

Figure 1.13 Power bipolar transistor characteristic: (a) second breakdown characteristic; (b) safe operating area

increase as the collector-emitter voltage (V_{CE}) increases. After a point A on the characteristic the device will go into saturation and the current will remain substantially constant until point B, when avalanche breakdown, or first breakdown, occurs. This causes a rapid rise in current until point C when a second breakdown effect develops. This results in rapid local heating of the silicon die, the collector-emitter voltage collapses, and the current escalates, destroying the transistor. There is a series of curves for different base currents and these give rise to individual second breakdown points, which all lie on a locus, as shown. As the duty cycle of the transistor decreases it runs cooler so that it can work on a wider second breakdown locus.

Power transistors must be operated in a mode such that second breakdown is avoided. This is done using the safe operating area (SOA) curves, of the type shown in Figure 1.13(b). Although these curves are for a device rated at a peak current of I_M and a voltage of V_M, the transistor cannot be run at this current and voltage simultaneously. For low values of V_{CE} the current can increase to I_M, where it is limited by the current-carrying capability of the bonding wire and the metallisation tracks used on the silicon. As V_{CE} increases so also does the power dissipation, so that eventually I_C will need to be decreased. For large values of V_{CE} the value of I_C is reduced still further in order to prevent the occurrence of second breakdown effects. The SOA of the transistor increases as the duty cycle reduces, since both the dissipation and second breakdown effects are now lower.

1.5.3 Construction

Several different construction techniques are used for power transistors, each giving some advantage, such as high voltage ratings or speed of operation. Figure 1.14 shows a few of these.

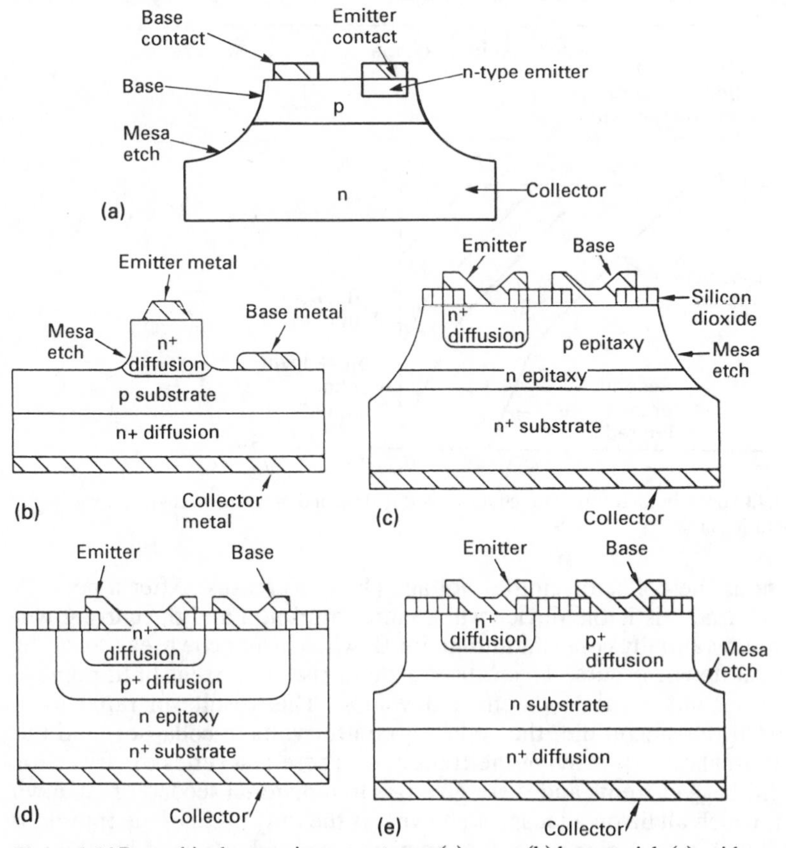

Figure 1.14 Power bipolar transistor structures: (a) mesa; (b) hometaxial; (c) epi-base; (d) planar epitaxial; (e) triple diffused

In the mesa structure the junction area is reduced by a mesa etch, so that junction capacitances are decreased. This technique also allows the edges of the transistor, mainly the collector-base junction, to be defined and passivated, which prevents contamination, so avoiding high electric fields.

The hometaxial structure is easy to produce so that it is cheap and rugged. It has good overall voltage ratings but suffers from relatively long switching times.

The epi-base transistor generally has a relatively low voltage rating, but this can be increased by adding high-resistivity collector epitaxy. This causes the collector voltage to be shared by the base and collector epitaxy layers, but the voltage rating is still less than for hometaxial. The process uses shallow emitter diffusions and a narrow-base epitaxy, so it has higher speed and current-handling capability than a hometaxial transistor for the same emitter area.

The planar epitaxial arrangement has all its junctions protected during fabrication by an oxide layer, so it is capable of being designed for very low leakage current. The diffusion process also enables very narrow base widths to be built, thus giving higher speeds and lower saturation voltage drops than hometaxial or epi-base transistors. The disadvantage of the planar epitaxial device is that it is not very resistant to the effects of second breakdown.

The triple-diffused transistor is better than planar on second breakdown, but it is still not very resistant to this. Speed and saturation voltage drops are similar to planar epitaxial, but the construction is more expensive and has a higher leakage current. The process is also difficult to control, so that a wide distribution in parameters is obtained between batches.

In bipolar transistors operating at high frequencies the current is forced

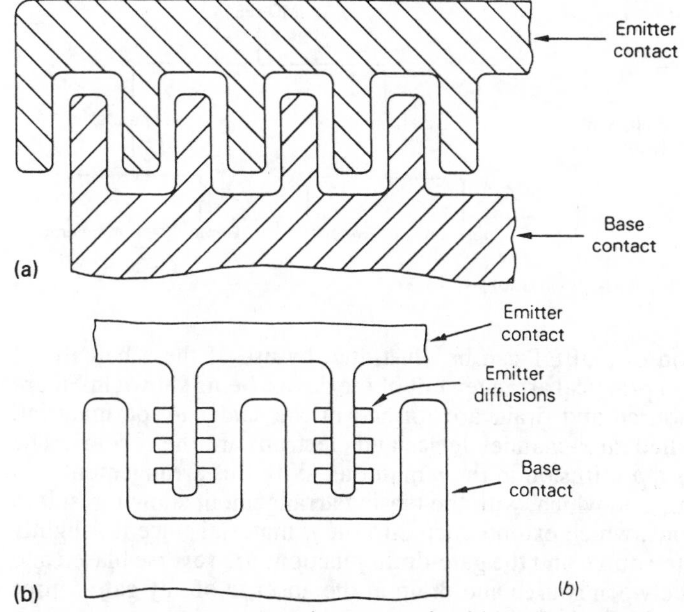

Figure 1.15 High-frequency transistor geometries: (a) interdigitated; (b) overlay

out towards the edges, so the ratio of emitter periphery to area is large. Two techniques are use to overcome this, as illustrated in Figure 1.15. In the interdigitated structure the base and emitter are interleaved and formed on the silicon die. Diffusions for both these are under the metal contact area shown. In the overlay device the emitter metal contact is formed over the base rather than adjacent to it. Emitter diffusions are segmented and spread out from the emitter metal contact, whereas the base diffusion lies below and along the length of the base metal contact.

1.6 Unipolar transistors

1.6.1 Principles of operation

Unipolar transistors, unlike bipolar, have a conduction mode which is dependent on only one type of carrier, which may be holes or electrons. Unipolar transistors are also known as field effect transistors (FET) and they are capable of several different operation modes, as illustrated in Figure 1.16. The junction field effect transistor (JFET) can have a *p* or an *n* channel, but it can only operate in depletion mode, as explained later. The metal oxide semiconductor field effect transistor (MOSFET) can also be *p* or *n* channel, and both these may be designed to operate in enhancement or depletion mode.

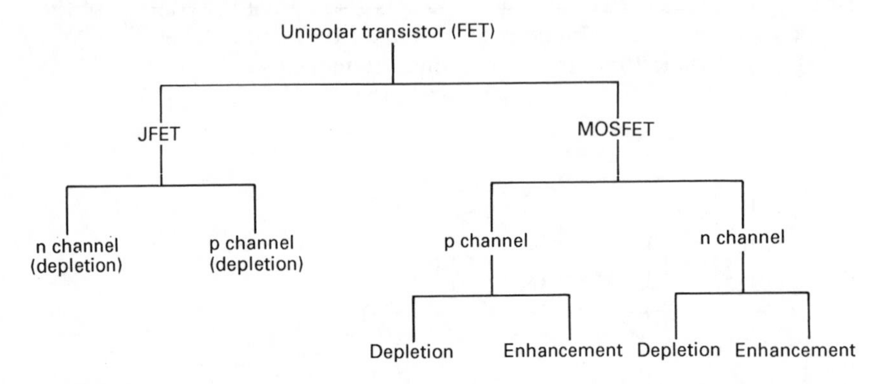

Figure 1.16 Unipolar transistor variations

The operation of a JFET can be illustrated by use of the schematic of Figure 1.17(a), a practical arrangement of the device being shown in Figure 1.17(b). The source and drain are formed in the same *n*-type material, hence this is called an *n*-channel device and electrons are the carriers. The gate is made as a *p* diffusion in the *n* material. With this arrangement *p–n* junctions are formed which, with the biasing arrangement shown, result in depletion regions, which extend deep into the *n* material since it is lightly doped. The gate-source and the gate-drain junctions are reverse biased and current flows between source and drain in the absence of any gate-source voltage. The device is therefore known as depletion mode.

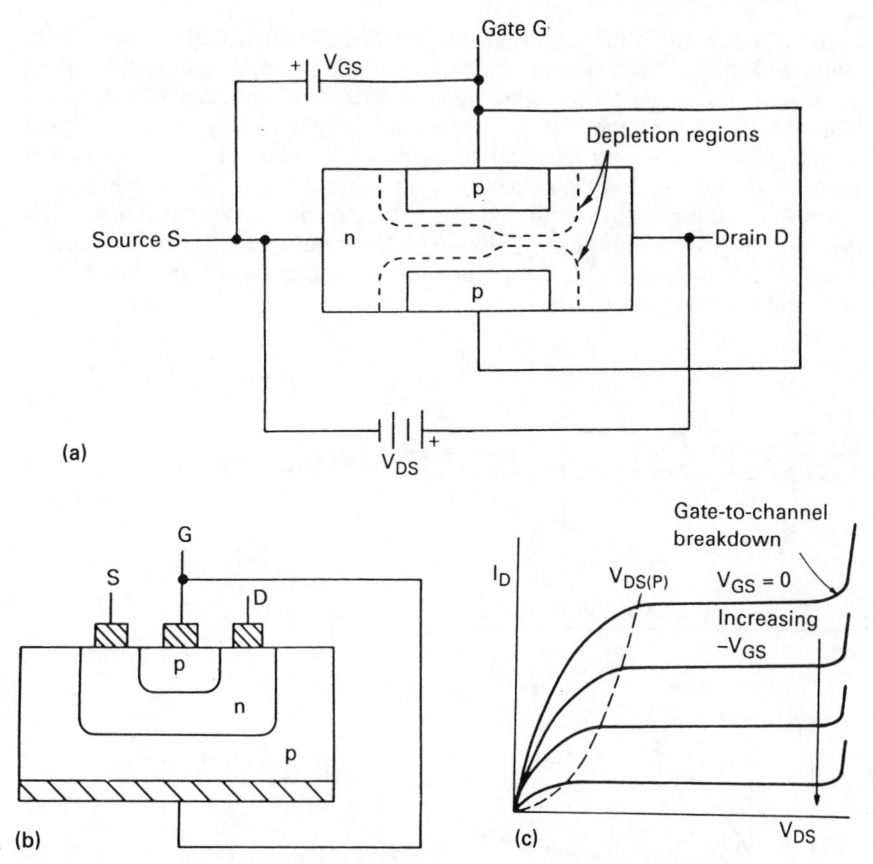

Figure 1.17 Junction field effect transistor (JFET): (a) *n*-channel representation; (b) *n*-channel practical representation; (c) static characteristic

Figure 1.17(c) shows the drain voltage and current curves for a depletion-mode device and it is seen that, with the gate-source voltage held constant, the drain current increases as the drain-source voltage increases. However increasing the drain-source voltage causes the depletion region to extend further, until the two regions in Figure 1.17(a) meet, resulting in the pinch-off state. The pinch-off voltage is shown on Figure 1.17(c) as $V_{DS(P)}$. Increasing the drain-source voltage beyond the pinch-off value does not cause any significant increase in drain current, but the point along the channel at which the depletion regions meet moves nearer to the source. The drain current is maintained by the electrons being swept through the depletion layer, similar to the process in the base of a bipolar transistor. Eventually the breakdown region is reached at high values of voltage. Making the gate-source voltage negative lowers the value of drain-source voltage at which pinch-off occurs.

Figure 1.18(a) shows the construction of a MOSFET. The source and drain diffusions are separated by the gate region so no current flows in the absence of a gate voltage. This device is therefore called enhancement mode. For a depletion mode MOSFET a narrow *n* channel would be formed under the gate such that current flowed when there was no gate voltage. Figure 1.18(b) shows the characteristic for an enhancement-mode transistor. As the gate-source voltage is increased a point is reached, called the threshold voltage (V_T), when an inversion layer (the doped semiconductor reverses its polarity) is formed under the gate connecting the source to the drain and resulting in current flow. The value of this threshold voltage is determined by the impurity concentration in the semiconductor, the amount of charge in the gate oxide, the type of metal used for the gate, and temperature. As temperature increases, the threshold voltage decreases.

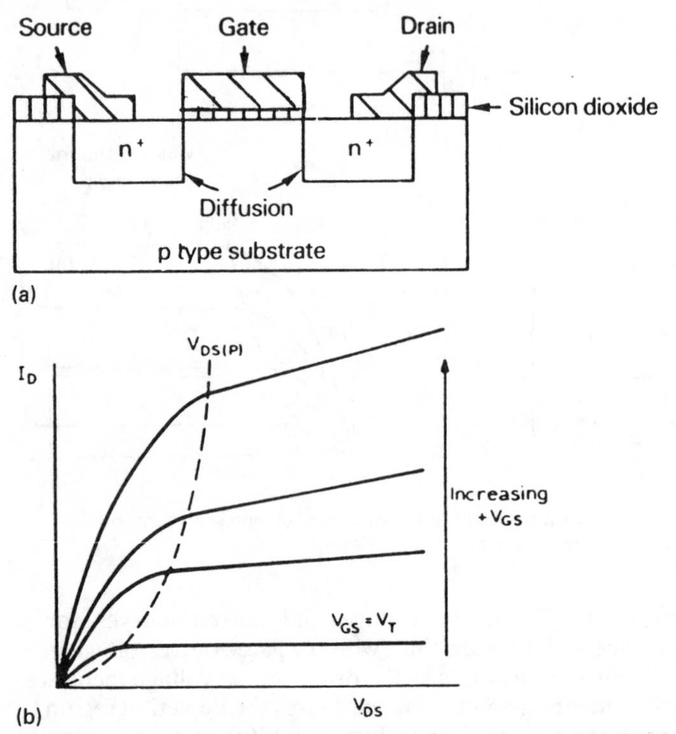

Figure 1.18 Metal oxide semiconductor field effect transistor (MOSFET): (a) construction of an *n*-channel device; (b) static characteristic of an *n*-channel enhancement-mode device

Increasing drain-source voltage causes a rise in the drain current. However, the resulting increase in the voltage drop causes a reduction in the channel conductivity. This reduction has the same effect as a constriction in the inversion layer, resulting in a knee in the curves of Figure 1.18(b). When the drain-source and gate-source voltages are equal the voltage between the gate and drain ends of the channel is zero and this is the pinch-off condition ($V_{DS(P)}$). Further increase in the drain-source voltage causes the drain depletion layer to increase and the end of the channel to move towards the source. The drain current is kept substantially constant by the electrons being swept through the depletion layer.

Figure 1.19 shows some of the symbols possible for unipolar transistors. The arrows point from the *p* to the *n* regions. The depletion-mode device shows a solid line between source and drain, since current flows in the absence of any gate voltage, and the MOSFET symbol indicates the oxide layer between gate and source-drain.

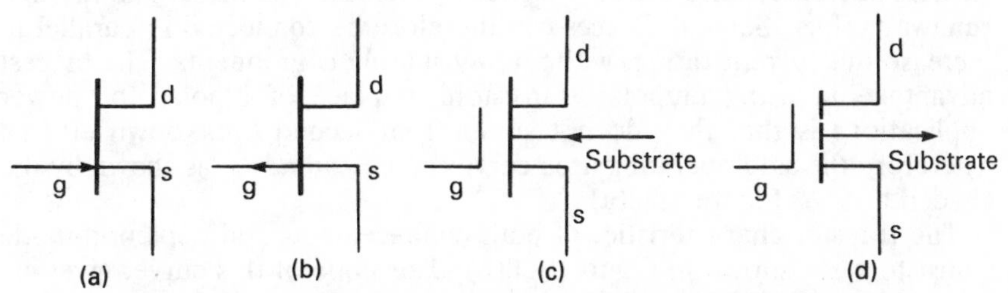

Figure 1.19 Unipolar transistor symbols: (a) JFET *n* channel; (b) JFET *p* channel; (c) MOSFET *n* channel depletion mode; (d) MOSFET *p* channel enhancement mode

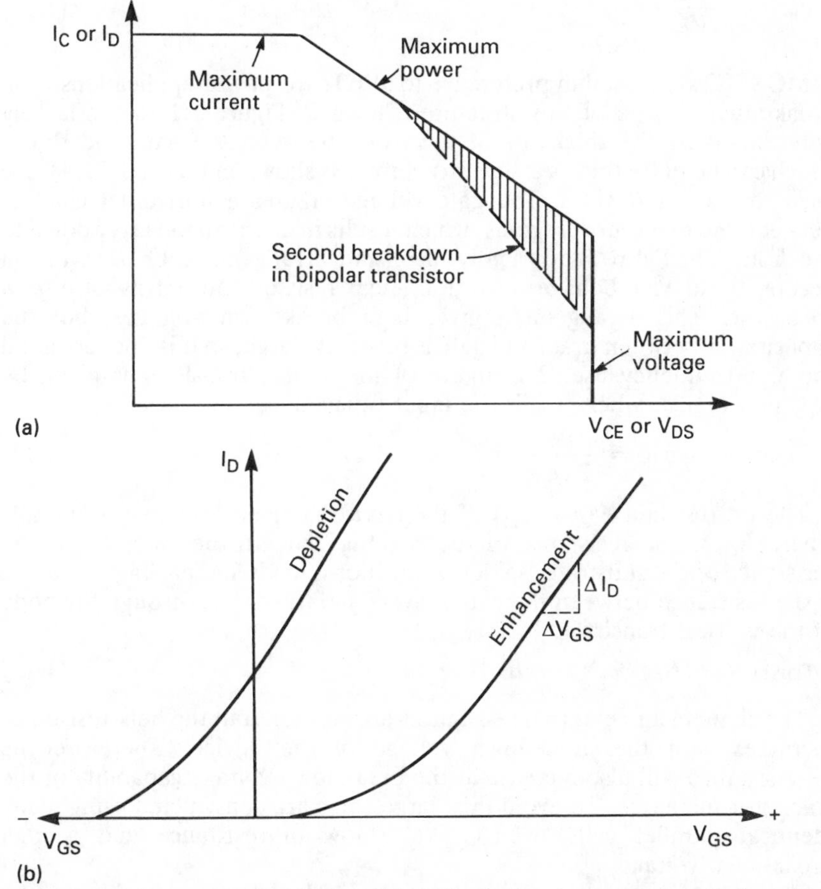

Figure 1.20 Unipolar transistor characteristics: (a) safe operating area and comparison with a bipolar transistor; (b) transfer characteristic

1.6.2 Characteristics and construction

Unipolar transistors are majority carrier bulk semiconductors and are free from minority carrier storage times, so they are inherently faster than bipolar devices. They are also voltage controlled so that their gain is much higher. As the temperature in a unipolar transistor increases, the majority carriers decrease, since the bulk resistivity increases, so there is no thermal runaway effect. Several devices can therefore be connected in parallel to increase the current rating, without any sharing components. The biggest advantage in using unipolar transistors in place of bipolar for power applications is that they do not suffer from second breakdown effects. Therefore the safe operating area curve can be extended, as shown by the shaded area of Figure 1.20(a).

The transfer characteristics of both enhancement- and depletion-mode transistors are shown in Figure 1.20(b). The slope of this curve, given by equation (1.11), is known as the transconductance or mutual conductance, and determines the amplification factor.

$$g_{\mathrm{m}} = \left.\frac{\Delta I_{\mathrm{D}}}{\Delta V_{\mathrm{GS}}}\right|_{V_{\mathrm{DS}}} \tag{1.11}$$

MOSFETs are used in preference to JFETs for power applications. The breakdown voltage of this structure, shown in Figure 1.18(a), is largely determined by the thickness of the oxide between the gate and drain, which can be quite thin. Vertical structures, as shown in Figure 1.21(a), are used for power devices. The gate voltage creates a horizontal channel between the two source regions, which results in an inversion layer down to the drain. The figure shows a polysilicon gate arrangement. Once inversion occurs, the device behaves as a non-linear resistor and not as an *n–p–n* transistor. This arrangement gives high breakdown voltages, but the capacitance between drain and gate is relatively large, so it is not very good for high-frequency use. The figure of merit of a transistor is given by equation (1.12), where C_{in} is the input capacitance:

$$\text{Figure of merit} = \frac{g_{\mathrm{m}}}{2\pi C_{\mathrm{in}}} \tag{1.12}$$

The on-resistance ($r_{\mathrm{DS(ON)}}$) of the device is given by equation (1.13), where R_{ch} is the resistance of the channel beneath the gate; R_{ex} is the resistance of the substrate, solder connections, leads and package; and R_{bk} is the resistance between the two p layers and the drain, through the body of the vertical transistor.

$$r_{\mathrm{DS(ON)}} = R_{\mathrm{bk}} + R_{\mathrm{ch}} + R_{\mathrm{ex}} \tag{1.13}$$

The channel and external resistances are constant but the bulk resistance increases with the breakdown voltage of the device. Therefore the on-resistance will also increase as the breakdown voltage capability of the transistor increases. To avoid this, large chips are constructed using many identical parallel cells, which give a low on-resistance and a high breakdown voltage.

The shorter the gate channel in a unipolar device, the higher the transconductance and the lower the parasitic capacitance. The maximum frequency of operation (f_{max}) is given by equation (1.14), where W_c is the length of the channel in micrometres and V_c is the carrier velocity in cm/s:

$$f_{max} = \frac{V_c}{2\pi W_c} \tag{1.14}$$

Figure 1.21(b) shows a short-channel arrangement which is formed by *p* and *n* diffusions, hence it is often called a double-diffused (DMOS) transistor. As the drain-source voltage is increased the depletion layer spreads into the drain drift region since it is lightly doped. Since the drain drift region is increased the breakdown voltage is higher, but the drain-source resistance also increases. This can be reduced by using several parallel cells on the semiconductor die.

Figure 1.21(c) shows part of the short-channel transistor of Figure 1.21(b), with an illustration of the parasitic components. The action of the parasitic transistor is usually avoided by a metal contact which bridges the source and the body, i.e. the base and the emitter of the parasitic transistor. Therefore theoretically this transistor should be off, but since the ohmic *p* layer puts a resistance into the base of this transistor it is possible to turn it on, for example via the parasitic capacitance (C_p) which can destroy the device. The parasitic capacitance is formed by the *p* base and the *n* drift region. The parasitic diode is useful when switching inductive loads.

Most power unipolar transistors are currently of the short-channel vertical type, and are made from many identical cells on the chip. A popular arrangement, the V-groove MOSFET or VMOS device, is shown in Figure 1.21(e). Its operation is identical to the DMOS structure except that it is a vertical structure, and there are now two parallel paths down to the drain. The metal gate shown in the figure is also usually replaced by a silicon gate.

1.7 Thyristors

1.7.1 Principles of operation

The thyristor, also known as a silicon-controlled rectifier (SCR), is a four-layered semiconductor device with three terminals, as illustrated in Figure 1.22(a). If the anode is connected to a positive supply, with respect to the cathode, then junctions J_1 and J_3 are forward biased and J_2 is reverse biased. The *p–n–p–n* structure can conveniently be represented by the *p–n–p* and *n–p–n* transistors, as shown in Figure 1.22(b). There are now clearly two possible gate connections G_1 and G_2, and the thyristor should be able to be turned on by putting current into G_2 or taking it out of G_1. The latter technique has poor gain and is not normally used, so the terminal is not brought out of the thyristor case.

If I_{CO1} and I_{CO2} are the leakage currents of the two transistors and α_1 and α_2 their gains, then the anode current of the combination is given by

$$I_A = \frac{\alpha_2 I_G + I_{CO1} + I_{CO2}}{1 - (\alpha_1 + \alpha_2)} \tag{1.15}$$

Usually the gains of the transistors are low, so that in the absence of any gate current the overall anode current is a little more than the leakage currents. When the sum of the gains approaches unity, however, the

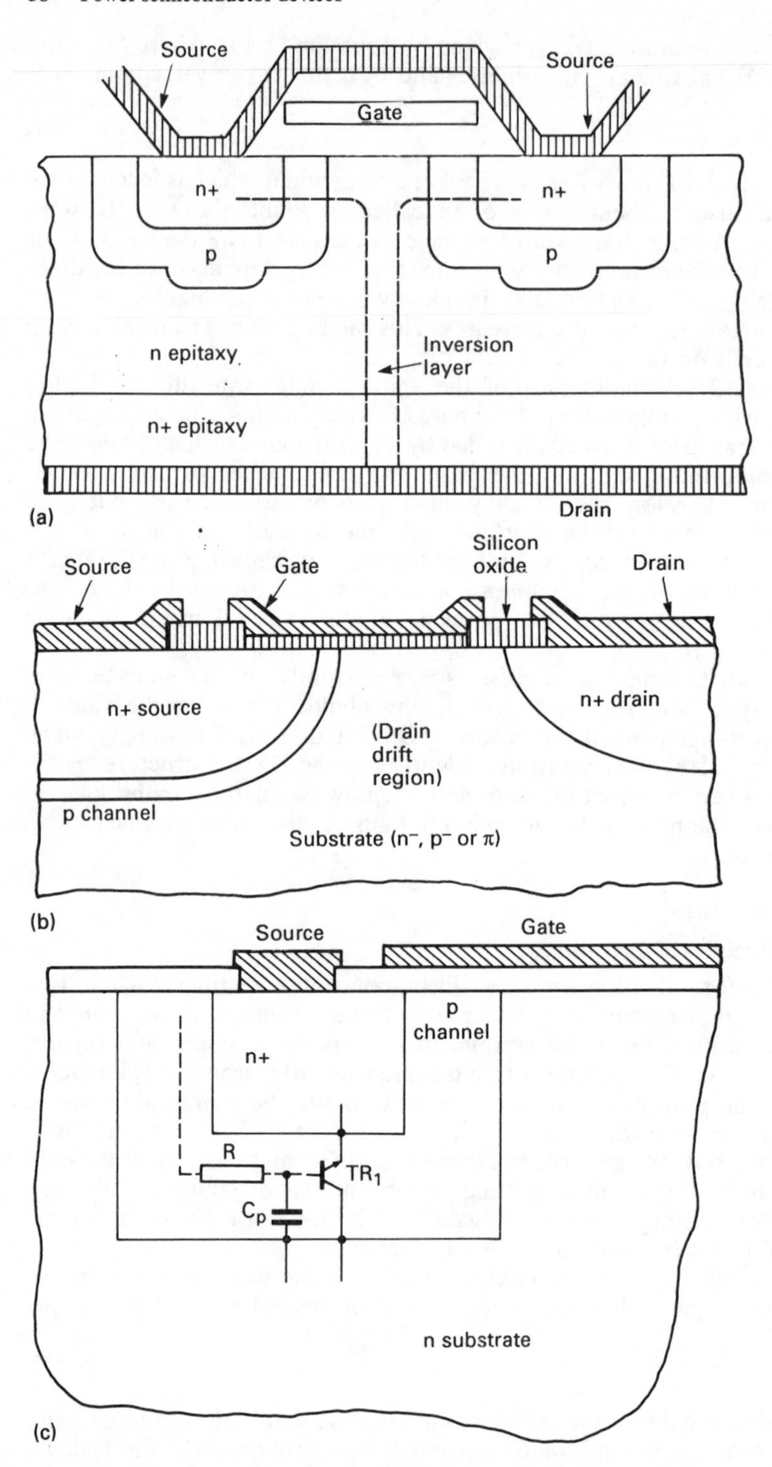

Figure 1.21 Unipolar transistor structures: (a) a vertical MOSFET; (b) double-diffused MOS (DMOS); (c) part of semiconductor structure showing parasitic components;

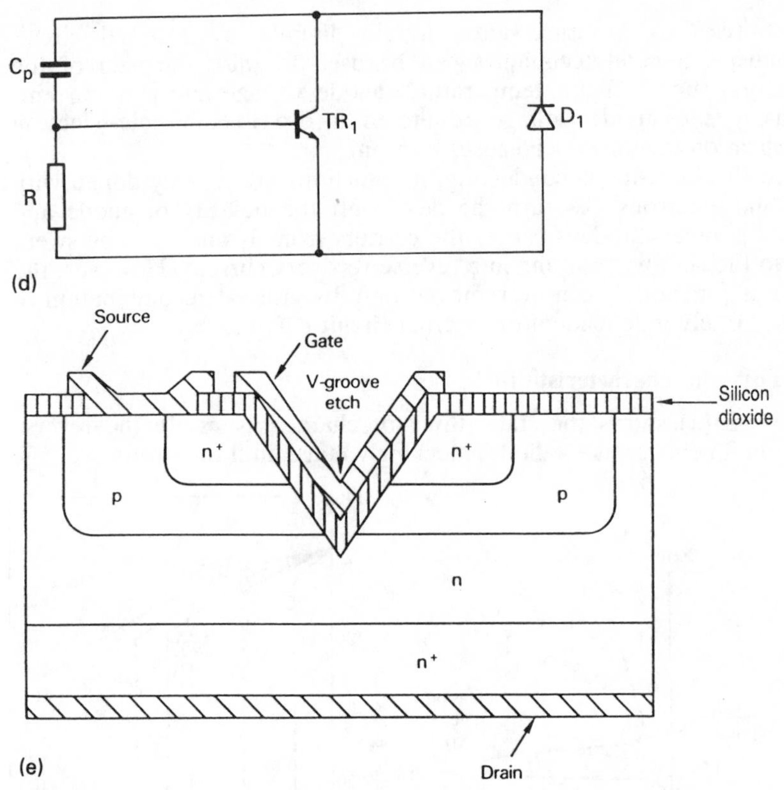

Figure 1.21 continued (d) equivalent circuit of parasitic components in (c); (e) V-groove MOS (VMOS)

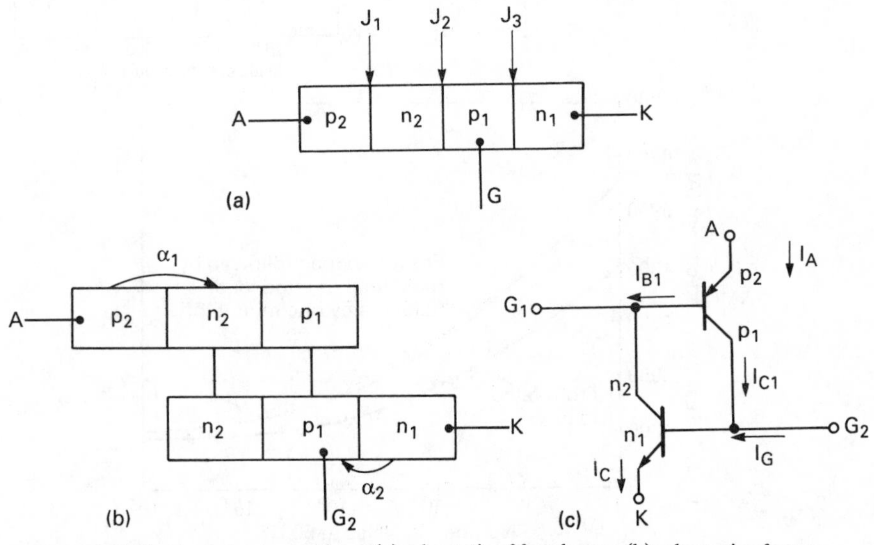

Figure 1.22 Thyristor representations: (a) schematic of four layers; (b) schematic of two transistors; (c) two-transistor analogy

current rises to a large value, usually limited by external circuit impedances. Several techniques can be used to raise the gain of the transistors, such as light, temperature, anode voltage and gate current. Increasing the anode voltage results in a rise in the hole–electron multiplication factors which causes turn-on.

Once the thyristor is conducting, its junctions are heavily doped with holes and electrons. To turn the device off the polarity of anode and cathode is reversed. This causes the carriers from J_1 and J_3 to be swept away to the supply, resulting in a reverse recovery current. However, the charge at junction J_2 can be removed only by gradual recombination, a process largely independent of external circuit conditions.

1.7.2 Thyristor characteristics

Figure 1.23(a) shows the static thyristor characteristics. In the reverse direction it behaves like a diode, blocking voltage until the reverse voltage

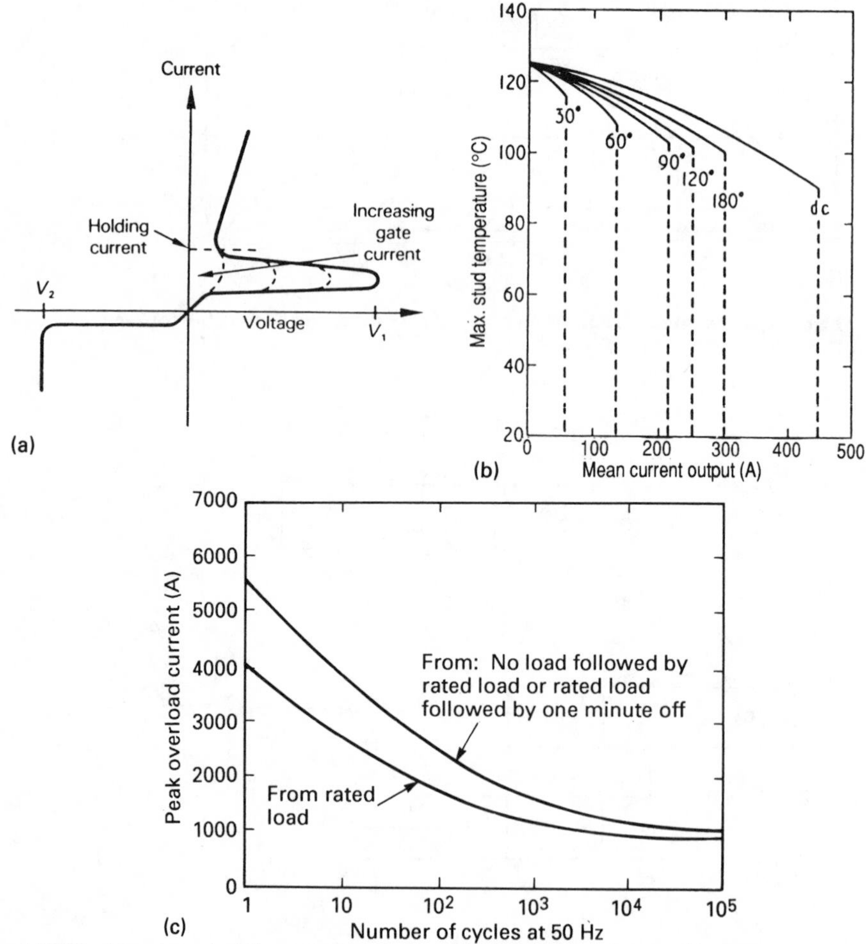

Figure 1.23 Thyristor characteristics: (a) static characteristics; (b) current curves; (c) surge rating;

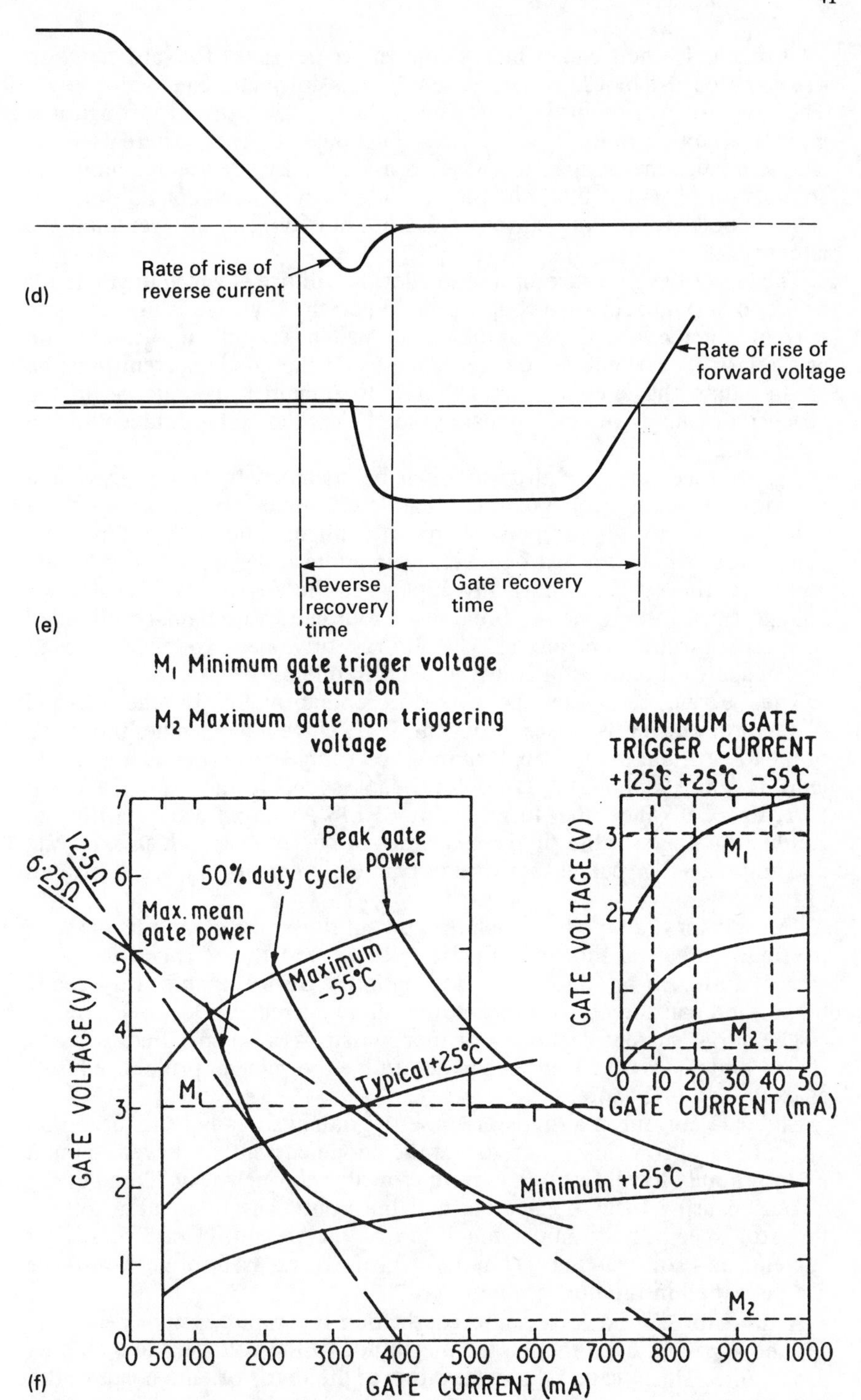

Figure 1.23 continued (d) current waveform during turn-off; (e) voltage waveform during turn-off; (f) gate characteristic

V_2 is reached, when avalanche breakdown occurs. In the forward direction the thyristor also blocks voltage until it breaks down into conduction at V_1. The thyristor will go into conduction so long as the current through it is greater than a value called the latching current. Thereafter its characteristics are similar to those of a diode, the device remaining in conduction provided the anode-to-cathode current does not fall below a value called the holding current. This holding current is lower than the latching current.

The larger the gate current, the smaller the voltage at which the thyristor breaks down into forward conduction. When the thyristor is fired its gate current is made very large, so the device switches rapidly into conduction at a relatively low anode voltage. Once again the anode current must be greater than the latching current for it to remain in conduction in the absence of gate drive, and it must not fall below the holding current or the device will turn off.

The voltage rating of thyristors can be specified by three terms, the repetitive blocking voltage, the breakdown voltage or peak repetitive voltage, and the non-repetitive peak voltage. These three may be considered in the forward or reverse directions. The repetitive blocking voltage is that which is normally applied to the device. The breakdown voltage causes the device to break over, but causes no damage provided the power through it is limited. The non-repetitive peak voltage can result in damage to the thyristor if it is applied frequently.

The current rating of the thyristor is determined by the thermal dissipation which this causes through it. RMS current determines the rating of the device, but in most applications the average current delivered to the load is more important. Therefore data sheets usually give average currents, but since the form factor (RMS/Average) varies with the conduction angle of the thyristor a series of curves exist, which show the maximum average current for various conduction angles. This is illustrated in Figure 1.23(b).

The shorter the time for which the current flows, the greater its possible overload value, as shown in Figure 1.23(c), and this is called its surge current rating. If rated load was flowing in the device, then it will already be hot and can therefore carry a lower surge current.

The surge current capability of the thyristor is also specified as its I^2t rating, and this is primarily used to determine the value of protective fuses, as explained in Chapter 5.

The gate current in a thyristor causes a gradual spread of the turned-on area of the silicon chip. Therefore if the anode current is allowed to build up too rapidly it will result in current crowding through a small area of the device, causing localised heating and burn-out. The di/dt rating of the thyristor specifies the maximum value of the permitted rate of rise of current, and construction techniques which can be used to increase these are described in the next section.

A thyristor can be forced into conduction if a rapidly rising voltage is applied across it, even though the magnitude of the voltage is less than its peak value. This is called the dv/dt rating of the thyristor, and it causes the thyristor to turn on due to current flow through the capacitance associated with the depletion layer (C_d) of the reverse-biased junction, given by

$$i = C_d \frac{dv}{dt} \tag{1.16}$$

Techniques which improve this rating are also illustrated in the next section.

The switching characteristic of a thyristor determines its switching losses and maximum operating frequency, much as it did for the transistor. The shape of the turn-on curve is very similar to Figure 1.12, where the current through the device rises as the anode-to-cathode voltage falls. The time for 10% turn-on, measured from the application of gate drive, is called the delay time, and that between 10% and 90% is the rise time. The sum of the delay and rise times is the turn-on time of the thyristor. The turn-on time is reduced if a steep rising gate pulse is used and the power in this drive is increased.

The turn-off waveforms of a thyristor are shown in Figures 1.23(d) and 1.23(e). During forward conduction all the junctions of the device are forward biased. To be able to block voltage the charge carriers must be removed, and this is usually done by applying a reverse voltage across the device, a process known as commutation. This causes holes and electrons to migrate from the centre junction to the end junctions, until all the carriers at the centre have recombined. A reverse recovery current flows during this process, and it starts to decrease, with a corresponding increase in reverse voltage across the thyristor, when recombination has been completed. The time between the start of the reverse recovery current and when it has fallen below a specified value, say 20% of the peak, is called the reverse recovery time. The magnitude of this time is largely determined by the amount of forward current which was flowing in the thyristor prior to turn-off and the rate of decay of this current.

A further time, called the gate-recovery time, is now needed before the thyristor is capable of again blocking forward voltage. This time increases at high junction temperatures and with an increase in the rate of reapplied forward voltage (dv/dt). The turn-off time of the thyristor is the sum of the reverse-recovery and gate-recovery times. Several techniques exist for reducing the turn-off time, such as adding gold doping which will decrease the minority carrier lifetime, but this will now increase the voltage drop across the device when it is in forward conduction.

The gate characteristics of a thyristor are important in the design of thyristor drive circuitry. These are given by means of the curves shown in Figure 1.23(f). The spread in the gate-to-cathode diode characteristic is given by the three curves at −55°C, +25°C, and +125°C, so that for any load line, such as the two shown for 6.25 Ω and 12.5 Ω, the operating point can be fairly widely spread. These must lie outside the box, bounded by the minimum gate voltage and gate current required to turn the thyristor on, whilst at the same time it must be below the relevant maximum power-dissipation curve. The shorter the gate duty cycle, the larger the permitted gate drive, and high-power pulse firing is often used for thyristors to ensure rapid turn-on.

In the example shown in Figure 1.23(f) to turn on all devices at 25°C would require a current and voltage exceeding 20 mA and 3.0 V. The graph is also useful in determining the gate drive impedance. For instance, suppose the thyristor is driven from a source of V through a resistance of

$R\,\Omega$. The peak gate dissipation would occur when the gate characteristic is such that it has a voltage drop equal to $V/2$. The dissipation is then equal to $V^2/4R$. For a duty cycle of $x\%$ equation (1.17) would hold, where $P_{G(AV)}$ is the mean gate power:

$$R = \frac{x\,V^2}{400\,P_{G(AV)}} \tag{1.17}$$

Therefore for V equal to 5 V and $P_{G(AV)}$ equal to 0.5 W, R equals 12.5 Ω for a 100% duty cycle and 6.25 Ω for a 50% duty cycle. R is the minimum value of the trigger source impedance. Plotting the above two load lines on Figure 1.23(f), they are seen to be tangential to their respective maximum gate power curves. Gate source impedance is usually chosen to give an operating load line between this line and the area bounded by the minimum firing gate voltage and current.

1.7.3 Thyristor construction

Several different thyristor structures are used to achieve various performance parameters. Figure 1.24(a) shows a conventional arrangement in which the edges of the silicon chip are bevelled to reduce stress at the junctions and so enable the voltage rating to be increased. An alternative technique for increasing voltage rating is to increase the thickness of the control layer, but this also results in an increased voltage drop across the thyristor.

Figure 1.24(b) shows what is known as the shorted emitter thyristor structure. It is used for applications which require a high dv/dt rating. In this structure the current generated by a rapidly rising voltage flows directly to the cathode, so that only a small proportion of it crosses the *p–n* junction as gate current.

It was mentioned earlier that when a thyristor turns on the initial conducting area is localised around the gate lead and then spreads relatively slowly over the whole silicon chip. The velocity of spread is about 50 to 100 m/s and since a modern, 2000 A mean rated thyristor would have a chip size of about 100 mm the turn-on time of the whole chip is relatively long, resulting in current crowding and a limit on the d*i*/d*t* rating of the thyristor. Where a large value of this rating is required special constructional techniques must be used. The most direct is to use an interdigitated gate structure, as shown in Figure 1.24(c), so that several areas around the chip periphery are triggered simultaneously. Unfortunately this also means that a much larger gate current is required since the gates are, in effect, connected in parallel. This disadvantage is overcome by the regenerative gate and amplifying gate structures.

The amplifying gate works on the principle shown in Figures 1.24(d) and 1.24(e). A low-power auxiliary thyristor is used to trigger the main high-power thyristor. The auxiliary thyristor is triggered by an external source, the power for the main thyristor being derived from the load which is being driven by the main thyristor.

The regenerative gate structure of Figure 1.24(f) makes use of a phenomenon known as 'emitter lip' resistance, which acts as an

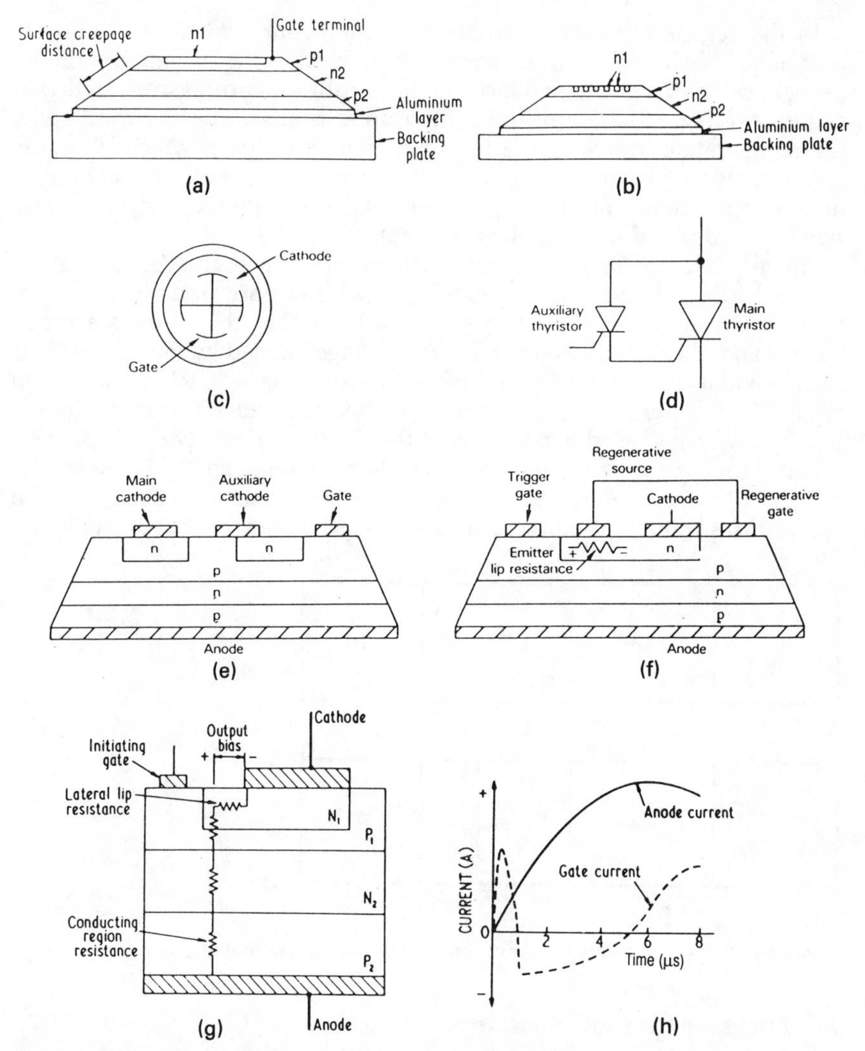

Figure 1.24 Thyristor configurations: (a) dice with bevelled edge; (b) shorted-emitter; (c) interdigitated gate; (d) amplifying gate schematic; (e) amplifying gate structure; (f) regenerative gate; (g) cell cross section showing effect of emitter lip resistance; (h) effect of emitter lip resistance on the current–time characteristic

impediment to the rapid turn-on of conventional thyristors. When a device is initially triggered most of the anode current tends to be squeezed through a small turned-on portion near the gate, and this effect is worsened by the fact that all thyristors have an 'emitter lip' between the cathode *n* layer and the cathode plate. This lip presents impedance to current flow, as shown in Figure 1.24(g), so that if the anode current is increasing rapidly a voltage is developed across it which opposes the gate signal. If the gate voltage is low, it may even be reversed, as in Figure 1.24(h), which would reduce the turn-on time still further.

In the regenerative gate arrangement of Figure 1.24(f) conduction is commenced by a signal on the trigger gate, which is the only gate terminal brought out of the package. Once anode-to-cathode current starts to flow it causes a voltage drop across the emitter lip resistance. This is picked up by the regenerative source and fed to the regenerative gate, which is usually an interdigitated arrangement. The external circuit needs to provide only a modest gate signal at the trigger terminal since the regenerative gate signals are derived from the load current.

All the thyristor structures described so far have been symmetrical, in that the forward and reverse blocking capabilities are optimised. Often, however, a thyristor is used in series with another thyristor or a diode which can provide the reverse blocking voltage needed by the circuit. It is then possible to use an asymmetrical thyristor (or ASCR), as shown in Figure 1.25, where a highly doped layer has been added near the anode junction, which stops the extension of the electrical field. The ASCR has a reduced forward voltage drop and lower turn-on time, and so lower losses, but also a lower reverse blocking voltage capability, so it is usually used in applications where it is connected in series with another rectifying device.

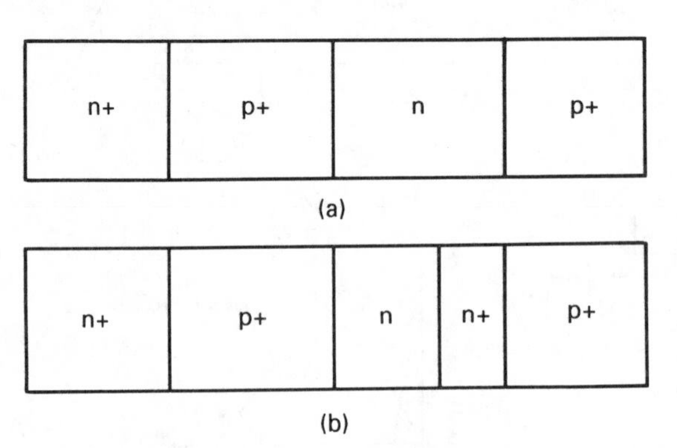

Figure 1.25 Comparison between a conventional and an asymmetrical thyristor

1.8 The gate turn-off switch

The gate turn-off switch (GTO) is similar in construction to a thyristor, having four layers, with three terminals, anode, cathode and gate. It is turned on by current flowing into the gate terminal, as for a conventional thyristor, but it can be turned off by taking current out of the gate, that is, with negative gate current. The operation of the GTO can be explained with reference to the two-transistor analogy of Figure 1.22(c). Here gate current I_G is shown as flowing into terminal G_2, but assuming that the two transistors are in conduction, then if gate current is taken out of this terminal, it will divert all the base current away from the *n–p–n* transistor, turning it off. If I_G is the value of this negative gate current, then turn-off will occur if it exceeds that given by

$$I_c \geqslant \left(\frac{\alpha_1 + \alpha_2 - 1}{\alpha_2}\right) I_A \qquad (1.18)$$

Defining the turn-off gain β_{OFF} of the GTO by equation (1.19), allows equation (1.18) to be rewritten as in equation (1.20):

$$\beta_{OFF} = \frac{I_A}{I_G} \tag{1.19}$$

$$\beta_{OFF} \leq \left(\frac{\alpha_2}{\alpha_1 + \alpha_2 - 1}\right) \tag{1.20}$$

Typical values of turn-off gain for a GTO are in the region of 5 to 10, and a large value of this parameter is desirable for an efficient device. It can be increased by making the gain of the *n–p–n* transistor (α_2) as close to unity as possible, by having a narrow base region (p_1) and heavily doping the emitter (n_1). At the same time the gain of the *p–n–p* transistor (α_1) must be kept low, by making its base (n_2) wide and adding gold doping, or using some other lifetime control technique. These modifications give a fast recovery device, but also result in the GTO having a higher voltage drop than a thyristor.

The symbol for the GTO is shown in Figure 1.26(a), being similar to that of a thyristor except for the gate, indicating the dual direction of current flow. The static characteristics for the GTO, shown in Figure 1.26(b), are also similar to that of the thyristor, once it has been turned on by sufficient

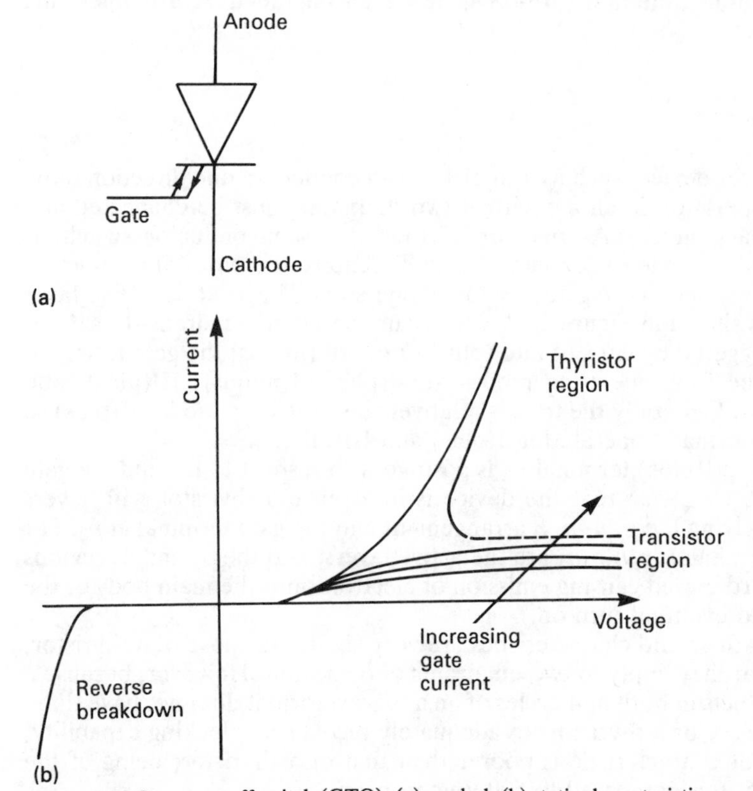

Figure 1.26 Gate turn-off switch (GTO): (a) symbol; (b) static characteristic

gate current. At low levels of gate current it operates in the transistor region, having a family of curves for increasing gate drive.

Most of the ratings and characteristics of a GTO are the same as those of a thyristor, except for the following. The GTO has a high forward blocking voltage rating, comparable to that attainable from a thyristor, but its reverse voltage rating is low, in the region of 10–20 V, and in this aspect it is similar to a transistor. The GTO also has a higher voltage drop and a higher latching current than a thyristor. This latter parameter means that the gate drive needs to be maintained for a longer period, during turn-on, to ensure that latching current is reached.

As expected, it is in the turn-off process that the GTO differs most from a thyristor. Prior to turn-off all the regions are heavily saturated, and the excess charge needs to be removed before turn-off can occur, resulting in storage time and fall time. The excess charge is first removed from the p_1 layer, in the vicinity of the gate terminal, and this recovered region then spreads over the whole junction. Current continues to flow by squeezing into parts of the junction not yet turned off, until eventually all the region recovers and the storage period ends. The effect of turn-off is therefore similar to the turn-on of a thyristor, where the current initially squeezes into the small turned-on region closest to the gate. Two parameters are quoted in GTO data sheets, which are not given for thyristors, the gate turn-off voltage (V_{GQ}) and the gate turn-off current (I_{GQ}). The turn-off time, turn-on time and turn-off gain are all degraded with temperature increase.

1.9 Triacs

A four-layer device, such as a thyristor, can conduct in one direction only, and for operation in an a.c. circuit two thyristors must be connected in a back-to-back mode. A triac, or TRIode AC semiconductor switch, is designed to be able to conduct in both directions, the onset of conduction being controlled by a gate, as for a thyristor. The triac is a five-layer device, as shown in Figure 1.27, which can operate in quadrants I or III. It can be triggered by current into (plus) or out of (minus) the gate terminal, so that the four operating modes are I(plus), I(minus), III(plus), and III(minus). Generally the triac is relatively insensitive in mode III(plus) so that it is normally operated in I(plus) and III(minus).

For mode I(plus) terminal L_2 is positive with respect to L_1, and the gate is positive to L_1, so that the device is in essence a thyristor with layers $p_1n_2p_2n_3$ giving the p–n–p–n arrangement and the gate terminal at p_2. For mode III(minus), $p_2n_2p_1n_1$ are the active layers, and the p_2 and n_3 regions are forward biased causing emission of electrons into the main body of the device and eventual turn-on.

Triac ratings and characteristics are very similar to those of a thyristor, except that they apply to two quadrants of operation. However, because a triac conducts in both half cycles of an a.c. waveform it does not have time, as in the case of a thyristor, to adequately recover its blocking capability, so its dv/dt characteristic is poorer than that of a thyristor, being of the order of 5 V/μs compared to 500 V/μs respectively.

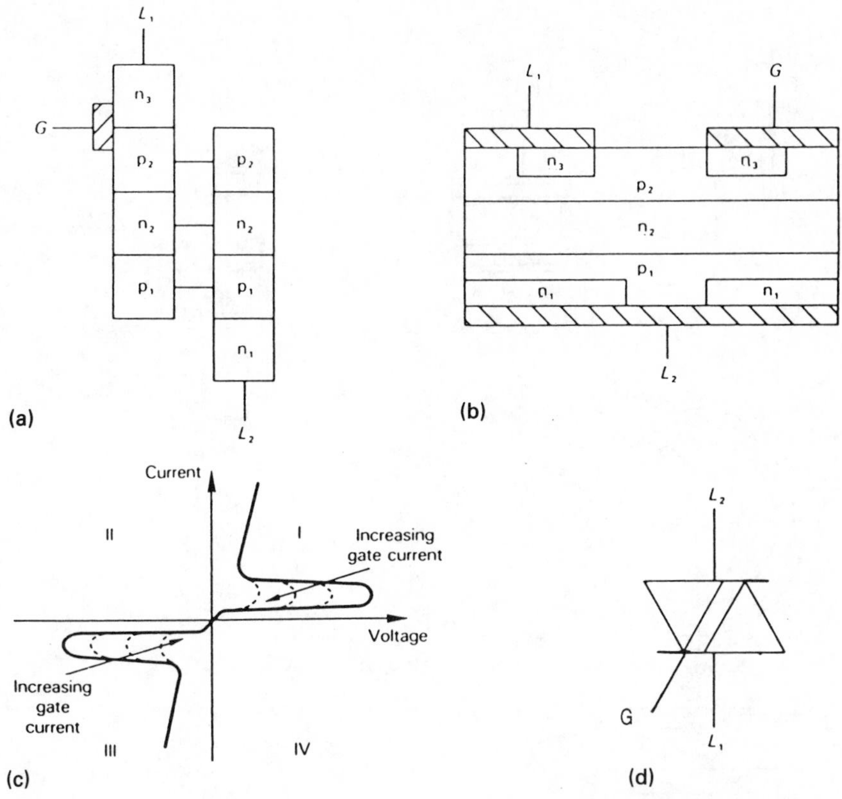

Figure 1.27 Triac: (a) five-layer representation; (b) construction; (c) characteristic; (d) symbol

1.10 Power semiconductor packaging

The package used with power semiconductor devices has to accomplish several functions, some of which are as follows:

(i) Provide a convenient method for electrical current to flow from the device.
(ii) Enable the heat generated in the silicon to be conducted away to the ambient, usually via a heatsink.
(iii) Give mechanical support to the semiconductor dice.
(iv) Protect the semiconductor dice from the chemical effects of the environment.
(v) Give adequate insulation between the gate, cathode and anode terminals of the device.

Many different packages have been used for power semiconductors, from the small metal can TO18, TO39, TO3 through to plastic TO92, TO220, TO218, and the large stud-mounted and 'hockey puck' devices. A few of the larger packages are illustrated in Figure 1.28. The relevant surface of the silicon dice, which can be the cathode or anode for a power

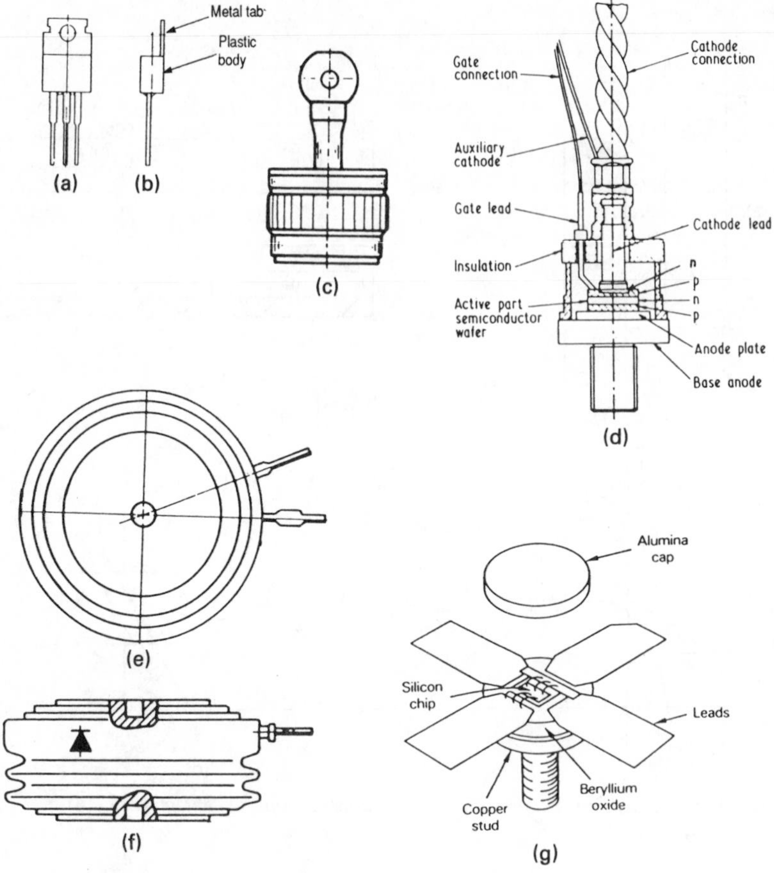

Figure 1.28 Power semiconductor packages: (a) and (b) TO220; (c) press fit; (d) stud; (e) and (f) hockey puck; (g) high-frequency

diode, the collector for a power transistor, and the anode for the thyristor, is coated with a multilayer structure of titanium, nickel and silver. Gold was used in the past but is becoming less popular. This dice is then attached to the semiconductor package header, made from a copper–nickel alloy, using soft or hard solders.

Soft solders, such as lead, silver, indium, antimony, or a mixture of these, are easier to use, but are not able to withstand repeated thermal cycles without fatigue. Hard solders, such as molybdenum, and eutectics like Au-Ge, Au-Sn, Au-Si, are able to withstand many more thermal cycles. It has been shown that the number of cycles to failure (N) for a silicon dice of diameter D, cycled through a temperature range $\mathrm{d}T$, is given, for hard-soldered joints, by

$$N = \left(\frac{K}{D\,\mathrm{d}T}\right)^2 \tag{1.21}$$

where K is a constant.

Connections to the emitter and base of a transistor, or the gate and cathode of a thyristor, are usually made by ultrasonic bonding using aluminium or aluminium–magnesium wire. Aluminium wire is needed for the gate terminal of a thyristor to ensure a non-rectifying contact.

For very large devices, such as the hockey puck package shown in Figure 1.28, the silicon dice is held in contact with the package using pressure only, usually in the form of springs or washers. This is known as compression bonding, and the absence of all solders makes this arrangement specially suitable for large thermal stresses. Therefore, this technique is commonly used for silicon dice greater than 25 mm in diameter.

Plastic packages such as TO22O are relatively inexpensive, easy to mount, and the epoxy used gives a good hermetic seal.

The stud-type encapsulation is shown in Figure 1.28, for a thyristor, with the dice hard soldered into the package, although compression bonding could be used instead. The dice is soldered to the base and encapsulated in ceramic, usually a top metal being used which is welded to the header.

An important consideration in high-frequency applications is the package, which must have low stray capacitance and good cooling properties, Figure 1.28(g) showing one type of package which can be used for high-frequency transistors. Twin-emitter leads are used, which gives symmetry of board layout when devices are combined for greater power. The leads are low-inductance strip lines, which can interface to microstrip lines used in UHF–VHF equipment. Beryllium oxide forms the dice insulator, since it has good thermal conductivity and the dice is located onto a copper stud which is bolted to a heatsink.

1.11 Voltage-reference diodes

Voltage-reference diodes are two-layer devices which are used primarily for providing a stable voltage reference, or for overvoltage surge suppression. The device is capable of passing a large current whilst maintaining a high voltage across it, hence it dissipates considerable amount of power, placing it in the category of a power semiconductor. Figure 1.29(a) shows the symbol for a voltage-reference diode. In the forward direction it behaves like an ordinary diode, whilst in the reverse direction it functions as a reference diode.

The operation of the voltage-reference diode is based on two distinct effects, zener and avalanche, although these devices are often referred to collectively as zener diodes. If the *p* and *n* regions of the diode are heavily doped, then the depletion region between them will be narrow. Applying a low voltage in the reverse direction across the device will now cause a high electric field to be formed across the junction, given by the ratio of this voltage to the depth of the depletion layer. When this field exceeds a critical value, equal to 3×10^{-5} V/cm, electrons will gain sufficient energy to break away from their bonds (see Figure 1.5) causing a large current to flow. This is known as zener breakdown and the characteristic is shown in Figure 1.29(b). The zener breakdown occurs below about 5 V and has a fairly gentle knee in its characteristic curve.

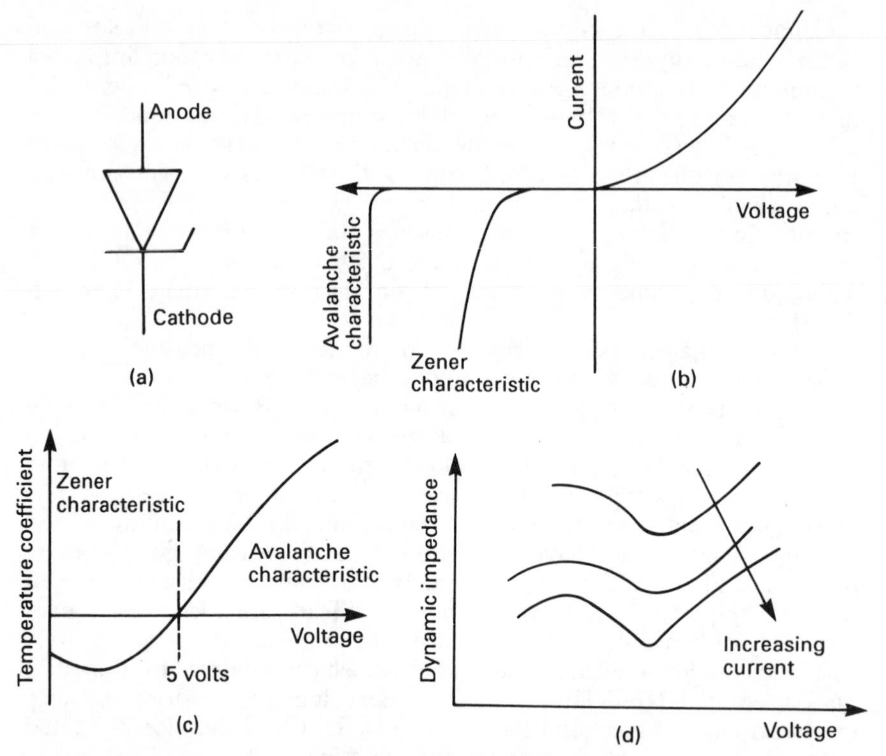

Figure 1.29 Voltage-reference diode characteristics: (a) symbol; (b) static characteristic; (c) temperature coefficient of reference voltage; (d) dynamic impedance

If the *p* and *n* layers of the diode are lightly doped, then they will have a wider depletion layer between them, so that the critical field strength for the zener effect cannot be reached, even under relatively large reverse voltages. However, the wider space charge region now allows the electrons to reach sufficient velocity, if the distance is greater than the mean free space for electrons at a given temperature, such that, on impact, they have enough energy to knock out other electrons from their crystal structure. These new electrons then free further electrons, so that the current builds up to a high value very quickly. This is known as the avalanche effect and, as shown in Figure 1.29(b), it has a very sharp knee in its characteristic, occurring at voltages in excess of about 7 V. In between 5 V and 7 V a combination of the zener and avalanche effects takes place.

Voltage-reference diodes are available in the range of 4–200 V by control of the doping profiles. Diodes below about 10 V are usually made by alloyed junctions, whilst above this the junctions are diffused into an epitaxial layer. The tolerance in the breakdown voltage is usually ±5%, but it is possible to select to tighter limits. The zener voltage decreases with temperature, whilst the avalanche voltage increases with temperature, so that the temperature coefficient of reference voltage follows a curve similar to that shown in Figure 1.29(c). By combining zener and avalanche diodes it is possible therefore to obtain devices with very good temperature characteristics. The same effect is obtained by using an avalanche diode in series with a forward-biased conventional diode, since a forward-biased junction has a negative temperature coefficient of voltage. These devices are often referred to as stabistors.

As seen from the characteristics of Figure 1.29(b), the reference voltage is not strictly constant, but varies with the current through the device. There is therefore an impedance Z_d associated with this, called the dynamic impedance and given by

$$Z_d = \frac{\Delta V_Z}{\Delta I_Z} \qquad (1.22)$$

where ΔV_Z and ΔI_Z are the voltage and current at any point on the characteristic. Dynamic impedance curves are illustrated in Figure 1.29(d).

Diodes used for voltage transient suppression need to have a very fast turn-on time, whereas normal voltage-reference diodes have a parasitic inductance, which causes the turn-on to be relatively slow. Cellular technology is now used, similar to that of a power MOSFET, to give a fast device which can handle about 200 A, with a non-repetitive power dissipation, without a heatsink, of 2500 W for 1 ms. For repetitive operation a heatsink is required.

1.12 Power Darlington

One of the disadvantages of using power transistors is their low gain at high current levels, which requires a large base current, placing stringent requirements on the base drive circuitry. This is one of the reasons why thyristors, which need a fraction of the gate power for turn-on, are so popular. The power Darlington, named after its inventor, also overcomes this disadvantage by using a combination of two transistors, one to provide the power output and one for the drive input. Figure 1.30 shows a variety of such Darlingtons, in which transistor TR_1 provides the base drive to the power output transistor TR_2. The overall gain of the combination is now equal to the product of the gains of the two individual transistors. However, the disadvantage is that the voltage drop is also high, being equal to the collector emitter saturation voltage of the output transistor plus the base emitter saturation voltage of the driver transistor.

Additional components are often added to the two-transistor pair, as in Figure 1.30(d). Resistors R_1 and R_2 prevent leakage current amplification, whilst diode D_1 is used to speed up the turn-off, by removing the charge carriers stored in the base of the output transistor. Diode D_2 protects the output transistor by preventing reverse voltages across it. The Darlington pair, with all the external components shown in Figure 1.30(d), can be made from discrete components, although it is more usual to build them as a monolithic structure into a semiconductor dice.

1.13 Photothyristors

Photothyristors have the same basic structure as conventional thyristors and function in similar modes, except that they are triggered by light

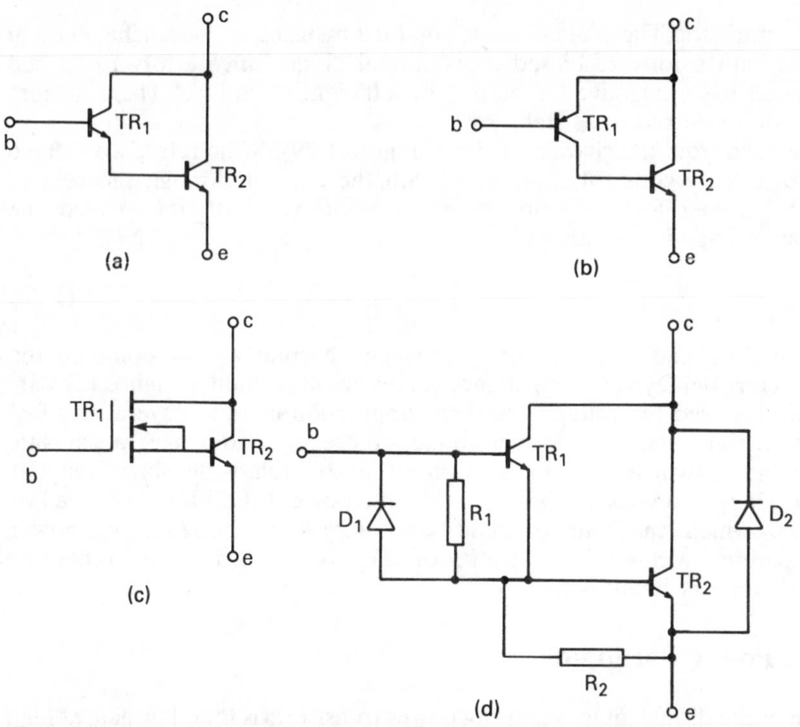

Figure 1.30 Darlington transistor arrangements: (a) conventional; (b) composite; (c) MOS-bipolar; (d) conventional with bias components added

instead of gate current. The symbol for the photothyristor, also called a light-activated SCR (LASCR) or a light-triggered thyristor (LTT), is shown in Figure 1.31(a), and it is seen that the gate terminal is sometimes brought out so that a combination of light and gate current can be used to control the device. The advantage of using light triggering is that the device is now insensitive to electrical signals, which can cause faulty operation in electrically noisy environments.

Figure 1.31(b) shows the construction of a photothyristor. In the absence of light, junction J_2 is reverse biased. When illuminated, hole–electron pairs are created in the vicinity of J_2 and are swept across to the anode and cathode. This acts as a triggering current, and if it is large enough it will turn on the thyristor. Usually a bias can be applied on the gate lead to vary the threshold light level for turn-on.

The photothyristor is made from thin layers to enable greater light penetration, but this also results in a lower blocking voltage. It is designed with a larger junction to increase light sensitivity, but this has the effect of making it more sensitive to variations in temperature and voltage, and in increasing the turn-off time. Resistors can be added between the gate and cathode to reduce its susceptibility to noise and dv/dt effects, but this degrades its sensitivity to light triggering. Commercially available photothyristors rival conventional devices, having voltage and current ratings in excess of 6 kV and 3 kA.

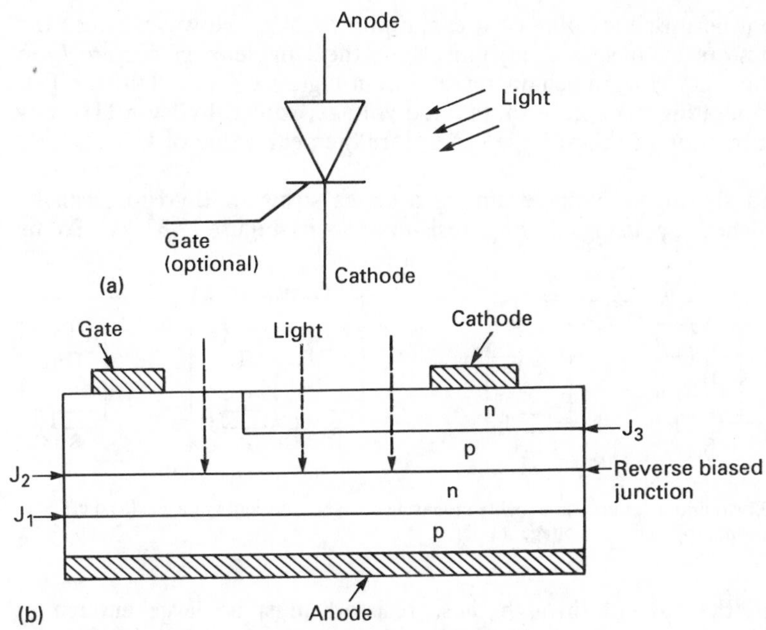

Figure 1.31 The photothyristor: (a) symbol; (b) construction

1.14 Series and parallel operation

Semiconductors are available today with the ability of controlling many megawatts of power, therefore it is only in very specialised applications that several devices need to be connected in series or in parallel, in order to increase the voltage or current rating, respectively, over that available from a single device. The techniques for doing so are illustrated in this section with respect to the thyristor, although these apply equally to any of the other power semiconductors described in the book.

1.14.1 Series operation

Figure 1.32(a) shows two thyristors connected in series to share a voltage V. If the peak-rated voltage of each thyristor is V_{pk}, it is hoped that the

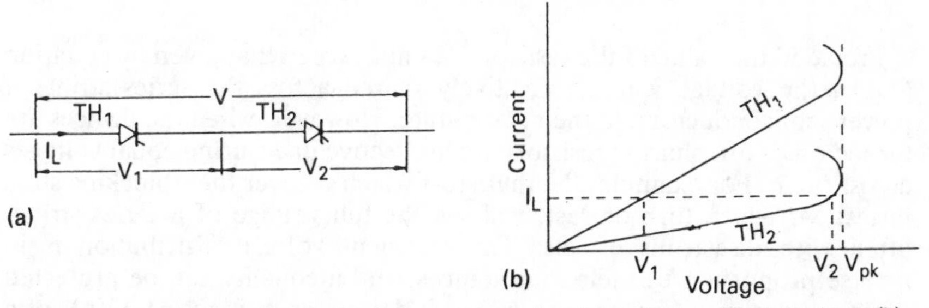

Figure 1.32 Unequal voltage distribution in a series string: (a) series-connected thyristor; (b) spread in device characteristics

maximum permissible value of V can approach $2V_{pk}$. However, since the two thyristors are in series they must share the same leakage current I_l, so that if they had blocking characteristics as in Figure 1.32(b), thyristor TH_2 will be operating very close to its rated voltage, whilst thyristor TH_1 only blocks a fraction of this voltage. Therefore the peak value of V is severely limited.

Forced sharing of voltage among a series string of thyristors can be accomplished by using sharing resistors, as in Figure 1.33(a). To be

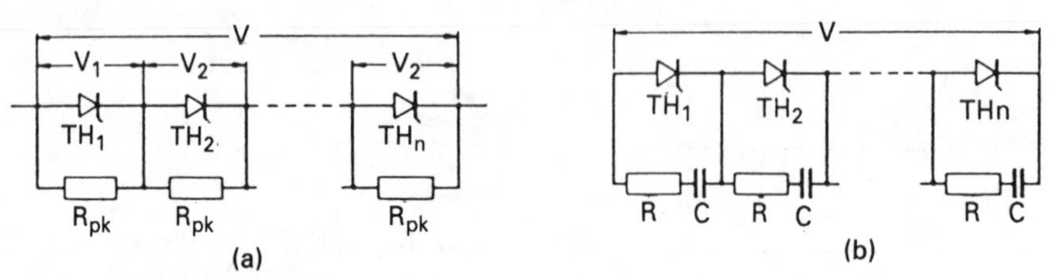

Figure 1.33 Additions for greater equality in sharing d.c. and transient voltages: (a) d.c. voltage sharing; (b) transient voltage sharing

effective, the current through these resistors must be large enough to swamp the inequality in the thyristor leakage currents. If R_{pk} denotes the maximum value of this resistance, in the limiting condition the worst case of unequal sharing occurs when one thyristor, say TH_1, has negligible leakage whereas the remaining have maximum leakage current, I_l. If there are n cells in series, then equations (1.23) and (1.24) can be written down, where V_1 is the peak-rated voltage of the thyristors and V is the maximum voltage which can be safely applied to the series string when sharing resistor R_{pk} is used:

$$V - V_1 = V_2(n - 1) \tag{1.23}$$

$$V_1 = \left(I_l + \frac{V_2}{R_{pk}}\right) R_{pk} \tag{1.24}$$

Solving equations (1.23) and (1.24) gives

$$R_{pk} = \frac{nV_1 - V}{(n - 1)\, I_l} \tag{1.25}$$

Provided the value of the resistor does not exceed that given by equation (1.25), the voltage will be effectively shared across the series string of power semiconductors, in their static state. However, when the devices are turned on or off, sharing resistors are ineffective in ensuring equal voltages across them. For example, the thyristors which recover their blocking state fastest, or which turn on last, will see the full voltage of a series string, often with disastrous effects. This transient voltage distribution is in inverse proportion to their capacitances, and inequality can be protected against by using a capacitor across each device, as in Figure 1.33(b), of a value greater than the device capacitance. This slows down the rate of change of voltage across the device during turn-off and turn-on. A low-valued resistor is also now required, in series with the capacitance, to limit the large discharge currents of the capacitors, which would otherwise occur when the thyristors are turned on. The value of the capacitance C can be obtained from empirical equation (1.26), where I_F is the peak forward current through the thyristor before commutation occurs and V_1 is the maximum voltage rating of the thyristors:

$$C = \frac{5I_F}{V_1} \tag{1.26}$$

In a normal application the sharing components for static and dynamic operation are combined, so that each power device would have across it both the sharing components shown in Figures 1.33(a) and (b).

1.14.2 Parallel operation

Several power semiconductor devices, connected in parallel, all have the same voltage across them when they are conducting, but due to unequal characteristics they carry different currents. This is illustrated for a thyristor pair in Figure 1.34. Once again it is possible to swamp the inequality of the device voltage drops by series-connected impedances, such as inductors or resistors, or by current-sharing reactors. Inductances are effective only during transient conditions, when the current is changing. Current-sharing reactors, where the imbalance in different parts of the reactor forces equal current flow in the various parallel arms, are expensive and bulky. Resistors are cheaper, but dissipate power.

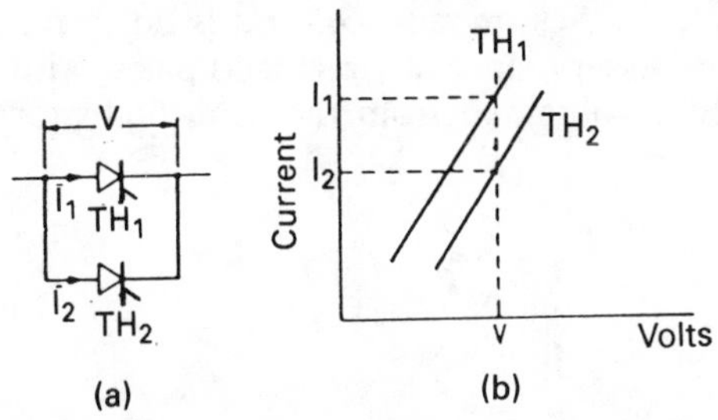

Figure 1.34 Unequal current distribution in paralleled thyristors: (a) paralleled thyristors; (b) spread in forward characteristics

Figure 1.35 shows two thyristors connected in parallel with sharing resistors. If V_1 and V_2 are the maximum and minimum voltage drops across the thyristors and I_1 and I_2 the permitted variations in the shared current, then equation (1.29), which gives the desired value of the sharing resistor R, can be derived from equations (1.27) and (1.28):

$$V = I_1R + V_1 \tag{1.27}$$

$$V = I_2R + V_2 \tag{1.28}$$

$$R = \frac{V_2 - V_1}{I_1 - I_2} \tag{1.29}$$

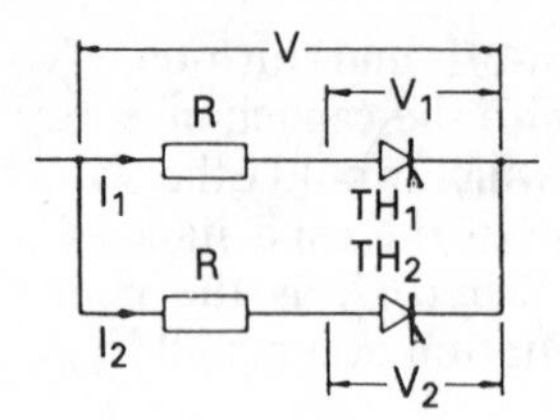

Figure 1.35 Series resistance for forced current sharing

The series resistors force d.c. more equally through paralleled power devices, but fail to compensate for unequal turn-on times, latching currents or holding currents. For instance, in a parallel-connected set of thyristors, if one device turns on before the rest, it will be forced to momentarily carry the full load current. This need not be serious so long as the di/dt or surge rating of the thyristor is not exceeded. If required, a small reactor can be connected in series with each device, to compensate for unequal turn-on times.

If the latching currents of a parallel set of thyristors differ, it could mean that for pulse firing some thyristors will reach their latching currents before the end of the pulse whereas others will not. This will mean that only these devices will conduct and support the total load current, and clearly the firing signal must be long enough to prevent this from occurring. Similarly, if the thyristor currents decrease momentarily, devices with holding currents below this value will turn off. When the load current again increases to its full load value the off thyristors will not turn on unless refired, and some devices may be overloaded. It is therefore essential to ensure that the load current does not fall to too low a value.

Although series connection is common, paralleling power devices is inefficient, due to the loss across the series impedance, and is not often used. In the instances where it is necessary, specially matched pairs, with equal voltage drops, can normally be obtained from the semiconductor manufacturers.

Chapter 2

Thermal design

2.1 Introduction

The parameter which has the greatest impact on the design of power semiconductor devices is the power which they generate and that which can be dissipated through it. This power dissipation results in temperature rises and bond fracture, due to uneven expansion between the silicon dice and its joints within the package. Special precautions must be taken to dissipate this heat to the ambient in order to prevent excessive temperature rises within the device.

This chapter describes the losses which occur within power semiconductors, their thermal characteristics, and the techniques which are used to cool the devices.

2.2 Power losses in semiconductors

There are four main sources of power loss within a semiconductor, as follows:

(i) Power loss during forward conduction. For a diode this is given by the product of the current through the device and the forward voltage drop across it. The same applies for a thyristor, but since the conduction angle can now be varied the power loss curves are usually given in data sheets as in Figure 2.1, where the average current over the whole conduction cycle is used. For a transistor the forward conduction loss is given by the product of the collector current and voltage, to which is added the base dissipation, equal to the product of the base current and voltage. Usually the base losses are small compared to the collector loss.

(ii) Leakage loss, when the power semiconductor is blocking voltage in the forward or reverse direction. This can occur when a diode or thyristor is reverse biased, or when a transistor or thyristor is forward biased but not turned on. These losses are usually small in comparison with the forward conduction losses.

(iii) Switching losses which occur during turn-on or turn-off of the power semiconductor. Although relatively small, these losses can become appreciable when the device is being operated at high frequencies.

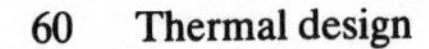

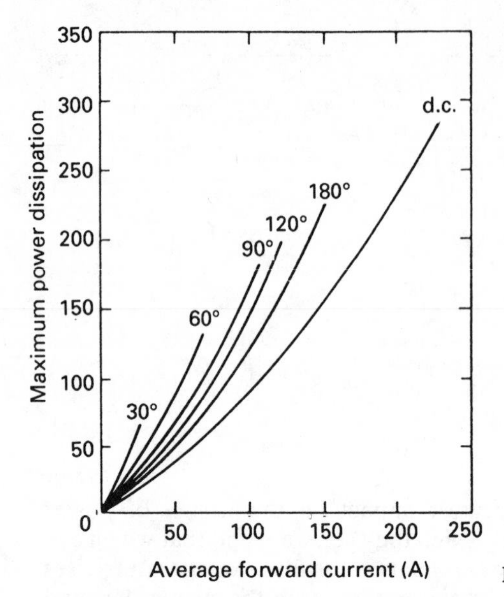

Figure 2.1 Power loss curves for a thyristor

They can occur due to the gradual turn-on and turn-off processes, which enable large amounts of currents to flow whilst the voltage across the device is still high. Figures 1.8(c) and 1.8(d) show one example of the turn-on phenomenon, whilst Figures 1.8(e) and 1.8(f) show an example of a turn-off characteristic. These turn-on and turn-off losses give rise to power spikes in the overall power-dissipation curve, as in Figure 2.2, the mean power loss due to switching being seen to be relatively small at low frequencies.

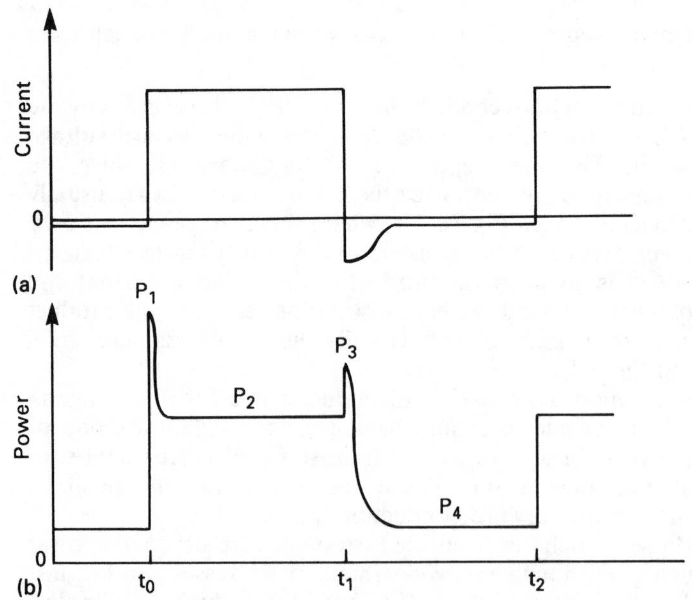

Figure 2.2 Power losses in a semiconductor during a typical cycle: (a) current; (b) power loss

(iv) Losses in the control terminal of the power semiconductor. The loss in the base of a transistor has already been considered as part of the forward conduction loss in item (i) above, since it is always present whilst the device is on. However, devices such as thyristors need only be pulse fired, so the gate power loss can be separated from the forward conduction loss. Four parameters are now usually defined; the peak gate power P_{GM}, i.e. the maximum value of the product of the forward gate current and voltage which is permissible; the average gate power $P_{G(AV)}$, which is the maximum value of the forward gate current and voltage averaged over one cycle; and peak reverse gate power P_{GMR} and average reverse gate power $P_{GR(AV)}$, which are the corresponding reverse values. Figure 1.23 shows the peak gate power curves within which the gate drive locus needs to stay.

2.3 Semiconductor thermal characteristics

A power semiconductor mounted on a heatsink can be analysed by analogy with electrical circuits in which the flow of current is replaced by heat transfer and the electrical impedances by thermal resistances. The unit of heat transfer is measured in joules per seconds or watts, and the unit of thermal resistance is in degrees centigrade per watt. Therefore if Q is the thermal power in watts being dissipated within a device, and $\mathrm{d}T$ is the temperature difference across the device in degrees Centigrade, then the thermal resistance R_{th} of the device is given by

$$R_{th} = \mathrm{d}T/Q \qquad (°\mathrm{C/W}) \tag{2.1}$$

A complex thermal circuit, such as a power semiconductor mounted on a heatsink, can be broken into its separate parts and then analysed using equation (2.1). Figure 2.3 shows the equivalent circuit of such an assembly.

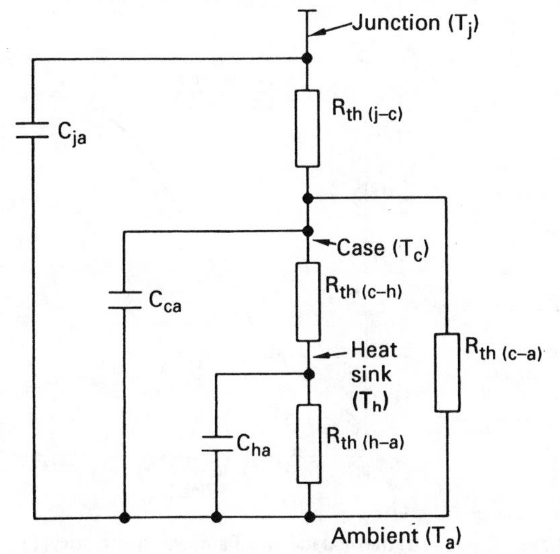

Figure 2.3 Equivalent circuit of a semiconductor device mounted on a heatsink

If T_j and T_c are the temperatures of the semiconductor junction and its case, and $R_{th(j-c)}$ the thermal resistance between junction and case, then for a power flow of Q W between junction and case the thermal resistance is given by

$$R_{th(j-c)} = (T_j - T_c)/Q \tag{2.2}$$

Similarly, the other thermal resistances between case and heatsink, and heatsink and ambient, can be obtained. Figure 2.3 also shows the thermal capacitances (C_{ja}, etc.) which can generally be ignored in any r.m.s. calculation and are only used for transient analysis. The thermal resistance between case and ambient is usually large compared to that through the heatsink, so that it too can be ignored. The equivalent circuit therefore simplifies to three elements in series, and for this total system the thermal resistance between semiconductor junction and ambient is given by equation (2.3) and the temperature rise by equation (2.4):

$$R_{th(j-a)} = R_{th(j-c)} + R_{th(c-h)} + R_{th(h-a)} \tag{2.3}$$

$$T_j - T_a = Q\,R_{th(j-a)} \tag{2.4}$$

So far, the discussions have dealt exclusively with instances in which there is steady state power loss in the semiconductor. Often, however, only intermittent operation is required, and Figures 2.4(a) and 2.4(b) show the effect of a step increase in power on the junction temperature. The power device, along with any heatsink used, presents a finite thermal mass so that the junction temperature increases gradually. Since thermal resistance is defined as the ratio of the rise in temperature to the power increase, this impedance will build up with time, as in Figure 2.4(c), and this is referred to as the transient thermal resistance ($R_{th(t)}$). It is generally difficult to calculate the transient thermal resistance accurately for an assembly, and it is measured experimentally and published as a graph in data sheets. It

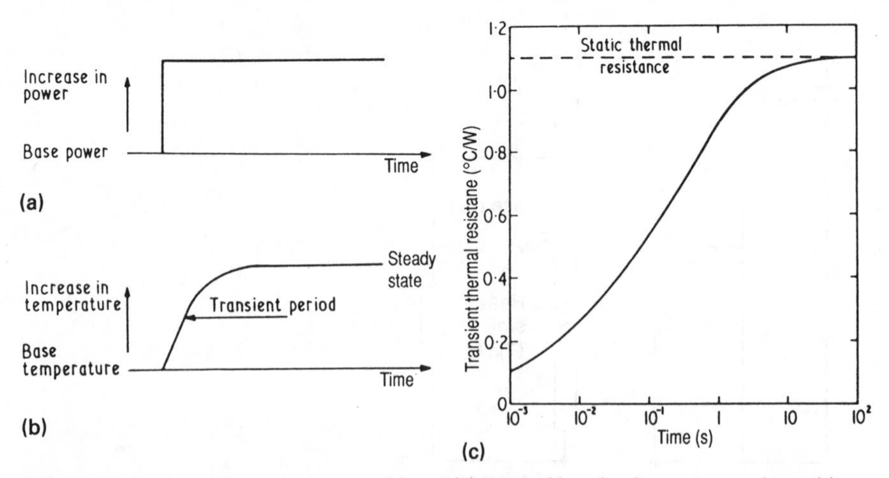

Figure 2.4 Transient thermal resistance: (a) and (b) thermal inertia of a power transistor; (c) transient thermal resistance curve

should be noted that this information applies for a fully conducting cell, such as when operating into fault conditions, and it cannot be used for power devices which are in the process of being turned on, since now the turned-on area is still spreading.

Figure 2.5 shows some examples of power pulses, and how the temperature rise can be calculated under these conditions. The single pulse shown in Figure 2.5(a) can be considered to be made up of two pulses, P_m

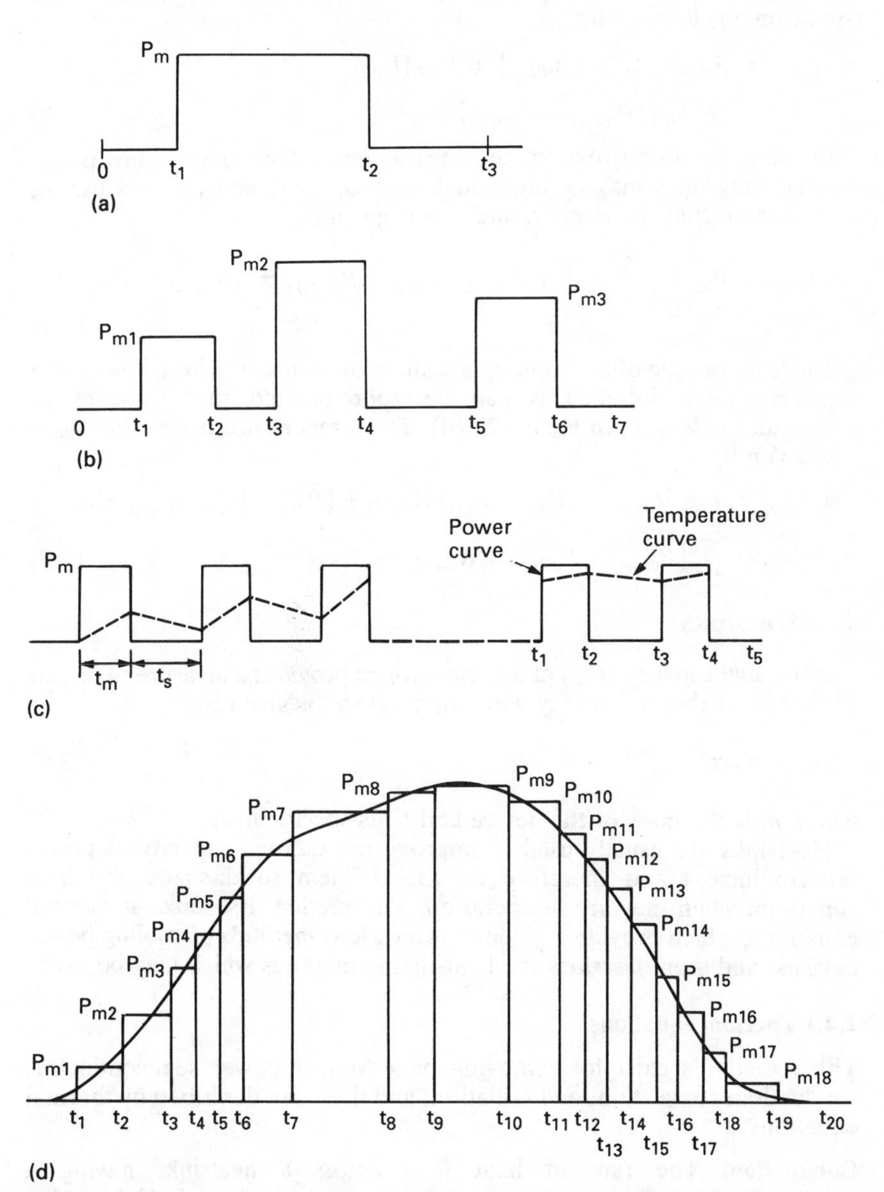

Figure 2.5 Illustration of power pulses: (a) single pulse; (b) multi-pulse; (c) pulse train; (d) irregular pulse

starting at t_1 and $-P_m$ starting at t_2. The junction temperature rise at time t_3 is then given by

$$\begin{aligned} dT_{(t_3)} &= P_m R_{th(t_1)} - P_m R_{th(t_2)} \\ &= P_m \left[R_{th(t_1)} - R_{th(t_2)}\right] \end{aligned} \tag{2.5}$$

The case of multiple-power pulses, shown in Figure 2.5(b), can be considered as a series of single superimposed pulses, and the temperature rise at time t_7 is given by

$$\begin{aligned} dT_{(t_7)} = {} & P_{m1} \left[R_{th(t_1)} - R_{th(t_2)}\right] + P_{m2} \left[R_{th(t_3)} - R_{th(t_4)}\right] \\ & + P_{m3} \left[R_{th(t_5)} - R_{th(t_6)}\right] \end{aligned} \tag{2.6}$$

In the pulse train situation, shown in Figure 2.5(c), the last few pulses are the only ones making individual contributions, and the rest can be averaged, so that the temperature rise is given by

$$dT_{(t_5)} = P_M \left[\frac{t_m}{t_s + t_m}\right] + P_M \left[\{R_{th(t_1)} - R_{th(t_2)}\} + \{R_{th(t_3)} - R_{th(t_4)}\}\right] \tag{2.7}$$

The last example of transient operation to be considered here is than of a non-rectangular pulse. This can be approximated into a series of rectangular pulses, as in Figure 2.5(d). The temperature rise at time t_{20} is now given by

$$\begin{aligned} dT_{(t_{20})} &= P_{m1} R_{th(t_1)} + (P_{m2} - P_{m1}) R_{th(t_2)} + (P_{m3} - P_{m2}) R_{th(t_3)} + \ldots \\ &= \sum \{P_{m(n)} - P_{m(n-1)}\} R_{th(t_n)} \end{aligned} \tag{2.8}$$

2.4 Heatsinks

The thermal capacity (C_{th}) of a device, which provides a measure of its rate of change of thermal energy with temperature, is given by

$$C_{th} = C.m \tag{2.9}$$

where m is the mass of the device and C its specific heat.

Heatsinks are usually used to improve the thermal capacity of power semiconductors and therefore to enable them to dissipate the heat generated when they are in operation. This section first looks at thermal equations, which provide a guide to the various methods of cooling power devices, and then describes the heatsinking methods which may be used.

2.4.1 Thermal equations

Three methods exist for removing heat from a power semiconductor, conduction, convection, and radiation, and these are described by thermal equations.

Conduction. The rate of heat flow across a heatsink, having a cross-sectional area of a, a thickness of d, and a thermal conductivity of k_T, is given by equation (2.10), where dT is the temperature difference across the heatsink and P_c is the rate of heat flow in watts. The thermal resistance of the heatsink (R_{th}) is then given by equation (2.11):

$$P_c = \frac{k_T a \, dT}{d} \tag{2.10}$$

$$R_{th} = \frac{dT}{P_c}$$

$$= \frac{d}{k_T a} \tag{2.11}$$

Convection. Convection may be due to natural air flow or forced air flow. Forced convection is further dependent on whether the air flow is laminar or turbulent. At low air velocities the flow is laminar and this changes at higher velocities to turbulent, the actual point of changeover being dependent on the design of the heatsink.

The heat flow from the heatsink having a cross-sectional area of a, a vertical length of l, and a temperature difference above the surrounding of dT, is given empirically by

$$P_n = k_n a \frac{(dT)^{1.25}}{l^{0.25}} \tag{2.12}$$

where the constant k_n has a value of about 1.37.

If the heatsink is now cooled by forced air, having a velocity of v_a, then the heat flow for laminar and turbulent air flow are given by equations (2.13) and (2.14) respectively, where the constants have the approximate values $k_{fl} = 3.9$ and $k_{ft} = 6.0$:

$$P_{fl} = k_{fl} a \, dT \left(\frac{V_a}{l}\right)^{1/2} \tag{2.13}$$

$$P_{ft} = k_{ft} a \, dT \frac{V_a^{0.8}}{l^{0.2}} \tag{2.14}$$

Radiation. The heat loss due to radiation is dependent on the emissivity of the heatsink (ε) and its temperature difference above the ambient. The maximum value of emissivity is unity, that of a black body radiator. If T_1 and T_2 are the temperatures of the surface of the heatsink and that of the surrounding air, then the heat loss in watts due to radiation is given by

$$P_r = k_r a \varepsilon (T_1^4 - T_2^4) \tag{2.15}$$

where the constant k_r is approximately equal to 5.7×10^{-8}.

Thermal analysis using the equations given above can result in errors up to 25% since many factors affect the actual heat-dissipation properties of heatsinks. The errors arise due to:

(i) The mix of heat transfer modes and the difficulty of predicting the actual heat transfer path. Heat radiating from adjacent bodies also grossly affects the final result.
(ii) The variation in power dissipation between semiconductors of the same type, even when these come from the same batch. Power

dissipation will vary due to differences in clamping pressure between the device and its heatsink, and these are difficult to predict.

(iii) Many of the constants used in the thermal equations are actually low-order variables and choosing the right value often needs judgement and experience.

Analysis of forced air-cooled systems gives less accurate results than analysis of natural air-cooled equipment because:

(i) There are differences in flow over interior and exterior surfaces.
(ii) It is not possible to calculate the air velocity at each point in the flow path.
(iii) Thermal analysis usually assumes symmetrical shapes, e.g. cylinders and spheres, and in practice these shapes rarely occur.

Performance can be improved by mounting heatsinks vertically within an enclosure with openings at the top and bottom, to create a chimney effect. Several devices can also be mounted on a common heatsink, since this would result in a higher temperature differential between heatsink and ambient, and so improve its efficiency, although the devices now run hotter and upstream components will be working at a higher case temperature. The advantage of mounting devices on the same heatsink is that there is good thermal coupling, as is needed when they are being operated in parallel.

2.4.2 Construction of heatsinks

Heatsinks are available in a variety of shapes and sizes to accommodate the many different package types used for power semiconductors. For example, Figure 2.6(a) shows a heatsink for a hockey puck construction

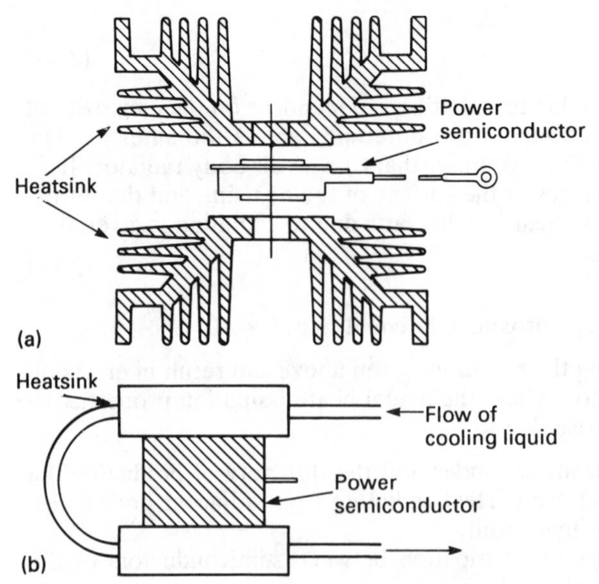

Figure 2.6 Examples of heatsinks: (a) air cooled; (b) liquid cooled

which is air cooled, whilst Figure 2.6(b) gives an arrangement for the same device cooled by liquid, as described in section 2.5.

Heatsinks are usually made from aluminium alloy extrusion, aluminium being a good conductor of heat, malleable so that it can be readily shaped, easy to extrude and made with a smooth surface finish. Aluminium gives a heatsink which is inferior to that made from copper, but it is much cheaper. Heatsinks are designed with a large surface area, for radiation and convection of heat, and the weight is minimised. The heatsink may be left bright, but coloured matt surfaces are more efficient. Black is not necessarily the best colour since at the temperatures being considered heat radiation occurs in the infrared region and all enamels, varnishes, anodised surfaces and oil paints have high emissivities regardless of colour.

Heatsinks are usually designed with fins, the greater the number of fins, the larger the area for convection cooling, but if the fins are too close together there is less heat radiation, so a compromise is needed in the heatsink design. Forced air-cooled heatsinks are three to four times more efficient than natural cooled systems. Radiation effects are now negligible, and since the rate of air flow is less dependent on temperature the thermal resistance is less variable, so the thermal system can be assumed to be linear. The air flow is also more independent of heatsink fin spacing and should be designed to create turbulence over the surface of the fins and break up any layer of static air.

Electrical isolation is often needed when a device is mounted on a heatsink, and this can be obtained by using isolating washers. Several materials are used for these washers; beryllia is the most expensive but has the highest thermal conductivity and dielectric strength, followed by hardened anodised aluminium washers which have good thermal conductivity and dielectric strength. Mica washers were very popular but they suffer from the fact that they can crack and peel and, because they are transparent, it is easy for two to become stuck together, which would cause an increase in the thermal resistance. High-temperature plastics such as Kapton and Mylar have lower dielectric strength than mica but they are cheaper, and since they are coloured their shading gives a visual indication if two are stuck together. An alternative to using an isolating washer is to spray the heatsink during manufacture with an electrical insulation material.

The interface between the case of the component being cooled and the heatsink has a relatively low thermal resistance compared to other parts of the system. However, the resistance can increase during assembly by a factor of 10 times unless care is taken to minimise it. This is done by keeping the mating surfaces clean, by applying adequate mating pressure and by using a thermal grease between them. This grease, or heatsink compound, is a silicone material filled with heat-conductive metal oxides. The grease must not dry out, melt or harden even after operating for long periods at high temperatures such as 200°C. Figure 2.7 illustrates how the grease helps to even out the temperatures in the semiconductor package. Without the grease a slight bump, bow or dust particle causes the temperature at A to be higher than that at B. The thermal grease replaces the air and has a much lower thermal resistance so that the temperatures in the copper tab, and therefore the semiconductor, are more even. Figure

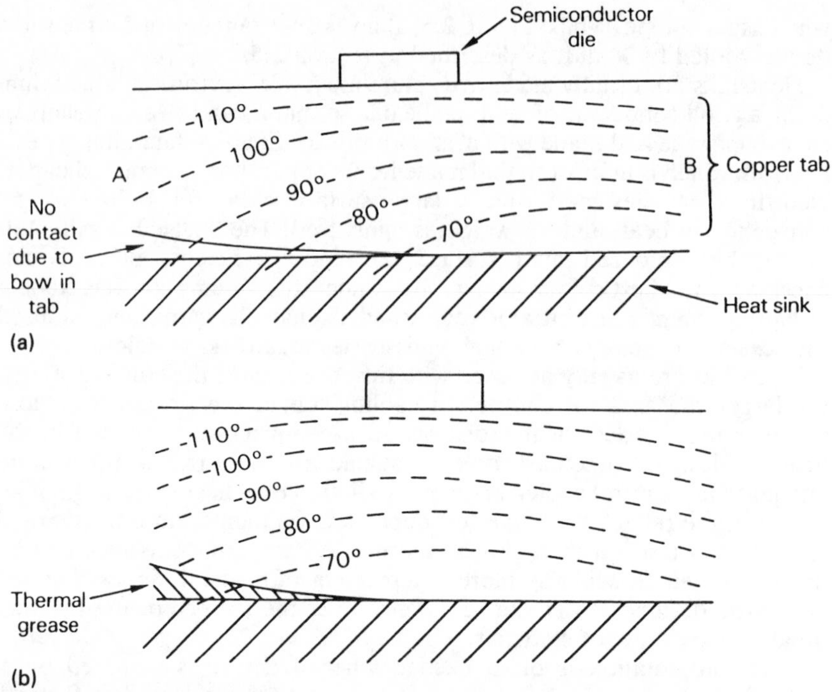

Figure 2.7 Temperature contours at the heatsink–power device interface: (a) without thermal grease; (b) with thermal grease

2.8 compares the thermal resistance of some commonly used components; diamond is used to cool some specialised devices operating at high frequency.

Figure 2.9 illustrates curves which may be used to determine the thermal resistance of a heatsink in any application. The left-hand set of curves gives the power dissipation through the power semiconductor, and is similar to

Material	Thermal resistance (°C cm/W)
Diamond	0.02 to 0.1
Copper	0.3
Aluminium	0.5
Solder	2.0
Thermal grease	130
Mica	150
Mylar	400
Still air	3000

Figure 2.8 Comparison of thermal resistances of some typical materials

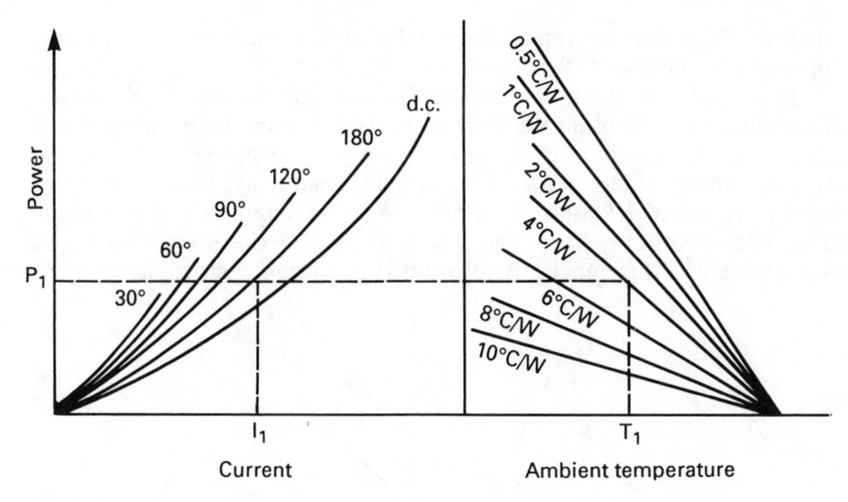

Figure 2.9 Curves used to calculate heatsink thermal resistance

Figure 2.1, whereas the right-hand curves are for a heatsink and give its thermal resistance. Such a series of curves would be obtained, for example, in a forced air-cooled heatsink operating under a variety of air speeds. If I_1 is the current through the power semiconductor, then P_1 is the power dissipated through it, and for an ambient operating temperature of T_1 the heatsink must have a thermal resistance of 4°C/W.

2.5 Liquid cooling

Liquids can be used to cool power semiconductors, and this method is more effective than air cooling. Several methods exist for liquid cooling:

(i) Interconnecting a series of individual, specially designed heatsinks, each of which carry a single device. These are all linked together by pipes through which the cooling liquid flows.
(ii) Mounting the power devices on a common liquid-cooled structure, such as a hollow bus bar, the components being electrically insulated from each other.
(iii) Immersing the power devices into the cooling liquid, the components sometimes having small heatsinks fitted to their bodies.

The liquids used can be water or oil. Water has a high speed of flow but can cause electrolytic corrosion and can also freeze. De-ionised water is often used to which has been added a suitable anti-freeze agent. Oil is more viscous than water and can be inflammable. However, it does not permit the flow of electrolytic currents and devices can be immersed directly into it.

Liquid-cooled systems can have a thermal resistance below 0.01°C/W, which compares with figures of 0.25°C/W for natural air-cooled systems and 0.1°C/W for forced air cooling. The added advantage of liquid cooling is that the heat can be removed to a location remote from the power semiconductor before it is dissipated.

A heat pipe is a device which is sometimes used with power semiconductors to conduct the heat away from a component, mounted in an inaccessible position, to a larger, remote dissipater. A metal bar conducts heat very inefficiently: for example, conducting 1 kW of heat in a solid copper rod of 1.5 cm diameter over a 30 cm length would give about 800°C difference between its ends. A heat pipe of the same dimensions would give a 2°C difference, therefore it is much more efficient.

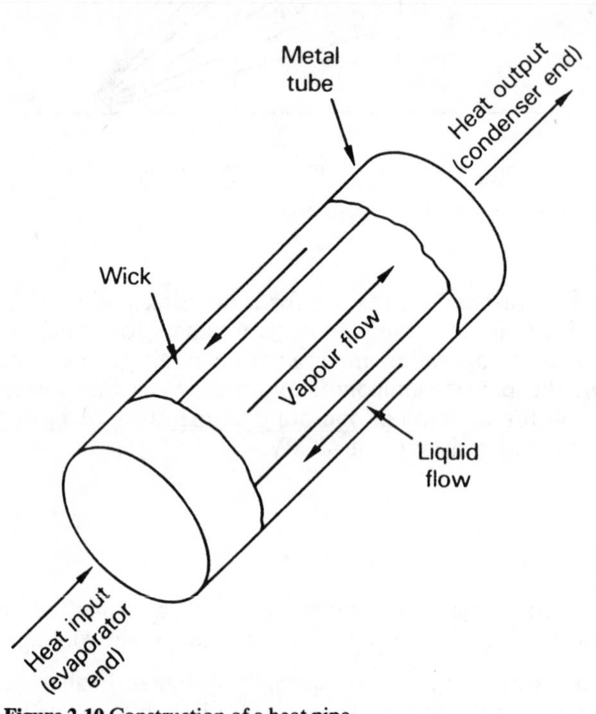

Figure 2.10 Construction of a heat pipe

Figure 2.10 shows the construction of a heat pipe. A hollow metal tube is sealed at both ends and its walls are lined with a wick material. The inside of the tube contains a small quantity of a working fluid which is in a partial vacuum so that it boils at a lower temperature than it would at atmospheric pressure. The component to be cooled is attached to the evaporator end of the tube. The working fluid vaporises and heat is absorbed in converting the liquid to vapour. The vapour travels towards the condenser end of the tube and this end is cooled externally by a heatsink. The vapour gives up its heat at this end as latent heat and condenses. The condensed fluid is returned along the wick by capillary action to the evaporator end. When the vapour condenses it tends to increase the vacuum so that more vapour is drawn from the evaporator end. A heat pipe is typically 0.3–1 cm in diameter and up to 50 cm long, with a variety of shapes to suit the equipment layout, operating in the range +20°C to +200°C, although the range −200°C to +600°C can be covered if necessary.

Two equations are used in the design of a heat pipe. The maximum heat transport capability Q_M of the pipe is given by equation (2.16), where d is the inside diameter of the pipe, l is the length from condenser to evaporator, and K is a constant which is dependent on the heat pipe geometry, the wick material, the working fluid and the orientation of the pipe. The maximum temperature difference between the evaporator and condenser for a power of Q is given by equation (2.17), where a_E is the evaporator area, a_c the condenser area, and b a constant dependent on the configuration of the pipe. Equation (2.17) is used to obtain an approximate design for a heat pipe which must then be tested and modified:

$$Q_M = \frac{K d^2}{l} \tag{2.16}$$

$$dT = \frac{Q}{b}\left[\frac{1}{a_E} + \frac{1}{a_c}\right] \tag{2.17}$$

Chapter 3

Power semiconductor control components

3.1 Introduction

This chapter examines components which are capable of handling relatively low levels of power, but which are used to control high-power semiconductors. These components often take the form of low-power equivalents of the power components themselves, such as diodes, transistors, and light-triggered thyristors, although in this chapter only two types of components are considered; those which provide trigger pulses and are often used in oscillator circuits, and those which are used to isoldate the drive circuits from the main power devices.

In addition to these components, which transmit the pulse to the power semiconductors, logic devices are widely used as part of the overall control electronic function, which includes semi-custom-integrated circuits and microprocessors.

3.2 Power semiconductor control requirements

Power semiconductors which have a control terminal and are therefore capable of being controlled are transistors, thyristors, gate turn-off switches and triacs. These devices vary in that a transistor needs a control signal at its base during its entire conduction period, whilst the other components are turned on by trigger pulses on their gate terminals. The transistor, however, is similar to a gate turn-off switch in that it can be turned on and off by a control signal, whilst thyristors and triacs depend on a momentary break in the load current for turn-off.

Irrespective of the type of semiconductor device, they all require a turn-on signal having a shape similar to that of Figure 3.1. A high-amplitude pulse V_1 is essential to provide the overdrive needed during the turn-on period, so that the turn-on time is reduced and the switching losses in the power semiconductor minimised. The pulse must have a short rise time since this also affects the turn-on delay time. The initial overdrive is usually kept to a short period since otherwise the power dissipation in the control section of the power semiconductor, and in the control circuitry, would be excessive. After time t_1 this pulse is reduced to a lower value V_2 such that the dissipation is within the requirements of the power

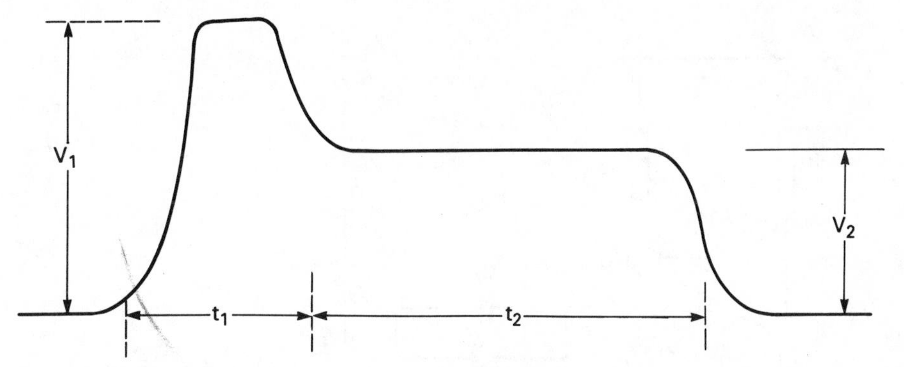

Figure 3.1 Desired control signal waveform for a power semiconductor

semiconductor. The shorter the pulse duration, the higher can be the power in the pulse, as shown in the gate control characteristic for a thyristor in Figure 1.23.

The overall control drive period (t_1+t_2) must be long enough to suit the power semiconductor device. For a transistor this means the total conduction period of the device. For a pulse-triggered power semiconductor the drive must be present until the current through the device reaches a critical value, called the latching current. If the load is inductive or the voltage across the device is rising slowly, then an extended pulse is required, so that it is common to control these components with a train of pulses. Oscillations on the pulse need to be minimised since they reduce the effective pulse duration, and a negative period may turn the power semiconductor off.

Pulse-triggered semiconductors such as thyristors need to be protected from spurious turn-on, and this is often done by connecting a low impedance such as a resistor between the gate and cathode terminals, or by applying a slight negative voltage to the gate terminal when it is to be non-conducting. A positive voltage on the gate terminal is also to be avoided when the device is reverse biased since it increases its leakage current and hence its dissipation.

A gate turn-off switch is turned on and off by means of a pulse on its gate, the turn-off pulse having a reverse polarity to the turn-on pulse. Various circuits exist for this, one being shown in Figure 3.2. With transistor TR_2 off transistor, TR_1 is on so that gate drive is provided to the GTO. Capacitor C_1 charges and its voltage is clamped by the zener diode. When TR_2 turns on it turns off TR_1 and discharges C_1 applying a reverse gate current through the GTO, turning it off.

3.3 Trigger devices

This section describes semiconductor devices which, although low-power devices themselves, are used to control high-power semiconductors. They are commonly called trigger devices, and they cover the unijunction transistor and its variants, the silicon unilateral and bilateral switch, and the diac.

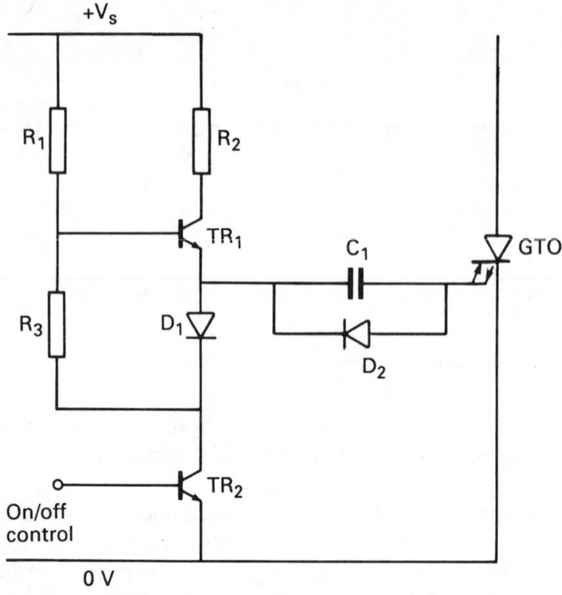

Figure 3.2 Control circuit for a gate turn-off switch (GTO)

3.3.1 Unijunction transistor

The unijunction transistor is a three-terminal, two-layer, semiconductor device, which exhibits a negative resistance region, enabling it to be used as an oscillator and trigger device. Figure 3.3 shows the construction, symbol and characteristic of the component. In its simplest arrangement it consists of an *n*-type silicon bar with two ohmic contacts for the base 1 and base 2 terminals. The emitter terminal is a *p*-type aluminium wire which is alloyed onto the silicon bar to form a *p–n* junction. The symbol for the UJT shows base 1 and base 2 terminals at right angles to indicate that these are non-rectifying ohmic contacts, but the emitter is represented by an arrow since it is a rectifying *p–n* junction, the arrow pointing from the *p*-type emitter to the *n*-type base.

The resistance between the two bases, which consists of the ohmic resistance of the silicon, results in a uniform voltage drop when the diode is biased as in Figure 3.3(d), and this results in the emitter being reverse biased by ηV_{BB}, where η is known as the intrinsic stand-off ratio of the transistor. Therefore only a small leakage current will flow, as shown by the device characteristic of Figure 3.3(e), the emitter junction being reverse biased and the interbase resistance being in the region of 5–10 kΩ. When the voltage exceeds the peak point voltage V_P given by

$$V_P = \eta V_{BB} + V_D \tag{3.1}$$

where V_D is the forward voltage drop of the silicon diode formed by the emitter base region, current starts to flow and minority carrier injection occurs into the base, increasing the conductivity between the emitter-base 1 and causing current to flow between the two bases.

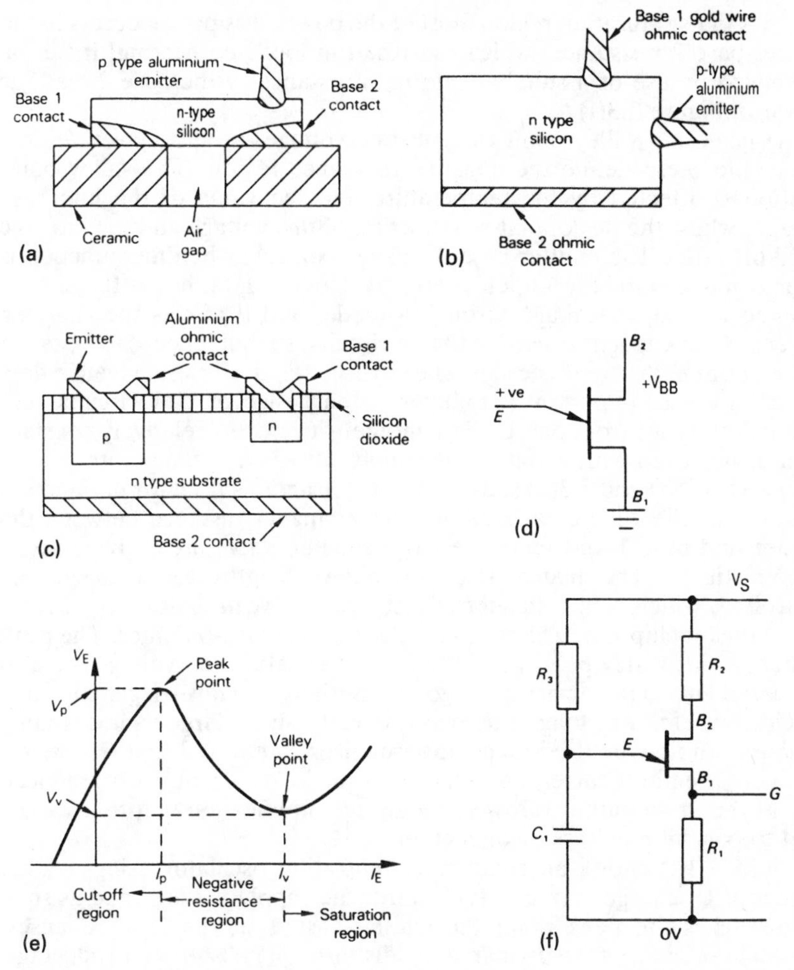

Figure 3.3 Unijunction transistor: (a) series bar assembly; (b) cube assembly; (c) planar diffused assembly; (d) symbol and biasing; (e) characteristic; (f) oscillator circuit

The increasing current is accompanied by a reduction of voltage, causing further emission from the emitter, resulting in regenerative action or a negative resistance effect. This is shown clearly in the characteristics of Figure 3.3(e). Beyond the valley point the saturation region is reached, where further increase in conduction is limited by hole–electron recombination, so the emitter voltage must be increased in order to increase the emitter current.

Several parameters are given in UJT data sheets. The interbase resistance r_{BB} is measured between the two base terminals with the emitter open, and its value varies slightly with applied voltage and with temperature, data sheets normally providing a curve showing the variation of this resistance with temperature. The emitter-base 1 resistance r_{B1} decreases with increasing emitter current, although the emitter-base 2 resistance r_{B2} is not significantly affected by this current change. When the device is in its saturation region most of the power dissipation occurs in the emitter-base 2 resistance, which can result in localised heating; it can be minimised by use of a suitable external resistance in the base 2 lead, as shown in Figure 3.3(f).

The peak and valley point currents and voltages are also given in data sheets and these define the negative resistance region. The valley point location is affected by the temperature and the value of the interbase voltage, whilst the peak point is a function of this voltage and the intrinsic stand-off ratio. The emitter-base 2 leakage current, when this junction is reverse biased with base 1 open, is also given in the data sheets. It is similar in value to that of leakage through a diode, and it affects the charging current of any capacitors used in timing circuits, so that it needs to be taken into account in the circuit design. The intrinsic stand-off ratio, given in data sheets, is a very important parameter in the design of UJT circuits, and although it varies from one UJT to another, it remains relatively constant for a device even with variations in supply voltage and temperature.

Figures 3.3(b) and 3.3(c) show two other structures used for unijunction transistors. The cube arrangement gives a smaller distance between the emitter and base 1 and therefore has a smaller active area, giving faster turn-on times. The planar structure allows lengths to be accurately controlled, which results in shorter distances between emitter and base 1 and a smaller chip size. This again results in faster turn-on times. The peak point current, valley point current and emitter saturation voltage are also decreased and so the device gives good sensitivity and low trigger currents, which is useful for long time-delay circuits since large-valued timing resistors can now be used, and capacitor sizes can be reduced. However, the average emitter current, which is often the load current, is also reduced so that the drive output is lower, requiring amplification before it can be used to control power semiconductors.

Figure 3.3(f) shows an elementary relaxation oscillator using a UJT. Capacitor C_1 charges through R_3 towards the supply voltage, and as soon as it reaches the peak point the emitter-base 1 of the UJT collapses, allowing the capacitor to discharge rapidly through resistor R_1, producing a positive spike across it. When the voltage falls to the valley point the UJT recovers and the capacitor again begins to charge through its resistor. The train of positive pulses at point G can be used to trigger a power semiconductor, as will be described in later chapters.

3.3.2 Complementary and programmable UJT

There are two variations of the unijunction transistor which, although they have a different construction, exhibit very similar negative resistance characteristics and are also widely used to control power semiconductor devices. These are the complementary unijunction transistor (CUJT) and the programmable unijunction transistor (PUT), shown in Figure 3.4.

The complementary unijunction transistor is a four-layer device consisting of a *p–n–p/n–p–n* arrangement with internal biasing resistors, all built into a silicon planar monolithic die. The transistor pair is normally off, but will turn on when the emitter goes more negative than the base 1 terminal (B_1) by a value given in equation (3.1). Once in the conduction

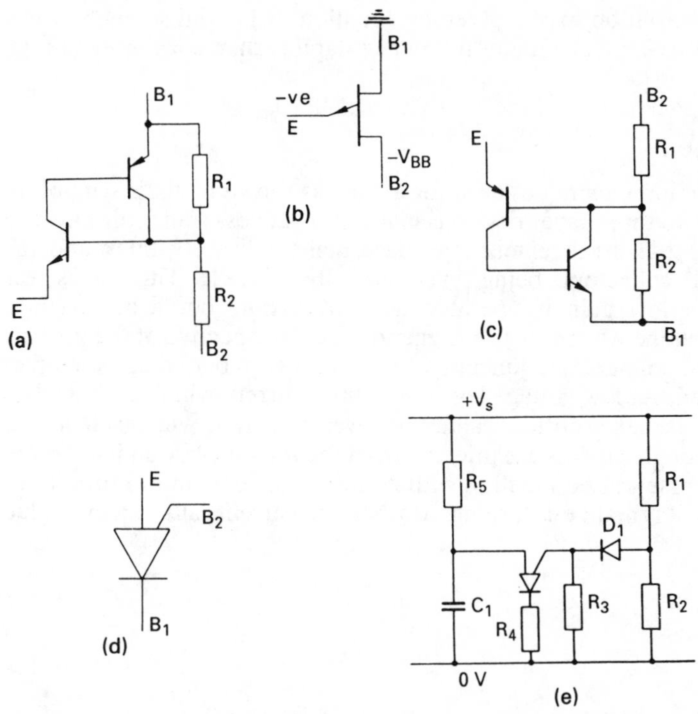

Figure 3.4 Complementary and programmable unijunction transistors: (a) complementary unijunction transistor (CUJT); (b) CUJT symbol; (c) programmable unijunction transistor (PUT); (d) PUT symbol; (e) a PUT oscillator circuit

mode, the device exhibits a negative resistance characteristic, similar to that of Figure 3.3(e). Since the applied voltages and currents are of polarity opposite to that of the conventional UJT this device is called a complementary UJT.

The intrinsic stand-off ratio of the CUJT is determined by the ratio of the two resistors R_1 and R_2 and since these are situated on the same silicon die as the rest of the components the device can be designed to exhibit a tight tolerance in this parameter, good stability over a wide temperature range, and low saturation voltage.

The programmable unijunction transistor is also a four-layer device but it has its resistors R_1 and R_2 external to the silicon die so that they can be selected by the user. This enables parameters such as the intrinsic stand-off ratio, interbase resistance and the peak and valley point currents, to be programmable. By choice of suitable external components these parameters can be made less susceptible to temperature variations than a conventional UJT. As seen from its symbol, the PUT is basically a thyristor with an anode gate.

Figure 3.4(e) shows the PUT used in a relaxation oscillator circuit similar to that of Figure 3.3(f). External resistors R_1 and R_2 determine the value of the intrinsic stand-off ratio for the device and timing circuit R_5C_1 can be modified to vary the oscillator frequency. Diode D_1 compensates for the temperature variation in V_D, given by equation (3.1), and so enables this circuit to maintain much greater frequency stability than conventional UJT systems.

3.3.3 The diac

Two construction techniques exist for a diac, as shown by their symbols of Figure 3.5. The three-layer device is made as a gateless transistor having a *p–n–p* or *n–p–n* arrangement, the base being relatively thick and the emitter and collectors being symmetrically placed. This gives the component a low gain if operated as a transistor, but a symmetrical breakover voltage when run as a trigger device. Irrespective of the polarity of the applied voltage, one junction is always forward biased and the other reverse biased. At low voltage levels very little current will flow, but when the voltage reaches a critical value the reverse current will reach such a value that enough carriers are injected from the forward-biased junction to flood the reverse-biased junction with minority carriers, and to turn it on. The device will remain on, turning off when the current falls to a low value again.

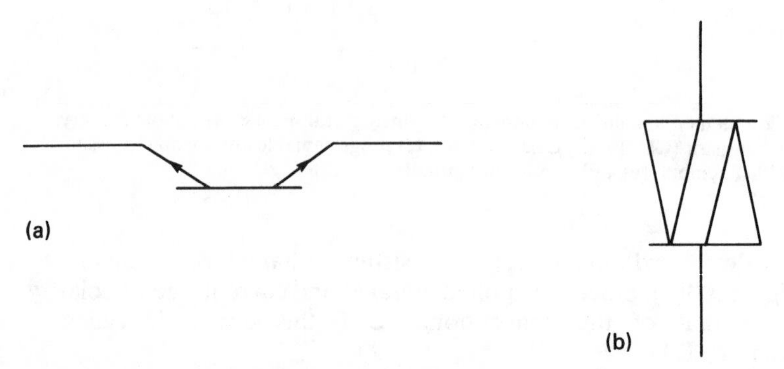

Figure 3.5 Diac symbols: (a) three-layer device; (b) five-layer device

The three-layer diac has a typical breakdown voltage of about 30 V with a breakback voltage of about 8 V and carrying a current of 1 A for a short duration. Much better characteristics, such as a lower breakback voltage, can be obtained by using a five-layer structure, which essentially consists of a triac, as shown in Figure 1.27(b), in which the gate has been omitted.

The characteristics of a diac are as shown in Figure 1.27(c) for the case where the gate current is zero. Being bi-directional devices, diacs are very useful for firing triacs, and often the two components are built in the same silicon die, the diac being formed in the gate of the triac. The device is now referred to as a quadrac. Figure 3.6 shows a full-wave phase-control system which has found extensive use for domestic applications such as heater controls, light dimmers and motor speed variation. It is perhaps the simplest of circuits, containing three components, the triac and diac being available as a single unit.

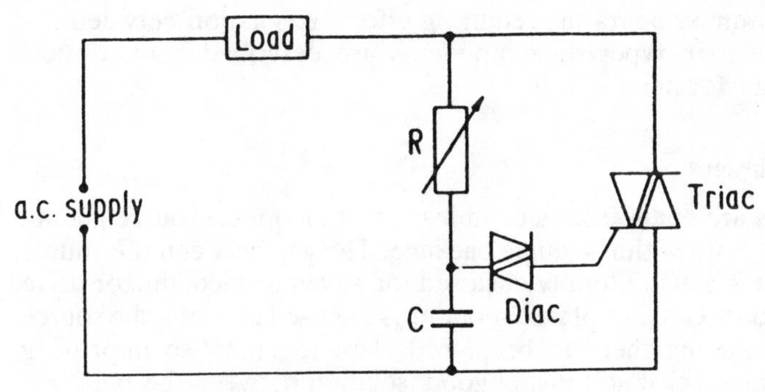

Figure 3.6 A simple diac phase-control circuit

3.3.4 Silicon unilateral and bilateral switches

The silicon unilateral switch (SUS) is primarily an integrated circuit in which the trigger point of the two transistors is determined by a zener, denoted by Z_1 in Figure 3.7(a). The device is normally off until the breakdown voltage of the zener diode is reached, when the *p–n–p* transistor turns on providing base drive to the *n–p–n* transistor and so turning the two devices on. This trigger point is determined by the value of the zener voltage and is typically 5–10 V. Two SUS components can be connected in a reverse arrangement within the same case, to provide a bi-directional trigger device called a silicon bilateral switch (SBS). Figure 3.7 shows the symbols for the SUS and SBS trigger devices.

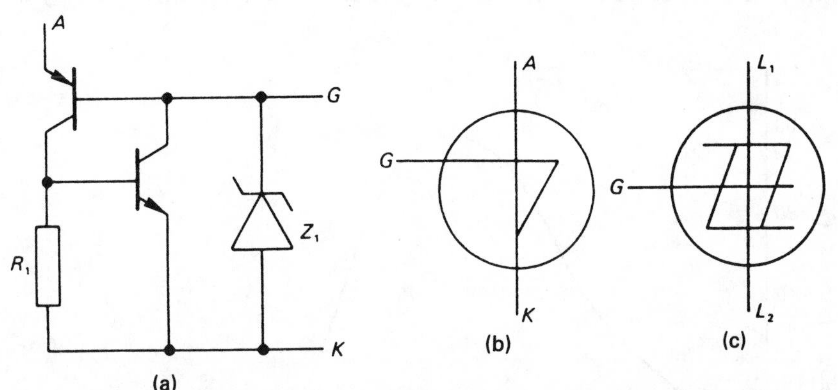

Figure 3.7 Silicon unilateral and bilateral switch trigger devices: (a) silicon unilateral switch construction; (b) silicon unilateral switch symbol; (c) silicon bilateral switch symbol

3.4 Isolating components

This section describes two classes of components which have been used to isolate the power semiconductor from the low-power circuit which is controlling it. Often the power semiconductor regulates megawatts of power, whilst the low-power circuit is handling a few watts and is connected to a human operator, requiring effective isolation between the two parts. Two main types of components are described here, optical couplers and transformers.

3.4.1 Optical couplers

Optical couplers are made from a combination of an optical source and an optical detector, both within a single package. The gap between the source and the detector is usually totally enclosed for power semiconductor drive circuit applications. Glass or plastic separators are used between the source and detector, enabling them to be placed close together, so improving coupling efficiency whilst still giving good isolation between the two.

Light-emitting diodes (LEDs) are invariably used as the optical source, although many different types of LEDs are in use, the most common being GaAlAs, which has an emission in the near infrared (750–850 nm) and GaAs having an emission in the infrared (940 nm). The LED material chosen needs to match the spectral response of the silicon photodetector, shown in Figure 3.8, as closely as possible and also have good efficiency in terms of light emission for current input. GaAlAs is the most popular material since its band gap can be varied relatively easily, to modify the emissions in the range 650–900 nm, by varying the gallium-to-aluminium ratio. GaAlAs also needs a low drive current, so that it is well suited to being driven directly from low-power logic circuits.

The parameters of most interest in optical couplers are the isolation between source and detector, the input–output current transfer ratio and the speed of operation. The isolation resistance is of the order of $10^{11}\,\Omega$, and is usually higher than the leakage resistance between package pins on the printed circuit board. Another way of expressing this isolation is by the

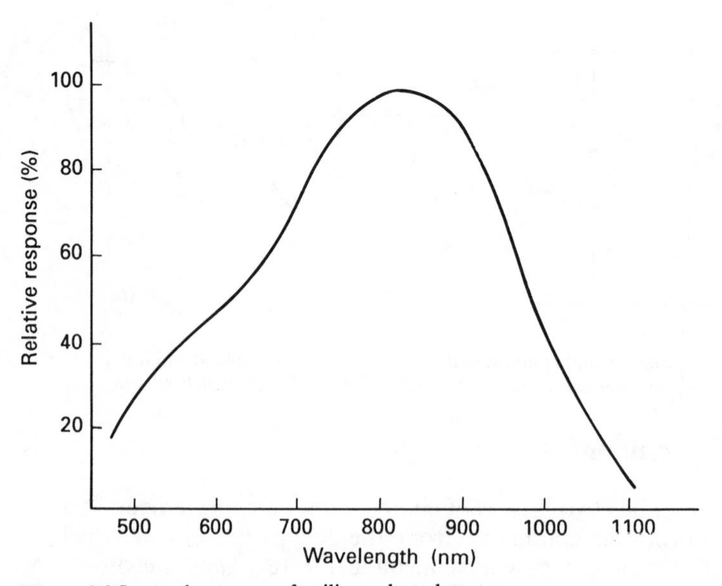

Figure 3.8 Spectral response of a silicon photodetector

maximum voltage which can be applied between input and output without breakdown. If breakdown occurs it can form a resistive path due to carbonised moulding on the surface, or it can result in a short circuit caused by molten lead wires bridging the lead frames of the emitter and detector. For high isolation voltages the moulding is usually designed such that the input and output pins are brought out from separate sides of the package. It is also important to minimise the parasitic capacitance through the dielectric between input and output.

The current transfer ratio is given as the ratio of the output current to the input current of the source, when the detector is biased in a specified way. This ratio is determined by several factors, such as the level of current into the source and detector saturation. Generally, an LED is used as a source and the light output of the device falls with time, giving a decrease of transfer ratio. The operating speed of the coupler defines how fast it can be switched and is usually specified in terms of its maximum operating frequency.

Many different types of output detectors are used in optical couplers, a few of these being shown in Figure 3.9. The phototransistor coupler tends to be low cost with an operating speed of typically 100–500 kHz and a minimum current transfer ratio between 10% and 100%. Photo Darlington devices have a transfer ratio between 100% and 600% but this is difficult to predict accurately due to the wide variation in the gain of the Darlington stage. The operating speed is relatively low, being typically between 2 kHz and 10 kHz.

For high currents, photothyristor and phototriac output stages are used. The current into the LED, which is needed to trigger the thyristor or triac, is now an important parameter. Since the coupling efficiency between the LED and photothyristor is low it is important that the thyristor is designed to have a high gate sensitivity. This usually requires careful process control

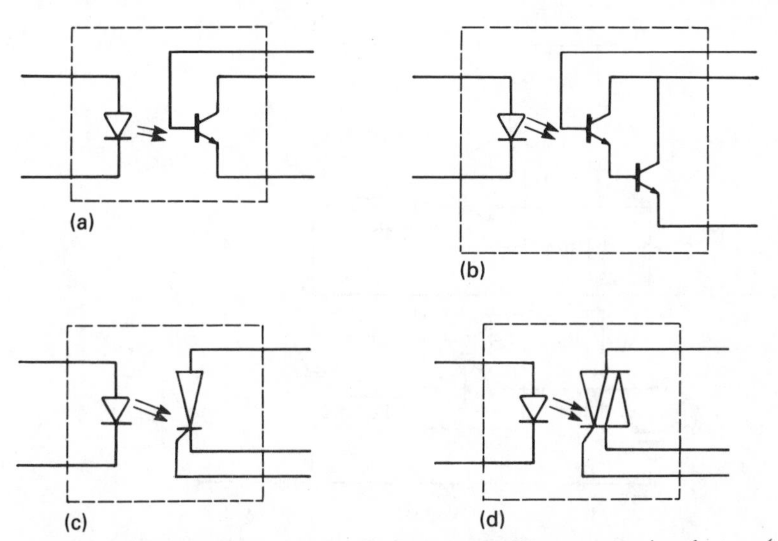

Figure 3.9 Examples of optical couplers used for controlling power semiconductors: (a) transistor output; (b) Darlington output; (c) thyristor output; (d) triac output

in order not to degrade other parameters, such as voltage rating. Photothyristor and phototriac couplers require of the order of 20–30 mA to trigger and can provide between 200 and 300 mA of output current. The thyristor turn-on time is typically 2–20 μs.

Figure 3.10 shows a few examples using optical couplers. A separate low-voltage supply is used in Figure 3.10(a), so the voltage rating of the optical coupler output stage is relatively low. The circuit of Figure 3.10(b) does not require a separate gate power supply, since it derives this from the

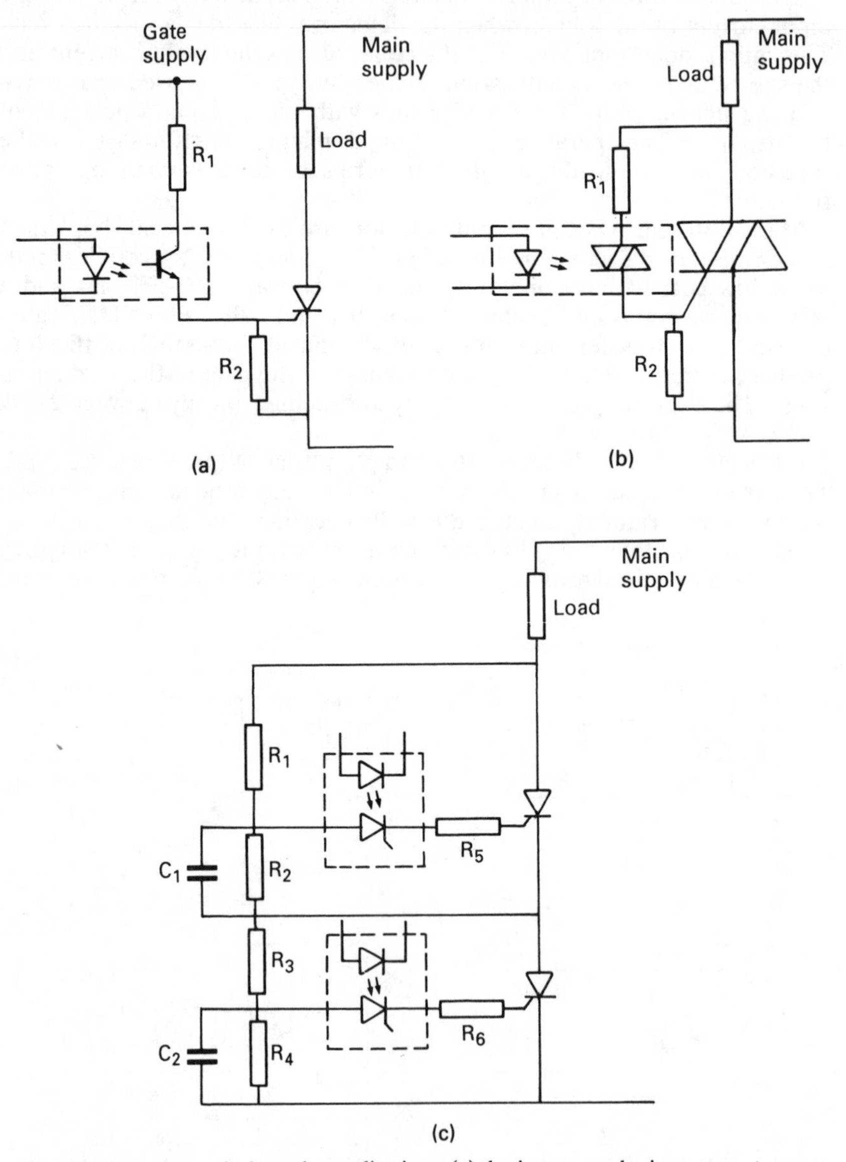

Figure 3.10 Typical optical coupler applications: (a) thyristor control using a separate power supply; (b) triac control using the load supply; (c) series thyristor control

same source as the main power semiconductor. However, now the output stage will see the full load voltage when the power triac is off, so it must be rated for this value. Many optical couplers have a zero crossing detector built into the package, enabling the power triac to be switched at the zero crossing points of the a.c. mains, so minimising the generation of radio frequency interference. The circuit of Figure 3.10(c) shows two series-connected thyristors which are simultaneously fired. The blocking voltage of the optical coupler output stages is reduced by the use of a voltage-dividing resistor chain across the power thyristors, although now these resistors provide a leakage path for the load current, which bypasses the main power devices.

3.4.2 Isolating transformers

Many different types of transformers are used in power circuits, such as those for d.c. power supplies, current transformers, autotransformers and pulse transformers. Pulse transformers provide the isolation between the low-power control circuit and the power semiconductor, and these are considered in this section, although the parameters of all transformers are very similar.

The peak secondary voltage of a transformer will change between the no-load and full-load currents, due to voltage drops in the secondary and primary windings, even if the primary voltage is kept constant. This is called the regulation of the transformer, and is defined as the change measured as a percentage of the full-load voltage. The efficiency of the transformer is an indication of how well it converts the input power into output power, the difference between the two being dissipated as transformer losses, generating heat.

Transformers need to withstand high voltages between the secondary and primary. These voltages result in corona, dielectric failure, surface creepage and flashover between points. Corona, being partial discharge within the transformer, can destroy the insulation and also generates radio frequency interference, which affects adjacent equipment and circuits. It increases with the magnitude of the applied voltage or frequency.

Flashover is arcing between parts of the transformer and creepage is flashover across the surface of the insulation, both of which can result in high voltages in secondary circuits. The dielectric strength of the insulation between the primary and secondary windings is usually measured as a maximum withstanding voltage per unit thickness of insulation. Solids have a higher dielectric strength than liquids and gases. In a transformer, the presence of gas adjacent to a solid insulator presents a weakness in which corona can be generated, and this limits the maximum voltage which can be applied across the insulation system. It is therefore important to avoid air gaps in series with the insulation. When several insulating materials are used in series, the stress in each is inversely proportional to its dielectric constant. Therefore the insulation with the lowest constant has the highest stress, and this is usually air or gas. When a direct voltage is applied to the material the voltage drop is mainly due to its resistivity, so the material with the highest resistivity has the highest stress.

Transformers can be shielded electrostatically and electromagnetically.

Electrostatic shielding reduces the voltage transfer through the inter-winding capacitances. It is needed to prevent the transfer of transient voltages or high-frequency noise from the power input circuit to the secondary circuit. The shield is usually a grounded metallic plate between the primary and secondary windings. Electromagnetic shielding is used to attenuate the magnetic field which leaks from the magnetic circuit of the transformer and induces voltages in adjoining circuits. Placing a magnetic shield around the transformer is not usually very effective, since most of the stray lines of flux from the transformer would be perpendicular to the shield and would pass through it. A better solution is to separate the transformer and adjoining sensitive circuits and to orientate them to minimise pick-up. The adjoining circuits can also be shielded by layers of thin, high-permeability material, which are usually interleaved with layers of non-magnetic material, such as copper.

Pulse transformers must be capable of passing a square wave, or a pulse having a short rise and fall time, without appreciable distortion of the waveform. Figure 3.11 shows a typical output voltage waveform from a pulse transformer and indicates the terms used to describe it.

A pulse transformer should be small, able to handle pulses with a short rise time and having a high pulse width-to-pulse rise time, called the span ratio. It should also be capable of resolving adjacent pulses in a high pulse-repetitive frequency application. Transformers are available in many sizes, several devices often being mounted inside a dual-in-line package.

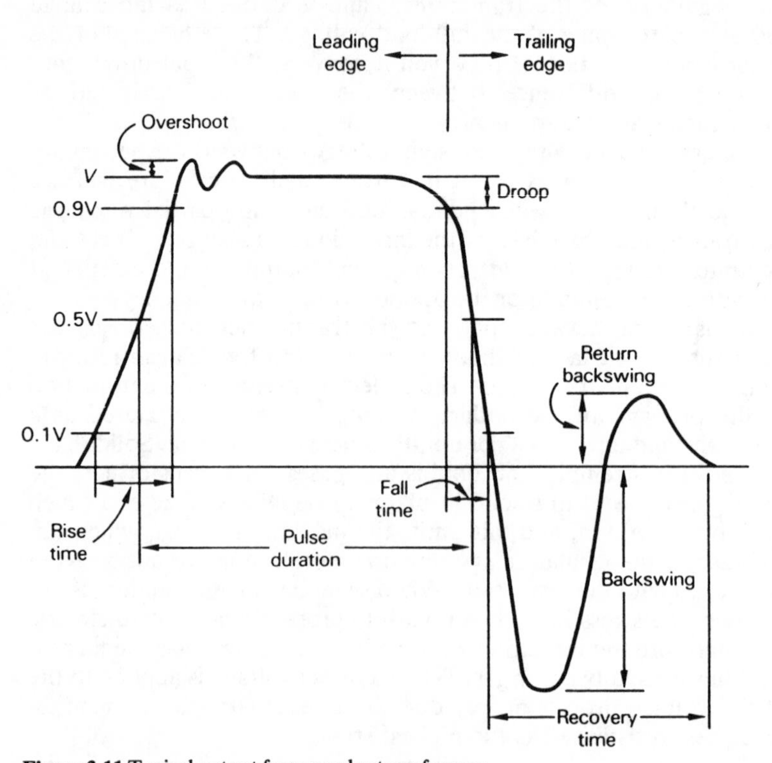

Figure 3.11 Typical output from a pulse transformer

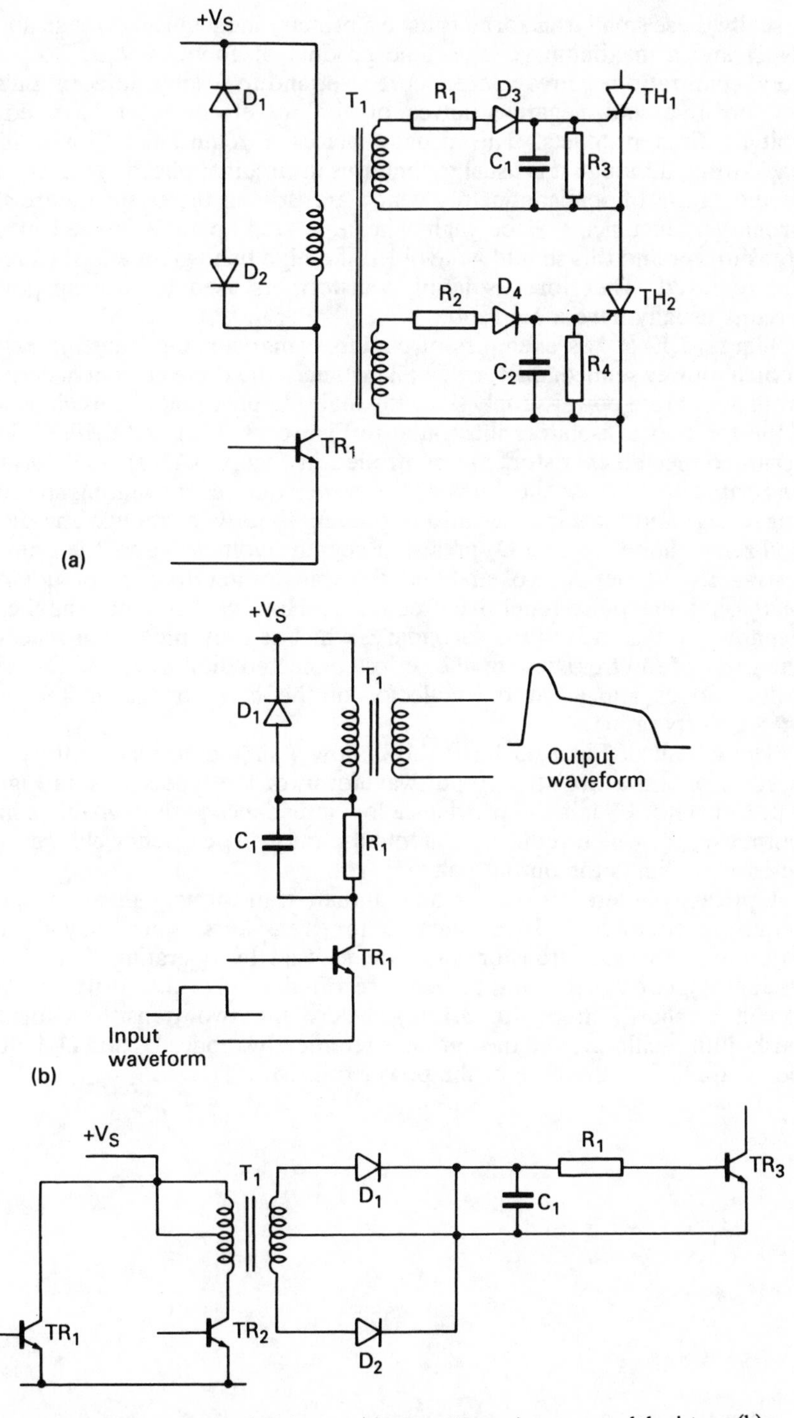

Figure 3.12 Transformer isolation circuits: (a) controlling series connected thyristors; (b) a two-step output waveform; (c) controlling a power transistor

Usually these small transformers have a primary inductance less than about 2 mH and a maximum voltage–time product of about 10 V μs. To get a good span ratio requires a larger core size, and to resolve adjacent pulses the positive and negative halves of the waveform must have equal voltage–time products. This usually places a requirement for a high backswing, although it is usual to limit this in circuit applications, as it adds to the rating of semiconductors which are driving the transformer. The primary inductance will be high if a large step-up ratio is used in the transformer and this should be avoided if a high bandwidth and span ratio are required. Therefore, isolating transformers used for driving power circuits usually have a 1:1 ratio.

Figure 3.12 shows examples of circuits using isolating transformers to control power semiconductors. In all instances the drive current is derived from a separate power supply since the main supply cannot be used, as was done for optical isolators illustrated in Figures 3.10(b) and 3.10(c). Two series-connected thyristors are controlled in Figure 3.12(a), the resistors and capacitors across the gates of the power devices preventing spurious triggering, and the capacitor also improving their dv/dt rating. The diode and zener diode, D_1 and D_2 present a negative voltage across the primary during its off period, so enabling the transformer to recover quickly, increasing the pulse-repetitive frequency. However, this also applies a negative voltage across the secondary, which is prevented from reaching the gates of the thyristors by the series-connected diodes D_3 and D_4. This voltage does appear at the collector of the drive transistor TR_1 and increases its rating.

The circuit of Figure 3.12(b) shows how a simple arrangement can be used to obtain a two-step output waveform, of the type shown in Figure 3.1. Capacitor C_1 initially provides a low-impedance path to enable a high output pulse, which reduces in amplitude once the capacitor charges and resistor R_1 limits the output voltage.

Optical couplers are better suited than transformers in applications requiring continuous drive, such as for transistors, since they do not saturate. However, transformers can be used by operating them in an oscillatory circuit, the output being rectified to give d.c. drive. Such a circuit is shown in Figure 3.12(c), where the two transistors form a push–pull oscillator, and the output is rectified by diodes D_1 and D_2 before being applied to the base of the power transistor TR_3.

Chapter 4

Electromagnetic compatibility

4.1 Introduction

Power electronic circuits, by switching large amounts of current at high voltages, can generate electrical signals which affect other electronic systems. These unwanted signals give rise to electromagnetic interference (EMI), also known as radio frequency interference (RFI), since they occur at higher frequencies. The signals can be transmitted by radiation through space or by conduction along cable.

Apart from emitting EMI, the control circuit of power systems can also be affected by EMI generated by its own power circuitry, by other circuits or by natural phenomena. When this occurs the system is said to be susceptible to EMI. Any system which does not emit EMI above a given level, and is not affected by EMI, is stated to have achieved electromagnetic compatibility (EMC).

This chapter first describes some of the concepts behind electromagnetic compatibility, followed by a description of the sources of EMI in power electronic systems. The method by which the effects of EMI can be minimised, using circuit techniques and shielding, to achieve EMC are then considered, followed by an introduction to the regulatory position on EMI. Finally the principles involved in the measurement of EMI are explained.

4.2 EMC concepts

There are three elements to any EMC system, the source of the EMI, the media through which it is transmitted, and the receptor, which is any system that suffers adversely due to the received EMI. Therefore electromagnetic compatibility can be achieved by reducing the EMI levels from the source, blocking the propagation path of the EMI signals, or by making the receiver less susceptible to the received EMI signals.

The source of the EMI is primarily any system where the current or voltage changes rapidly (for example, the breaking of current by relay contacts, arcing of motor commutators, high-frequency switching such as the rapid turn-on and turn-off of a thyristor).

EMI can be radiated through space, as electromagnetic waves, or it can be conducted as a current along a cable. Conduction can take the form of common-mode or differential-mode currents. For differential mode the currents are equal and opposite on the two wires and are caused primarily by other users on the same lines. Common-mode currents are almost equal in amplitude on the two lines, but travel in the same direction. These currents are mainly caused by coupling of radiated EMI to the power lines and by stray capacitive coupling to the body of the equipment.

Emissions can be classed as broadband and narrowband. In broadband emission the signal bandwidth is greater than the reference bandwidth and the pulse-repetition frequency is less than that of the reference bandwidth. The reference bandwidth, for EMC purposes, may be considered to be the equipment being interfered with, or a test receiver. For a narrowband emission the signal bandwidth is less than the reference bandwidth and the pulse-repetition frequency is greater than that of the reference bandwidth. Figure 4.1 illustrates the difference between broadband and narrowband emissions. Broadband emissions are caused by low-frequency repetitive pulses or individual impulses of electrical or magnetic state changes, such as in switching or commutation. This results in many spectral lines separated by a frequency less than the receptor bandwidth. The frequency components may be coherent, i.e. harmonically related in frequency, or incoherent, i.e. random, such as noise. Narrowband emissions are caused by high-frequency components separated in frequency by greater than the receptor bandwidth.

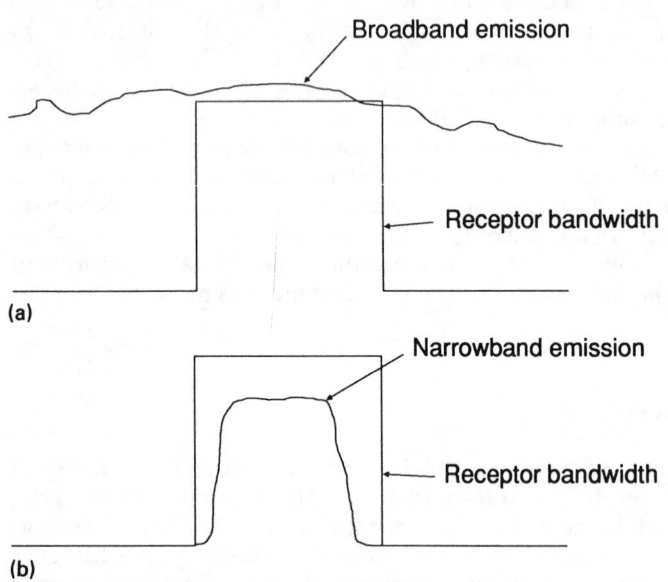

Figure 4.1 Types of emissions: (a) broadband; (b) narrowband

4.3 Sources of EMI

EMI sources can be broadly divided into two categories, natural and man made. Naturally caused EMI below 10 MHz is mainly due to atmospheric noise resulting from electrical storms. Above 10 MHz they are primarily as a result of cosmic noise and solar radiation. Lightning, which is caused by a sudden discharge of high potential between earth and a cloud, or a cloud to cloud, can result in interference to ground or airborne equipment, and damage if a direct hit occurs. Peak currents can exceed 50 kA with rate of rise in the region of 100 kA/μs, giving field strengths in nearby conductors of greater than 200 kV/m. The voltage induced in antenna systems, having a large physical length, is especially high and these are also prone to direct strikes.

Man-made EMI can be intentional or unintentional. In both cases it is the variation of the voltage and current which produces EMI, whose magnitude depends on the value of the current, the length of the conductors, the rate of change of voltage and current, and the physical position of the conductors relative to each other and any earth plains. Examples of intentional EMI are radar, radio, television, and pagers. Companies near airfields suffer most from EMI resulting from radar, usually in the frequency ranges 600 MHz, 1000 MHz and 10 000 MHz. Field strengths approach 200 V/m, with buildings giving only a low level of shielding.

The most common source of radio interference, and the one which is on the increase, results from mobile radio. A variety of frequencies are used but the power levels do not exceed about 50 W. However, the source, such as police cars patrolling the street, can be very close to the equipment being interfered with, so the field strength can exceed 20 V/m.

Unintentional man-made interference is caused by sources such as switches, relays, motors, and fluorescent lights. The inrush current of transformers during turn-on is another source of interference, as is the rapid collapse of current in inductive elements, resulting in transient voltages. Integrated circuits also generate EMI due to their high operating speeds and the close proximity of circuit elements on a silicon die, giving stray capacitive coupling elements.

The generation of EMI in power semiconductor circuits can be illustrated by the circuit of Figure 4.2, in which the power semiconductor is

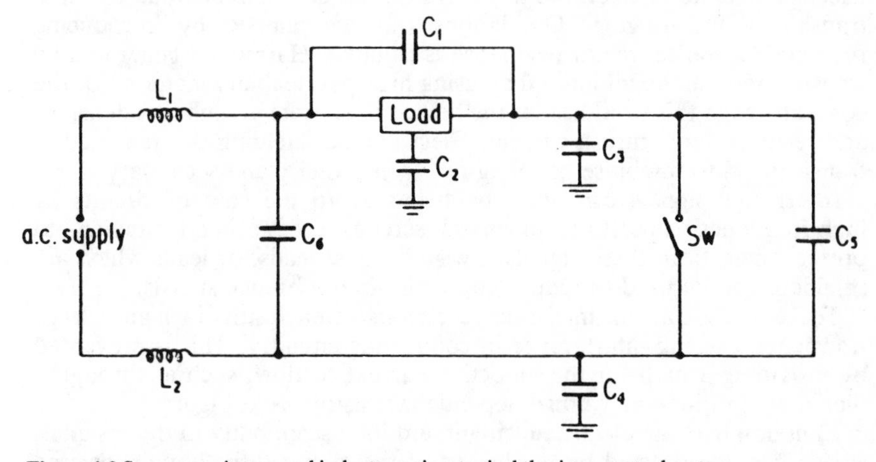

Figure 4.2 Stray capacitance and inductance in a typical thyristor control system

shown as a switch S_W. In this circuit L_1 and L_2 are series line impedances, which could be part of the supply transformer leakage reactance and the self inductance of cables between this and the load. C_1 and C_2 are stray capacitances across the load and from the load to earth. C_3 and C_4 are also stray capacitances to earth and C_5 and C_6 appear between the lines. With S_W open, the capacitors are charged to the peak instantaneous supply voltage, and when the switch closes the capacitors discharge. This creates an oscillatory system which produces a wide spectrum of unwanted frequencies, the magnitude of the EMI produced being determined by the peak energy stored in the capacitors at the instant the switch closes.

An especially strong source of EMI is nuclear electromagnetic pulse (NEMP), which results as a by-product of a nuclear explosion. The frequency spectrum of NEMP covers a wide range from 10 kHz to 10 GHz and is therefore difficult to protect against. The EMI resulting from NEMP depends on the type of nuclear explosion: high altitude, air burst or surface burst. In all cases the explosion releases high-energy gamma radiation which collides with air molecules releasing free electrons, called Compton electrons, resulting in a Compton current.

In a high-altitude explosion the Compton electrons spiral around the earth's magnetic field and produce large current loops. The resulting fields exceed 50 kV/m and rise times are less than 10 ns. Because of the height of the explosion the EMI field has a large coverage, for example an explosion at a height of 500 km in the centre of the USA will cover the whole of North America. Therefore equipment which must continue to function in the event of a nuclear explosion must be unaffected by NEMP.

Air-burst explosions result in a small vertical dipole current and relatively weak radiated fields. Surface burst, on the other hand, gives a large vertical dipole current due to the asymmetry of the air–earth interface, but has local coverage only.

4.4 Circuit design for EMC

Electromagnetic interference is generated in power circuits due to rapid transitions and ringing. Oscillations can be damped by introducing resistance if the source of resonance is isolated. Harmonics generated by transformers can be minimised by using high-permeability material for the core, although this would cause the device to operate at high flux densities and result in large inrush current. Electrostatic shielding is often used in transformers to minimise coupling between primary and secondary.

Interfering signals can often be bypassed to the case of circuits by high-frequency capacitors, or metal screens used around circuitry to protect them from these signals. Twisted signal leads, or leads which are shielded, can be used to reduce coupling of interference signals.

The collapse of flux in inductive circuits often results in high-voltage transients, causing interference in connecting circuitry. This is prevented by providing a path for the inductive current to flow, such as through a diode, zener diode or voltage-dependent resistor, as in Figure 4.3.

Emission from an electronic circuit and its susceptibility to these signals is significantly affected by the layout of the circuit, usually on a printed

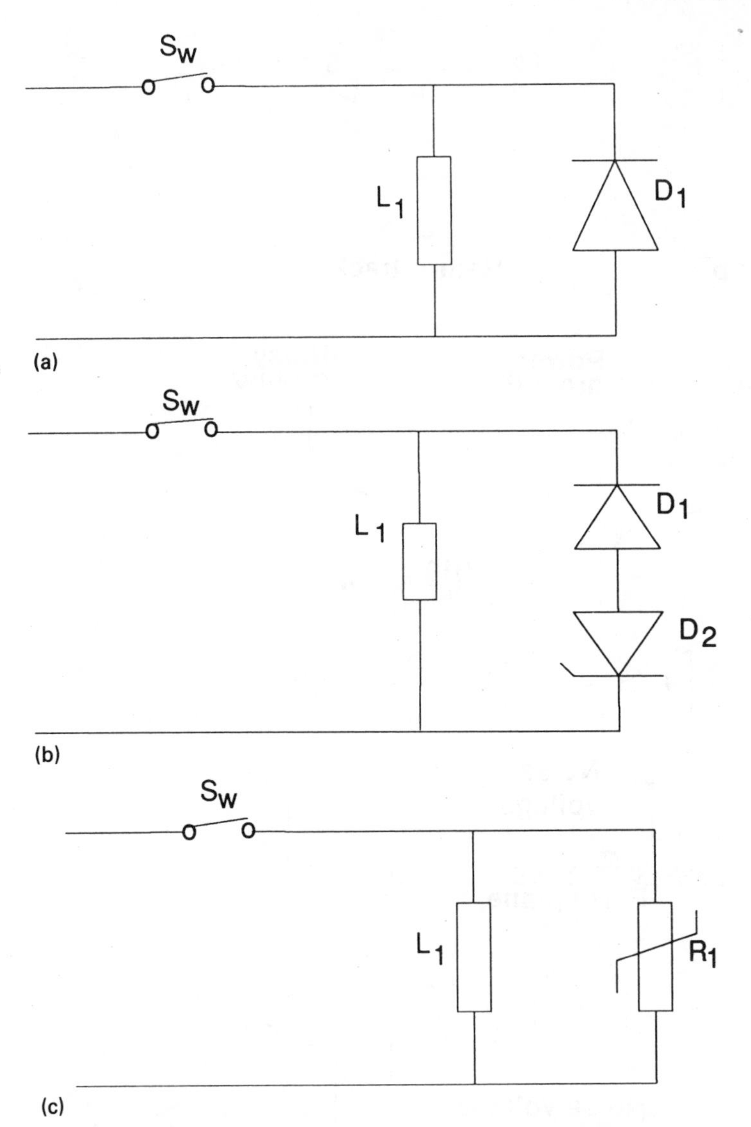

Figure 4.3 Suppression of voltage transients across an inductive load: (a) diode; (b) zener; (c) voltage-dependent resistor

circuit board operating at high frequencies. In these instances earthing strategies become important, as shown in Figure 4.4. In the common earth return arrangement the return current from all the devices passes through the return track to the earth point. The current from elements 2 and 3 causes a voltage drop of V_p to appear at the input to element 1, due to the impedance Z_p of the return track. This signal forms an unwanted interfering signal at the terminal of element 1, and its effect is especially severe if elements 2 and 3 are heavy-current devices or element 1 has a high sensitivity.

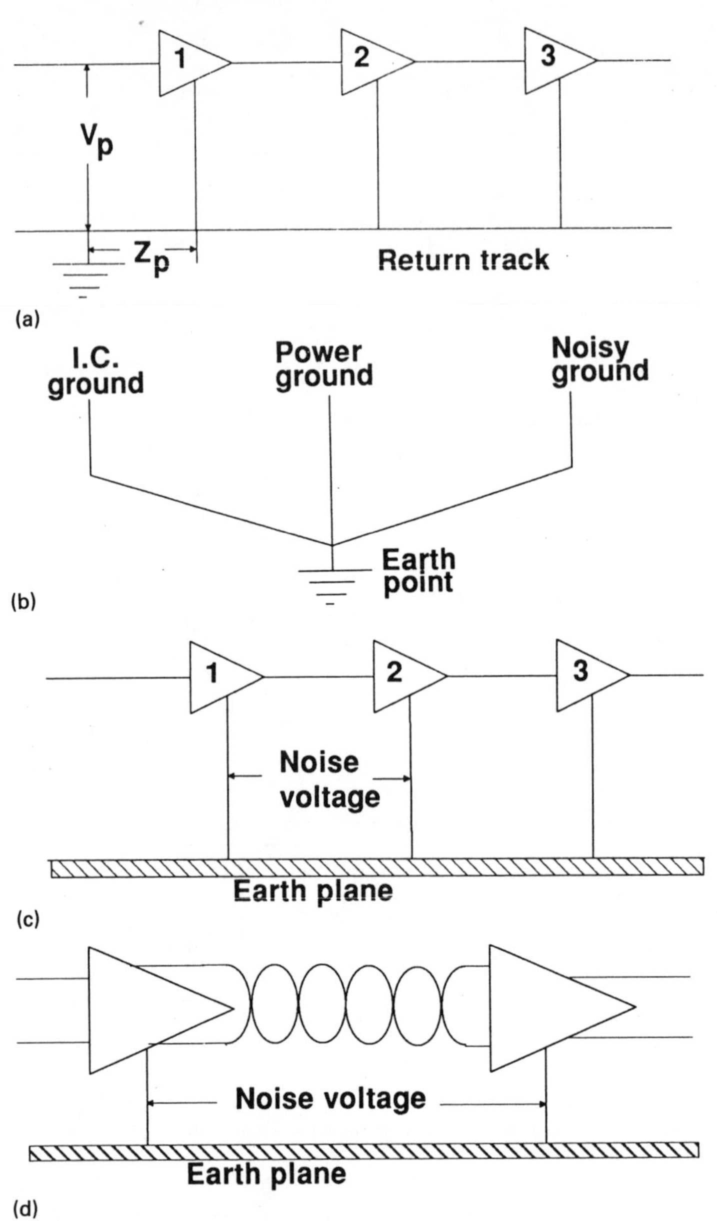

Figure 4.4 Earthing strategies: (a) common earth return; (b) separate earth returns; (c) multipoint ground; (d) balanced differential circuit

Several techniques can be used to overcome the problem of a common earth return, illustrated in Figure 4.4(a). Increased track widths can be used so as to reduce track impedance. Heavy-current components can be placed closest to the earth point. Alternatively, separate earth returns can be used for the various elements in the circuit. For example, Figure 4.4(b) shows three separate earth returns, each carrying a different level and type of current. This is especially useful in circuits which have a very different mix of circuit elements.

The signal ground should have a low impedance to handle large signal currents, and this is usually done by making the ground plane large. The inductance is reduced by placing the signal current-carrying conductor close to its ground return. Single-point grounds, as illustrated in Figures 4.4(a) and 4.4(b), are difficult to maintain at a low impedance and are not suitable for frequencies above about 10 MHz. Generally, above these frequencies multipoint grounds are used, as in Figure 4.4(c). However, care is now needed to prevent the occurrence of ground loops, which can generate fields which interfere with the signal.

Interference noise induced in circuits can be common mode, in which the voltage is induced between common terminals of the transmitting and receiving circuits, and differential mode, where the currents are caused to flow in opposite directions in the source and return lines. For the single-ended circuit shown in Figure 4.4(c) the noise voltage which appears in the signal loop will interfere with the signal, either adding or subtracting from it. The function of the differential amplifier, in the balanced differential circuit of Figure 4.4(d), is to sense the differential signal and to reject the common-mode signal. Therefore this circuit can tolerate a substantial amount of common-mode noise.

The impedance of power supplies used with electronic circuits should be as low as possible, and common-mode impedance coupling must be avoided. Several techniques are used:

(i) Decoupling the circuit elements using high-frequency capacitors. It is important in these instances to keep the leads short to minimise their inductance, and for critical applications integrated circuit holders can be used which have space in their body for mounting a decoupling capacitor, so that it is attached close to the pins of the integrated circuit.
(ii) Increasing the cross section of the power supply tracks, to reduce their impedance.
(iii) Keeping high-voltage and power return rails as close to each other as possible.

Multiple pins are often used on the printed circuit board connectors to transmit the power into and out of the circuit. This enables a network of power supply rails and earth returns to be built up to carry current to the high-usage devices. Often these tracks divide the board into areas containing high-, medium- and low-frequency circuits, which are therefore effectively segregated from each other.

4.5 EMI shielding techniques

The previous section described design techniques which could be used to reduce a system's susceptibility to EMI whilst also reducing the amount of EMI which it generated. The present section describes methods for protecting against EMI reaching a given system, or for preventing EMI which has been generated within a circuit from reaching other systems. The concepts involved in radiated EMI shields are first introduced, followed by a description of the factors which need to be considered in the design of the shields and protection against conducted EMI.

4.5.1 Radiated EMI shielding concepts

EMI can be caused by electrical or magnetic fields and a shield is any material which is placed in the path of the field to impede it, although this material is usually a conductor. The effectiveness of the shield is determined by the distance between the source of EMI and the receiver, the type of field, and the characteristic of the material used in the shield, including the number and type of discontinuities in it.

Shielding is effective in attenuating the interfering fields by absorption within its body, or by reflection off its surface. If E_u is the strength of the unshielded electrical field and E_s is the subsequent strength of this field after it has passed through a shield, as shown in Figure 4.5(a), then the shielding effectiveness is given by

$$S = 20 \log \frac{E_u}{E_s} \tag{4.1}$$

Similarly, if H_u and H_s are the corresponding unshielded and shielded magnetic fields, then equation (4.2) gives the shielding effectiveness:

$$S = 20 \log \frac{H_u}{H_s} \tag{4.2}$$

The shielding effectiveness can also be expressed in terms of the absorption loss (A_l) and reflection loss (R_l), as in equation (4.3), where all the terms are in decibels. The factor B_l is introduced to take account of multiple reflections, as described later:

$$S = A_l + R_l + B_l \tag{4.3}$$

Conductors have poor dielectric characteristics, so fields, whether electrical or magnetic, will suffer absorption loss when going through them. The amount of absorption is determined by the skin depth (δ) of the material, where skin depth is the distance into the conductor at which the field is attenuated by an amount equal to $1/e$. Skin depth decreases with the conductivity (γ) and permeability (μ) of the material, and with frequency (ω), as given by

$$\delta = \left[\frac{2}{\omega \mu \gamma}\right]^{1/2} \tag{4.4}$$

The field strength decreases exponentially as it passes through the material, so that the shielded field strength for the arrangement of Figure 4.5(a) is given by equation (4.5) and the absorption loss by equation (4.6):

$$E_s = E_u \, e^{-d/\delta} \tag{4.5}$$

$$A = 8.69\,(d/\delta) \qquad \text{(dB)} \tag{4.6}$$

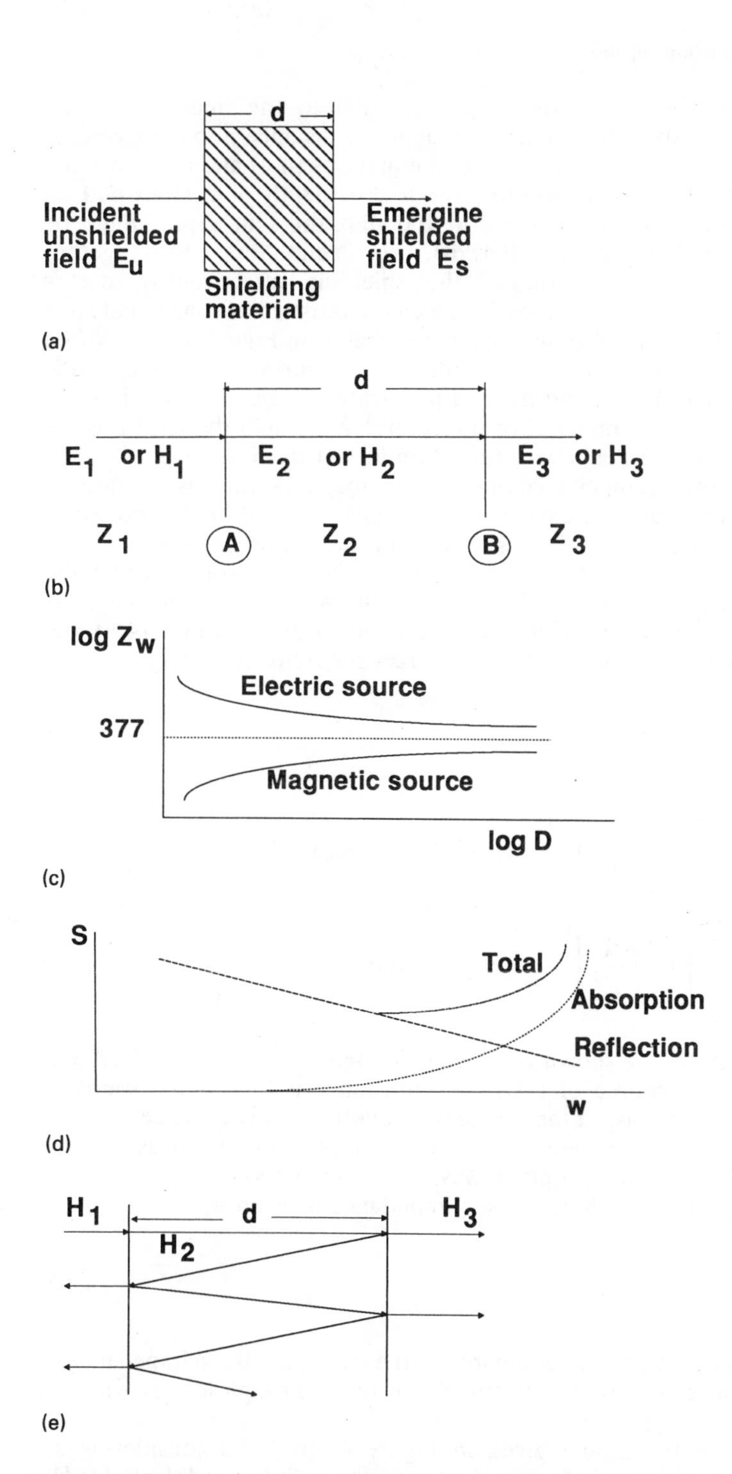

Figure 4.5 Shielding concepts: (a) shielded and unshielded fields; (b) reflection calculation model; (c) plot of electromagnetic wave impedance for an electric and a magnetic source; (d) variation of shielding effectiveness with frequency; (e) illustration of multiple reflections

Therefore the absorption loss is proportional to the thickness of the material and equals 8.69 for each skin depth of distance into the shielding material. As an example into the use of equations (4.4) and (4.5), a 1 mm aluminium sheet at 1 kHz would have a skin depth of 2.7 mm and an absorption loss of 3.2 dB, whereas at 100 MHz the skin depth would be 8.3 μm, and the absorption loss 1047 dB. Reflection loss, which together with absorption loss determines the shielding effectiveness, occurs whenever there is a discontinuity in the characteristic impedance between the shield and its surrounding. This is illustrated in Figure 4.5(b). When the field reaches the point of discontinuity, at surface A, some of the incident field is reflected and the rest penetrates, to be attenuated by the absorption loss in the material of impedance Z_2. When the field reaches surface B it will exit, again with reflection loss at this interface.

The characteristic impedance of an electromagnetic wave is the ratio of the electrical to magnetic fields. This impedance depends on the properties of the shield material and the distance from the source of interference to the measurement point. If D is the distance from the source and λ the wavelength, then, defining K by the value shown in equation (4.7), the impedance of the electromagnetic wave is given by equations (4.8) and (4.9), for a magnetic and an electric source, respectively:

$$K = \frac{2\pi D}{\lambda} \tag{4.7}$$

$$Z_W = 120\pi \left[1 + \frac{1}{K^2}\right]^{-1/2} \quad \text{(magnetic source)} \tag{4.8}$$

$$Z_W = 120\pi \left[1 + \frac{1}{K^2}\right]^{1/2} \quad \text{(electric source)} \tag{4.9}$$

These equations are shown graphically in Figure 4.5(c), which illustrates that close to the source of EMI the impedance is high if the source is electrical, whilst it is low if the source is magnetic. As the distance from the source increases, the impedance reaches asymptotically towards the free space wave impedance of a plane wave, of 120 π, or 377 Ω.

For a conductor the characteristic impedance is given by

$$Z_c = [\omega\,\mu/\gamma]^{1/2} \tag{4.10}$$

This impedance is very low, compared to the free space wave impedance of 377 Ω, therefore boundary reflections can result in high losses, which is desirable.

Returning to the model given in Figure 4.5(b), and considering an electric field for the present, even if absorption loss is ignored the fields E_1, E_2 and E_3 will be different, the difference being due to reflection losses at the two interfaces A and B. The relationships between these fields are equations (4.11) and (4.12), when multiple reflections, as described later, are ignored:

$$E_2 = \frac{2\,Z_1\,E_1}{Z_1 + Z_2} \tag{4.11}$$

$$E_3 = \frac{2\,Z_2\,E_2}{(Z_2 + Z_3)} = \frac{4Z_1\,Z_2\,E_1}{(Z_1 + Z_2)\,(Z_2 + Z_3)} \tag{4.12}$$

If Z_2 is a shield placed in air, then Z_3 equals Z_1, and equation (4.12) reduces to that given by equation (4.13). This is further simplified if the shield is a conductor and Z_2 is small, as shown in equation (4.14):

$$E_3 = \frac{4\,Z_1\,Z_2\,E_1}{(Z_1 + Z_2)^2} \tag{4.13}$$

$$E_3 = \frac{4\,Z_2\,E_1}{Z_1} \tag{4.14}$$

Using equation (4.14) the reflective component of the shielding effectiveness, as in equation (4.1), is given by

$$R_{\mathrm{E}} = 20\log\left(\frac{Z_1}{4\,Z_2}\right) \tag{4.15}$$

This indicates that the smaller the conductor's characteristic impedance, the greater the effectiveness of the shielding.

The characteristic impedance of a conductor increases with frequency, being given by proportionality (4.16), so that the shielding effectiveness is as in proportionality (4.17). However, absorption loss decreases with frequency since the skin depth decreases, so that the total shielding effectiveness, as indicated by equation (4.3), varies as in Figure 4.5(d). Generally, a conductive sheet gives very good electric field shielding, except for very thin coatings:

$$Z_2 \propto \omega^{1/2} \tag{4.16}$$

$$R_e \propto \log \omega^{-1/2} \tag{4.17}$$

For an electric field most of the reflections occur at boundary A, in Figure 4.5(b), giving high reflection loss and low penetration. For a magnetic field most of the reflections occur at boundary B, and there is therefore high penetration. Low-frequency magnetic fields have low reflection loss and depend on absorption loss for their shielding effectiveness. For a magnetic field equations (4.11) to (4.14) are equally applicable, where the electric fields E_1, E_2 and E_3 are replaced by magnetic fields H_1, H_2 and H_3. The reflective component of the shielding loss is given by

$$R_{\mathrm{H}} = 20\log\left(\frac{Z_1}{4\,Z_2}\right) \tag{4.18}$$

ignoring multiple reflections, which is similar to that of the electric field of equation (4.15).

Because very low reflections occur in a magnetic field at the air–conductor interface, the field inside the conductor is high. In fact, using the magnetic equivalent of equation (4.11), H_2 will be twice the value of the incident field H_1, for Z_2 much smaller than Z_1. If the absorption losses in the conductor are low then most of this energy will be reflected at the conductor–air interface, and it will result in multiple reflections within the conductor. This is illustrated in Figure 4.5(e), where a small amount of field is shown to leak away at each reflection. Eventually, in the ideal case of zero absorption loss, half the field will emerge to the right and half to the left, so that there are no reflection losses. To allow for this multiple-reflection phenomenon a term B_l is introduced into the shielding effectiveness equation (4.3).

The value of the correction term B_l is given by

$$B_l = 20 \log (1 - e^{-2d/\delta}) \quad (4.19)$$

It is always negative to show that reflection loss predictions without considering multiple reflections are too optimistic. B_l can be omitted if the absorption loss is high, greater than about 10 dB, so that multiple reflections are minimised, although for low-frequency magnetic fields, and thin shield, this term is almost always needed.

4.5.2 Shield design

The shielding effectiveness is often less than that predicted by the equations given in the previous section due to discontinuities in the shield, such as seams and holes. These discontinuities impede the flow of induced current in the shield, which is responsible for generating a field opposing the interfering field. Wherever possible, the location of the discontinuity must be such as to minimise its effect on these currents. For example, the seam should be located such that the circulating current does not have to flow across it, as in Figure 4.6(b), but flows parallel to it, as in Figure 4.6(a).

The joints between the different parts of the shield represent its weakest point, and the transfer impedance of these joints is a useful concept for characterising them. If a current I flows on the outside of the shield, as in Figure 4.6(c), and it induces a voltage V on the inside of the shield, then the transfer impedance is equal to the ratio V/I. The impedance at a seam is due to the contact impedance and the surface impedance of any gasket which is used. The contact impedance can be represented by a parallel-connected RC circuit whilst the gasket impedance is a series-connected RL circuit. Therefore if there is poor contact resistance this will usually show up as a drop in transfer impedance as the frequency increases.

The contact resistance at a joint must be kept as low as possible, and this can be done by applying sufficient pressure; by using materials with low contact resistance; by having a large surface area of contact; and by avoiding corrosion. Figure 4.6(d) shows how simple shaping techniques

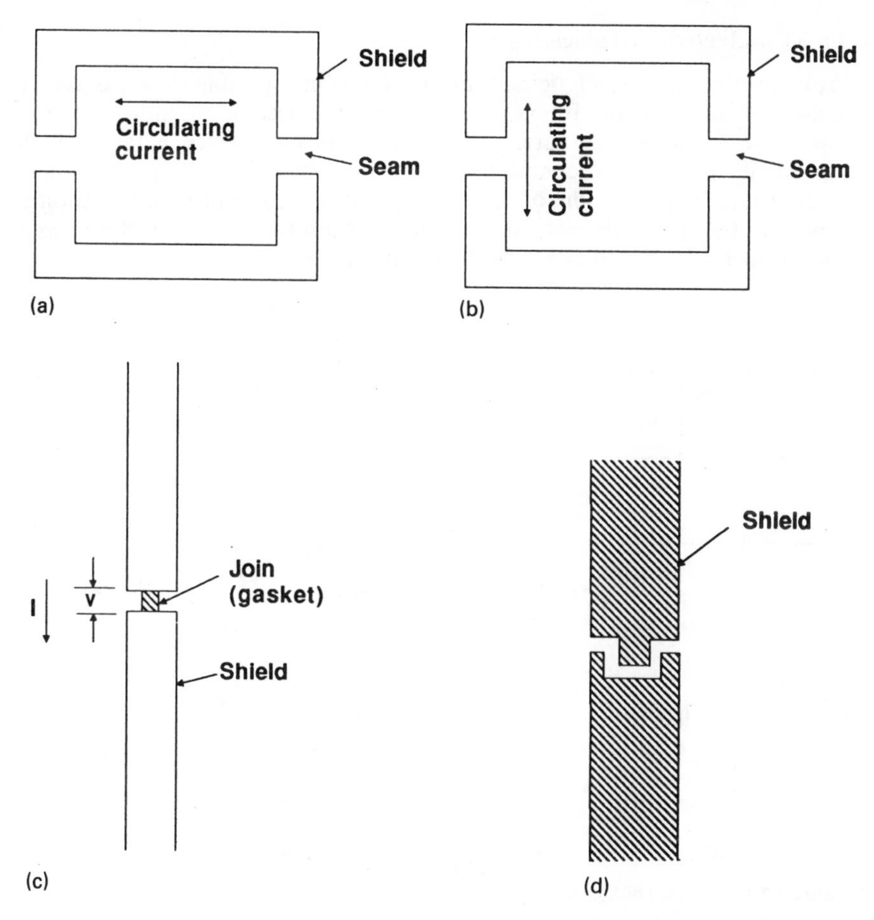

Figure 4.6 Seams in shields: (a) and (b) position of seam relative to circulating current; (c) illustration of transfer impedance of a seam; (d) reducing transfer impedance by increasing contact area

can be used to increase the contact area of a joint, and corrosion can be reduced by not placing dissimilar materials in contact with each other.

Screened cables, brought into the screened enclosure, should have their screens terminated in connectors which totally enclose their screens and provide a solid connection from the screen to the shielded wall of the enclosure. Both ends of the cable should be so connected, and the finish used on the enclosure must not be such as to make it difficult to provide a good connection.

The low-frequency performance of a shield is determined by the electrical properties of its walls and joints, whereas the high-frequency performance is much more influenced by the gaps and apertures which exist in its surface. For effective screening there should be no gaps which are greater than one hundredth of the lowest wavelength of the interfering radiation.

4.5.3 Conducted EMI shielding

Protection against EMI being transmitted along a cable is achieved by means of suppression filters, which consist basically of inductive and capacitive elements. A variety of such filters exist since a type which suppresses interference completely from one system may be quite useless in another. The location of the filter is also important, and it should generally be placed directly at the source of interference, and the output and input leads should never be bundled together.

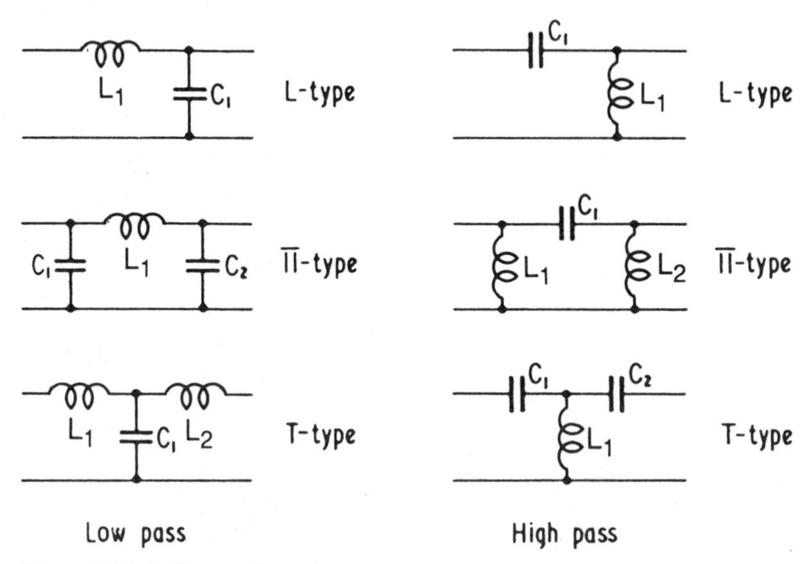

Figure 4.7 Basic filter configuration

Figure 4.7 shows some of the basic filter arrangements which are used for suppressing conducted interference. Where large attenuation is required it is preferable to use several stages of smaller filters, rather than one large system. Low-pass filters are commonly used in a variety of modifications, as shown in Figure 4.8. The arrangement of Figure 4.8(a) is perhaps the most basic, and although it is effective in suppressing symmetrical interference it is not suitable for asymmetrical interference, which is caused by leakage between lines and earth. The circuit of Figure 4.8(b) is now used, where the polarity of the windings in the split inductor L_1L_2 are such that the d.c. components of current through them are cancelled, so that the core size is reduced. Figures 4.8(c) and 4.8(d) introduce progressively more attenuation into the interfering signal path, at the expense of greater complexity. Generally, arrangements of the type shown in Figure 4.8(d) should not be used if S_W is a semiconductor switch, since the surge current through the switch when it turns on, caused by the discharge of the capacitors, may destroy it.

Components chosen for the filters should also be carefully designed. The inductors must have low stray capacitors, so they should not be multilayer,

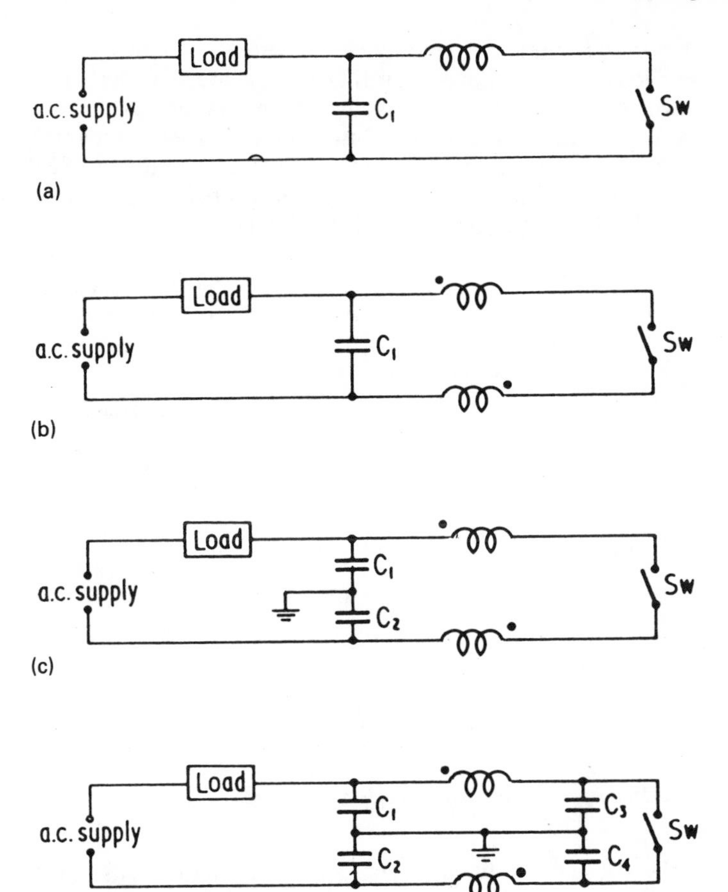

Figure 4.8 Modifications to the basic filter arrangements: (a) basic L-type; (b) and (c) L-type; (d) π-type

and the capacitors must have low series inductance. The filter should be enclosed in a screened box, which is connected to the wall of the shielded enclosure, so that interference signals cannot become coupled from the noisy side to the quiet side of the system.

4.5.4 EMC standards

Most countries have their own standards organisations which regulate their RFI policies, examples of these being the BSI in the UK, the FCC in the USA, and the VDE in West Germany. The military also have their own requirements, referred to as MIL-STD and DEF standards, although the needs of commercial and military equipment are different. The military require that their equipment continue to perform in battlefield conditions, whilst commercial interests centre on the need to protect radio, television, telecommunication and domestic systems.

Figure 4.9 shows the spread of frequencies covered by a few typical standards, as expected, the military operating over a much wider range than the commercial. In order to ensure some commonality of standards between different countries the IEC (International Electrotechnical Commission) set up a subcommittee in 1934, called the CISPR (International Special Committee on Radio Interference), to prepare guidelines on EMC. These are now followed by most countries.

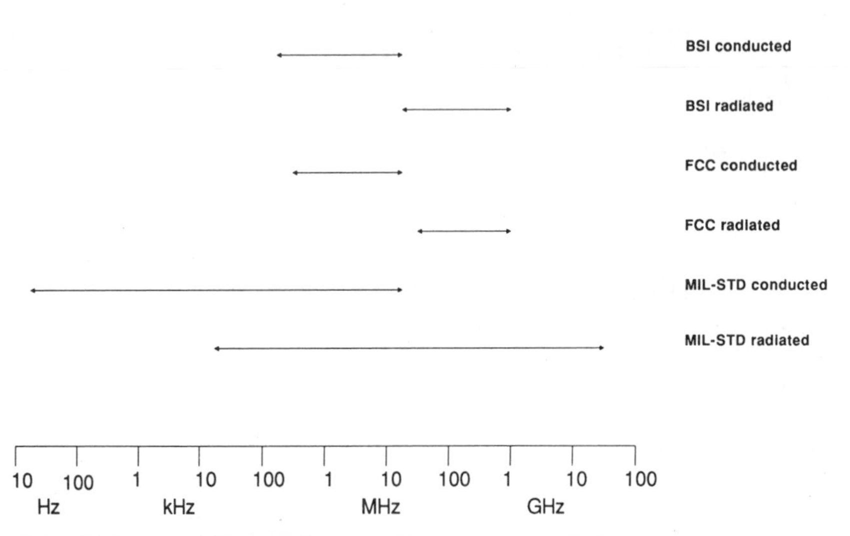

Figure 4.9 Spectrum of frequencies covered by typical EMC standards

In the UK the BSI maintained BS 6527, which defined its EMC requirements, although this has now been replaced by a European standard, EN 55022, which was ratified by all CENELEC countries in June 1986, and is the system to which most major European countries are working. EN 55022 covers information technology equipment, i.e. receiving data equipment such as data input lines and keyboards; processing data equipment, such as computation and storage; and data output equipment. This standard grants two classes of approval, Class A and Class B. The less stringent approval class, Class A, is intended for commercial users, on the assumption that commercial equipment is usually better protected against RFI. It uses a protection distance of 30 m, and it can be too liberal for domestic users. Class B is used for domestic equipment, where the protection distance is reduced to 10 m. Tables 4.1 and 4.2 provide the limits for conducted and radiated RFI, as specified in EN 55022, the quasi-square and average limits being measured by a quasi-square and average detector respectively.

The FCC administers the use of the frequency spectrum in the USA and its rules span many areas, for example Title 47 of its 'Code of Federal Regulations' covers telecommunications and has four volumes, each made up of many parts. Volume 1, Part 15 specifies emissions from radio frequency devices, such as radio and TV receivers, low-power radio-controlled appliances, and computing devices. The FCC also operates two

Table 4.1 EN 55022 conducted interference limits

Frequency range (MHz)	*Limits* dB (MV)			
	Class A		*Class B*	
	Quasi-peak	*Average*	*Quasi-peak*	*Average*
0.15–0.50	79	66	66–56	56–46
0.50–5	73	60	56	46
5–30	73	60	60	50

Table 4.2 EN 55022 radiated interference limits

Frequency range (MHz)	*Quasi-peak limits* dB (μV/m)	
	Class A (test distance 30 m)	*Class B (test distance 10 m)*
30–230	30	30
230–1000	37	37

classes of approvals. Class A is for commercial users, and to gain approval equipment manufacturers do their own tests and keep the results on file. The FCC may ask to see these results or to conduct some of the tests. Warning labels need to be attached to the equipment stating that it complies with FCC Class A and that its use in a residential area could cause problems. FCC Class B has a more stringent requirement and is intended for domestic equipment. As before, manufacturers conduct the tests, but the FCC approves all results and keeps these on its own files. Once again

Table 4.3 FCC conducted interference limits

Frequency range (MHz)	*Maximum r.f. line voltage* (μV)	
	Class A	*Class B*
0.45–1.6	1000	250
1.6–30	3000	250

Table 4.4 FCC radiated interference limits

Frequency range (MHz)	*Field strength* (μV/m)	
	Class A (test distance 30 m)	*Class B (test distance 3 m)*
30–88	30	100
88–216	50	150
216–1000	70	200

the FCC may repeat some of the tests, and labels must be attached to the equipment stating that it meets FCC Class B. Tables 4.3 and 4.4 show the FCC limits for conducted and radiated interference, as specified in Volume 1, Part 15, Subpart J.

4.5.5 EMC measurement

EMC testing can be done on the bench top, in an open field site, in an anechoic chamber, or in a shielded room. Bench-top tests are generally only used to provide a quick check on the various components of the system, which are later verified by full testing. The open field site is used for commercial equipment, and shielded rooms usually for tests on military equipment. Semi-anechoic chambers are used for both military and commercial tests.

When checking the RFI generated from a piece of equipment a suitable pick-up device is required. The receiver must be sensitive, able to read low-level signals without introducing distortions. For tests on shielded enclosures both a source of RFI and a measurement device are required. Readings can be taken by tuning the generator and receiver at a series of single frequencies, although this is usually slow and can result in incomplete results, since critical narrowband frequencies may be missed. It is now better to use swept broadband techniques, as shown in Figure 4.10, where the tracking generator source sweeps synchronously with the input frequency of the spectrum analyser. The output from the spectrum analyser may be stored, manipulated and displayed, if required.

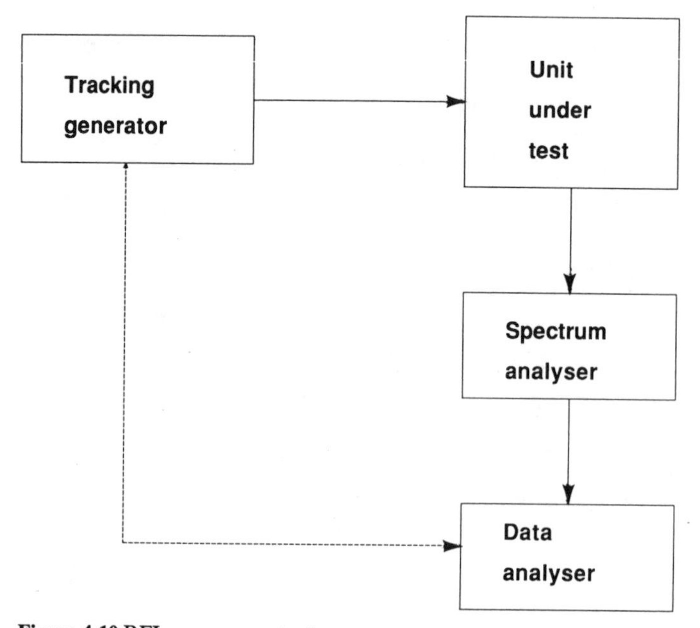

Figure 4.10 RFI measurement set up

The shielding effectiveness of an enclosure can be obtained by measuring the fields with and without the presence of the shield, and then applying equations (4.1) and (4.2). Separate antennas can be used for the electric and magnetic fields, the electric field tests usually going down to 1 MHz. For magnetic fields this should cover the range to 10 kHz, since the leakage of the magnetic field component, of the plane wave, is greater than the electric field component at low frequencies.

The RFI shield being tested can be radiated by means of an external source and the field inside the enclosure measured. This method gives acceptable results so long as the transmitting antenna is sufficiently far from the source to give wavefront uniformity at the shield and a correct wave impedance. At low frequencies this would require a large spacing and high transmitting power to effect the shield, so that it is more usual to use a smaller spacing, reducing the power requirement but accepting that the wave impedance is now not correct and may not be consistent. The other disadvantage of reducing the spacing between the shield and transmitting antenna is that the shield couples to the incident field of the antenna and would give false results.

The problem of transmitter power can be solved by putting the transmitting antenna inside the screened enclosure being tested, and measuring the field which escapes outside. However, localised current paths are now produced in the screen and these will influence the transmitter, giving results which are not representative of the plane-wave

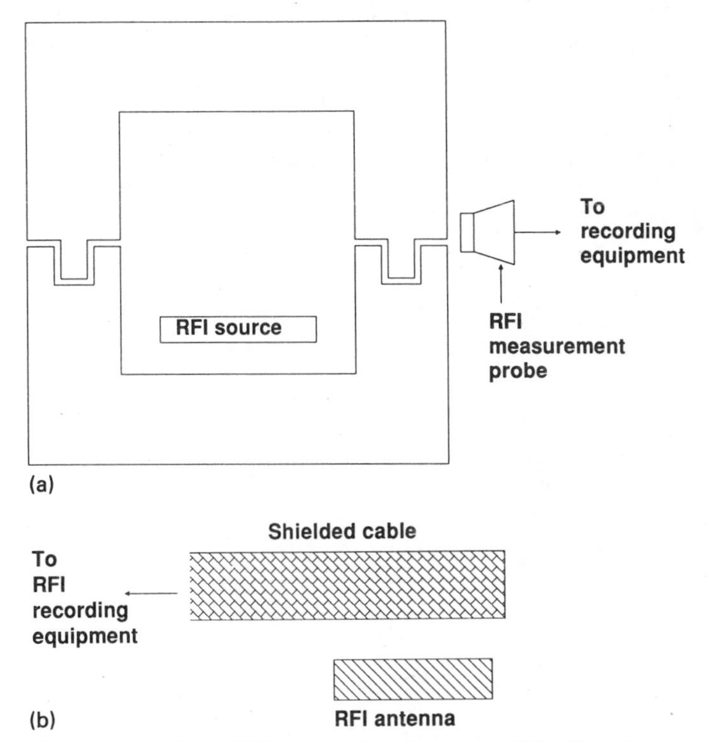

Figure 4.11 Bench-top RFI testing: (a) enclosure tests; (b) cable tests

shielding performance of the enclosure. When measurements are done in a shielded room the room would exhibit many resonance modes, which would mask the true performance of the shield over several bands of frequency. Generally, low-frequency measurement problems can be minimised by use of a Rhombic Simulator, which produces a plane wave over a wide frequency range, having a wave impedance very near to that of free space, given by equation (4.8), of 377 Ω.

Bench-top testing is useful for system troubleshooting at the component level before going on to total system tests in open field sites or screened rooms. These tests require a small antenna for localised shielding effectiveness tests, as shown in Figure 4.11(a). Cable shielding testing on the bench top, shown in Figure 4.11(b), are also very important, since problems here are a major cause of system noise, crosstalk and errors. The cable is used as the receiving antenna, being connected to a spectrum analyser, the signal coupled into the shielded and earthed cable being low for an effective shield.

Repeatability of EMC measurements is an important consideration, and problems which cause non-repeatable readings include the following:

(i) Electric field sensitivity of the receiving antenna, which picks up stray electric fields coupled to the test engineer and cables. Variation in this coupling is also caused by the orientation of the various elements. The problem can be minimised by reducing the electric field sensitivity of the antenna.
(ii) Screen room interactions, which should be reduced, or at least variations in the physical layout of the tests should be avoided in order to duplicate the interactions.
(iii) Poorly maintained and calibrated test equipment.
(iv) Not using correct assembly procedures for the equipment being tested, when it is dismantled and re-assembled, so that its performance varies, for example due to the variation in pressure on joining gaskets.

Chapter 5

Power semiconductor protection

5.1 Introduction

In spite of its relatively large power-handling capability, a power semiconductor can be easily destroyed due to voltage or currents which exceed its ratings. High currents cause localised or general heating, and excessive voltages can result in punch-through of the silicon wafer. Protection devices are often connected into power semiconductor circuits, but it must be remembered that this reduces the overall reliability of the system, since the component count has been increased, and these protection components can fail themselves, sometimes destroying the components which they are supposed to be protecting.

This chapter considers the factors causing failure in power circuits, and the protection techniques and components which can be used.

5.2 Causes of failure in power circuits

The mechanisms which result in failure within power circuits can be related to the power semiconductor device itself or to external factors within the system. Device-related factors are:

(i) The breakdown of the component, causing it to operate in a continuous on-mode. This could be as a result of a current overload or an overvoltage.
(ii) The power semiconductor operating in a blocking mode, so that other components in the system are overstressed. This mode may be a result of a failure in the component itself or due to a failure of the circuit controlling it.

Circuit-related failure-mode mechanisms are:

(i) Load related, where load short circuits cause excessive current surges through the controlling devices or switching the load generates voltage spikes.
(ii) Supply related, usually seen as voltage spikes on the a.c. or d.c. supply lines.
(iii) Other circuit-generated current or voltage overloads, such as voltage transients generated when the current changes suddenly within inductive circuits.

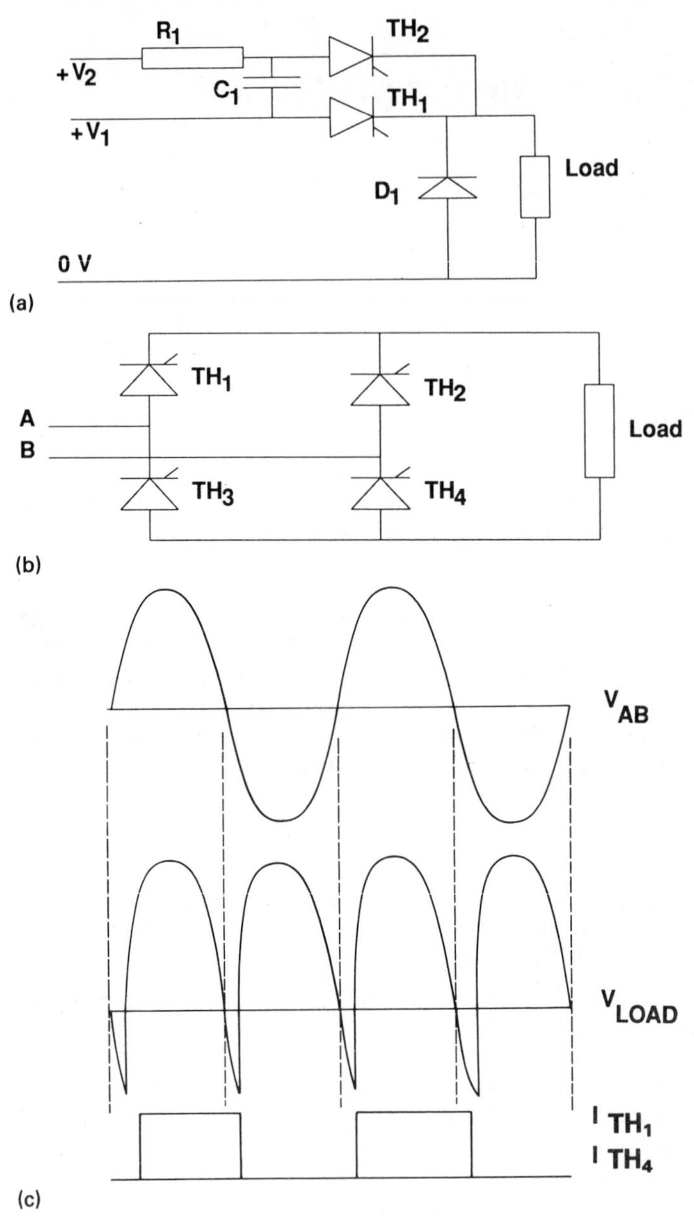

Figure 5.1 Effect of component failure in power circuits: (a) forced commutated circuit; (b) naturally commutated circuit; (c) waveforms for naturally commutated circuit

Figure 5.1 illustrates the effects of component-related failure in two common circuits. In Figure 5.1(a) thyristor TH_1 is the main load-carrying device and thyristor TH_2 is an auxiliary component, which discharges capacitor C_1, causing the main thyristor to turn off. Diode D_1 is a free-wheeling device placed across the load, to carry the inductive load current when the thyristors are off. If this diode were to fail to a short circuit it would cause the two thyristors to carry large amounts of current when they turn on, and if it failed to an open circuit the decay of load current would cause a high-voltage spike to be generated, destroying the semiconductors. Failure of thyristor TH_2 would result in TH_1 being permanently on, causing it to overheat.

In the bridge circuit, shown in Figure 5.1(b), the devices are controlled in pairs, TH_1 and TH_4 together and TH_2 and TH_3 together. The current and voltage waveforms are as shown in Figure 5.1(c), where each device conducts for half a cycle. If any component, say TH_3, now fails to an open circuit, then the other thyristors, in this case TH_1 and TH_4, will carry the full-load current continuously, causing them to overheat and perhaps be destroyed. However, if thyristor TH_2 was fired after TH_3 had failed, then the inductive load current would free-wheel through thyristors TH_2 and TH_4, causing overheating. If any of the thyristors fail to a short circuit then this short circuit is applied across the lines, resulting in a failure of the other devices. Failure of the a.c. supply line would cause the load current to free-wheel through the two arms of the bridge, assuming that the thyristors have been fired.

Circuit-related failure mechanisms can be due to a variety of reasons. Fault current, as a result of short circuits, can cause heating and loss of control of the power semiconductor switch, the current building up over several a.c. cycles to reach a steady state value determined by the circuit voltage and impedance. The high current can also result from discharge of circuit capacitances, such as snubbers used across the lines or capacitors used in power supplies, and is one of the main causes of di/dt failure.

Lightning strikes can cause a steep-fronted, line-to-ground, surge of voltage which destroys the power semiconductors, although the surge may be attenuated to some extent by output transformers. Lightning arrestors are usually placed across lines to guard against this. Transformer switching is another source of high-voltage transients. On switch-on the inrush current can result in oscillations within the resonant secondary winding, due to transformer leakage inductance and the distributed parasitic capacitance of the secondary winding. On switch-off the magnetising current is interrupted and this results in a collapse of the core flux, generating voltage transients on the secondary. This effect is greatest on light loads, when the primary current is passing through zero. Energising a step-down autotransformer results in the interwinding capacitance causing the primary voltage to be momentarily coupled through to the secondary, giving an overvoltage.

Overvoltages are also caused by discontinuous current operation in inductive circuits, the energy stored in the inductance causing a high-voltage spike. Current interruption can occur due to several causes, such as power semiconductors turning off too fast, or due to operation of protection devices such as circuit breakers and fuses. Voltage transients can be avoided by several techniques, for example:

(i) Switching the secondary of transformers, so as to avoid the inrush current which occurs when the primary is switched;
(ii) Ensuring that switching devices do not operate too quickly, or have

long arcing times, so that the inductive energy in the circuit is dissipated in the arc;

(iii) Using lightning and other surge arrestors close to the equipment being protected;

(iv) Placing a capacitor across the secondary of autotransformers, to divide the voltage coupled from the primary due to the stray winding capacitance.

5.3 Overvoltage protection

Three types of voltage suppressors are commonly used: those which store the transient energy of the overvoltage and then dissipate it later as heat; those which directly convert the energy of the overvoltage into heat, which is then dissipated; and those which convert the overvoltage into an overcurrent, and then use the techniques described in section 5.4 to protect against it.

Capacitors are commonly used to store the energy of transient overvoltages, which are then dissipated in resistors. Figure 5.2(a) shows an

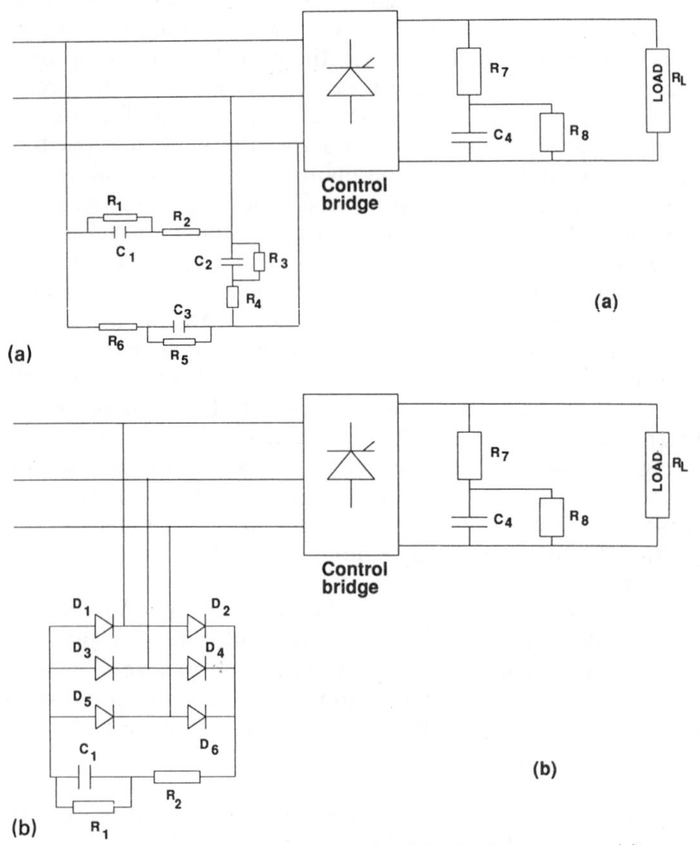

Figure 5.2 Transient overvoltage protection using R-C suppressors: (a) a.c. capacitor; (b) d.c. capacitor

arrangement in which R-C suppressors are connected across the three-phase supply lines, as well as across the d.c. load. The transient voltage is partly dissipated in the line impedances and series resistor, and the rest appears as an increase in capacitor voltage. The larger the value of capacitance, the lower this voltage rise. The resistors in series with the suppression capacitors limit the charging current and also damp down the oscillations, resulting from resonance between these capacitors and inductances in the lines, which can lead to overvoltages.

The capacitors shown in Figure 5.2(a) carry power frequency components of current and their presence can also affect the commutation behaviour when changing from one supply phase to the next. Figure 5.2(b) shows an alternative arrangement which uses an auxiliary low-power bridge, but which requires only one suppression capacitor and this can be electrolytic, since it only carries d.c., enabling it to be physically small. If the inductance of the capacitance is high it will not be able to suppress fast-rising transients, and in these cases it is usual to shunt the electrolytic capacitor with a much lower-valued low-inductance capacitor, such as ceramic. Resistor R_1 limits the charging current of the capacitor and resistor R_2 discharges the capacitor, for safety reasons, after the rectifier is de-energised.

Suppression components, which convert the voltage transient energy directly into heat, consist of devices with characteristics such as those shown in Figure 5.3(a), for unidirectional operation, and Figure 5.3(b) for bi-directional operation. Very little current flows through the device until the breakover voltage V_B is reached, after which time the current rises rapidly and the voltage is held substantially constant. These devices are available in a range of voltage ratings, which are fixed and cannot be adapted to changes in circuit operating voltage, as is possible with an R-C suppressor.

The voltage clamping device can be connected across the power supply lines, or across the components being protected, as in Figure 5.3. The device should have a flat voltage–current characteristic once breakover has occurred, but it should also be able to absorb high energies for short durations. The zener diode exhibits the characteristic shown in Figure 5.3(a) and it can be used as a suppressor. It has flat characteristic, i.e. low slope resistance, but it is only able to absorb small energy levels, so it is used in low-power applications. Transient overvoltages usually have a high voltage value but then only last for a short time, or they have a relatively low voltage but last for longer. Therefore the energy requirements are fairly well defined, and suppression devices are usually specified in terms of watt-seconds of energy dissipated.

Surge suppressors based on selenium are designed to handle high energy levels and have a low inductance, so that they can suppress steep-rising waveforms. Their slope resistance is higher than that of a zener diode so the voltage will rise with current. Their operating temperature range is also relatively low. Selenium suppressors are based on selenium diodes operated in the reverse direction, having a characteristic similar to that of Figure 5.3(a). Two devices can be connected back to back for bi-directional operation, as in Figure 5.3(b), and several cells can be connected for higher-voltage operation. Figure 5.3(c) shows an unidirec-

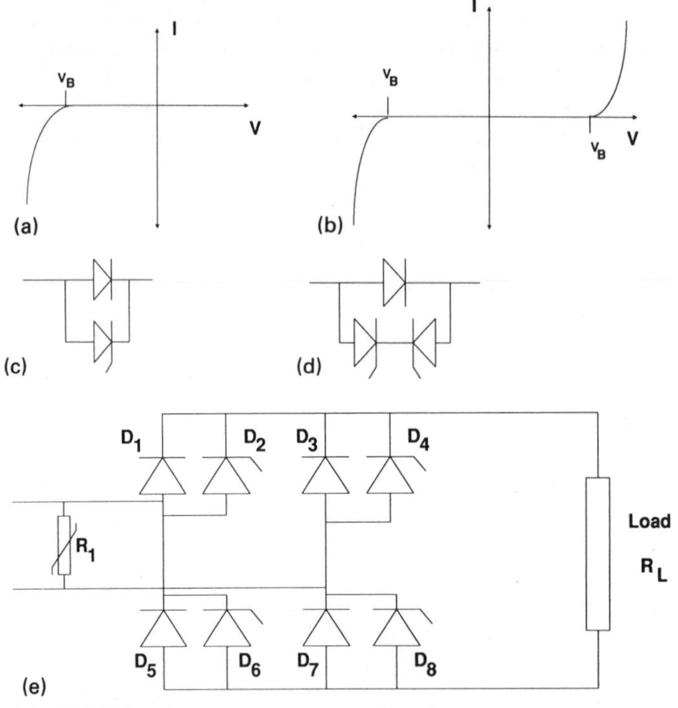

Figure 5.3 Transient overvoltage protection using surge suppression components: (a) unidirectional operating device; (b) bi-directional operating device; (c) protection of a diode; (d) protection of a thyristor; (e) bridge-circuit protection

tional cell connected to protect a diode, and Figure 5.3(d) a bi-directional cell for protecting a thyristor.

Avalanche diodes are also used for surge protection. They have a low slope resistance and a wide operating temperature range, but they are relatively costly and so are usually used to protect expensive components. Varistors, or voltage-dependent resistors, are made from oxides of metals or of silicon carbide. They have a bi-directional characteristic and a temperature range from about −25°C to +85°C. They are robust and low cost, with an operating voltage typically between 30 V and 100 V and a relatively high slope resistance. Figure 5.3(e) shows a circuit which uses a varistor across the a.c. lines and a selenium suppressor across each component.

Figure 5.4 illustrates a simple crowbar circuit, in which the overvoltage is converted into an overcurrent, which is then protected by the fuse F_s. An overvoltage appearing at the main thyristor TH_2 would cause zener diode D_1 to start to conduct, which would apply gate current to protection thyristor TH_1, turning it on. This thyristor applies a short circuit across the supply lines, blowing the protection fuse. The circuit can act very quickly, being limited primarily by the turn-on time of thyristor TH_1.

An R-C network, called a snubber network, is also commonly used to protect thyristors and triacs against dv/dt effects, which cause spurious

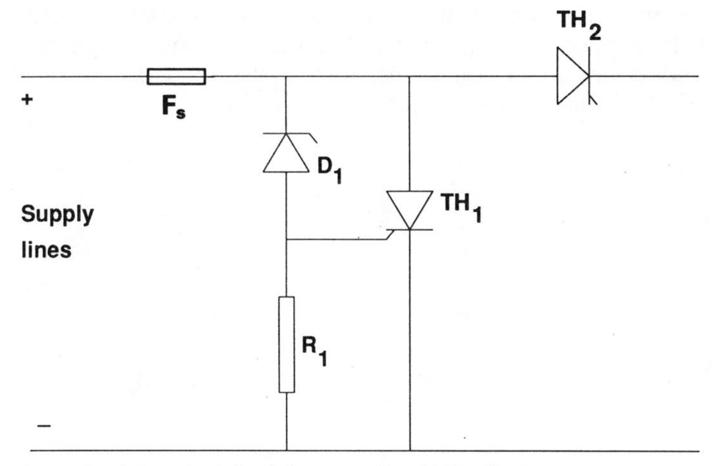

Figure 5.4 A 'crowbar' circuit for overvoltage protection

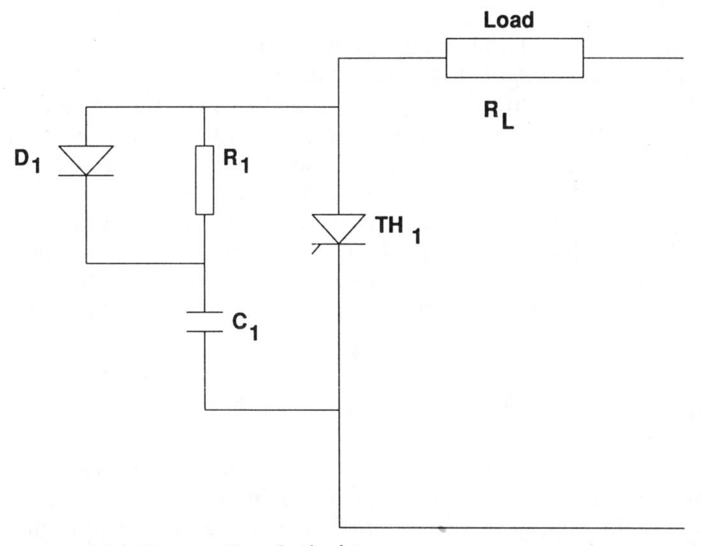

Figure 5.5 dv/dt protection of a thyristor

turn-on. This is shown in Figure 5.5, where resistor R_1 prevents the capacitor discharging directly into the thyristor when it turns on, giving large d*i*/d*t* currents. The capacitor charges via the circuit inductances and limits the transient voltage rise and its rate of increase. If the voltage is to reach a peak of V_o and $(dv/dt)_c$ is the critical rate of rise of voltage which the thyristor can tolerate, then the value of C_1 must be greater than

$$C_1 \geqslant \frac{0.632\ V_o}{R_L\ (dv/dt)_c} \tag{5.1}$$

The R-C snubber also softens the reverse recovery of the power semiconductor, so reducing the spikes generated. Diode D_1 may be added for improved dv/dt protection since the voltage drop across resistor R_1, when the capacitor is charging, will be seen by the power semiconductor. The value of the suppression capacitor and resistor are often empirically chosen, depending on circuit conditions, with R_1 being in the range of 10–100 Ω, and C_1 in the range 0.1–1 μF.

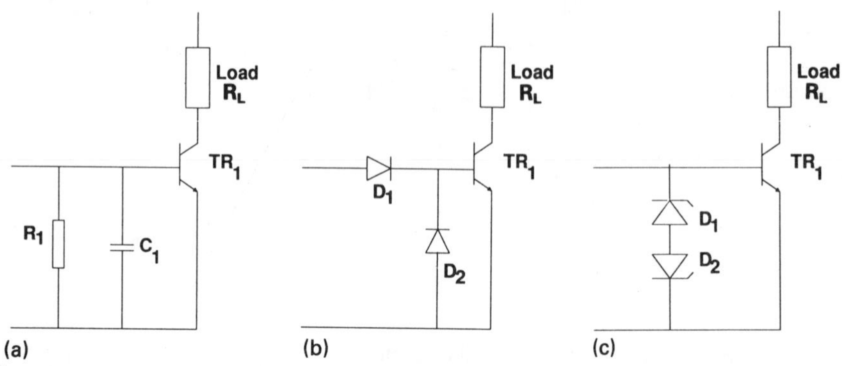

Figure 5.6 Control terminal protection: (a) R-C circuit; (b) diodes; (c) surge suppressors

Figure 5.6 shows some of the many circuits which have been used to protect the control terminals of power semiconductors, which can be the base of a transistor, as shown, or the gate of a thyristor or triac. Usually, inductive or capacitive coupling between the power and control circuits induces voltage transients into the control terminal, and this terminal can be isolated by transformers or optoisolators to minimise the effect and the leads twisted together or shielded. Apart from protecting the gate terminal, the devices affect the power device characteristics. For example, capacitance in the base or gate would increase the turn-on and turn-off times and, for a thyristor, would increase the dv/dt rating and the holding and latching currents.

5.4 Overcurrent protection

Several factors determine the choice of an overcurrent, or fault current, protection system for power circuits. The most important of these is, perhaps, whether the supply source is 'stiff' or 'soft.' A soft supply has series impedances, of sufficient amount to significantly impede the rate of current rise during a circuit fault, whereas a stiff supply results in a rapid rise in current, within a traction of a cycle.

If, when running from a soft source, there is a low probability of a sustained fault over several cycles, then the power device may be chosen such that the fault current is below its surge rating. Alternatively, the fault current can be sensed and the power semiconductors turned off before the current rises to dangerous proportions, or else an auxiliary crowbar thyristor can be fired to blow a fuse, as shown in Figure 5.7, which is similar to that of Figure 5.4 except that the fault current in resistor R_1 is used to turn on the crowbar thyristor TH_1. A third alternative is to use electromechanical contactors or thermal overload trips, which operate in 20–80 ms, so that the power semiconductor needs to carry the fault current during this period.

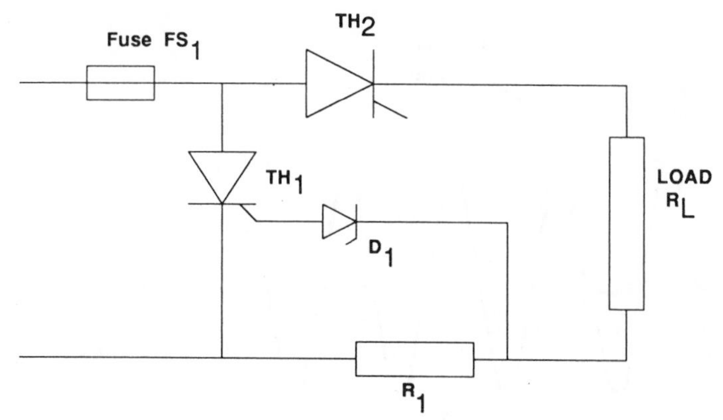

Figure 5.7 Overcurrent protection by a 'crowbar' circuit

The rapid build-up of fault current in a stiff supply means that circuit isolation must be quickly achieved. Figure 5.8 shows the waveforms for an a.c. circuit under conditions of short circuit or heavy overload, and a fuse is usually used to protect under these conditions, causing the current to be terminated at time t_2. This fuse current is shown in greater detail in Figure 5.8(c). At the commencement of a fault the current builds up rapidly, the rate of increase being limited by the relatively small circuit impedance. If no action were taken the current would rise to the peak prospective fault value, being limited only by line impedances, before reversing due to reversal of the supply. Long before this current is reached the fuse melts, at point A, the current increasing slightly to B before the energy, dissipated in the arc of the fuse, causes it to decrease to zero at C. It is important that the rate of decrease of current in the arc is not too severe, or it can give rise to high-voltage transients in circuit impedances, which would destroy the thyristors, and in most fuse designs the melting and arcing times are approximately equal.

A fuse consists of a metal element which carries the normal steady state load current, but overheats and melts if a fault current flows which is large enough and lasts for a sufficiently long time. If the current is i, the resistance of the fuse element R and the fuse melts after time t, then the energy needed to blow the fuse is i^2Rt. Since the resistance of the fuse is indeterminate it is usual to refer to the i^2t rating of the fuse.

Figure 5.9 shows the construction of a high-rupturing capacity (HRC) fuse. The fuse element is made from pure silver and usually has the V-notch structure shown, which gives a fuse with a greater r.m.s. current capability whilst having a reduced i^2t rating, so that it will melt quickly if a high fault current occurs. In low-current fuses it is difficult to make the constrictions in the element narrow enough, so plain wire is often used.

The body of the fuse must have good mechanical strength and be able to withstand thermal shock and the high temperatures which arise during normal running. Low-current and low-voltage fuses can use glass, but for high power levels ceramic, and sometimes silicon bonded glass fibre, is required to prevent the case from shattering during fuse rupture. The fuse body is often filled with quartz granules, which give rapid heat conduction

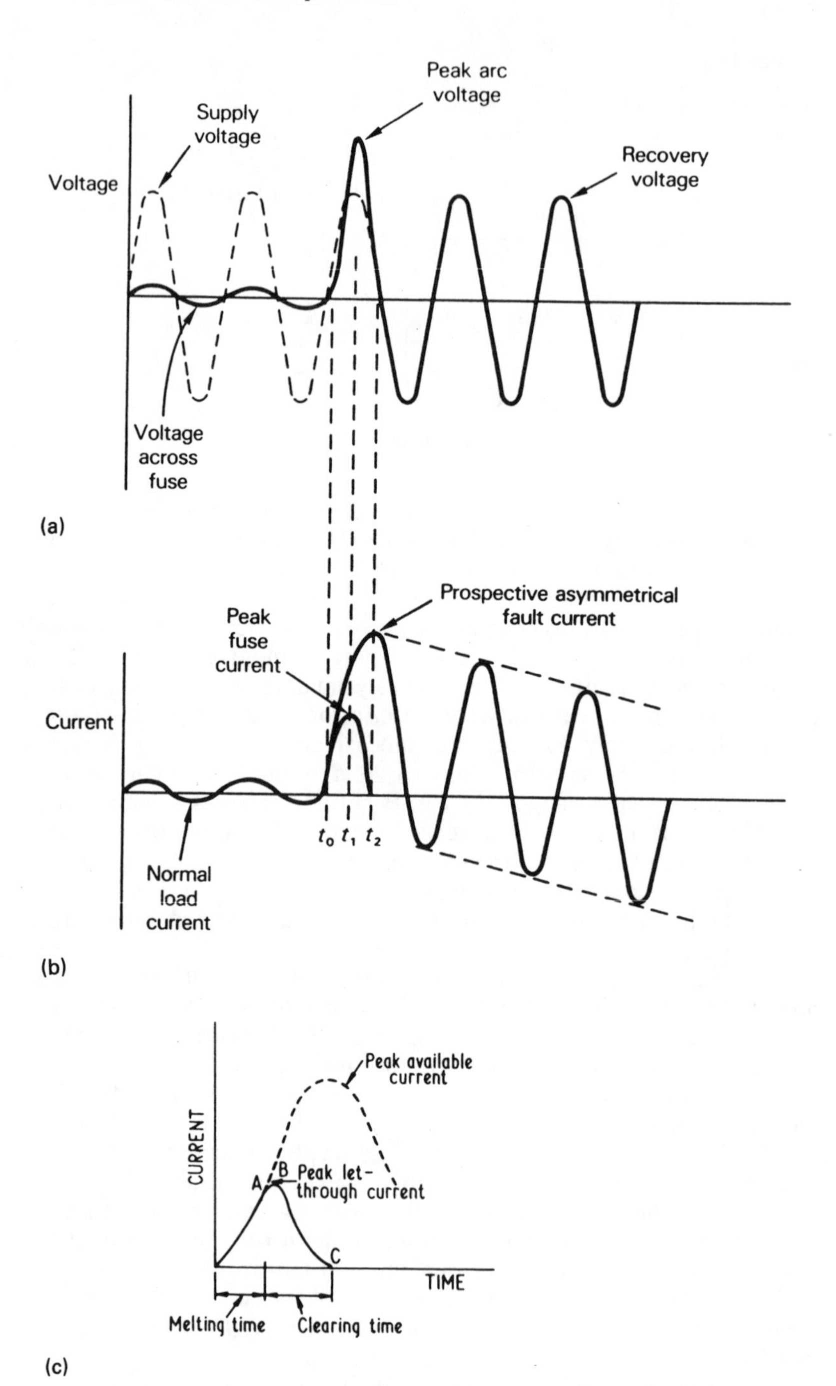

Figure 5.8 Fault current characteristics: (a) short-circuit voltage; (b) short-circuit current; (c) high-speed fuse characteristic

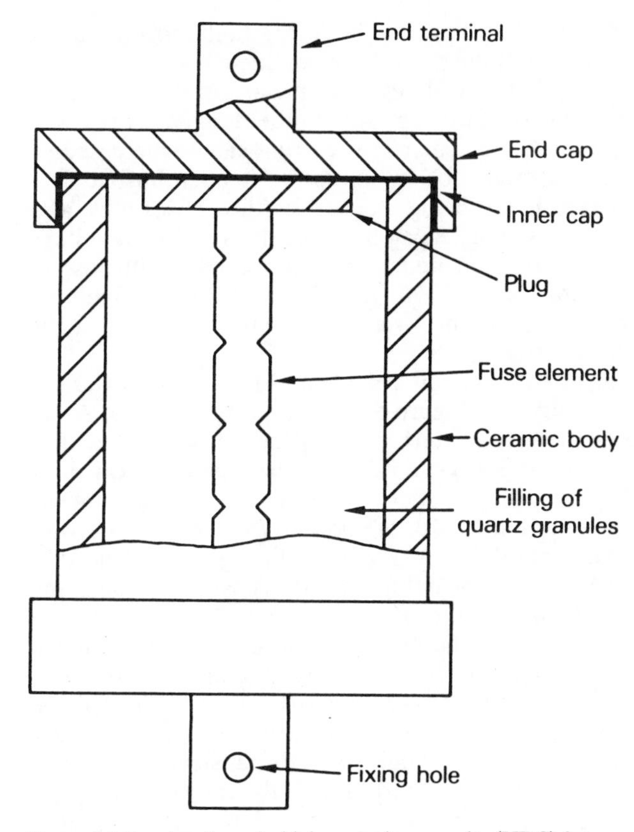

Figure 5.9 Construction of a high-rupturing capacity (HRC) fuse

from the fuse element to the case. The filler also extinguishes the arc by fusing with the resultant silver vapour to form a non-conductive material called fulgurite and distributes the high pressures generated in the fuse link, during overload conditions, over the ceramic body.

The end cap and end terminals must give good electrical contact and form a good fit with the inner cap. The material used is usually brass with a high copper content. The end terminal is welded, soldered, or riveted and soldered to the end cap. The construction used gives good electrical contact and isolates the element from external mechanical shock and vibration.

Power semiconductors and fast-acting fuses both exhibit the property of withstanding almost constant values of i^2t below about one cycle. This greatly simplifies the design of protective fuses, since all that is basically required is to ensure that the i^2t rating of the fuse is less than that of the semiconductor it is to protect, so that it will fail first. If the fuse is in the primary of a transformer and the semiconductor cell in the secondary, then the i^2t of the fuse and semiconductor must be related by

$$i^2t_{(\text{fuse})} \leqslant \left(\frac{V_{\text{sec}}}{V_{\text{prim}}}\right)^2 i^2t_{(\text{semi})} \tag{5.2}$$

For clearance times which last over one cycle the fault current curves must be kept below the rating of the power semiconductor, and even if the fuse operates within one cycle the peak let-through current must be below the rating of the semiconductor. Fuses should also be designed with a low ratio of i^2t to r.m.s. current rating. The r.m.s. current can be increased by cooling the fuse in an air stream, or by mounting it on a busbar or a heatsink, the i^2t characteristic not being affected by this. Fuses must also have a sufficient voltage-clearing capability, otherwise a high-voltage source can cause a continuous arc under fault conditions, with loss of protection. The arcing voltage should also be as low as possible to limit the peak voltage which occurs across the circuit elements, but this is at the expense of the i^2t characteristic.

The i^2t of a fuse is affected by circuit voltage and the prospective fault current, as shown by the curves of Figure 5.10 for a particular fuse. As an exercise into the use of these characteristics suppose a fuse is needed to protect a circuit operating at 300 V and 300 A, the asymmetrical peak current being 13.5 kVA. Using a 450 V, 400 A r.m.s. fuse, whose characteristics are shown in Figure 5.10, the normal i^2t is 800000 A^2s with an asymmetrical peak current of 200 kA. Figure 5.10(a) gives a reduced i^2t of 280000 at 300 V and drawing a line in Figure 5.10(b) passing through the 200 kA, 280000 A^2s point, the value of i^2t at 13.5 kVA is 150000. This is now the rating of the fuse under the circuit conditions described and it should be less than that of the power semiconductors being protected. It is possible to reduce the i^2t of the fuse by choosing one with a higher voltage rating, but the excess of fuse rating over supply voltage could give very abrupt arc quenching and generate high-voltage transients. A better method of reducing the fuse i^2t is to decrease the prospective peak

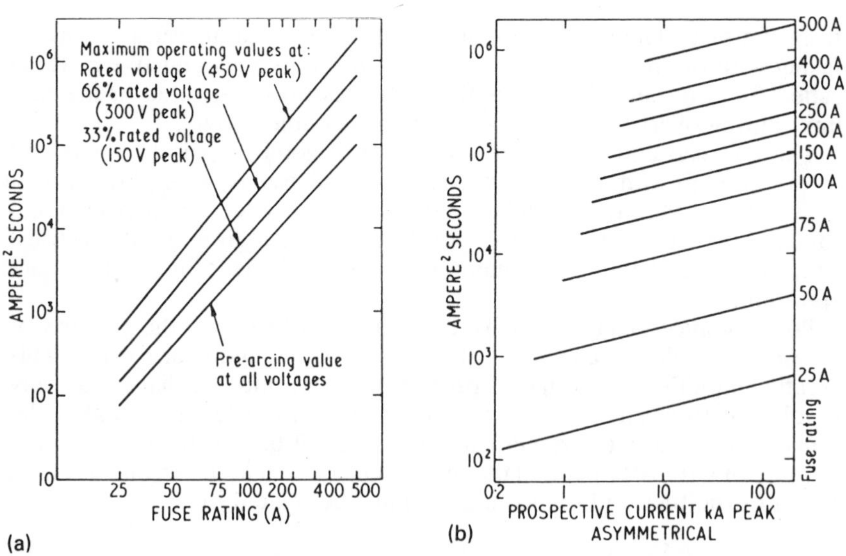

Figure 5.10 Variation of i^2t with circuit voltage and prospective current for a typical 450 V 400 A r.m.s. fuse: (a) circuit voltage; (b) prospective current

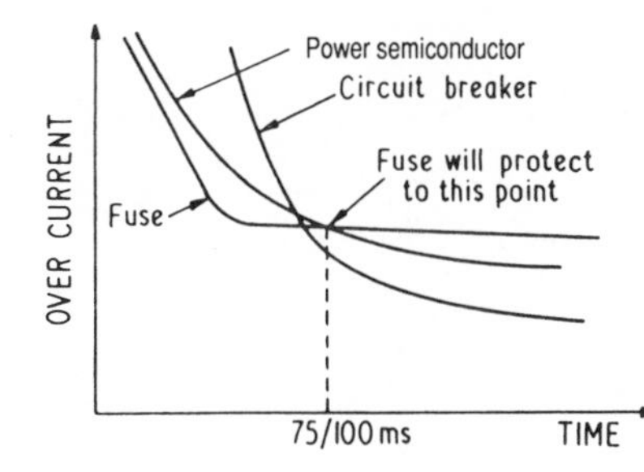

Figure 5.11 Characteristics of several protective devices

asymmetrical current by increasing circuit reactances. It is important to ensure that the fault current reaches a sufficient value to blow the fuse within a relatively short time, or the power semiconductor may be damaged. This is illustrated in Figure 5.11, where it is seen that for stiff supplies a fast-acting semiconductor fuse will protect a power semiconductor, whereas for soft supplies circuit breakers must be used.

The performance of a fuse can be improved by connecting several of them in parallel, the total steady state current being given by

$$i_{\text{total}} = i_{\text{one fuse}} \times N \times F \tag{5.3}$$

where N is the number of fuses in parallel and F is a factor which accounts for fuse mismatch, being typically 0.9.

The i^2t rating of the combination is given by

$$(i^2t)_{\text{total}} = (i^2t)_{\text{one fuse}} \times N^2 \tag{5.4}$$

Therefore if two fuses are connected in parallel each is required to have about half the steady state rating of one fuse, but the i^2t rating is improved by a factor of four. Fuses can be connected in parallel by having several elements in parallel within a single case.

As mentioned in section 5.2, rapid rates of current increase (di/dt) cause failure in power semiconductors. This current increase is due to low source impedances during a short circuit, or to discharge of suppression capacitors or recovery currents of other devices, as in Figure 1.9. Inductances may be added in the lines to limit the rate of rise of current within the power semiconductor, the inductors being air cored and therefore linear, or a saturable reactor.

Fuses are not effective in protecting power transistors, since these may come out of saturation as the fault current through them increases, due to insufficient base drive. This would limit the current, preventing the fuse from blowing quickly, whilst causing high dissipation across the power transistor, leading to failure. In these instances current sensing is used, the base drive being removed to turn off the transistor, or else a crowbar device, such as a thyristor, is fired to blow a line fuse, as shown in the simple circuit of Figure 5.7.

Part 2

Circuits

Chapter 6

Power semiconductor circuits – a résumé

6.1 Introduction

Chapters 1–5 have described components which are used in the design of power semiconductor circuits. The remaining chapters introduce the various circuits in which these power semiconductors may be used in order to regulate the power to the load. Circuit principles are described since in many cases several different types of power components may be used to perform the same functions.

Because of the diversity of power semiconductor circuits, the present chapter introduces the principles involved and these are then described further in following chapters.

6.2 Power switches

This book is primarily concerned with the use of power semiconductors in a switching mode, and therefore this is the basic type of power semiconductor circuit which may be used. The switching characteristics of the different components have already been described in Chapter 1, some devices needing continuous drive on their gate terminal when conducting, whilst others being triggered by pulses of current. Furthermore, some devices can be turned off by their control terminal, whilst others require the load current to be momentarily interrupted for turn-off. Chapter 7 describes some typical switching circuits and applications.

6.3 A.C. line control

In this application the power semiconductors are used to regulate the power flowing from an a.c. source to an a.c. load. Figure 6.1(a) shows two thyristors, connected in anti-parallel, which are used to control the power from the a.c. supply across lines A and B to the load across C and D. Instead of two thyristors a single triac could be used, although power semiconductors which may be turned off by their gate terminals, such as transistors and gate turn-off switches, are usually used in d.c. rather than a.c. applications.

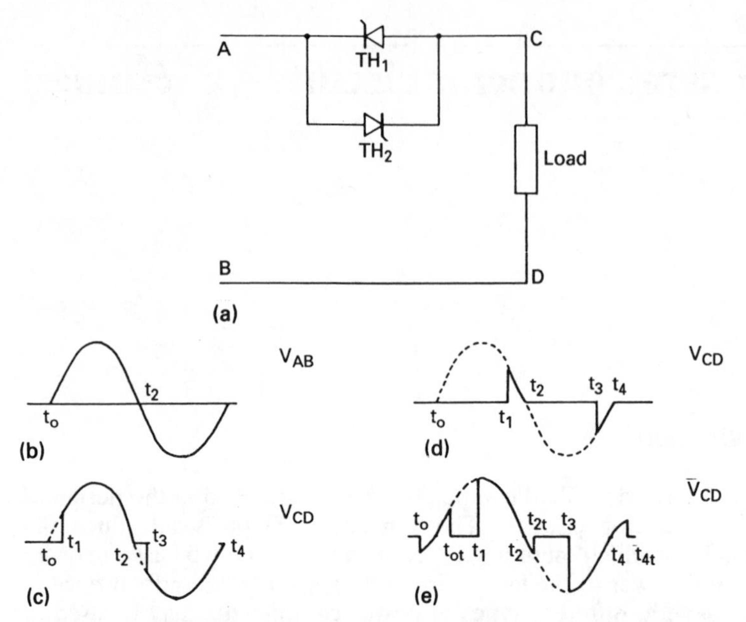

Figure 6.1 Single-phase a.c. line control: (a) circuit arrangement; (b)–(e) waveforms

Figures 6.1(b) to 6.1(e) show the circuit waveforms. The input voltage V_{AB} is shown in Figure 6.1(b), in which line A is positive to B over the first half cycle and negative during the second. Output voltage waveform for a resistive load is shown in Figure 6.1(c), where at time t_0 the supply voltage goes positive but the load voltage is zero since both thyristors are off.

At time t_1 thyristor TH_1 is turned on and since it presents a very low impedance the voltage at C is almost the same as that at A and the load voltage jumps to practically the a.c. input value. When the line voltage reverses, at time t_2, thyristor TH_1 is reverse biased and its current decays to zero, so that it turns off. Once again the output voltage is zero, until thyristor TH_2 is fired at time t_3, when it rises to the negative line voltage. The thyristor firing times, t_1 and t_2, are variable, and Figure 6.1(d) shows a longer delay from the zero a.c. voltage point than Figure 6.1(c), illustrating the principle of variable mean a.c. load voltage by control of the thyristor firing angles during an input cycle.

If the load is inductive the thyristors will not turn off directly the supply reverses, and the waveform for an inductive load is shown in Figure 6.1(e). Thyristor TH_1 is fired at time t_1 and conducts until time t_2, during which period it is forward biased, the current flowing from the supply to the load. After time t_2 the supply voltage reverses, but since the energy stored in the inductive load cannot be dissipated instantaneously, it forces TH_1 to remain conducting, whilst current flows against a positive potential from the load to the supply. This is known as regeneration. At time t_{21} the load current decays to a sufficiently low value to enable TH_1 to turn off, the load voltage then remaining at zero until the next thyristor is fired at time t_3.

A.C. line control circuits are described further in Chapter 8.

6.4 Controlled rectification and inversion

Power rectifiers are used where rectification of an a.c. supply to a d.c. load is required, although these devices cannot be controlled in relation to their turn-on point in an a.c. cycle. Where this control is required thyristors are almost invariably used due to their unidirectional conduction properties, and a typical circuit is shown in Figure 6.2, where the load is assumed to be resistive or inductive with a free-wheeling diode connected.

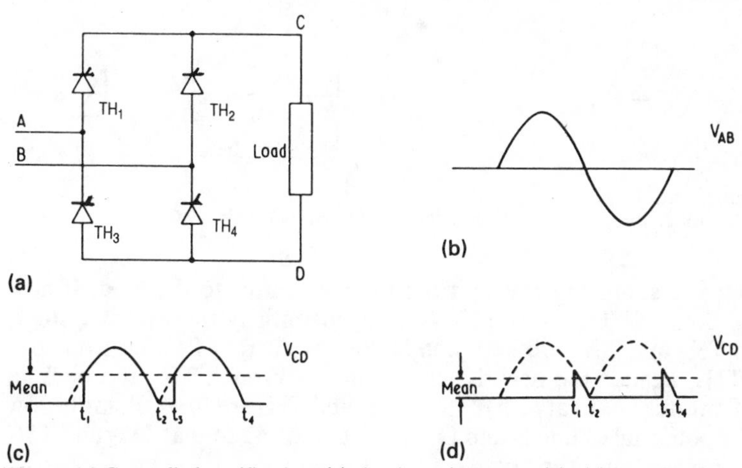

Figure 6.2 Controlled rectification: (a) circuit arrangement; (b)–(d) waveforms

Figure 6.2(b) shows the a.c. line voltage and Figure 6.2(c) the load voltage. At time t_1 thyristors TH_1 and TH_2 are fired so that the load voltage is almost equal to that of the supply. When the polarity of the input reverses at t_2 these thyristors turn off and the load voltage is zero, until at time t_3 thyristors TH_2 and TH_3 are fired. Even though the input line polarity has reversed, end C of the load is again connected to the most positive side, so that the load voltage is maintained unidirectional. Figure 6.2(d) illustrates how the mean value of the d.c. load voltage can be changed by altering the firing point of the thyristors during the half cycle.

If the load is inductive and is not bypassed by a free-wheeling diode, then the main thyristors will not turn off exactly at the zero input voltage points. As seen in the case of a.c. line control, the energy stored in the load during the rectification period forces the main thyristors to remain in conduction, current being fed back from the load to the negative supply. This is regeneration, or inversion as it is better known here, since current flows from a d.c. load to an a.c. source. Clearly, the maximum inversion period is 90° since only as much energy can be taken out of the load as is put in. However, if the load includes an independent d.c. source, as in Figure 6.3, then inversion over a full 180° is possible.

When the firing points of the thyristors in Figure 6.3 are delayed the output voltage of the bridge is not zero, but follows the negative contours of the line voltage, since the devices are kept conducting by current provided by E_B. In Figure 6.3(c) up to time t_2 thyristors TH_2 and TH_3 were

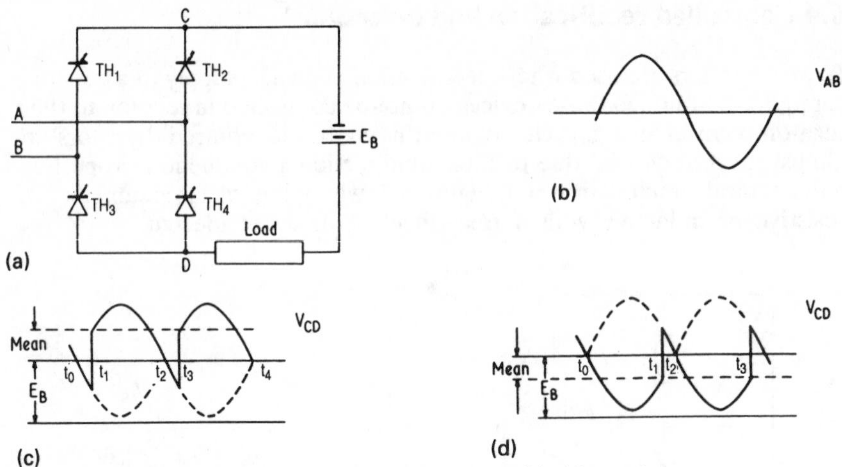

Figure 6.3 Controlled inversion: (a) circuit arrangement; (b)–(d) waveforms

conducting and supplying energy from the a.c. source to the load. If now the firing point of TH_1 and TH_4 is delayed until t_3 then from t_2 to t_3 thyristors TH_2 and TH_3 are kept conducting by current flowing from E_B, through TH_3, against the negative line voltage, through TH_2 and back to E_B. This continues until at t_3 thyristors TH_1 and TH_4 are fired. Point C now rises to the potential of line B and D falls to that of A, so that TH_2 and TH_3 are reverse biased and turn off.

In Figure 6.3(c) the mean output voltage is positive so that the a.c. lines give out more energy than they receive from the d.c. source. If the firing point of the thyristors is delayed beyond the 90° point, as in Figure 6.3(d), the mean voltage is made negative, there now being a net transfer of power from the d.c. source to the a.c. lines, i.e. an overall inversion.

Controlled rectification and inversion circuits are described in greater detail in Chapter 9.

6.5 Direct a.c. frequency converters

To convert from one frequency to another, one can either go directly from a.c. input to a.c. output or pass first through an intermediate energy storage stage, usually in the form of d.c. The former is a direct a.c. frequency converter, called a cycloconverter, and the output frequency is always less than that of the input. In the case of the d.c. link frequency converter, also called an inverter, the intermediate d.c. stage buffers the input and output, so that the output frequency is not related to the input and can be either greater or less than it.

Figure 6.4 shows a typical cycloconverter with its waveforms, assuming a resistive load. Examination of this figure will show that the two thyristor groups TH_1, TH_2, TH_7, TH_8, and TH_3, TH_4, TH_5, TH_6 form two bridges, similar to that of Figure 6.2, feeding the load in opposite directions.

To obtain a frequency reduction between input and output of 3:1 the following firing sequence can be observed. At t_0 when line A is positive,

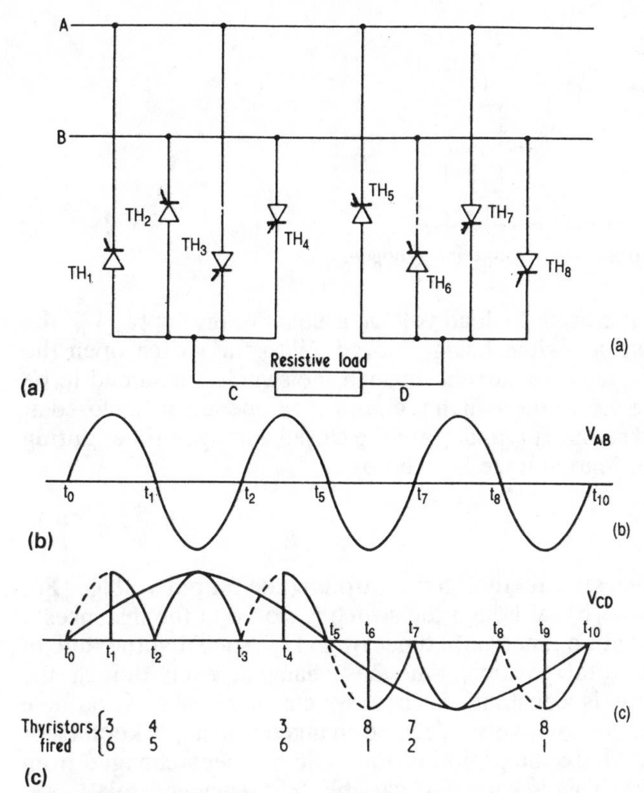

Figure 6.4 Direct a.c. frequency converter (cycloconverter): (a) circuit arrangement; (b) and (c) waveforms

TH$_3$ and TH$_6$ are fired, at t_2 thyristors TH$_4$ and TH$_5$ are fired, and at t_3 thyristors TH$_3$ and TH$_6$ are fired. These four thyristors are known as the positive group, since they make the load voltage positive, irrespective of the polarity of the input voltage. The negative group of thyristors, TH$_1$, TH$_2$, TH$_7$, TH$_8$, are fired over times t_5, t_7 and t_8 of the input voltage cycle. The firing point of the thyristors shown in Figure 6.4 have been delayed, so that the mean a.c. voltage is reduced with frequency. This is often necessary when the load is iron cored, such as a motor, so that the flux is maintained at the most optimum value.

Direct a.c. frequency changers are described in greater detail in Chapter 10.

6.6 D.C. line control

In the previous sections the power semiconductor was used to regulate the amount of energy between an a.c. input and an a.c. load. The circuit shown in Figure 6.5 controls a d.c. load operating from a d.c. supply, and the power semiconductor is shown as a switch. Such a circuit is frequently referred to as a chopper, and Figure 6.6 shows the output voltage waveforms.

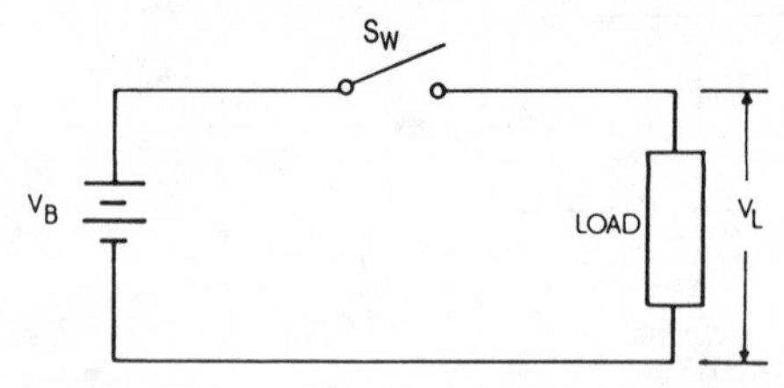

Figure 6.5 D.C. line control circuit arrangement (chopper)

When the switch is closed the load voltage is equal to the supply V_B, the voltage drop across the switch being ignored. With the switch open the voltage is zero if the leakage current through the switch is assumed to be negligible. In Figure 6.6(a) the switch is closed at t_1, opened at t_2, closed at t_3, opened at t_4, and so on. If t_c and t_o are the closed and open times during one cycle, the mean load voltage is given by

$$V_L = V_B \frac{t_c}{t_c + t_o} \tag{6.1}$$

This equation suggests a method for controlling the output voltage. For instance, in Figure 6.6(b), although the switch is closed at the instances t_1 and t_3 as before, it is not opened until times t_{21} and t_{41}, such that the ratio of the closed time to total periodic time has changed, even though the operating cycle time is constant. Such a system is known as variable mark-space control. Alternatively, t_o can be changed whilst t_c is kept fixed, as in Figure 6.6(c), where the period of the cycle has been changed from that in Figure 6.6(a). This is known as variable-frequency control.

No matter what method of control is used, equation (6.1) shows that the minimum load voltage is zero when t_c is zero, and is equal to V_B when t_o is zero, that is, when the switch is continuously closed. Such a system is called a step-down chopper since it always controls voltages below that of the supply.

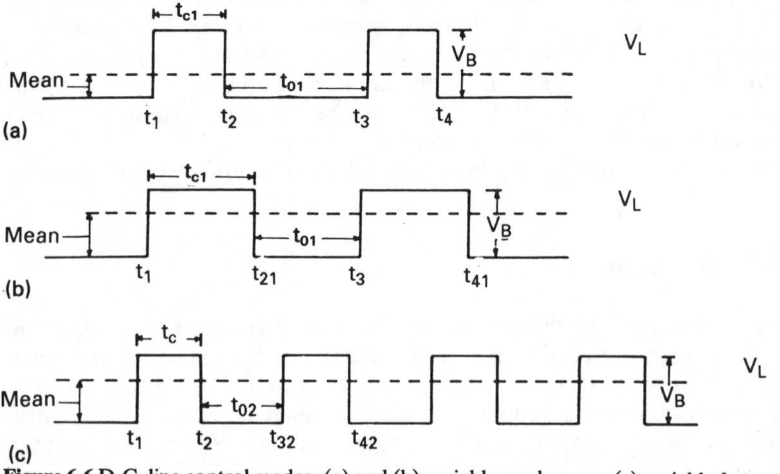

Figure 6.6 D.C. line control modes: (a) and (b) variable mark space; (c) variable frequency

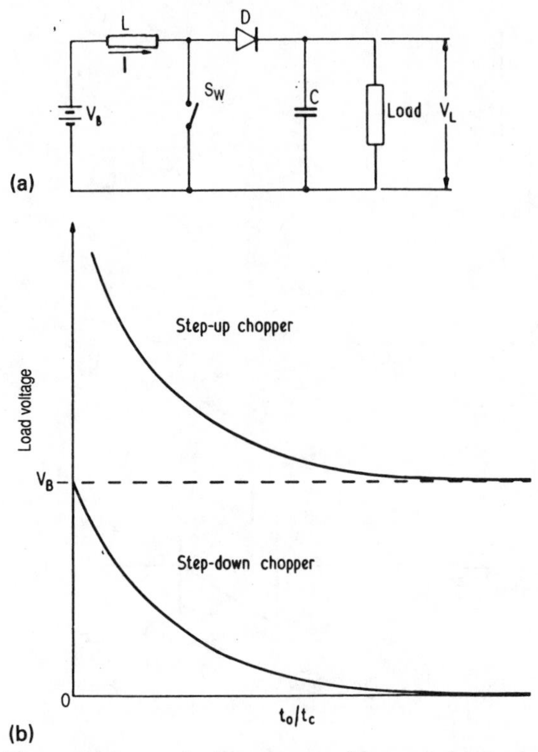

Figure 6.7 Step-up d.c. line converter (chopper): (a) circuit arrangement; (b) characteristic

Figure 6.7(a) illustrates a step-up chopper. When the switch is closed current flows in inductor L storing energy, and when the switch is open this energy transfers to the load, boosting its voltage above that of the supply. To analyse the circuit, L is assumed to be large enough to maintain the current I substantially constant, and the load voltage is assumed to be smooth. If the switch is closed for time t_c and open for t_o then, equating the energy transfers in the inductor during these periods, the load voltage is given by

$$V_B I t_c = (V_L - V_B) I t_o$$

$$\therefore \quad V_L = V_B \frac{t_c + t_o}{t_o} \tag{6.2}$$

Equations (6.1) and (6.2) are plotted in Figure 6.7(b), which illustrates the two different chopper types.

The switches shown in Figures 6.5 and 6.7 are usually transistors, gate turn-off switches or, for very high-power operation, thyristors. Since the supply is d.c. it is no longer sufficient, for thyristor switches, to turn them on and expect conduction to cease naturally when the supply polarity reverses, as in a.c. line control. They must be forcibly turned off, and such a system is referred to as forced commutation to differentiate it from a.c. line commutation, also called natural commutation.

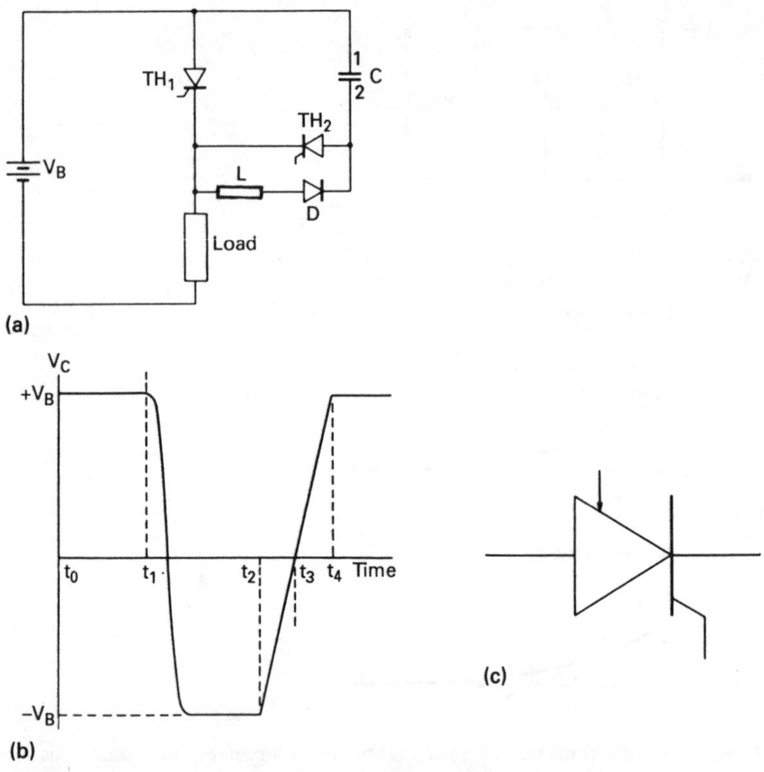

Figure 6.8 Forced commutation for a chopper circuit: (a) circuit arrangement; (b) waveform across the commutation capacitor; (c) forced commutated thyristor symbol

There are several methods for forced commutation of thyristors, and these are described in Chapter 11. Figure 6.8(a) shows an example and Figure 6.8(b) gives the voltage waveform across the commutation capacitor C. At time t_0 capacitor C is charged to the supply voltage with plate 1 positive, due to a previous cycle, and TH_2 is fired on the first switch on to ensure that this occurs. At time t_1 the main thyristor TH_1 is turned on to commence the load cycle. This also discharges C through L and D such that it resonates, recharging to V_B with plate 2 positive, resonant losses being ignored. To turn thyristor TH_1 off at any time t_2 thyristor TH_2 is fired, causing C to discharge through the load and recharge with plate 1 positive. Provided the time for which TH_1 is reverse biased, which is t_3 to t_2 in Figure 6.8(b), and exceeds its turn-off time, the thyristor will remain off until it is refired by a gate signal.

To differentiate forced commutated thyristors from those which are naturally turned off they can be drawn with the addition of an arrow, as in Figure 6.8(c). This arrow then represents all auxiliary thyristors, diodes, inductors and capacitors used in the forced commutation process.

It should also be noted that the loads in the chopper circuits shown have been assumed to be resistive. If they were inductive it would be necessary to add a free-wheeling diode across them to carry the inductive current and so prevent voltage transients during switch-off, which could destroy the semiconductors. D.C. line control circuits are described further in Chapter 12.

6.7 D.C. link frequency converters

A d.c. link frequency converter, or inverter, operates from a d.c. supply, so that if the input is a.c. it must first be rectified. Figure 6.9 shows one form of inverter circuit which could use any power semiconductor switch such as a transistor, GTO or thyristor, although a thyristor has been illustrated here. Note that all the thyristors are forced commutated by external circuitry which is not shown, and the load is again assumed to be resistive. At time t_0 thyristors TH_1 and TH_2 are turned on, and the load voltage rises to the positive value of the d.c. supply if the losses across the devices are ignored. At time t_3 thyristors TH_1 and TH_2 are turned off and thyristors TH_3 and TH_4 are fired, causing the load voltage to swing to the negative value of the d.c. supply. The half cycle is repeated at t_6 by turning off TH_3 and TH_4, and refiring TH_1 and TH_2 to give the waveforms shown in Figure 6.9(b). Voltage control of the load, as frequency changes, can be obtained by using the mark-space technique, illustrated earlier in Figure

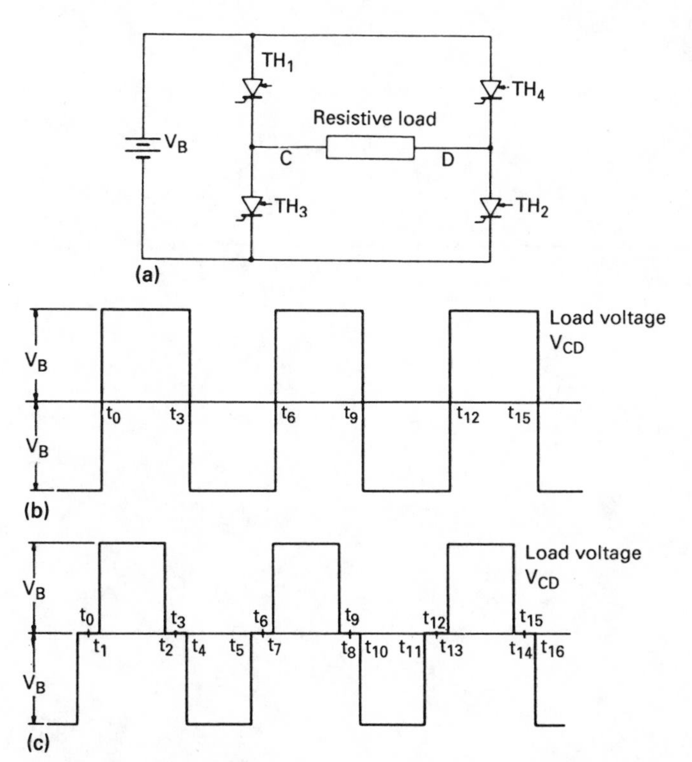

Figure 6.9 D.C. link frequency converter (inverter): (a) circuit arrangement; (b) and (c) waveforms

6.6 and shown again in Figure 6.9(c). At t_0 all four thyristors are maintained non-conducting. At t_1 thyristors TH_1 and TH_2 are turned on. At t_2 all thyristors are again off until TH_3 and TH_4 are fired at t_4. It must again be emphasised that this description applies for a resistive load only, which has been adopted to explain the principles of operation. D.C. link frequency changers are described in greater detail in Chapter 13.

The differences between direct a.c. and d.c. link frequency changers can again be summarised as follows:

(i) In the direct a.c. converter the semiconductor switches are all naturally commutated, whereas for a d.c. link converter they must be turned off by their gate control or forced commutated.
(ii) The d.c. link converter operates from a d.c. source so that its output is independent of the frequency and waveshape of the input, whereas the direct a.c. converter follows the envelope of the a.c. input.
(iii) The d.c. link converter can give an output range of frequencies which is infinitely variable. The direct a.c. converter, on the other hand, can only change the frequency in discrete steps of the input, and can only operate up to a maximum frequency equal to that of the input.

Chapter 7

Static switches

7.1 Introduction

The most basic applications of a thyristor are those in which its use is essentially that of a switch connecting a load to its supply, these circuits generally being referred to as static contactors, due to the absence of moving parts. In this chapter typical d.c. and a.c. contactor systems will be examined and reference made to electronic control and protection circuits, which often enable them to perform a variety of functions.

A static contactor has several desirable features when compared to its mechanical counterpart. There are no moving parts to wear or burn out; it is mechanically robust and noise free in operation; it can operate very quickly, in less than 0.1 ms, compared to about 50 ms for a mechanical contactor; and it is readily adaptable to sophisticated electronic control. This can include zero-level switching, such that the contactor closes at the zero voltage point of the supply waveform and so prevents interference in the lines; gradual build-up in voltage applied to the load to prevent current surges; and rapid isolation and switch-off in the event of a fault.

However, the static contactor is not used as frequently as its mechanical counterpart, and for some very good reasons. It is generally more expensive and physically larger; it is more prone to failure due to current and voltage overloads; and it does not provide complete isolation between supply and load due to leakage current through the power semiconductors. There are several applications where it is frequently used, for example in explosive atmospheres where arcing of contacts would be dangerous; in environments where a conventional contactor would generate intolerable interference, such as in computer installations; when the contactor is mounted in inaccessible conditions where they could not be maintained, since a static contactor needs no maintenance; where it is required to operate frequently, since static contactors do not wear out; and when the contactor is required to respond to voltage signals generated by some other process, such as a fault detector, since static contactors can be readily controlled by electronic signals.

Contactors can be used to operate from a.c. or d.c. supplies. Generally, mechanical contactors can have the same construction for both these sources, only their ratings being affected. Since some power semiconductors, such as thyristors and triacs, can be turned off only when the current through them falls below a certain value, static contactors using these devices differ in circuit detail when operating on a.c. or d.c.

7.2 A.C. contactors

A.C. contactors operate from single-phase or three-phase lines. Generally, the power semiconductors used for these applications are triacs or thyristors, and although thyristors are shown in all the illustrations in this section two back-to-back connected thyristors can be replaced by a triac, if required.

Figure 7.1(a) shows the simplest and most frequently used arrangement for a full-wave single-phase contactor. With thyristors 1 and 2

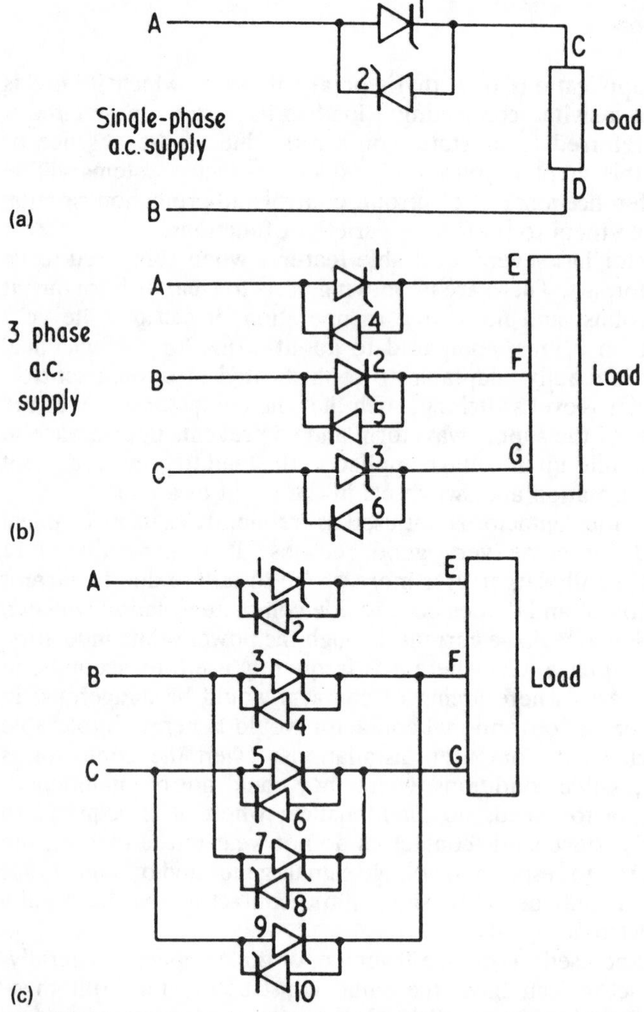

Figure 7.1 A.C. contactors: (a) single-phase; (b) single-phase unidirectional; (c) single phase reversing

non-conducting the load is effectively isolated from the supply lines A and B, if the leakage of the devices is ignored. When thyristors 1 and 2 are fired they can be considered as short circuits and the load is connected to the supply.

Figure 7.1(b) shows a three-phase static contactor. Only one thyristor and a reverse diode is required in each line, since one of the lines must conduct in the forward direction before any load current can flow. As in single-phase contactors, with no thyristor gate drive the load is isolated or off, and when the thyristors are turned on the load is connected to the supply.

If it is required to reverse the supply to the load, for instance where the load is a motor and its direction of rotation is to be changed, the circuit of Figure 7.1(c) must be used. With thyristors 1 to 6 conducting and thyristors 7 to 10 off, line A feeds E, B feeds F and C feeds G. To reverse the supply to the load, thyristors 1, 2 and 7 to 10 are on and thyristors 3 to 6 are off. Now A feeds E, B feeds G and C feeds F. It is necessary to use back-to-back thyristors, or triacs, in all the lines since if, for instance, 10 was a diode then firing thyristor 3 would cause a short circuit across lines B and C.

The devices in Figure 7.1 must all be rated to withstand the peak line voltage. For 240 V single-phase and 440 V three-phase supplies this means peak voltages of 340 V and 630 V respectively. To allow for line transients it is usual to employ 500 V and 800 V devices. Additionally, surge suppressors must be used between lines and across devices if the transients are troublesome. In this context it is perhaps worth noting that the circuit of Figure 7.1(b) is protected against reverse voltage transients by the diodes. Overvoltage in the forward direction could cause a thyristor to break over into conduction for a half cycle, but now it need not be damaged and the load may be such as not to be appreciably affected by the half cycle of power.

The current rating of the devices can be calculated as follows, with reference to the single-phase circuit. Assuming the line current to the load to be given by $I_{pk}\sin\theta$ when the thyristors are conducting, then the r.m.s. line current is given by

$$I_{rms} = \left[\frac{1}{\pi}\int_0^{\pi} I_{pk}^2 \sin^2\theta \, d\theta\right]^{1/2}$$

$$= \frac{I_{pk}}{\sqrt{2}} \tag{7.1}$$

The thyristors carry current for only half a cycle each, so their mean rating is given by $I_{T(av)}$ in equation (7.2), and this can be simplified to equation (7.3) by using equation (7.1):

$$I_{T(av)} = \frac{1}{2\pi}\int_0^{\pi} I_{pk}\sin\theta \, d\theta$$

$$= \frac{I_{pk}}{\pi} \tag{7.2}$$

$$I_{T(av)} = \frac{\sqrt{2}}{\pi} I_{rms}$$

$$= 0.45\, I_{rms} \qquad (7.3)$$

Therefore each thyristor must have a mean current rating of 0.45 times the r.m.s. line load current. Equation (7.2) applies equally well to the thyristors and diodes of the three-phase circuits in Figure 7.1, since these also carry half a cycle of the line current each.

7.3 D.C. contactors

Once conduction has been initiated in a thyristor or a triac it remains on until the current decays to zero. This is accomplished naturally in an a.c. circuit, owing to the reversal of the supply voltage, but for d.c. contactors this voltage reversal must be artificially provided across the power semiconductor, for a time in excess of its rated turn-off time.

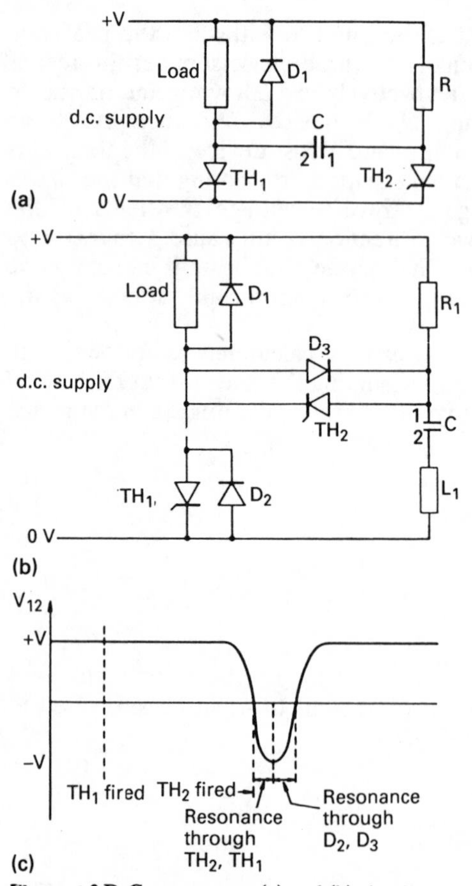

Figure 7.2 D.C. contactors: (a) and (b) circuit arrangements; (c) waveform across capacitor C in Figure 7.2(b)

Figure 7.2(a) shows a simple d.c. contactor circuit in which thyristor TH_1 is the main power semiconductor, which is fired to connect the load across the supply. This also permits capacitor C to charge through resistor R with plate 1 positive. To turn the contactor off thyristor TH_2 is fired, which connects C across thyristor TH_1, reverse biasing it. In effect, C provides an auxiliary path for the load current and consequently commences charging with plate 2 positive. Provided the time during which TH_1 is reverse biased is greater than its turn-off time it will remain off. When TH_1 is next fired thyristor TH_2 turns off since C is connected across it. This contactor circuit has two major limitations:

(i) It is inefficient, since thyristor TH_2 is conducting during the off period and energy is lost in resistor R. This resistor can be made large valued, but then the operating frequency is reduced.
(ii) After thyristor TH_1 is turned on sufficient time must be allowed for C to charge to the supply voltage, through R, before thyristor TH_2 can be fired to turn it off. Therefore the contactor has a top limit to its operating frequency, and the greater the value of R, the lower this frequency.

The contactor shown in Figure 7.2(b) is capable of much higher operating frequencies, even under conditions of high load impedance. If the load is substantially constant, diode D_2 may be omitted. Initially C charges through D_3 to the d.c. supply with plate 1 positive. Thyristor TH_1 is now turned on to supply load current. During this period C would discharge, owing to the leakage currents of TH_2 and D_3, had this not been compensated for by the current through R_1. Therefore this resistor can have a large value since it has only to pass the leakage current of two devices. To turn TH_1 off, TH_2 is fired. Capacitor C resonates with L through TH_1 and TH_2 and the voltage on C reverses, plate 2 being positive. Once again C resonates with L through D_2 and D_3, turning TH_1 and TH_2 off. Figure 7.2(c) shows the voltage on capacitor C assuming negligible resonant losses.

The mean current of thyristor TH_1 is given by the maximum load current I_{pk}, with a peak current, due to resonant discharges, given by

$$I_{TH(pk)} = \frac{V}{\sqrt{(C/L_1)}} \tag{7.4}$$

where V is the supply voltage.

The peak voltage of thyristor TH_1 is given by

$$V_{pk} = V + I_{pk}\sqrt{(L_1/C)} \tag{7.5}$$

The maximum turn-off time seen by TH_1 is given by

$$t_{OFF} = \sqrt{(L_1C}\left\{\frac{\pi}{2} - \sin^{-1}\left[\frac{I_{pk}}{2V}\sqrt{\left(\frac{L_1}{C}\right)}\right] + \cos^{-1}\left[\frac{I_{pk}}{V}\sqrt{\left(\frac{L_1}{C}\right)}\right]\right\} \tag{7.6}$$

The mean current rating of thyristor TH_2 is low, the peak value being given by $V\sqrt{(C/L_1)}$. Its voltage rating must exceed V.

Diode D_1 passes a peak current of I_{pk} and sees a voltage of $2V+I_{pk}\sqrt{(L_1/C)}$.

The mean current of diode D_3 is also small, the peak current being $V\sqrt{(C/L_1)}$ at a voltage of V.

Alternative commutation circuits may be used for static switching, and these are illustrated in Chapter 11.

7.4 Control and protection circuits

If the load for an a.c. contactor were resistive, it would be sufficient to fire each power semiconductor (thyristor or triac) with a single pulse at the start of each a.c. cycle, the pulse width being larger than the turn-off time of the thyristor. The waveforms for the single-phase contactor are shown in Figure 7.3 and are equally applicable to three-phase circuits. For inductive loads Figure 7.3(e) shows that there can be up to 90° lag between voltage and current. Therefore a pulse to thyristor TH_1 at the start of the positive half cycle, as shown in Figure 7.3(b), would be ineffective in turning the device on, since the load current is still flowing through TH_2 so that TH_1 is reverse biased. Only when this current has decayed to zero will TH_1 be able to conduct.

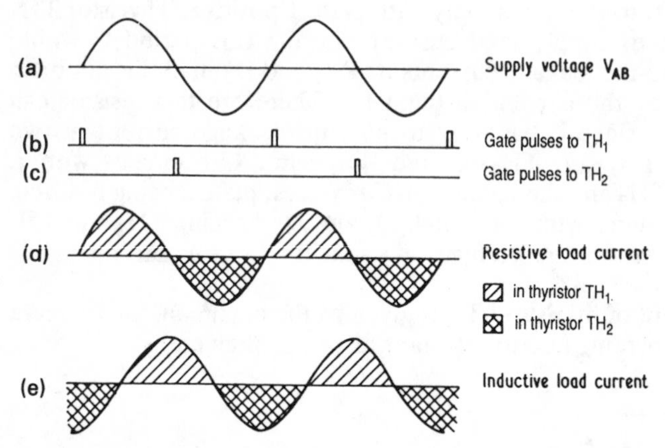

Figure 7.3 Circuit waveforms for the single-phase a.c. contactor of Figure 7.1(a)

To enable the a.c. contactor to work on inductive loads it is therefore necessary to maintain continuous gate drive for 90°. However, it is far simpler to fire TH_1 and TH_2 simultaneously and continuously for the duration of the contactor-on period. This eliminates any requirements for synchronising the gate pulse to the supply, the only disadvantage being that the gate pulses are now applied to a thyristor when it is reverse biased, so increasing its leakage current and hence dissipation. Since the reverse voltage is only equal to the forward voltage drop of the other conducting thyristor this dissipation increase is likely to be minimal. Figure 7.4 shows methods for obtaining continuous isolated drive to the power semiconductors. Optical couplers can be used with a relatively simple current source,

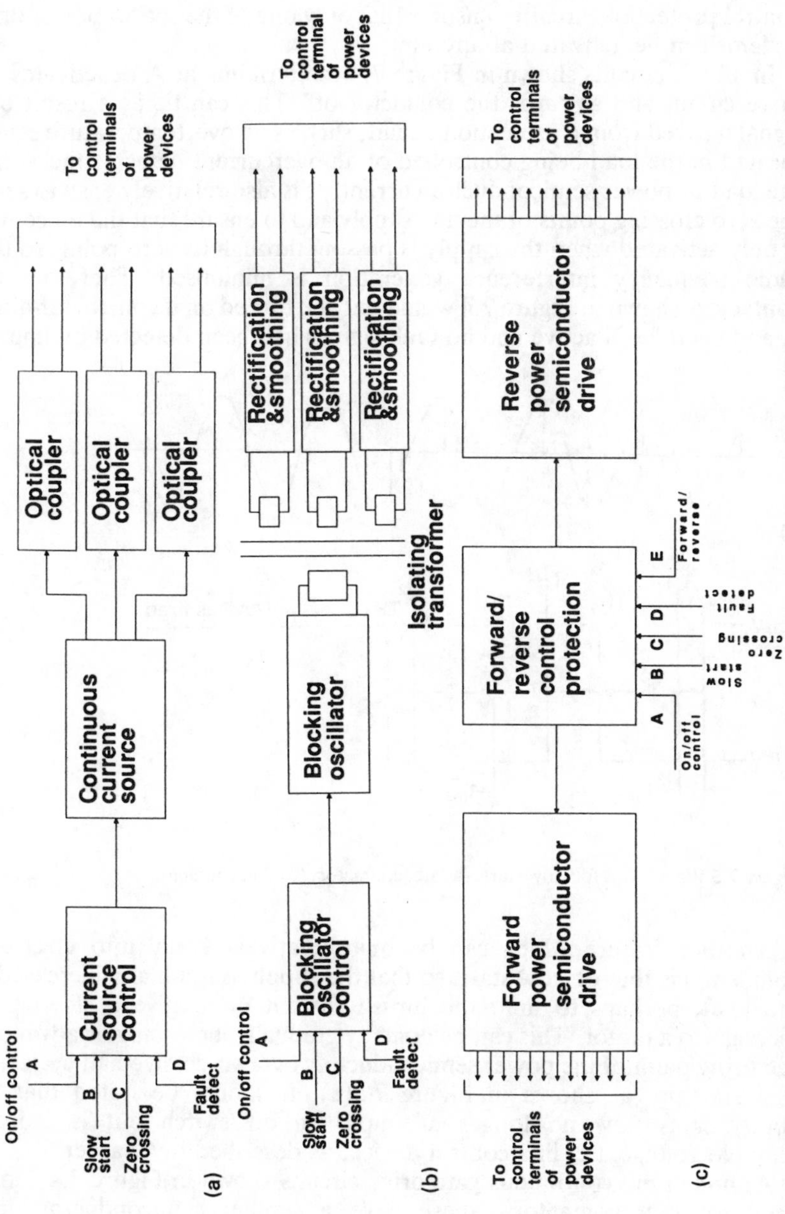

Figure 7.4 Three-phase a.c. contactor control: (a) optical couplers; (b) isolating transformers; (c) forward/reverse control

but since transformers saturate if operated in a d.c. mode they need to be driven at high frequency by an oscillator, the a.c. output from the secondaries being rectified and smoothed before being applied to the gates of the power semiconductors. Figure 7.4(c) shows a modification for driving the reversing contactor of Figure 7.1(c), where the forward/reverse control protection circuitry ensures that only one of the sets of power drive systems can be activated at any time.

In all the circuits shown in Figure 7.4 the terminal at A deactivates the drive circuit and so turns the contactor off. This can be as a result of a signal derived from a protection circuit, such as an overtemperature sensor located in the load being controlled or an overcurrent detector measuring the load or power semiconductor current. It is also relatively easy to sense the zero crossing points of the a.c. supply and to ensure that the on control is only activated when the supply is passing through its zero point, so that radio frequency interference generation is minimised. Therefore the contactors shown in Figure 7.4 would only be turned on if control terminals A and C are both active and no fault signals had been detected on line D.

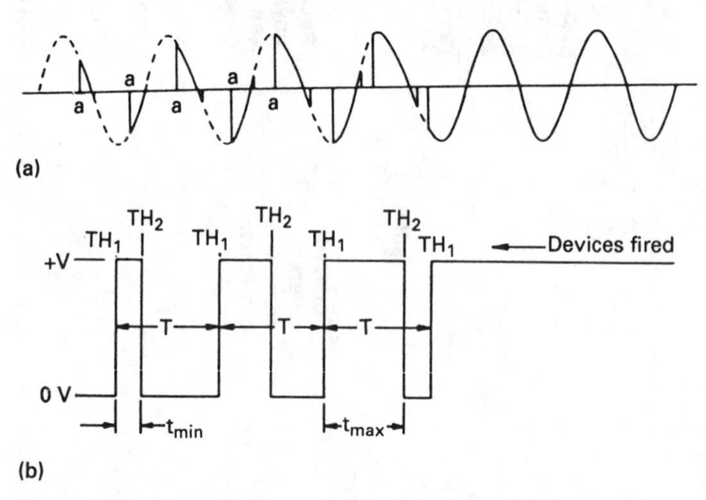

Figure 7.5 Waveforms for slow start: (a) a.c. contactor; (b) d.c. contactor

Another feature which can be built relatively easily into electronic contactors is that of slow start, so that the supply is gradually increased to the load, perhaps to limit the inrush current or to give a slow speed increase to a motor. This can be done by gradually increasing or advancing the firing point of the power semiconductors over successive half cycles, on first start-up, as shown in Figure 7.5(a). It should be noted that the contactor is now no longer a simple on–off switch but is more a variable-voltage a.c. line control device, as described in Chapter 6.

Although the continuous gate drive circuits shown in Figure 7.4 can be used for d.c. contactors, these have a simpler semiconductor drive requirement since the gate pulse need only be maintained for the length of time needed to allow the load current to rise above the device latching current. Simpler transformer-coupled pulse circuits can now be used, as in

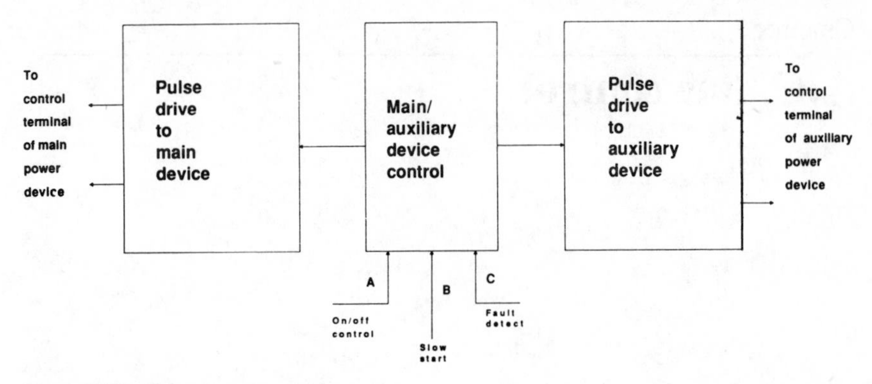

Figure 7.6 D.C. contactor control

Figure 7.6, the on/off control terminal ensuring that the main thyristor is fired for the start of the on-cycle and the auxiliary thyristor for the start of the off-period. As before, protection signals, such as overtemperature and overcurrent, can be fed in to turn the contactor off, and slow start can be incorporated as in Figure 7.5(b) by gradually increasing the mark-to-space ratio of the load voltage.

Chapter 8

A.C. line control

8.1 Introduction

The power flowing in an a.c. line can conveniently be controlled by series resistors, although this method is inefficient, so that for all but the lowest power levels transformers or variacs are used.

The thyristor a.c. line regulator was introduced in Chapter 6. This is relatively efficient since it works on the principle of blocking unwanted power, rather than dissipating it across the control device, and it also gives a system which is physically smaller and lighter than conventional methods. This is especially true when the power being handled is large, as is evident when comparing a 1 kV 1 kA thyristor a.c. regulator with an equivalent auto-transformer.

Static switching, described in Chapter 7, is a method of a.c. line control, although in this case the control is either on or off, there being no facility for continuously varying the amount of power flowing from the supply to the load. These variable systems are described in the present chapter. Phase-control techniques give the simplest a.c. regulating system and are best known. However, there are three other methods by which thyristor control of a.c. lines is possible, namely a.c. chopper control, integral half-cycle regulation and synchronous tap changing. These are described in the sections which follow.

8.2 Phase control

8.2.1 Single-phase circuits

Figure 8.1 shows three typical single-phase control systems with their circuit waveforms. The two-thyristor circuit of Figure 8.1(a) is the simplest, and if a triac is used in place of the two thyristors shown only a single power component is required. The circuit of Figure 8.1(b) uses two extra diodes compared to the basic two-thyristor system. This increases its cost and reduces the efficiency since for any conduction path there is now a series thyristor and diode. It has the advantage that the cathodes of the thyristors are commoned, so the gate drive circuit is simplified.

Referring to the circuit of Figure 8.1(b) and the waveforms given in Figure 8.1(d), at time t_0 input line A goes positive to B, and since thyristor

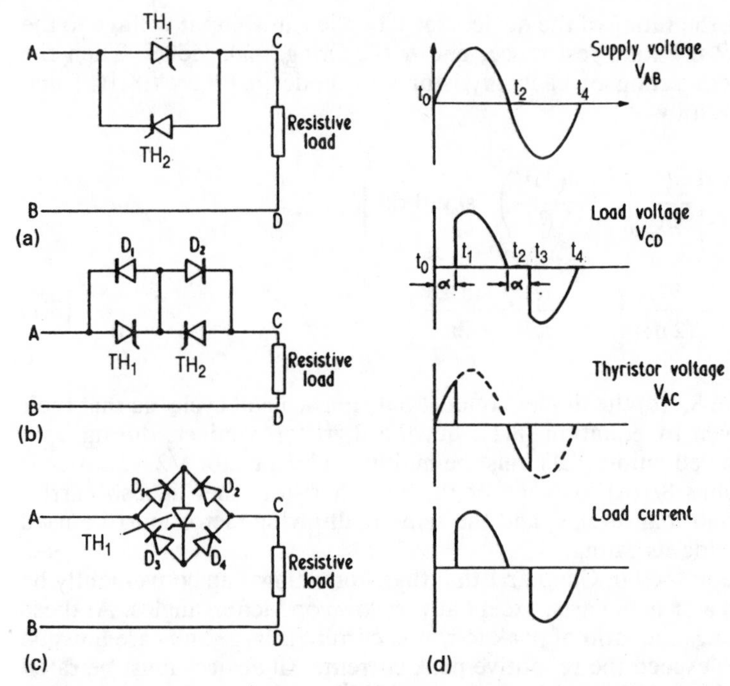

Figure 8.1 A.C. line phase control for a resistive load: (a)–(c) circuit arrangements; (d) waveforms

TH_1 is off no load current can flow and the load voltage is zero, the supply voltage now appearing across the section AC. When thyristor TH_1 is fired at time t_1 current flows through it and diode D_2 to the load. Similarly, in the negative half cycle, the firing of TH_2 is delayed by α from the zero voltage point. It is evident now why this system is called 'phase control', since it controls the phase, α in Figure 8.1(d), between the start of the supply voltage and the start of the load current, in order to vary the power flowing to the load.

Figure 8.1(c) shows an alternative system which uses only one thyristor and a diode bridge. The waveforms of Figure 8.1(d) still apply, although it must be kept in mind that the voltage across the thyristor is now never negative, due to the action of the diode bridge. Therefore with line A positive, thyristor TH_1 conducts from t_1 to t_2. At time t_2 the load voltage is zero and the thyristor must turn off. If this does not happen then, as soon as the supply reverses, the voltage across the thyristor will become positive, turning it on, and delay period t_2 to t_3 will be lost. This loss of control is most likely to occur on inductive loads.

Another disadvantage of the arrangement of Figure 8.1(c) is that there are voltage losses across three devices in any direction, two diodes and one thyristor, so the efficiency is lower than in the other two circuits of Figure 8.1. For high-voltage systems this may not be important and the circuit can often prove cheaper, since the diode bridge is lower cost than high-voltage thyristors, and the gate-firing circuit is also simplified since only one thyristor is used.

To derive the rating of the devices let V be the r.m.s. input voltage to the a.c. lines, R the load resistance, and α the firing angle delay. Then the r.m.s. current rating of each thyristor and diode in Figures 8.1(a) and 8.1(b) is given by

$$I_{T(rms)} = \left[\frac{1}{2\pi}\int_{\alpha}^{\pi}\left(\frac{\surd(2)V}{R}\right)^2 \sin^2\theta \, d\theta\right]^{1/2}$$

$$= \frac{V}{\surd(2)R}\left[1 - \frac{\alpha}{\pi} + \frac{\sin 2\alpha}{2\pi}\right]^{1/2} \tag{8.1}$$

For Figure 8.1(c) the diodes conduct only in each half cycle, so that their rating is given by equation (8.1), but the thyristor conducts during both half cycles so equation (8.1) must be multiplied by a factor $\surd 2$. If a triac is used in Figure 8.1(a) in place of the two thyristors then it also carries current in both half cycles, and the same multiplying factor must be used when obtaining its rating.

It was mentioned in Chapter 1 that thyristor ratings can conveniently be obtained on an r.m.s. basis, except at very low conduction angles. At these control settings the ratio of peak to r.m.s. current is large and care must be taken not to exceed the repetitive peak current. All devices must be rated to withstand a peak repetitive voltage of $V\surd 2$.

The r.m.s. load voltage and currents are similarly given by

$$V_L = \left[\frac{1}{\pi}\int_{\alpha}^{\pi} V\surd(2) \sin^2\theta \, d\theta\right]^{1/2}$$

$$= V\left[1 - \frac{\alpha}{\pi} + \frac{\sin 2\alpha}{2\pi}\right]^{1/2} \tag{8.2}$$

$$I_L = \frac{V}{R}\left[1 - \frac{\alpha}{\pi} + \frac{\sin 2\alpha}{2\pi}\right]^{1/2} \tag{8.3}$$

and it is seen from these equations that for $\alpha = 0$ the values of current and voltage are those of the sine wave input.

The load voltage and current waveforms shown in Figure 8.1(d) are rich in harmonics, and Figure 8.2 gives the Fourier analysis of the spectrum up to the seventh. Where load harmonics must be minimised a.c. chopper regulators are preferred, as described in section 8.3.

It was mentioned in Chapter 6 that for inductive loads the control thyristors do not cease conduction when the input voltage reverses, but are kept on by energy stored in the load, which is fed back to the supply. Figure 8.3 shows the modified waveforms for a series resistance–inductance load of power factor angle ϕ. Thyristor TH_1 in Figure 8.1(a) is fired at time t_1, which causes the load current to flow, the rising edge of the current being slower than for a purely resistive load. Thyristor TH_1 is kept conducting until time t_{21}, energy flowing from the load to the supply from

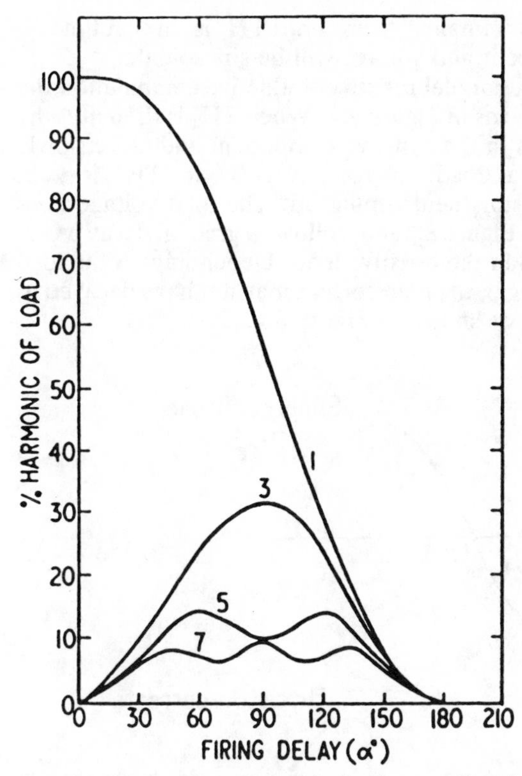

Figure 8.2 Harmonic content of single-phase a.c. line phase-control system with resistive load

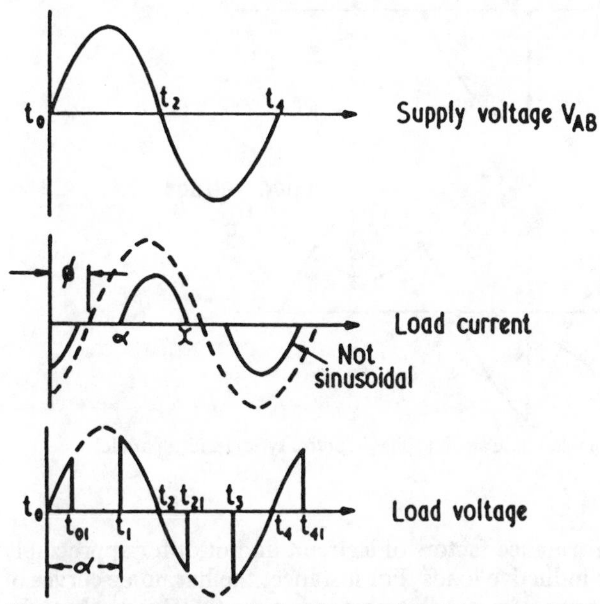

Figure 8.3 Waveforms for single-phase a.c. line phase-control system with a series resistive–inductive load

time t_2 to t_{21}. The load voltage remains at zero until TH_2 is fired at time t_3. Clearly if $\alpha \leqslant \phi$, the load current and voltage will be sinusoidal.

When the load consists of a parallel resistive–inductive combination the circuit waveforms are modified as in Figure 8.4. When TH_1 is fired at t_2 the load current rapidly increases in the resistive component and more slowly in the inductance. At t_3 the load voltage reverses but TH_1 is kept conducting until t_4, the thyristor then turning off. The load voltage does not fall abruptly to zero, as in Figure 8.3, but follows a gradual decay as the inductive energy is dissipated in the resistive load. Depending on the load power factor and the firing angle, the load current may not have decayed to zero before TH_2 is fired at t_5 as shown in Figure 8.4.

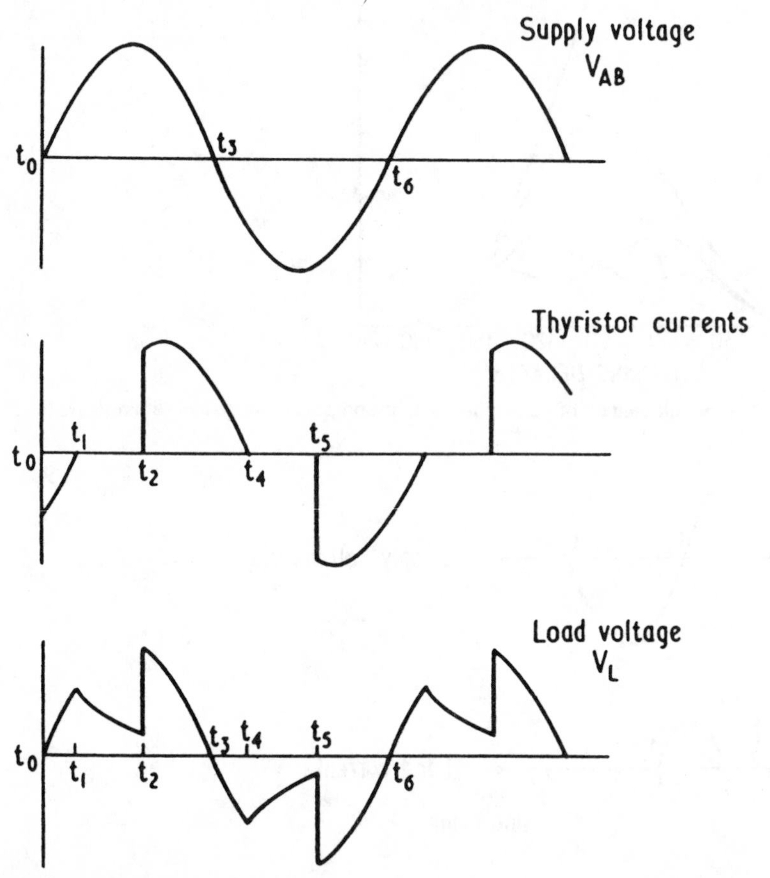

Figure 8.4 Waveforms for single-phase a.c. line phase-control system with a parallel resistive–inductive load

Generally, the performance factors of a circuit do not differ appreciably between resistive and inductive loads. For instance, the harmonic curves of Figure 8.2 are still approximately correct on inductive loads, but the harmonic content decreases with reducing power factors. The peak of the third harmonic will also now shift to slightly higher firing angles. The r.m.s. thyristor current and load voltage for a series resistance–inductance load are given by

$$I_{T(rms)} = \frac{V}{Z}\left[\frac{1}{2}\left\{\frac{x-\alpha}{\pi} - \frac{\sin(x-\alpha)}{\pi\cos\phi}\cos(\alpha+x+\phi)\right\}\right]^{1/2} \quad (8.4)$$

$$V_L = V\left[\frac{x-\alpha-\sin(x-\alpha)\cos(x+\alpha)}{\pi}\right]^{1/2} \quad (8.5)$$

Z is the load impedance and the cut-off angle x, in Figure 8.3, is given by

$$\sin(x-\phi) - \sin(\alpha-\phi)\exp\{-\cot\phi\,(x-\alpha)\} = 0 \quad (8.6)$$

Figure 8.5 shows plots of the variation of the r.m.s. load voltage for resistive and inductive loads.

As in most power semiconductor circuits, it is necessary to guard against two effects:

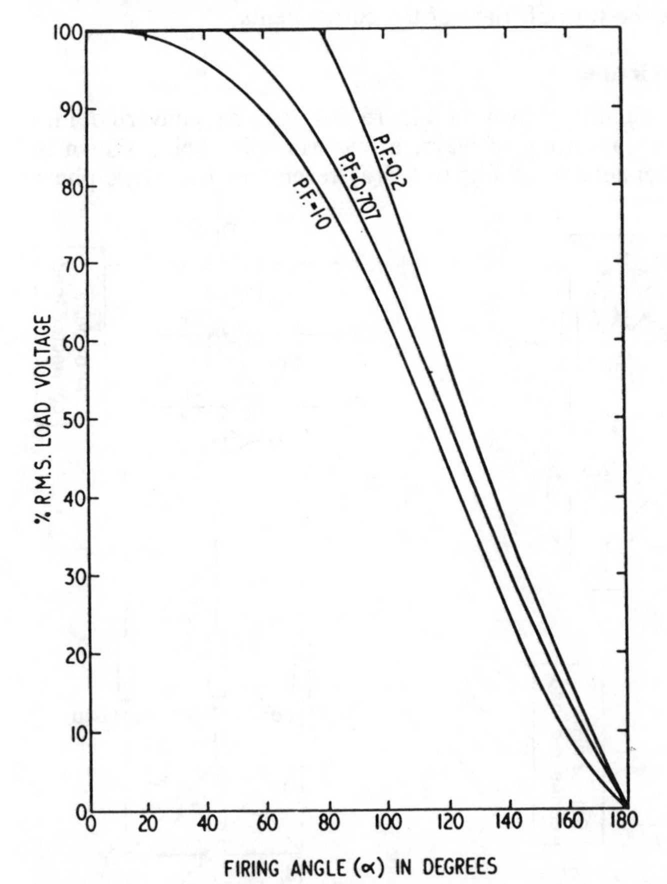

Figure 8.5 Variation of r.m.s. load voltage with firing angle, for a single-phase a.c. line phase-control system having a series resistive–inductive load

(i) d*i*/d*t*. When the load is capacitive a rapid rise in charging current can occur on initially turning on the power device. This d*i*/d*t* effect can destroy the semiconductors by causing local hot spots, even though the r.m.s. current rating has not been exceeded. When the load is a transformer stray capacitance across it would produce the same effect. In all such cases a linear or saturable reactor, in series with the power semiconductor, can be used to slow down the rate of current change.

(ii) dv/dt. Referring to Figure 8.1(a) and Figure 8.3, it can be seen that TH_1 conducts from period t_1 to t_{21}, the voltage across TH_2 being negligible during this period. At t_{21} thyristor TH_1 goes off and the voltage across TH_2 rises rapidly to the value of the input line voltage at this point. This dv/dt effect can cause the thyristor to switch on and conduct, even in the absence of a gate signal. The effect is more pronounced when triacs are used, since the device will have been conducting in the previous half cycle before it sees the dv/dt rise during its off period. It can be damped by R-C circuits across the power semiconductors, which reduce the rate of rise of voltage but also increase the turn-off time of the components.

8.2.2 Three-phase circuits

The single-phase circuits shown in Figure 8.1 can be converted into three-phase lines in a variety of ways, a few examples being shown in Figure 8.6. The connection of Figure 8.6(a) resembles the single-phase

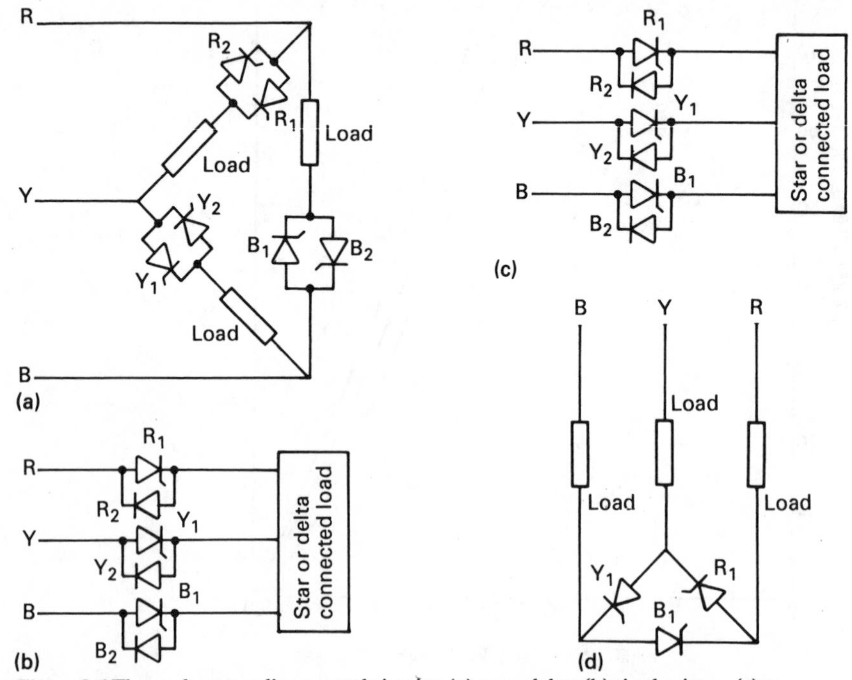

Figure 8.6 Three-phase a.c. line control circuits: (a) open delta; (b) six-thyristor; (c) thyristor/diode; (d) half-wave delta

system almost exactly and the same equations and circuit waveforms are applicable. It produces the lowest voltage harmonics over most of the control range, but is restricted for use with open-star connected loads. The circuit of Figure 8.6(b) is also similar in operation to the single-phase systems. Thyristors are fired in pairs, spaced 120° apart, in order to give the required phase sequence output, the load current flowing from a more positive to a negative phase. A typical sequence of conduction could be B_1 Y_2 R_1, Y_2 R_1 B_2, R_1 B_2 Y_1, B_2 Y_1 R_2, Y_1 R_2 B_1 and R_2 B_1 Y_2 over a 360° period.

The control system shown in Figure 8.6(c) uses one thyristor and a diode in each line. It has no counterpart in a single-phase circuit since the diode would then supply load current during a complete half cycle, although for three-phase systems the diode cannot conduct unless a thyristor in a more positive line has been fired. This circuit is simpler to use since only the thyristors require gate pulses, the diodes conducting automatically. The circuit shown in Figure 8.6(d) is the simplest, since it only uses three power components, but now each device has to carry a higher current than in the other circuits, the current flowing in the components for 240° in every cycle, at full load. Table 8.1 summarises the device ratings for the circuits shown in Figure 8.6.

Table 8.1 Device ratings for the three-phase circuits of Figure 8.6

Circuit number	*Thyristor voltage (percentage of a.c. line)*	*Thyristor r.m.s. current (percentage of a.c. line)*	*Full control delay angle (degrees)*
8.6(a)	141.4	40.8	180
8.6(b)	122.5	70.7	150
8.6(c)	122.5	70.7	210
8.6(d)	141.4	76.6	210

Figure 8.7 gives the variation of r.m.s. line voltage with firing angle for Figure 8.6(b). It is essentially similar to the single-phase operating curves but is limited to lower control angles. Another peculiarity of three-phase systems is that the neutral point voltage will vary with the conduction angle, owing to imbalance in the instantaneous phase voltages, and when a neutral line is present a current will flow. Figure 8.8 illustrates the waveform of this current for the arrangement of Figure 8.6(b) when a four-wire star-connected load is used. Clearly, the imbalance is the greatest at delay angles of 90°, which is illustrated in Figure 8.9, where the neutral current is then seen to be equal to the line current.

The harmonics generated in three-phase circuits have shapes similar to those of single-phase circuits although, as expected, they have a lower value and the harmonic present varies with the configuration. Figure 8.10 shows the plot of harmonics for the three-phase controller of Figure 8.6(b).

8.2.3 Control circuits

The essential features of a firing circuit for phase control are shown in Figure 8.11. The detector senses the zero voltage points of the input lines,

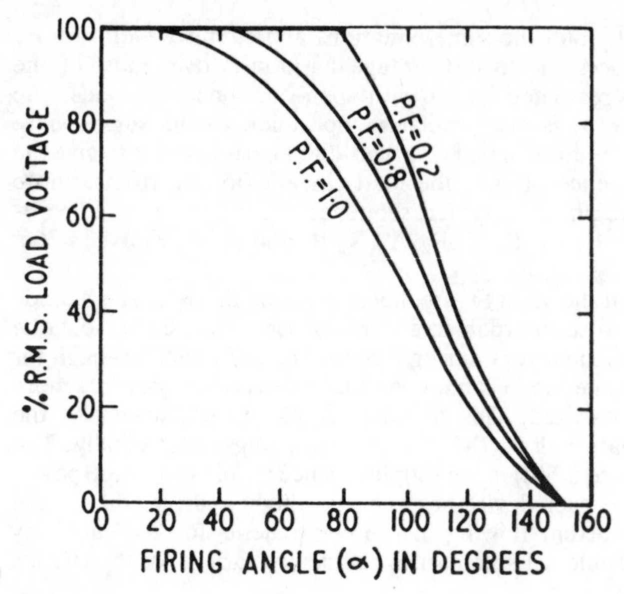

Figure 8.7 Variation of r.m.s. load voltage with firing angle for the three-phase control circuit of Figure 8.6(b)

or some other reference, the delay circuit providing a variable setting from this reference, after which the gate drivers are energised to turn the power semiconductors on.

The detector could be used to sense the voltage across the thyristors and begin the delay from the instant the previous thyristor goes off, although for inductive loads this can lead to asymmetry in the firing point. For example, in Figure 8.1(a) suppose that the load is inductive and that, due

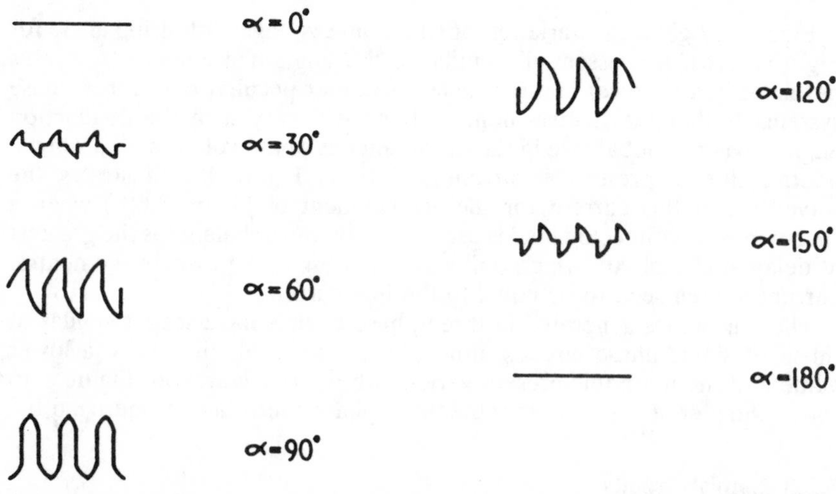

Figure 8.8 Neutral current waveforms for the three-phase circuit of Figure 8.6(b) with a star-connected resistive load

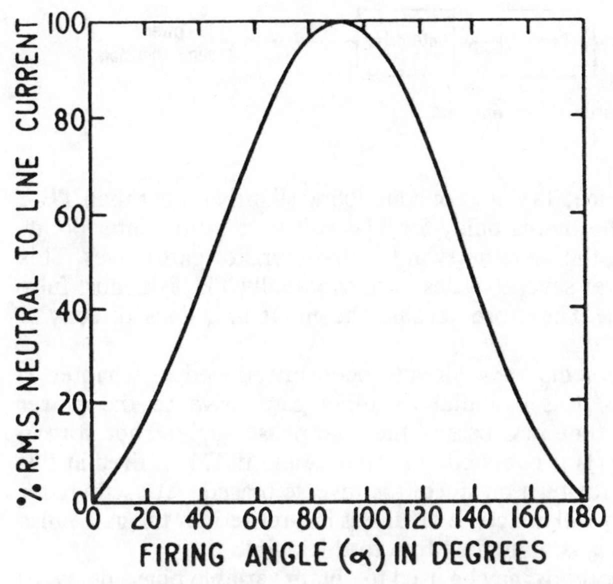

Figure 8.9 Variation of neutral current with firing angle for the three-phase circuit of Figure 8.6(b), with a star-connected four-wire resistive load

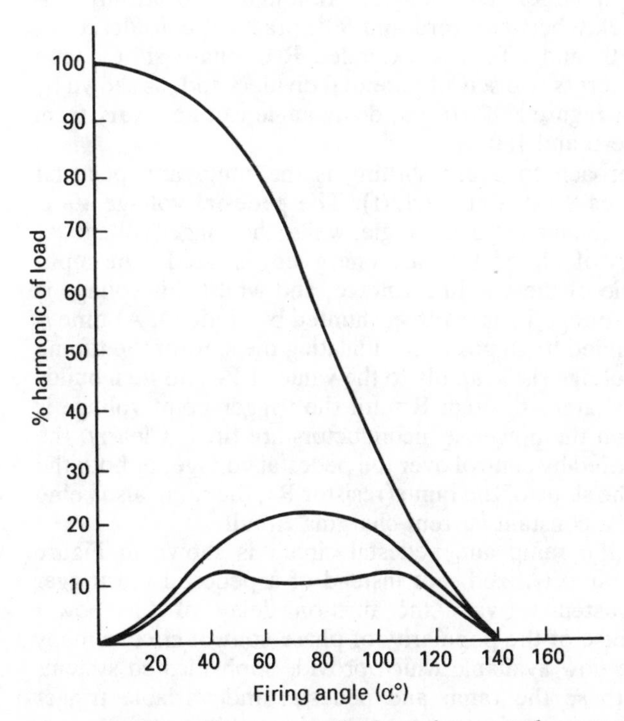

Figure 8.10 Harmonic content for the three-phase control system of Figure 8.6(b)

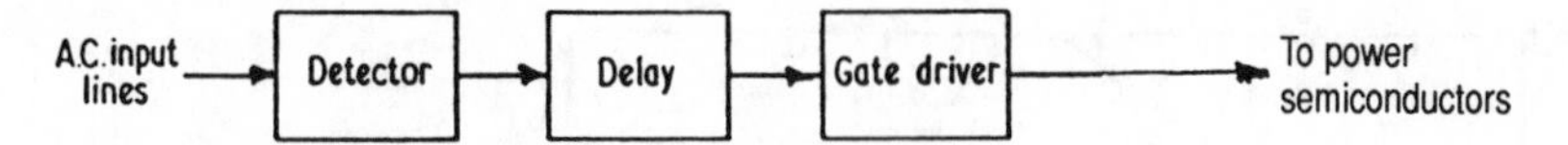

Figure 8.11 Phase-control circuit block diagram

to momentary imbalance, TH_1 is kept conducting slightly longer than TH_2. This will mean that the timing delay for TH_2 will start from a later point, still further increasing the asymmetry in the two-thyristor gate drives. This effect will build up over several cycles until eventually TH_1 is almost fully on and TH_2 fully off. Therefore sensing the input a.c. lines directly is preferred.

Gate drive requirements have already been introduced in Chapter 7. With inductive loads it is essential to apply gate drive to the power semiconductors for a time in excess of the load phase angle. Therefore in Figure 8.3 thyristor TH_2 is conducting at time t_0 and if TH_1 is fired at this point ($\alpha = 0$) it will not turn on since it is reverse biased. At t_{01} thyristor TH_2 goes off, but TH_1 will not come on unless it is refired, or the gate pulse which was applied at t_0 is maintained up to this point.

Several different methods may be used to obtain variable phase delay, as shown in Figure 8.12. In the simple R-C circuit of Figure 8.12(a) the voltage across the resistor leads that across the capacitor by 90°, as in the phasor diagram of Figure 8.12(b), so giving the delay α between the input and output voltages. As the resistance is increased the value of V_R increases, leading to a larger delay angle. Although theoretically this circuit could give a delay between zero and 90°, practical considerations limit it to between 10° and 80°. The extended R-C phase shift circuit compares the voltage across two sets of potential dividers and, as shown by the phasor diagram of Figure 8.12(d), the delay angle can now vary from theoretical limits of zero and 180°.

An alternative approach to phase shifting is the ramp and pedestal circuit shown in Figures 8.12(e) and 8.12(f). The pedestal voltage V_P is variable and is used to change the delay angle, whilst the trigger voltage, at which the gate drivers of Figure 8.11 are energised, is fixed. The input voltage V_{IN} is a sample of the a.c. line voltage, and whilst this voltage is negative the pedestal voltage V_P is in effect shunted by diode D. At time t_0 the line voltage is assumed to go positive, indicating the start of the timing cycle. The capacitor voltage rises rapidly to the value at V_P and then builds up more slowly as it charges through R until the trigger point voltage is reached at time t_1 when the power semiconductors are fired. Clearly, the trigger delay can be varied by control over the pedestal voltage, or both the pedestal voltage and the slope of the ramp (resistor R), the ramp also being made linear by use of a constant current-charging circuit.

A modification to the ramp and pedestal circuit is shown in Figure 8.12(g) in which the ramp is fixed, but instead of a pedestal the trigger point voltage is adjusted to vary the turn-on delay of the power semiconductors. Because of the popularity of phase-control circuits many integrated circuits are now available which provide sophisticated systems on a chip, and for these the ramp and pedestal and variable trigger threshold systems are easier to implement than phase-shift circuits.

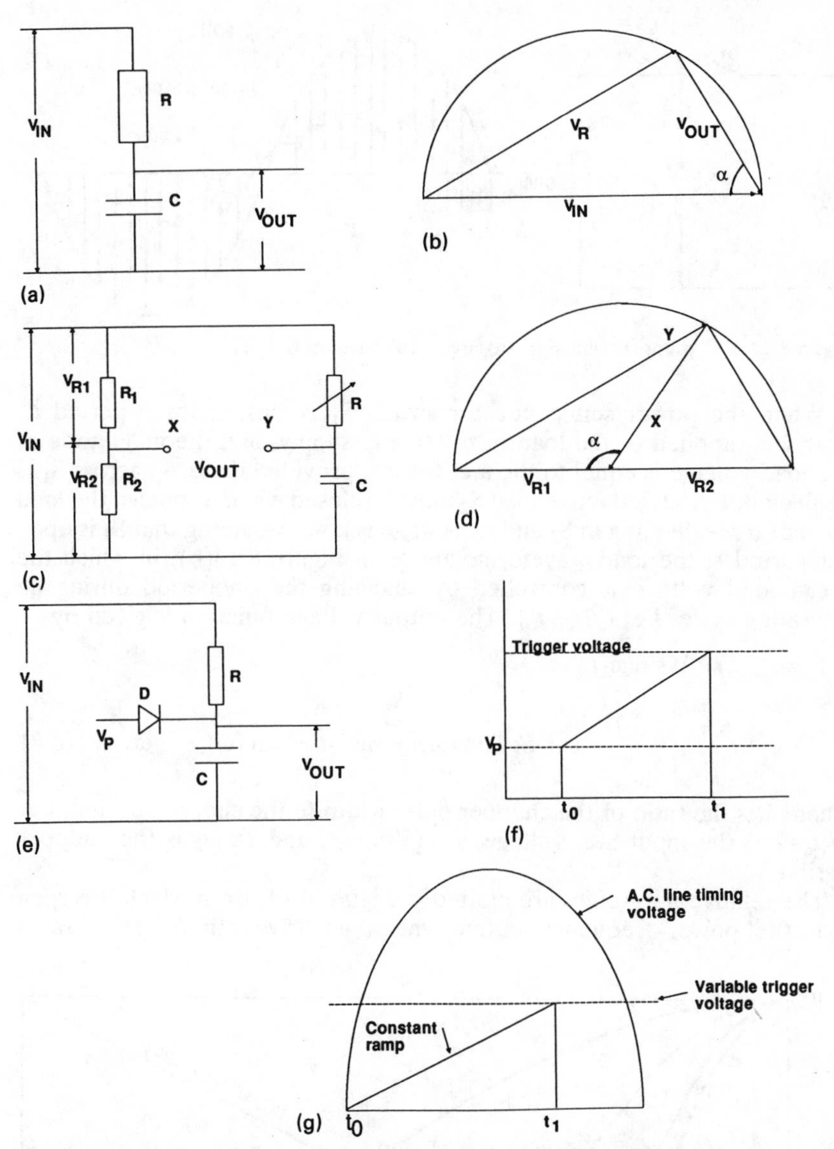

Figure 8.12 Phase-delay circuits: (a) and (b) simple R-C phase shift; (c) and (d) extended R-C phase shift; (e) and (f) ramp and pedestal; (g) variable trigger threshold

8.3 A.C. chopper regulation

Regulation of the a.c. power to a load by phase-control methods, as described in the previous section, has several disadvantages, one being the high harmonic content in the output, which is especially evident at large delay angles. An alternative to phase control is chopper regulation, which is illustrated in Figure 8.13.

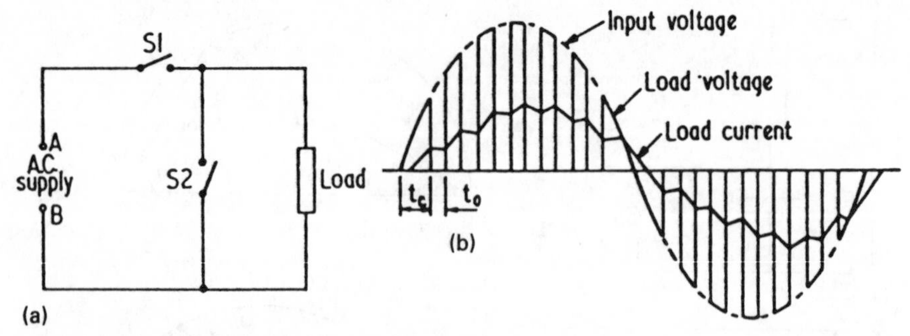

Figure 8.13 A.C. chopper regulation: (a) circuit; (b) waveforms

When the power semiconductor switch S_1 is closed, for a period t_c, power is supplied to the load from the a.c. supply, and the magnitude of the load voltage is equal to the a.c. source, provided the drop across S_1 is neglected. For an inductive load S_2 must be closed when S_1 opens, the load current free-wheeling in S_2 and its voltage is low. Assuming that S_1 is open for period t_o the load waveforms are as in Figure 8.13(b), in which the mean load voltage is controlled by changing the on period during an operating cycle, i.e. $t_c/(t_c+t_o)$. The output voltage function is given by

$$V = \sqrt{(2)}\, k\, V \sin \omega_1 t + \sum_n \left[\frac{\sqrt{(2)}V}{n} \sin nk \left\{ \sin(\omega_1 + n\omega_2)t - \sin(\omega_1 - n\omega_2)t \right\} \right] \quad (8.7)$$

where k is the ratio of the chopper pulse width to the chopper period, i.e. $t_c/(t_c+t_o)$, the input a.c. voltage is $\sqrt{}(V \sin \omega_1 t$, and $2\pi/\omega_2$ is the chopper period.

The relative harmonics are plotted in Figure 8.14, from which it is seen that the power frequency output varies linearly with k. The lowest

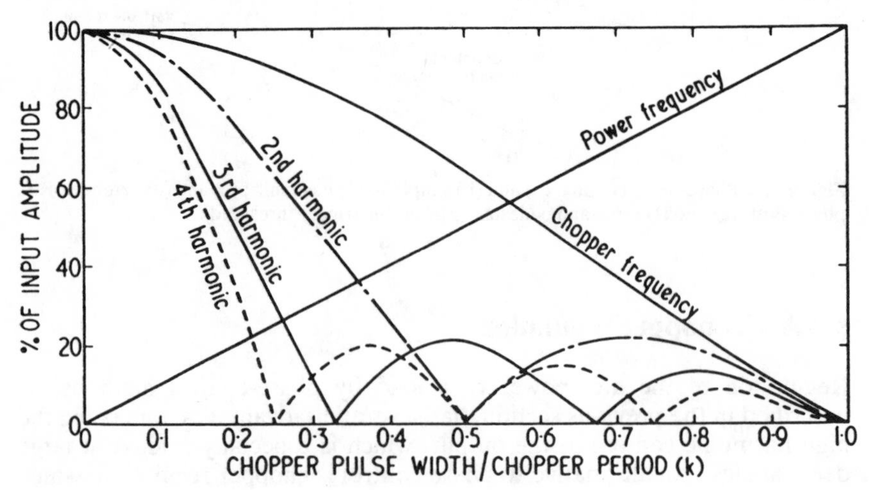

Figure 8.14 Harmonic content of a chopped sine wave

harmonic is approximately equal to the chopper frequency and can be filtered out relatively easily for high chopping frequencies.

A.C. chopper regulation has two basic advantages over phase-control techniques:

(i) The speed of response is faster. This is so since a chopper can be turned on or off at any instant in the a.c. cycle, whereas in phase-controlled circuits once the power switches have been turned on they will go off only when the supply voltage has reversed.

(ii) The harmonic content is lower, which is evident when comparing the load current waveforms given in Figure 8.13 with those in Figure 8.3. For low voltages the phase-control circuit produces a series of short current pulses, spaced 180° apart, whereas for chopper regulation the waveform can still be made to approximate to a sine wave, provided the chopping frequency is relatively high. Figure 8.15 shows the plot

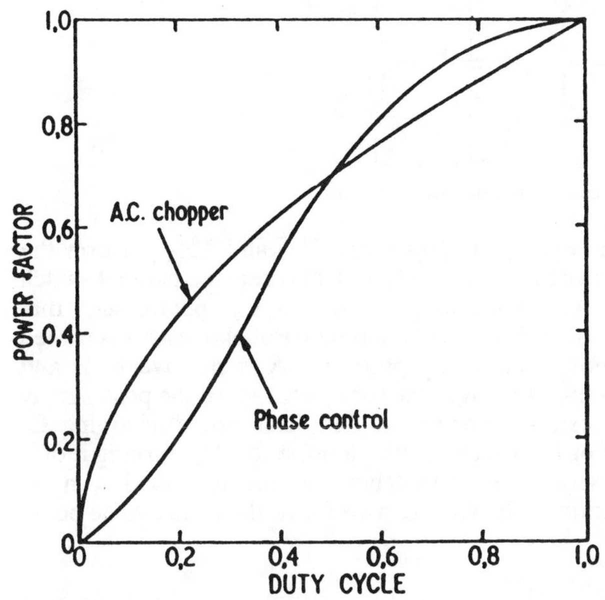

Figure 8.15 Power factor curves

of effective power factor, assuming a resistive load, against duty cycle, which is $t_c/(t_c+t_o)$ for chopper regulators and $(\pi-\alpha)/\pi$ for phase-controlled circuits. This graph shows that the chopper regulator has a higher power factor below 50% duty cycle, whilst above this value the phase-controlled waveform is superior. The difference at these higher voltage settings is not large, so that overall a chopper regulator gives a better performance.

The disadvantages of chopper regulators are principally:

(i) Higher cost due to more elaborate control and power circuitry.

(ii) Greater radio frequency interference generation, since the chopping is performed several times per cycle.

The gate turn-off switch is ideally suited for use in a.c. chopper regulators, since it can be turned off by control of its gate current, although transistors are used for lower power ratings. Where the power of a thyristor is required the device needs to be turned off before the end of a half cycle, i.e. it must be forced commutated. Several circuits may be used, as discussed in Chapter 11, Figure 8.16 showing one system which is ideally

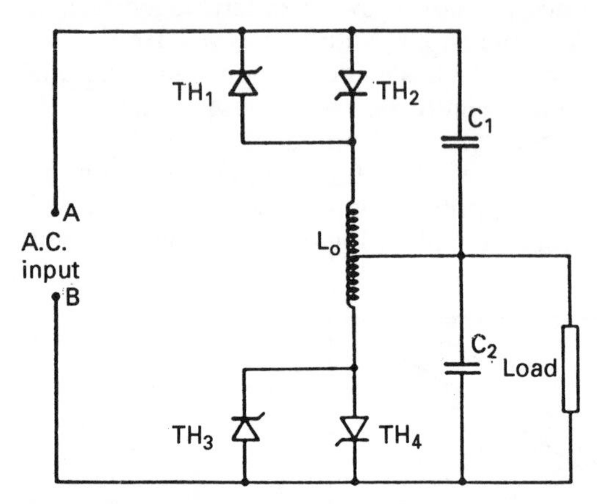

Figure 8.16 A.C. chopper regulator thyristor arrangement

suited to a.c. chopper regulation. Thyristors TH_1 and TH_2 perform the function of the series switch S_1, and TH_3 and TH_4 replace parallel switch S_2. Capacitors C_1, C_2 and centre-tapped inductor L_o operate such that when a series thyristor is fired its corresponding parallel thyristor is turned off, and vice versa. For instance, suppose line A is positive to B and thyristor TH_2 is conducting, so that capacitor C_2 charges to the peak supply voltage. To turn the series thyristor off TH_4 is fired, discharging C_2 through the lower half of L_o, which couples a pulse to TH_2, turning it off.

The rating of the series power switches, on resistive load, can be obtained by an integration of the voltage waveform, the r.m.s. value being given by

$$I_{T(rms)} = \frac{V}{R}\left[\frac{t_c}{2(t_c + t_o)}\right]^{1/2} \tag{8.8}$$

The rating of the shunt power switches depends largely on the power factor of the load. They are usually made comparable in size to the series devices, to allow for inductive loads operating on a 50% duty cycle.

A.C. chopper regulators are primarily used in applications which require a sine wave output, since their harmonic content is then more easily filtered out than in comparable phase-controlled circuits. Owing to the higher frequency harmonics involved, the filter section can be relatively simple, consisting essentially of a series band stop section, which removes the fundamental chopper frequency, followed by a low-pass section, which filters out higher-order harmonics.

The control circuit for a chopper regulator, which is designed to produce a stabilised sine wave output, is given in Figure 8.17. The a.c. supply is

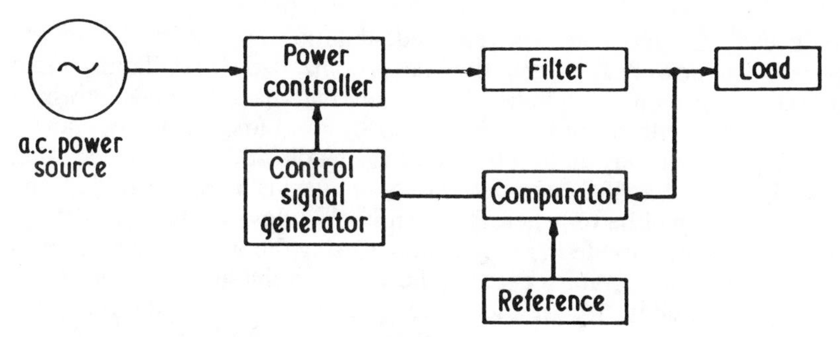

Figure 8.17 Control circuit for a.c. chopper regulator

chopped before being transformed to the required voltage level, and it is then filtered and fed to the load. The sensed voltage, fed back from the load to the comparator, may be a.c. or d.c.; if a d.c. signal is required the load voltage first being rectified and smoothed. A d.c. reference voltage also feeds the comparator, which produces an error signal to the control signal generator. This modifies the firing of the thyristors, so as to change the pulse width of the a.c. output and correct the error between the output voltage and its reference.

This control method produces an output voltage whose mean or r.m.s. value is monitored with reference to a desired input, having no control over the shape of the sine wave. The system also has a slow response speed owing to the delay introduced by signal rectification and smoothing. This disadvantage can be overcome by feeding a sine wave a.c. reference, derived from the a.c. power source, to the comparator, along with an a.c. signal fed back from the load. A comparison then occurs between instantaneous values of output and reference, such that if the output is greater, the parallel switch is operated, whilst if it is less, the series switch is closed to deliver more power to the load. The chopping frequency will now be variable. Apart from removing the delay in response speed, this system also causes the instantaneous output to follow the a.c. reference voltage to positive or negative limits which are determined by the gain of the overall system.

8.4 Integral half-cycle regulation

One of the disadvantages of phase control and chopper regulation techniques is that the power switch can be caused to turn on when there is a relatively large voltage across it, which results in a sharp increase in load current, with the possible generation of radio frequency interference. This effect can be minimised by the use of filters, but if the power levels being handled are large the filters can be bulky and expensive. Alternative techniques, known as zero voltage switching or integral half-cycle regulation, can then be used.

The principle of zero voltage switching consists in turning on a power switch at the beginning of a half cycle or not at all, the load voltage build-up then following the sine wave of the supply voltage. Therefore the main cause of interference is removed and no suppression filters are required. The power supplied to the load is controlled by regulating whole half cycles of the supply voltage, there being two main methods by which this is done, as illustrated in Figure 8.18. For burst firing the power switches are either fully on or fully off for the whole duration of the sensing period. For instance, if the temperature of an oven is being controlled, the power switches will be on when the oven cools below a reference setting, and power will be supplied to the heaters causing the oven temperature to rise. When this temperature exceeds the desired value the power switches turn off and remain off until a lower temperature limit is again reached.

Burst firing or on/off control systems are suitable for high-inertia loads, such as heating, but are not suitable for lighting or motor control since the operating frequency is too low. In these instances proportional control, also known as cycle syncopation, is used, the system working on a fixed period, measured in numbers of cycles or half cycles. Figure 8.18(b) illustrates a system in which this period is five cycles. To regulate the power the on-to-off duty cycle within this periodic time is varied. Once again this change can be in half cycles or in cycles, the important consideration being that the power device turns on as close to the start of the half cycle as possible.

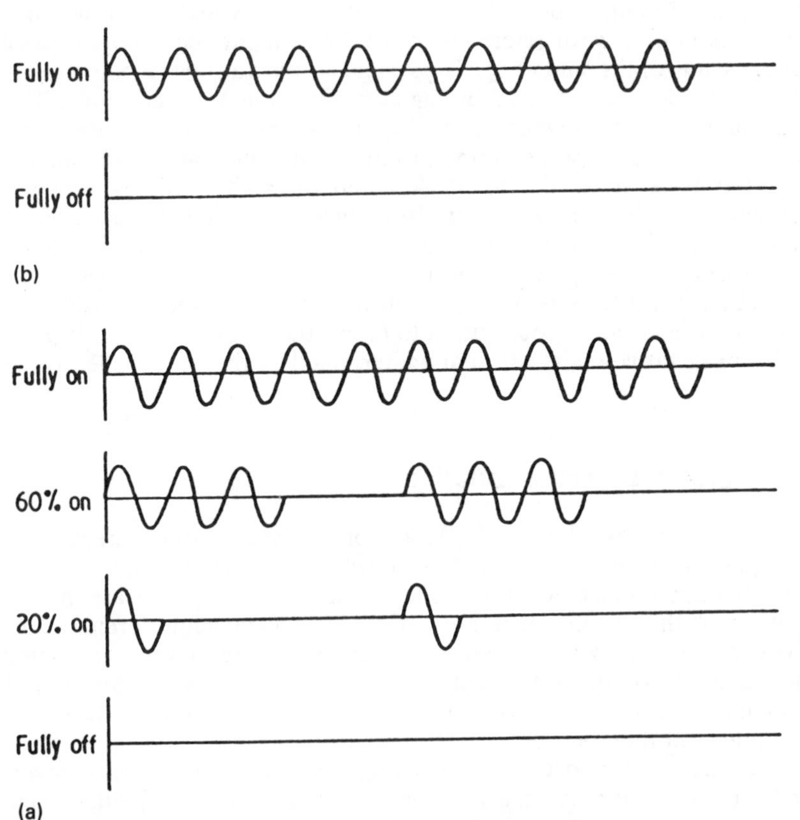

Figure 8.18 Operating modes for zero-voltage switching control

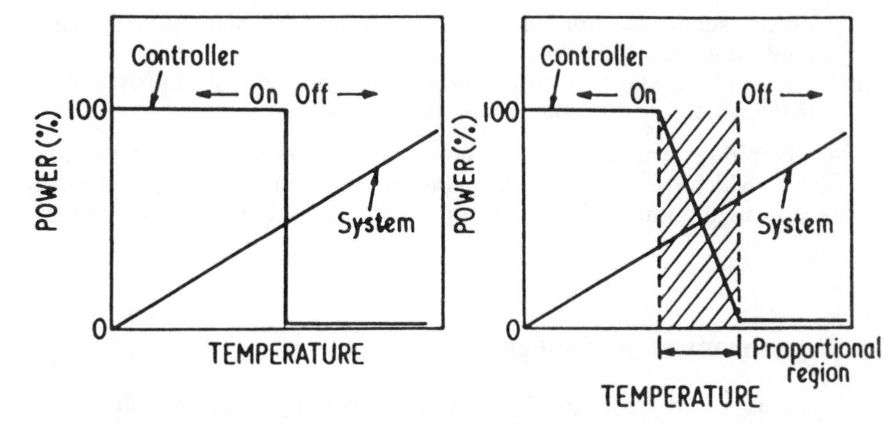

Figure 8.19 Control characteristics for zero-voltage switching systems: (a) burst firing; (b) proportional control

The difference between burst firing and proportional control is one of detail only, both being very similar in principle. Burst firing control systems have a large gain, as illustrated in Figure 8.19(a), so that the power switches are either off or on. Although satisfactory for many applications this high gain can in some cases lead to problems of instability, commonly known as hunting. To overcome this the system gain is reduced, as in Figure 8.19(b), so as to introduce a proportional band. Within this band the controller is able to anticipate the future state of the controlled parameter and to adjust the duty cycle of the thyristors accordingly. The width of the proportional band is important; if it is made too large the system will be sluggish and reach its final state slowly; if it is too narrow there will be large overshoots about the required setting, as in burst firing.

Irrespective of the method of control used, the mean power and r.m.s. voltage are dependent on the power switch duty cycle. If V is the r.m.s. input voltage and P is the power in the load with the supply uncontrolled, then the controlled voltage V_c and power P_c are given by

$$P_c = \frac{Pt}{T} \tag{8.9}$$

$$V_c = V\left[\frac{t}{T}\right]^{1/2} \tag{8.10}$$

where t is the power switch conduction period, expressed in units of time or number of half cycles or cycles of the input voltage waveform, and T is the operating period, in the same units as t.

There are many circuits that may be used to control thyristors operating in a zero voltage switching mode. They all have three basic sections:

(i) A power supply stage to drive all internal amplifiers and feed the gate energy to the power semiconductors.
(ii) A zero voltage-detecting stage, which senses the instant of zero supply voltage. This stage releases the power amplifiers for a short duration around this cross-over point so that they may trigger the

power semiconductors if required, or operate some other more continuous drive circuit.

(iii) An amplifier stage which magnifies the control signal to provide the drive needed to turn the power switches on.

As for phase control there are now several integrated circuit devices available which provide a convenient method for obtaining integral half-cycle regulation, using both burst and proportional control techniques.

8.5 Synchronous tap changer

In systems which have an output transformer and require only small adjustment of the voltage, synchronous tap changing is often preferable to ordinary phase control. Figure 8.20 shows a typical system and its output waveforms.

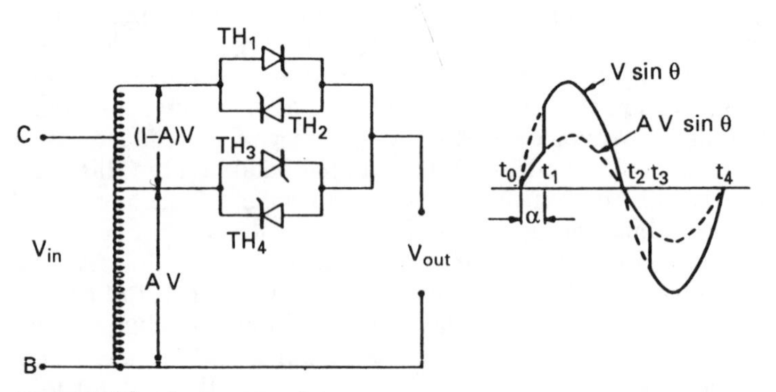

Figure 8.20 Synchronous tap changer

At time t_0 input line C goes positive and thyristor TH_3 is fired. After a delay α thyristor TH_1 is turned on, which reverse biases TH_3 and turns it off. The output voltage jumps to the new value of input, until at t_2 thyristor TH_1 goes off, assuming a resistive load. Thyristor TH_4 is fired at t_2 followed by TH_2 at t_3 to complete the negative half cycle. By moving through one complete tap-changing sequence synchronously with the supply voltage, natural commutation of the thyristor is possible. The values of A and α determine the output voltage as a function of input V. Although A is fixed for a given system, adjustment of the load is possible by varying α, the r.m.s. output voltage being given by

$$V_0 = \left[\frac{1}{\pi}\int_0^{\alpha}(AV\sin^2\theta)^2\,d\theta + \int_{\alpha}^{\pi}(V\sin\theta)^2\,d\theta\right]^{1/2}$$

$$= \frac{V}{2}\left[\frac{1}{\pi}\{(1-A^2)(\sin 2\alpha - 2\alpha) + 2\pi\}\right]^{1/2} \qquad (8.11)$$

The r.m.s. current rating of the thyristors depends on the conduction point α and is given by equation (8.12) for TH_3, TH_4 and by equation (8.13) for TH_1, TH_2, where R is the load resistance:

$$I_{T(rms)} = \frac{VA}{2R}\left[\frac{1}{\pi}\{2\alpha - \sin 2\alpha\}\right]^{1/2} \quad (8.12)$$

$$I_{T(rms)} = \frac{V}{2R}\left[\frac{1}{\pi}\{\sin 2\alpha - 2\alpha\} + 2\pi\right]^{1/2} \quad (8.13)$$

The voltage rating of devices need only be $(1-A)V$ since with one or other arm always conducting the transformer tap voltage appears across them.

This control method produces a lower distortion than traditional phase-control techniques, the Fourier series of the output being given by equation (8.14), the values of a_n and b_n being given by equations (8.15) and (8.16) respectively, for n equal to 3, 5, 7, and so on:

$$V_0 = \frac{a_0}{2} + \sum (a_n \cos n\omega t + b_n \sin n\omega t) \quad (8.14)$$

$$a_n = \frac{2V(1-A)}{\pi(1-n^2)}\left[\cos\alpha \cos n\alpha + n \sin\alpha \sin n\alpha - 1\right] \quad (8.15)$$

$$b_n = \frac{2V(1-A)}{\pi(1-n^2)}\left[\cos\alpha \sin n\alpha - n \sin\alpha \cos n\alpha\right] \quad (8.16)$$

From this equation the magnitude of any harmonic for a fixed value of α and A can be found. Harmonics may be reduced by increasing the number of tapping points at the expense of increased circuit complexity. Fundamental coefficients are given by equations (8.17) and (8.18), the amplitude of the fundamental being given by (8.19):

$$a_1 = \frac{2V(1+A)\sin^2\alpha}{2\pi} \quad (8.17)$$

$$b_1 = \frac{2V(1-A)\left[(\alpha \sin^2\alpha - \cos^2\alpha) - \sin\alpha \cos\alpha\right]}{2\pi} \quad (8.18)$$

$$\left[a_1^2 + b_1^2\right]^{1/2} \quad (18.19)$$

The operation of the circuit shown in Figure 8.20 becomes slightly more complicated on inductive load. For example, at t_2 thyristor TH_1 is conducting. Even though the line voltage now reverses, this thyristor will be kept on by the stored energy in the inductive load. To turn it off and so limit the output voltage to $AV\sin\theta$, thyristor TH_3 is turned on. This means, however, that provided TH_3 is still on, thyristor TH_2 cannot be fired at t_3, as would normally be done, or it would short-circuit through TH_3. Therefore although the control circuit for the thyristors is very similar to the usual phase-control systems described previously, it is now necessary to sense the load current as well as the zero voltage cross-over points and to inhibit the firing pulses if danger of short circuit exists.

Clearly, the system of Figure 8.20 can be extended to a greater number of taps if these were available on the output transformer. Generally, synchronous tap changing is used if an output transformer is available and only a limited range of control is required. Phase control is the most commonly used system and the simplest. Where load power being controlled is large and radio frequency interference requirements stringent, as for domestic heating applications, integral half-cycle control is preferable, provided the load has a relatively large inertia. A.C. chopper control finds application primarily on special systems where speed of response and minimisation of harmonics are essential.

Chapter 9

Phase-controlled rectification and inversion

9.1 Introduction

Controlled rectifier circuits had been in use long before power semiconductors were invented, the control element being the mercury arc rectifier. Although these are still found for very high-power applications, they have almost entirely been replaced by the semiconductor switch, primarily the thyristor.

Phase-controlled rectification and inversion was introduced in Chapter 6, and although the principle is fairly simple there are a bewildering number of different combinations which exist in practice, some of these being described in this chapter. All these circuits can usually be divided as follows:

(i) Bi-directional or unidirectional converters.
(ii) Each of the above converters can then be either push–pull or bridge in configuration, the push–pull arrangement requiring an input transformer.
(iii) All these converters can have any number of output pulses, the pulse number usually being related to the number of phases of the input a.c. supply. The higher the pulse number, the lower the ripple content in the output d.c. voltage.

Bi-directional converters are capable of inversion in addition to rectification, and this is often a desirable feature. Where it is not required unidirectional converters can be used, the resulting circuit being cheaper and having a smaller d.c. voltage ripple and input 'wattless' current content.

Push–pull circuits have the advantage that there is only one conducting device in series with the load, which was much more of an important consideration when mercury arc rectifiers were used, since they could have arc drops of 30 V or more. However, modern thyristors lose only a volt or two, so that this is important only when working from abnormally low supply voltages. Generally, bridge circuits have better transformer utilisation and are more frequently used.

The present chapter first describes the principles of unidirectional and bi-directional converters, followed by the effects of discontinuous load current and source reactance on their operation. The performance factors used in the analysis of these converter circuits are then obtained, and the chapter concludes with a description of gate-control and voltage-multiplication circuits, which is a special application of rectification.

9.2 Bi-directional converters

Perhaps the simplest bi-directional a.c. to d.c. converter is that shown in Figure 9.1. The thyristor is capable of supplying power only during half a cycle, when supply line A is positive with respect to B, so that this is a half-wave controller. On resistive loads the supply current follows the

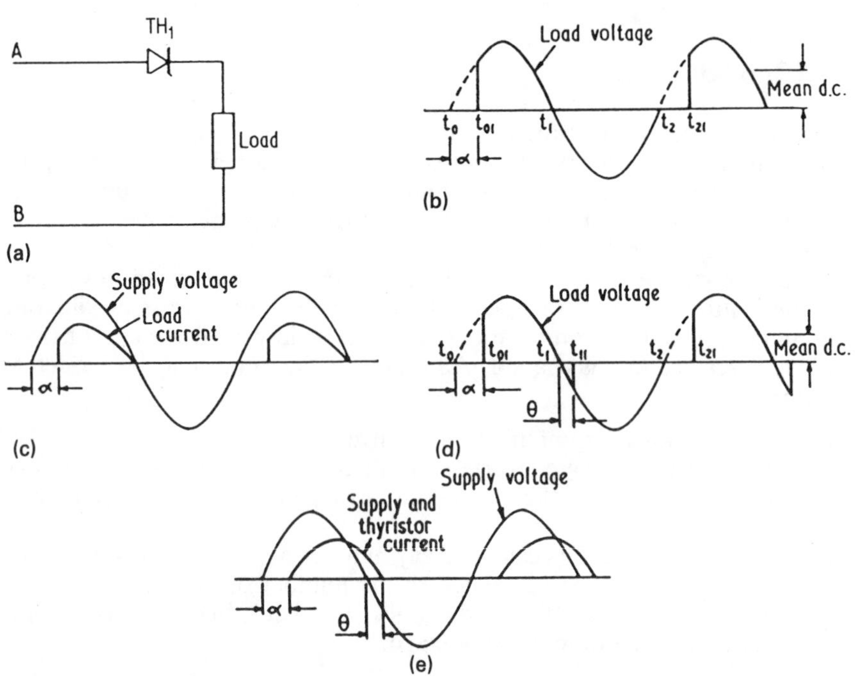

Figure 9.1 Half-wave bi-directional converter: (a) circuit arrangement: (b) and (c) resistive load waveforms; (d) and (e) inductive load waveforms

shape of the input a.c. voltage when the thyristor is conducting. When the thyristor is off the load current and voltage are zero, if the leakage through the device is neglected. Clearly, the delay angle α can be used to regulate the value of the mean d.c. output voltage. The operation of the circuit on inductive loads changes slightly. Now when the thyristor is fired, at t_{01} say, the load current will increase in a finite time through the inductive load. At t_1 the supply voltage reverses but TH_1 is kept conducting while the load energy stored during time t_{01} to t_1 is fed back to the supply. The load voltage goes negative, following the reverse half cycle of the supply voltage, and at t_{11} the load current falls to below the holding current of thyristor TH_1 and it goes off.

The half-wave circuit of Figure 9.1 is not normally used since it produces a large output voltage ripple, and is incapable of providing continuous load current. There are several ways in which it may be extended to full-wave operation, generally these systems falling into two groups:

(i) Push–pull converters, which require a tapped transformer input.
(ii) Bridge converters, where an input transformer may be used but is not an essential requirement for correct system operation.

Figure 9.2 shows the circuit of a push–pull bi-directional converter and Figure 9.3 gives its operating waveforms. It is assumed here that the load current is maintained at a constant d.c. value through the operating cycle. T_1 is a centre-tapped transformer whose turns ratio may be adjusted to give any primary to secondary voltage change. When terminal A of the a.c. supply goes positive TH_1 is forward biased and when the supply polarity reverses TH_2 becomes forward biased. When either thyristor conducts, the load voltage equals the instantaneous a.c. voltage across half the transformer secondary winding.

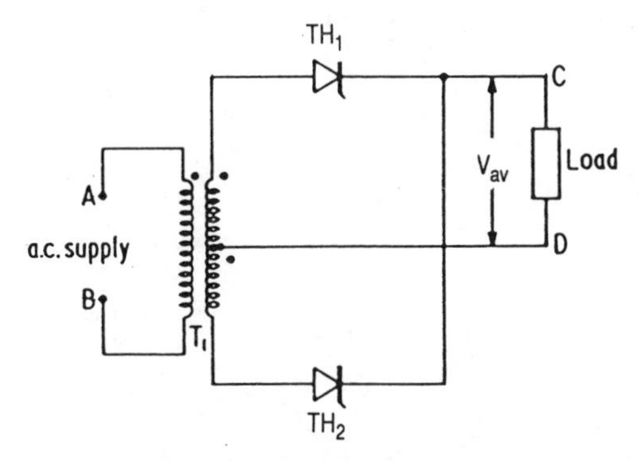

Figure 9.2 Push–pull two-pulse bi-directional converter

Referring to Figure 9.3(a), the thyristors are fired at the commencement of the a.c. cycle and, as far as the circuit is concerned, they behave exactly like diodes. The supply current is assumed square (ripple-free load current) and is composed of a fundamental and various higher harmonics. The fundamental of the current is in phase with the input voltage, so that the system behaves like a unity power factor load.

Figure 9.3(b) shows a delay of α between the start of a positive half cycle and the firing of the corresponding thyristor. Therefore before t_0 thyristor TH_2 was conducting, and when the supply voltage reverses at this point it is kept in conduction due to inductive load current, even though the voltage is negative. This is a regenerative period and power flows from the load to the supply. The voltage across TH_1 is positive and, referring to Figure 9.2, it is seen to be equal to the combined voltage across both halves of the secondary transformer winding, i.e. twice the load voltage. At t_{01} thyristor TH_1 is fired, the voltage across TH_2 now equalling that of the two halves of the secondary winding, and since it is negative this thyristor turns off. TH_1 now conducts up to t_{11} when thyristor TH_2 is refired. During t_{01} to t_1 power is fed from the supply to the load and from t_1 to t_{11} it is fed from the load back to the supply.

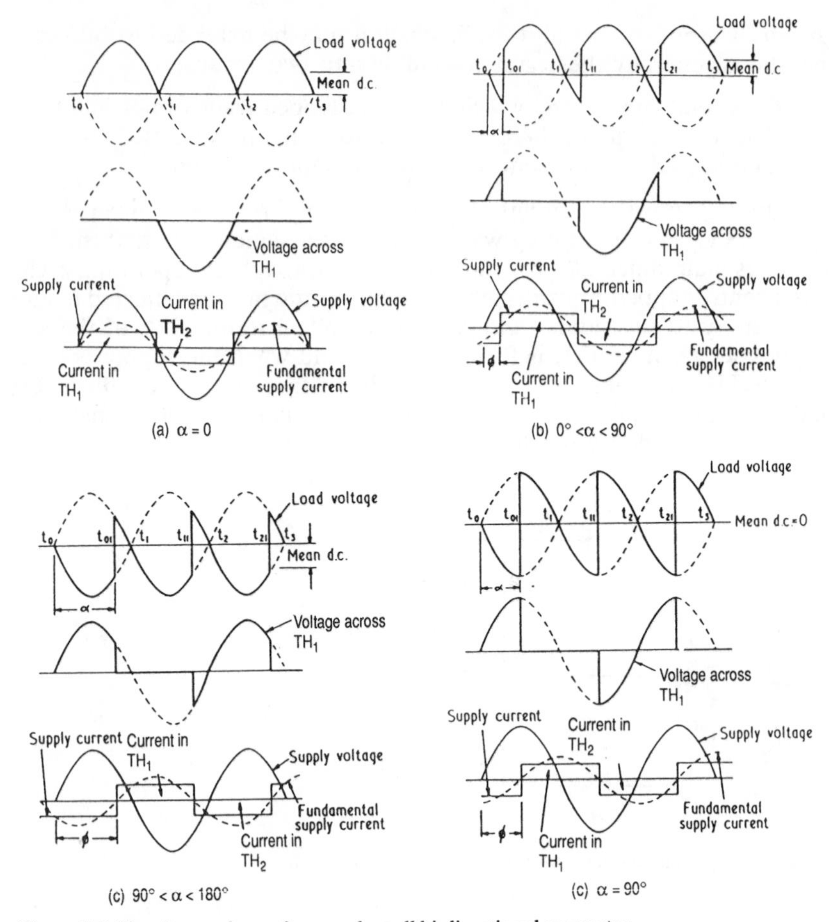

Figure 9.3 Circuit waveforms for a push–pull bi-directional converter

Figure 9.3 illustrates the circuit waveforms for progressively increasing delay angles, the following being noted from these curves:

(i) For delay angles up to 90° more power is fed into the load than is received back from the supply, and beyond this point there is a net transfer from the load to the supply. This cannot be maintained continuously unless the load is provided with a suitable d.c. power source.

(ii) Each thyristor conducts for a 180° period and the shape of the input current waveform is unchanged. However, the phase angle ϕ between the fundamental component of this current and voltage input changes such that $\phi = \alpha$. That this is necessarily true can be verified at $\alpha = 90°$, where there are now equal periods of power flow in both directions between the load and the supply. As far as the supply is concerned, therefore, the load is purely inductive and the phase angle ϕ must be 90°. A converter, even though itself consisting of a fixed value of resistance and inductance, will therefore present a varying load power factor as the firing angle is changed. For net rectification this power factor is lagging, although for inversion it changes to leading.

(iii) The mean d.c. load voltage decreases as the firing angle α is increased, and beyond 90° delay the voltage goes negative, reaching a peak negative value at 180°. Clearly, for a.c. to d.c. rectifier systems the negative voltage period is undesirable.

(iv) The value of the d.c. ripple voltage also increases as the firing angle is increased, up to 90° delay. Beyond this point the ripple in the negative voltage decreases as α is increased to 180°.

(v) The period for which a thyristor is reverse biased reduces progressively as the delay angle increases to 180°. A thyristor must, of course, be reverse biased for greater than its turn-off time in order to be successfully commutated. Therefore the maximum delay angle can never be raised to 180° and for practical systems it is normally limited to about 165° on 50 Hz systems. If a thyristor is not successfully commutated it will commence conduction the instant its anode voltage goes positive and so provide a complete half cycle of power to the load. There will therefore be an abrupt change in the converter operating mode from almost full inversion to full rectification.

Push–pull converter circuits are popularly used in applications which require an input transformer either for isolation purposes or for effective phase number increase. As will be seen later, the larger the number of input phases, the lower the d.c. voltage ripple and the higher the power which the converter can handle. However, when an input transformer is not essential a bridge system is often more economical, a single-phase bridge being shown in Figure 9.4. The operation of this bridge can be

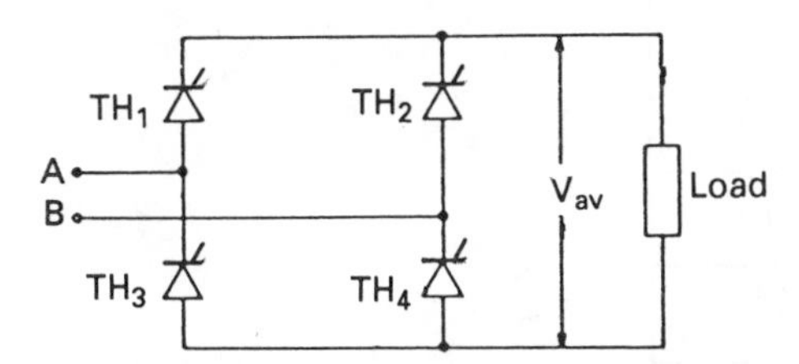

Figure 9.4 Bridge-type two-pulse bi-directional converter

followed by the waveforms of Figure 9.3, where TH_1, TH_4 and TH_2, TH_3 conduct in pairs. There are three points of difference between the push–pull and bridge converters, as follows.

(i) In a bridge system each thyristor must be rated to block the peak voltage across the a.c. inputs of the converter, so the peak load voltage and peak thyristor voltage are equal, whereas for a push–pull system it was seen that the thyristors must be rated for at least twice the peak load voltage.

(ii) A push–pull converter uses two devices compared to the four used for a bridge system, but their voltage rating is now doubled. For low-power systems the price of a thyristor is usually determined by its voltage rating, so that a bridge converter can prove cheaper, although this is not necessarily so when the current ratings of the devices increase. When considering costs it is also necessary to add the price of control systems, and since a push–pull converter uses fewer thyristors, and they have a common cathode, the cost of its drive circuitry should be less than for a comparable bridge converter.

(iii) In a push–pull converter there is only one thyristor in any conduction path between the supply and the load, whereas a bridge system has two series thyristors. Therefore the efficiency of a bridge converter would be expected to be lower than a comparable push–pull circuit although, since the thyristors normally have a drop of the order of one volt, this would only have a significant effect on very low voltage supplies.

Single-phase circuits are relatively simple in construction, but they are limited in power-handling capabilities and produce output voltage ripple which is much greater than that from three-phase systems. The circuits described so far can be termed two pulse, i.e. the ratio of the fundamental d.c. voltage ripple frequency to that of the input a.c. supply is two. The greater the pulse number, the lower the smoothing requirements of the circuit.

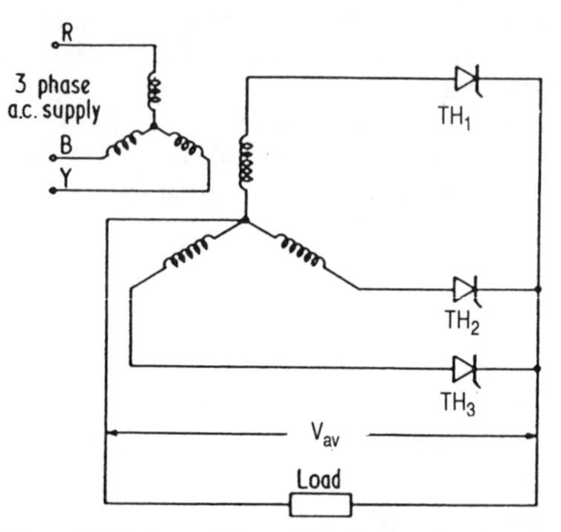

Figure 9.5 Push–pull three-pulse bi-directional converter

Figure 9.5 shows a three-pulse push–pull converter (also called a three-phase half-wave converter) and Figure 9.6 gives its operating waveforms. For zero firing angle delay the thyristors in the most positive phase conduct to the neutral line, the voltage across the thyristor being zero when it conducts and equalling the line voltage between it and the phase of the conducting device when the thyristor is off. Each thyristor is on for 120° and the supply current and voltage are as shown. The d.c. load current is assumed to be ripple free and the input a.c. current is seen to have a d.c. component equal to one third of the load current, but this magnetising current can normally be eliminated by zigzag connection of the input transformer.

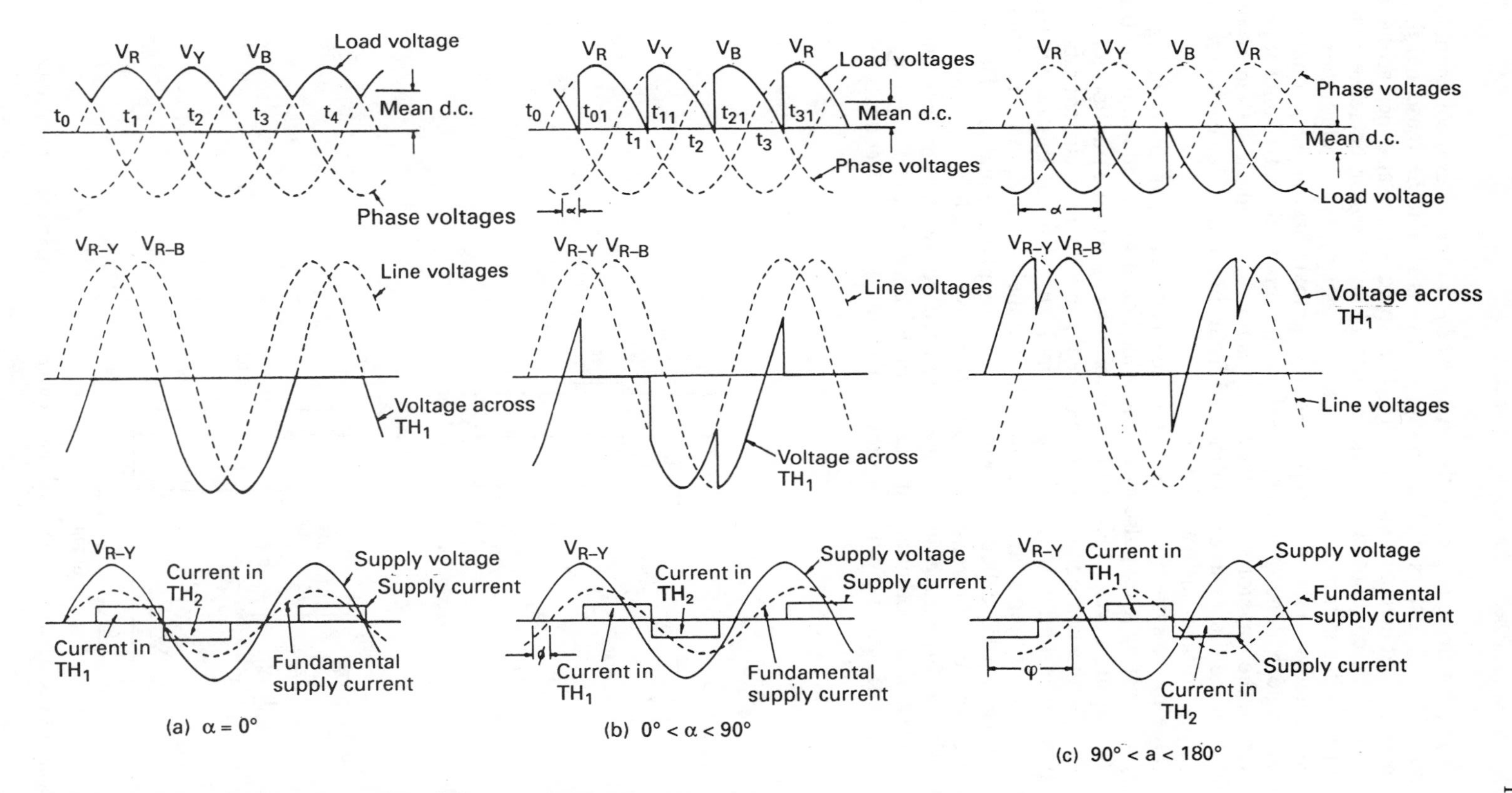

Figure 9.6 Circuit waveforms for a push–pull three-pulse bi-directional converter

When the firing angle is delayed by α beyond the natural commutation point of the thyristors, the mean d.c. voltage is seen to decrease and then to reverse (net inversion) for $\alpha > 90°$. The ripple voltage also increases and the negative commutation periods across the thyristor decrease, each thyristor still conducting for a 120° period, although, as expected, these are shifted relative to the input supply voltage so that phase angle ϕ equals the delay angle α. The d.c. component of the input current is unchanged, therefore the waveforms are essentially as for the two-pulse system expect that now the d.c. voltage is fabricated from three parts, for each cycle of the input voltage, instead of two.

Three-phase systems may be extended to six pulses, as shown in Figure 9.7, by creating a six-pulse system in which each line conducts to neutral for 60° during a cycle. This results in poor utilisation of the transformer and devices, with a consequent increase in their r.m.s. to mean current ratio, so that a much better solution is given in Figure 9.8. In this circuit two three-pulse systems are operated in parallel through an interphase transformer, also called an absorption coil or phase equaliser. This is a centre-tapped auto-transformer and its action is such as to cause it to absorb the instantaneous voltage difference across its windings and to produce a mean potential at its centre point. Therefore the d.c. voltage will contain a six-pulse ripple, as shown in Figure 9.9, the frequency of the

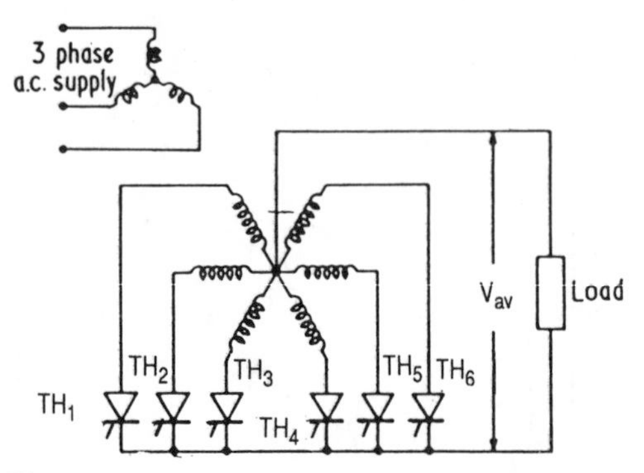

Figure 9.7 Push–pull six-pulse bi-directional converter

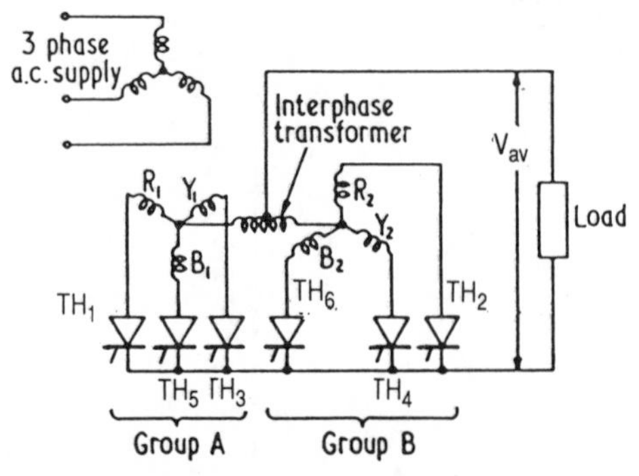

Figure 9.8 Push–pull six-pulse bi-directional converter using an interphase transformer

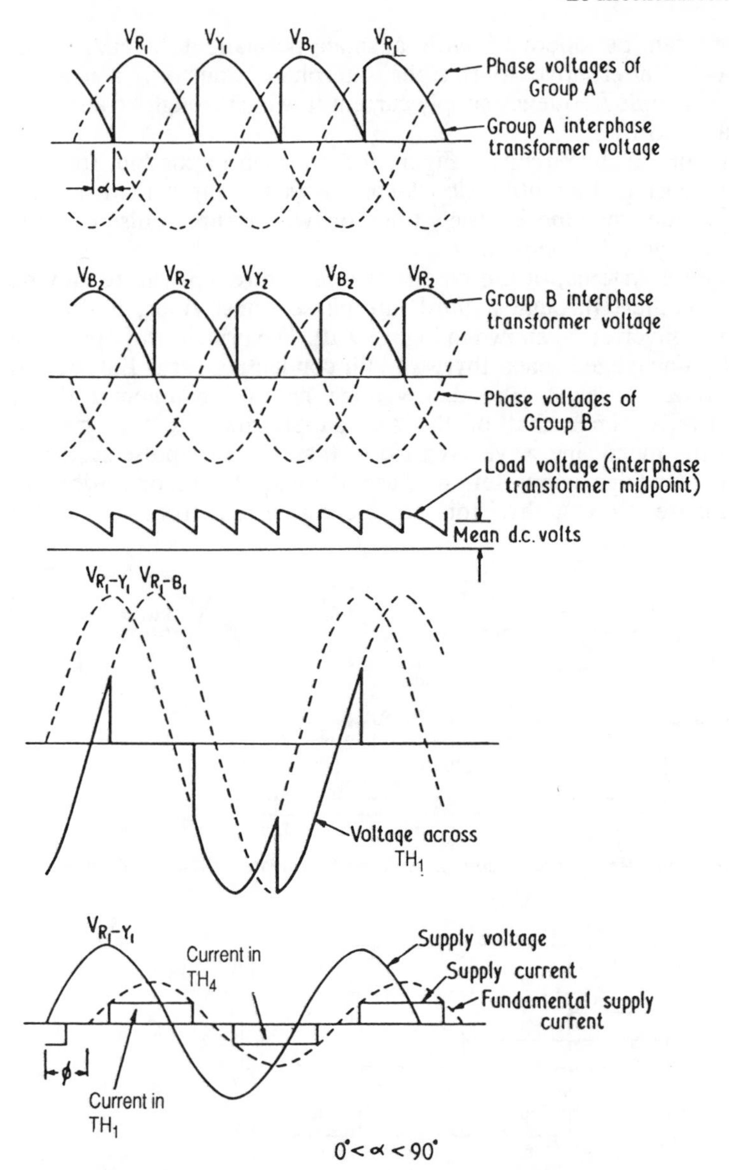

Figure 9.9 Circuit waveforms for a push–pull six-pulse bi-directional converter with interphase transformer

voltage across the interphase transformer being equal to three times that of the a.c. input.

An interphase transformer must be sufficiently fluxed to produce the sharing action between the two three-pulse systems, and on light loads this would not occur, so that the system then reverts to three-pulse operation, with a consequent shift in mean d.c. voltage. To prevent this change in voltage when the load falls below a critical value, known as the transit load, the rectifier can be operated with a small permanent bleeder load, although as an alternative to this the interphase transformer may be energised by a triple frequency supply current flowing through an auxiliary winding on its core.

The thyristors in the circuit of Figure 9.8 are seen to conduct for 120° each, so increasing their utilisation factor, with the input load current resembling a sine wave more closely than for two- or three-pulse circuits, and containing no d.c. component.

Two six-pulse systems, of the type shown in Figure 9.8, can readily be connected together, through a third interphase transformer, to form a twelve-pulse converter, as shown in Figure 9.10. The principle of operation is essentially unchanged, each thyristor still conducting for a 120° period during a cycle. However, the d.c. voltage has a fundamental ripple frequency twelve times that of the a.c. supply and the a.c. current approaches the mean sine wave even closer than for a six-pulse system.

Three-phase bridge converters are usually six-pulse in operation, as shown in Figure 9.11(a), thyristors conducting current from a line at a

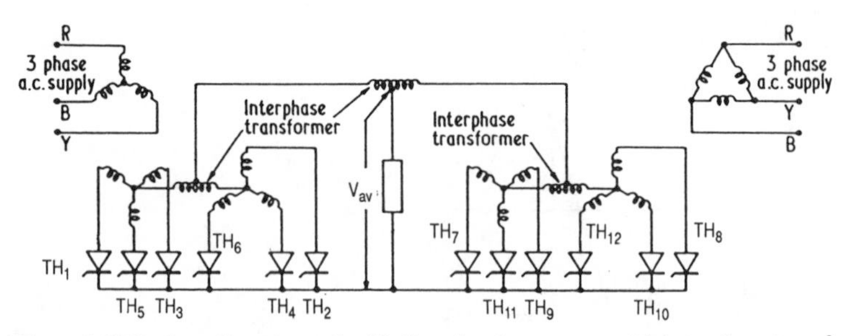

Figure 9.10 Push–pull twelve-pulse bi-directional converter with interphase transformer

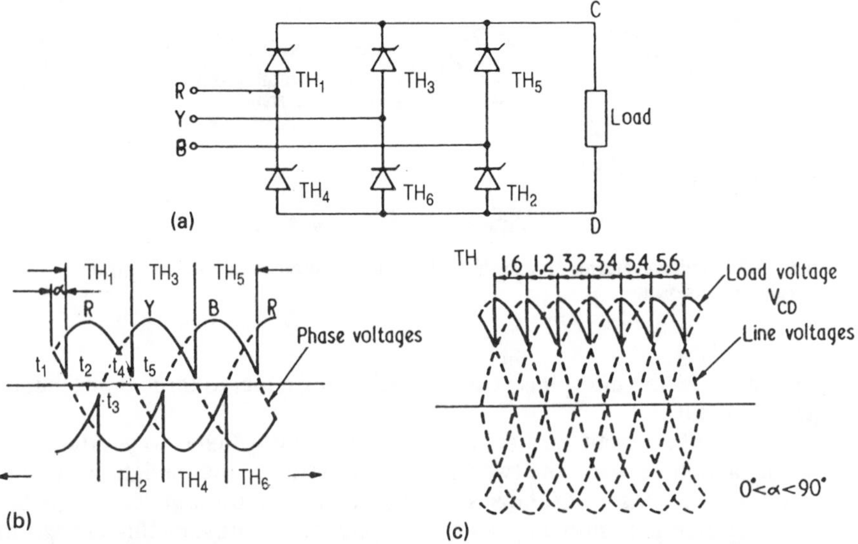

Figure 9.11 Bridge six-pulse bi-directional converter: (a) circuit; (b) and (c) waveforms

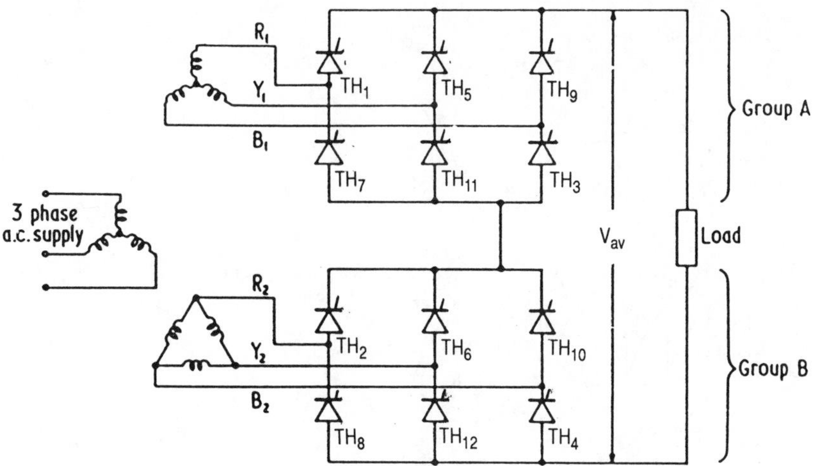

Figure 9.12 Bridge twelve-pulse bi-directional converter

more positive voltage to that at a lower potential, provided these have been fired. The circuit waveforms obtained are very similar to those shown in Figure 9.9, the thyristor conduction periods being shown in Figures 9.11(b) and 9.11(c). Thyristors conduct for 120° each and the load fundamental frequency is six times that of the a.c. supply. Similarly, two six-pulse bridge circuits can be operated, with suitable phase shifts, to give a twelve-pulse system, as in Figure 9.12. The thyristors again conduct for 120° each, the circuit waveforms being shown in Figure 9.13, which illustrates the reduction of the d.c. voltage and the a.c. current harmonics.

9.3 Unidirectional converters

Unidirectional converters are capable of passing power in one direction only, i.e. from the supply to the load. They can consist of a bi-directional circuit with the addition of a free-wheeling diode across the load, although often circuit modifications are made. Apart from circuit simplicity, unidirectional converters have other advantages, such as lower d.c. voltage ripple and a reduction of quadrature phase input current, which will be examined in this section with the help of typical circuits.

Figure 9.14 illustrates a half-wave circuit which has an additional diode D_1 connected across the load. The load waveforms given in Figure 9.14(b) can be contrasted with those obtained with the bi-directional circuit of Figure 9.1 and are seen to be identical on resistive loads. The operation of the circuit is as follows, assuming the load current is virtually ripple free. At t_{01} thyristor TH_1 is turned on and supplies power to the load. At t_1 the input voltage reverses, the inductive load tending to prevent any decay of load current and this now transfers from TH_1 to D_1. Therefore until TH_1 is refired at t_{21} the load current free-wheels in diode D_1, hence the term free-wheeling diode is applied to this device. If the voltage drop across D_1 is neglected the load voltage is zero during the free-wheeling period, so

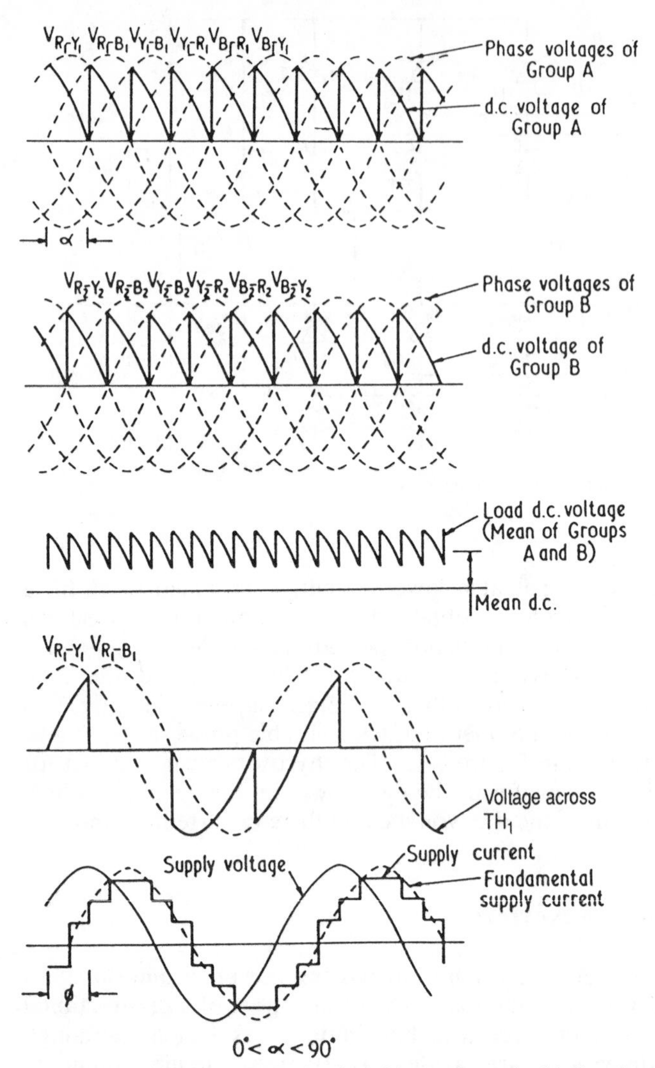

Figure 9.13 Circuit waveforms for the bridge twelve-pulse bi-directional converter of Figure 9.12

that the converter is now no longer capable of passing power back from the load to the supply.

As seen from the above discussion, it is possible to modify any bi-directional converter to unidirectional operation simply by connecting a free-wheeling diode across the load terminals. This is illustrated in Figure 9.15(a), which shows the modification applied to the circuit of Figure 9.4. Thyristors in diagonal arms of the bridge are fired in pairs, as before, at any delay angle α required, but whereas for a bi-directional converter these would have been maintained in conduction for a 180° period, until commutated by the firing of the opposite pair of thyristors, for

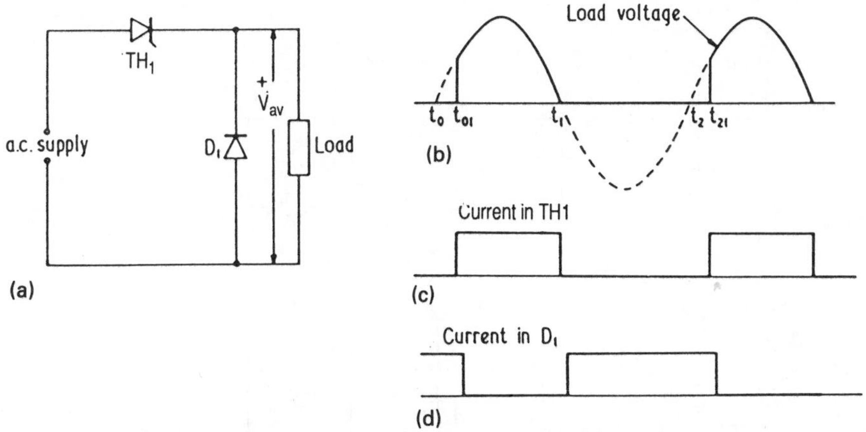

Figure 9.14 Half-wave unidirectional converter: (a) circuit; (b)–(d) waveforms

unidirectional circuits the thyristors turn off as soon as the supply voltage reverses, load current now being carried by the free-wheeling diode D_1.

Several modifications can be made to bi-directional circuits, apart from using free-wheeling diodes, to prevent them returning power to the load, which often simplifies the overall circuit arrangement. Figure 9.15(b) shows a system where half the bridge thyristors have been replaced by diodes, since diodes are cheaper than thyristors and do not need any associated gate drive circuitry, so that the overall cost of the converter has been reduced. Referring to the load voltage waveforms, the operation of the system can be explained as follows. At t_{01} thyristor TH_2 is fired, line A being positive to B so that load current flows via D_1 and TH_2. At t_1 the input voltages reverses, load inductance causing a free-wheeling current to flow in TH_2 and D_2 until time t_{11}, when TH_1 is turned on and current commutates to this thyristor. It is seen from this description that all devices carry current for 180°, irrespective of the firing angle delay α. This is convenient for calculating semiconductor ratings, but the system does have several limitations.

Suppose the converter is working with a low delay angle, thyristor TH_2 being fired at t_{01}, and it is desired to turn the bridge off. All thyristor gate pulses would be removed, causing the load current at t_1 to transfer to TH_2 and D_2. If the load inductance is large enough this current will not decay to zero during the half cycle interval, so that at t_2 thyristor TH_2 is still conducting. Since its anode voltage goes positive at this instance the thyristor will continue to conduct, with a zero delay angle, the load current flowing via TH_2 and D_1 for a complete half cycle. Depending on the load inductance, this state could be maintained continuously, the converter half-waving throughout, the only way to turn the bridge off being to refire the thyristors at appropriate instances and to increase α gradually, to reduce the load current. When this current reaches a value which is insufficient to keep the free-wheeling current on for 180° the thyristor gate pulses may be removed. This circuit limitation can be overcome by the addition of free-wheeling diode D_3 so that each thyristor is commutated at the end of a half cycle, when the diode conducts. If gate pulses are now

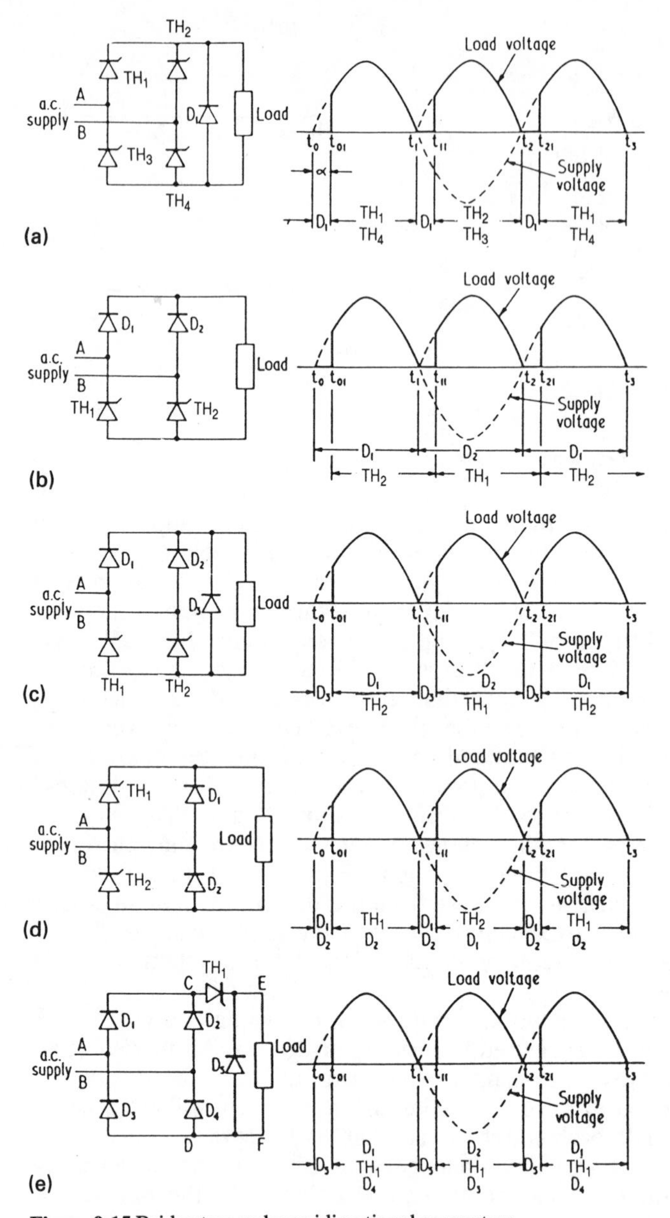

Figure 9.15 Bridge two-pulse unidirectional converters

removed, say from TH_2, then at time t_1 this thyristor will go off and, since none of the thyristors are refired, the load current will free-wheel in diode D_3 until it decays to zero.

An alternative system to Figure 9.15(c), which overcomes the half-waving effect and so gives a controller with a greater response speed, is shown in Figure 9.15(d), in which the number of devices have not been increased but they have been rearranged so that D_1 and D_2 give a free-wheeling path for load current. At t_{01} thyristor TH_1 is fired and load current flows via TH_1 and D_2 as before. At t_1 the current free-wheels through D_1 and D_2 so that thyristor TH_1 goes off, the system behaving as in Figure 9.15(c), where all conducting thyristors are commutated at the end of half cycles, except that now the conducting period for the bridge thyristors and diodes are $180° - \alpha$ and $180° + \alpha$ respectively, so that they could be unequally loaded depending on the load duty cycle.

Since thyristors are more expensive than diodes the converter given in Figure 9.15(e) sometimes proves economical. Free-wheeling diode D_5 is not necessary if the load is purely resistive. As seen, the system consists essentially of providing a fully rectified wave at CD and then regulating this with thyristor TH_1. Since the waveform across the thyristor therefore falls to zero only briefly every half cycle, a free-wheeling diode is essential, even for slightly inductive loads, to ensure successful device commutation. The operation of the circuit is readily followed by reference to the associated load voltage waveform.

Figure 9.16 shows the load voltage and current waveforms obtained from unidirectional circuits for various delay angles. Several features of these systems are evident from this figure and the above discussions:

(i) The load voltage has a lower ripple due to the absence of negative portions of the waveform.
(ii) The mean load voltage varies from a maximum to zero as the delay angle changes from 0° to 180°.
(iii) The power factor angle ϕ changes proportionally to the delay angle α, as for bi-directional converters.
(iv) Whereas for bi-directional converters the load current waveshape was unchanged as the delay angle varied, for unidirectional converters the load current period decreases with delay angle increase, so that at $\alpha = 90°$ the load current is unchanged in value from that at zero delay angle. However, since d.c. load voltage is zero there is now no net input power and all the a.c. current is quadrature component or wattless. For a unidirectional converter, on the other hand, the a.c. input current at very low d.c. output voltages is also very small, so that the quadrature component of the current has been reduced.
(v) As mentioned above, unidirectional converters are often cheaper than bi-directional ones.
(vi) There are no regenerative periods, so that a unidirectional converter cannot pass power from the d.c. to the a.c. side.

It is seen from items (i), (iv) and (v) that there is an advantage to be gained from using unidirectional converters in two-pulse systems, hence their popularity in applications which do not require regeneration. For systems with more than two pulses it will be seen later that the d.c. ripple frequency increases between bi-directional and unidirectional systems by a factor of two, so that, depending on the control range, when d.c. voltage filtering requirements are stringent bi-directional converters are sometimes preferred.

Bridge circuits can be converted from bi-directional to unidirectional operation by changing half the devices from thyristors to diodes, although the same rule does not apply to push–pull converters. The circuit of Figure

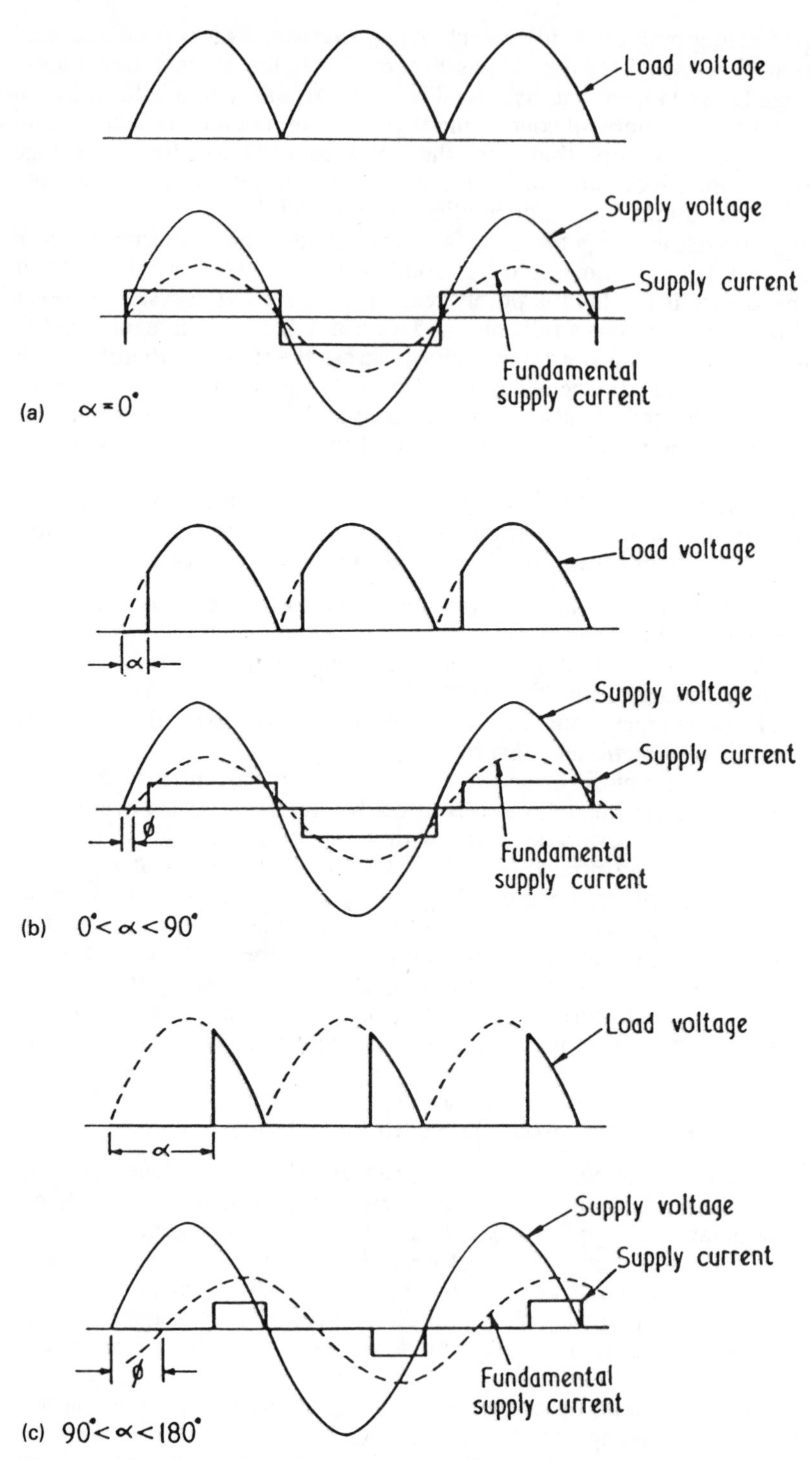

Figure 9.16 Circuit waveforms for a two-pulse unidirectional converter

9.2, for instance, will not operate correctly if one thyristor is replaced by a diode, all such systems requiring a free-wheeling diode for unidirectional operation, as shown in Figure 9.17.

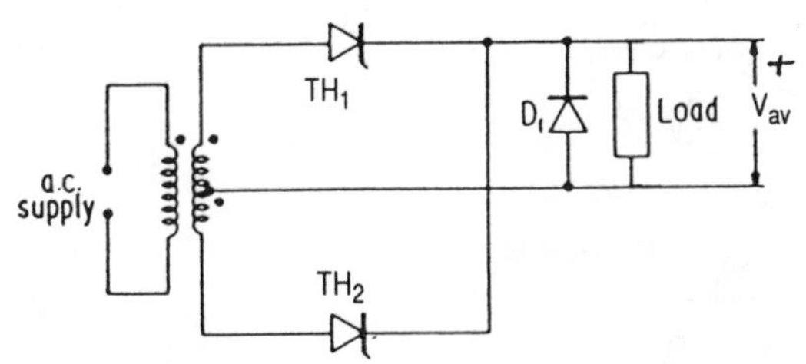

Figure 9.17 Push–pull two-pulse unidirectional converter

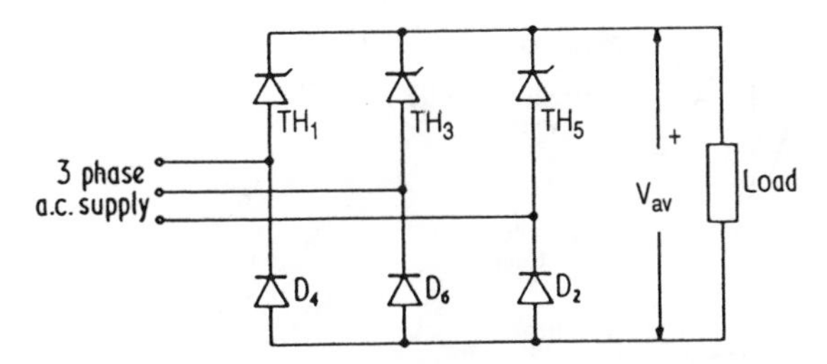

Figure 9.18 Bridge three-pulse unidirectional converter

Figure 9.18 shows a three-phase unidirectional converter in which half the thyristors of a bi-directional circuit have been replaced by diodes. It suffers from the same disadvantages as that of the single-phase circuit of Figure 9.15(b), but it is not possible to overcome these by a rearrangement of the components, as was done in Figure 9.15(d). Instead a free-wheeling diode must be used, as in Figure 9.15(c). The circuit waveforms for Figure 9.18 are shown in Figure 9.19. The top half of the bridge is controlled so that conduction will occur from the most positive phase in which a thyristor has been fired, the current in the bottom of the bridge commutating naturally from one diode to the next, depending on which is connected to the most negative phase. In Figure 9.19 for $\alpha = 0$ the output is a six-pulse waveform such as that obtained for Figure 9.11, but as the delay angle increases, the waveform changes to three-pulse, so that in Figure 9.19(b) the fundamental voltage ripple is three times that of the input a.c. At this stage there are as yet no free-wheeling current periods. The a.c. current has two durations of 120° whose phase relation to each other and to the supply voltage changes with α. Beyond the 60° delay point free-wheeling times increase, these leading to periods of zero voltage in the load waveform and a reduction of positive and negative portions of the a.c. current. In effect, the a.c. supply current waveform may be considered as being made up of a positive (thyristor) part and a negative (diode) part and as the delay angles increase the diode block remains stationary, since it is uncontrolled, whilst the thyristor block moves towards it. Beyond 60° delay

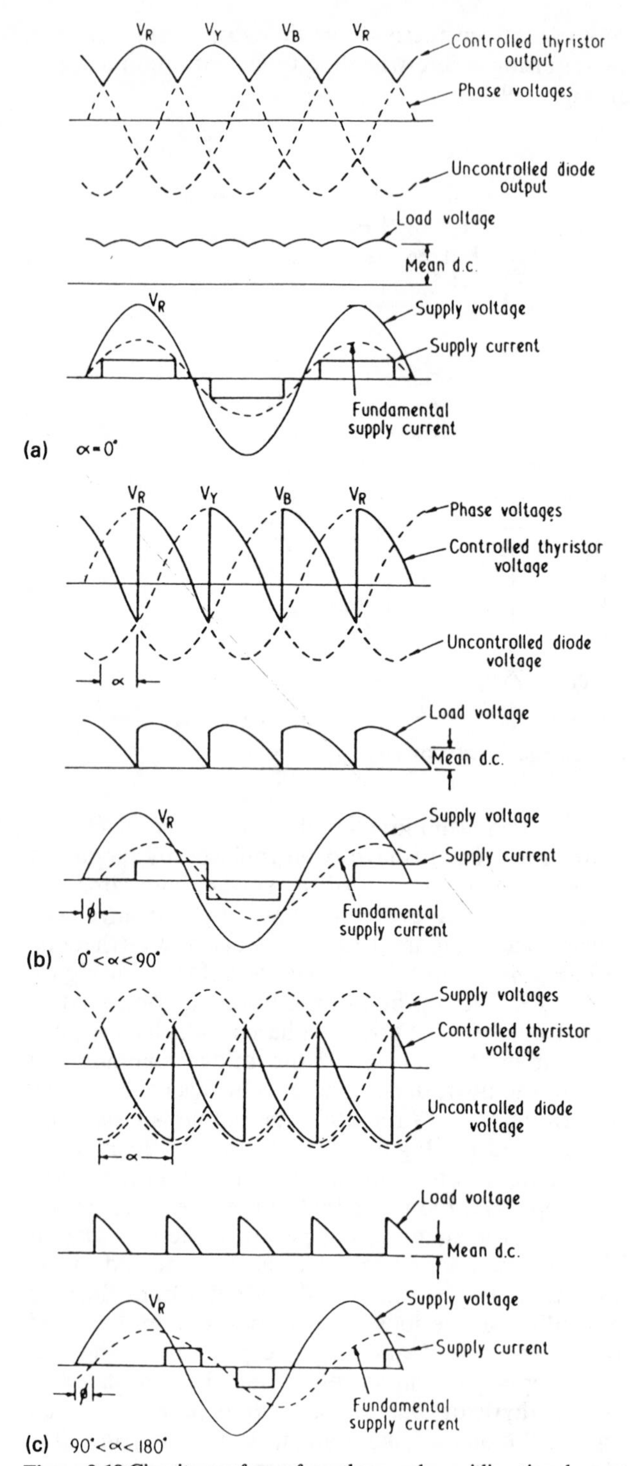

Figure 9.19 Circuit waveforms for a three-pulse unidirectional converter

the two overlap to an increasing extent, resulting in an increase in the free-wheeling period.

It has been stated above that an unidirectional wave has a lower d.c. voltage content, owing to the absence of negative portions of the waveform. However, for high pulse numbers it is seen above that there can be a reduction of ripple frequency by a factor of 2 from that of bi-directional converters. There is now obviously a compromise situation where, depending on the range of control voltage required, one converter would give lower ripple output than another. Ripple content from converter circuits are considered again in following sections.

9.4 Discontinuous load current

In the previous sections it has been assumed that, apart from resistive loads, the load current has been continuous during a cycle of operation, and if this is not the case then the load voltage waveform will be modified from those illustrated earlier. Generally, these systems are not easy to analyse since the output conditions are dependent on the load.

Figure 9.20 shows the waveforms from a bi-directional converter for various delay angles, where the load Q factor is finite. For delay α_1 the load voltage just dips to zero, although the load inductance maintains the current continuous. At α_2 the current is still continuous, so that the load voltage follows the contour of the a.c. input voltage and swings negative over certain portions. For larger delays of α_3 the load inductance is insufficient to maintain a prolonged regenerative period and the current decays to zero, and for a passive load its voltage will now be zero, as shown. Increasing the delay angle further reduces the d.c. voltage, but it

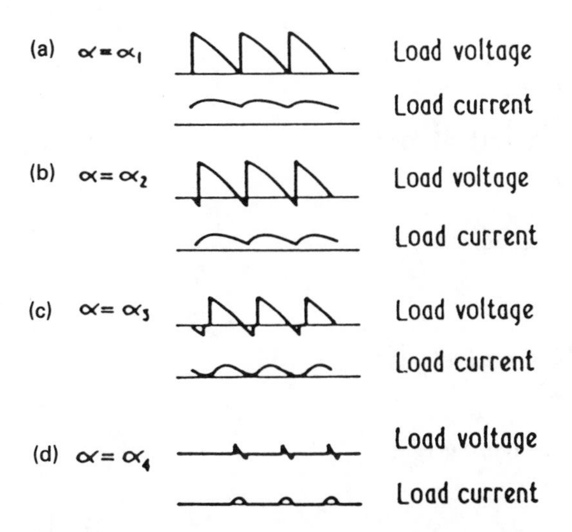

Figure 9.20 Load waveforms for a six-pulse bi-directional converter operating with a finite load inductance and possible discontinuous load current

also reduces the peak load current so that the regeneration period is shortened.

In Figure 9.20 the load has been assumed to be passive, whereas if it contained a d.c. source, such as a motor back e.m.f., the voltage during zero load current periods would rise to the value of this e.m.f. This is shown in Figure 9.21, which illustrates the effect of maintaining the converter firing angle fixed but of changing the load current, as would normally happen in a d.c. motor under variable torque conditions. The shape of the load waveform is seen to change. During continuous load current periods it follows that of the a.c. supply, but when the load current becomes discontinuous it rises to the value of the load back e.m.f. Since the mean voltage of the load varies with its waveform the effective load voltage has been changed although the firing angle has not. This is highly undesirable in many applications and illustrates the advisability of introducing external load inductances so as to maintain continuous load

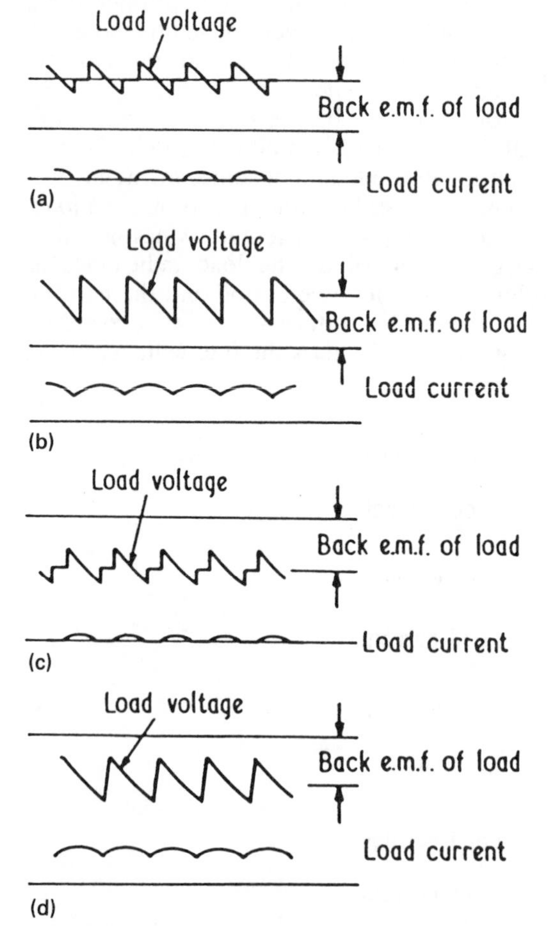

Figure 9.21 Load waveforms for a six-pulse converter operating with a back e.m.f. load: (a) and (c) light loads; (b) and (d) heavy loads

current. Figure 9.21(c) shows the effect of discontinuous load current during the inverting mode of the converter, assuming that the load has an internal back e.m.f. which provides the inverter voltage.

In order to ensure that the load current never becomes discontinuous it is possible to analyse the converter circuit to determine the minimum load inductance required under various firing angles, such an analysis depending on the value of load back e.m.f. and being involved, although it has been done for passive loads. Figure 9.22 shows the load voltage and

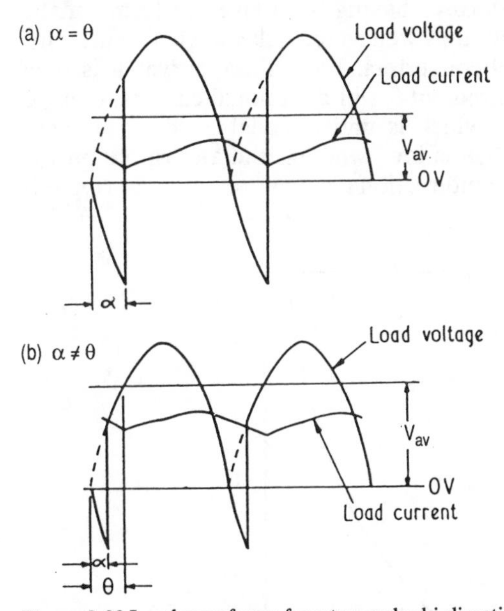

Figure 9.22 Load waveforms for a two-pulse bi-directional converter

current waveforms for a single-phase bi-directional converter, the load current reaching a minimum value at the intersection points of the instantaneous and mean d.c. waveforms. It can be shown that the value of critical inductance L_c required to prevent this current from falling below zero is related to the circuit resistance R and angular frequency ω of the supply by

$$\frac{\omega L_c}{R} = \frac{\pi}{2\cos\alpha}\left[\cos\theta + \frac{2}{\pi}\sin\alpha - \frac{2}{\pi}\cos\alpha\left\{\frac{\pi}{2} + \alpha - \theta\right\}\right] \quad (9.1)$$

For large firing angles, $\alpha = \theta$ so that equation (9.1) reduces to

$$\frac{\omega L_c}{R} = \tan\alpha \quad (9.2)$$

For unidirectional converters the analysis is complicated further due to zero voltage periods in the load waveform, and is found to be given by

$$\frac{\omega L_c}{R} = \theta - \alpha - \frac{\pi}{2} + \frac{\alpha + \sin\alpha + \pi\cos\theta}{1 + \cos\alpha} \quad (9.3)$$

As before, there are two operating conditions according to whether α is smaller or larger than a critical value (35.5°), which gives $\alpha = \theta$. In this case equation (9.3) reduces to

$$\frac{\omega L_c}{R} = -\frac{\pi}{2} + \frac{\alpha + \sin\alpha + \pi\cos\alpha}{1 + \cos\alpha} \tag{9.4}$$

Similar analysis can be made for three-phase converters.

Figures 9.23 and 9.24 show the plot of L_c/R for single- and three-phase converters, the abscissae in both cases having been given in terms of the percentage of maximum d.c. output as well as the delay angle required for unidirectional and bi-directional converters. From these curves it is seen that the critical inductance requirement for bi-directional circuits tends to infinity at low output voltages, which is understandable, since at these delay angles the mean d.c. voltage is low, whereas the a.c. ripple on the voltage is at its peak value. For unidirectional converters the a.c. ripple is

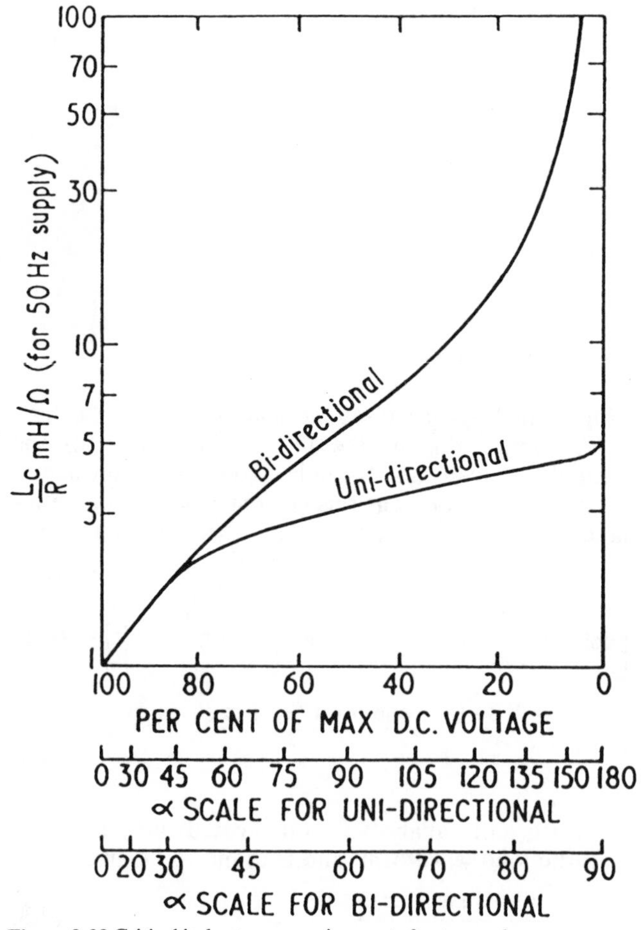

Figure 9.23 Critical inductance requirements for two-pulse converters

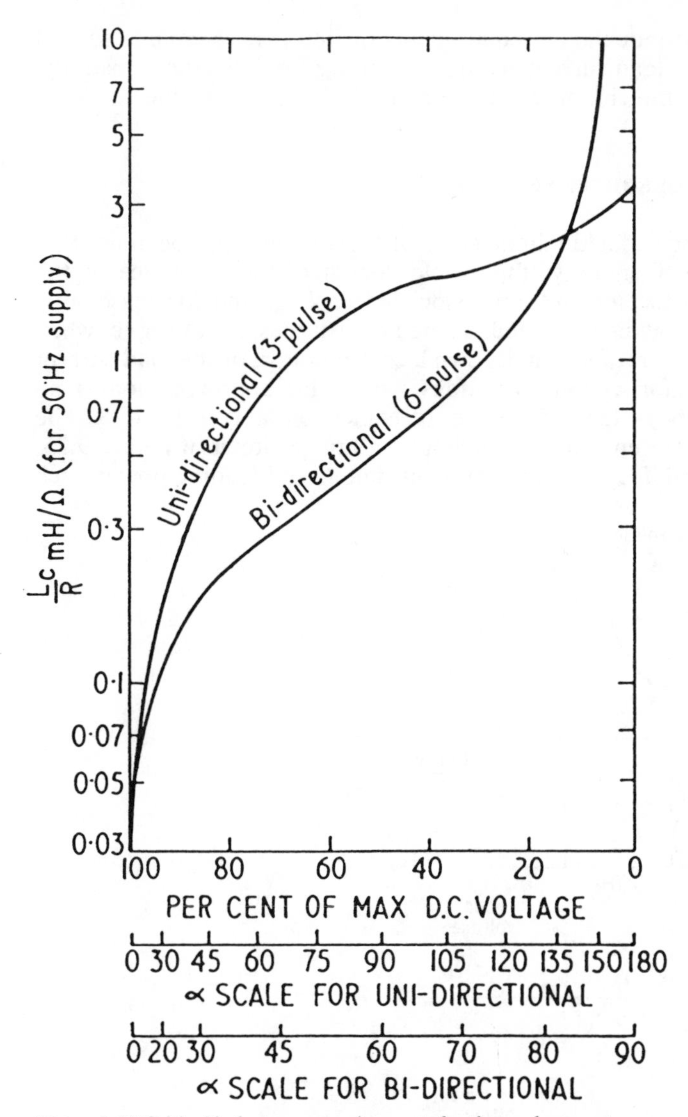

Figure 9.24 Critical inductance requirements for three-phase systems

also very small at delay angles approaching 180°, so that the critical inductance is a finite value. Figure 9.23 clearly shows the advantage of unidirectional converters in requiring smaller load inductance, but Figure 9.24 illustrates that the unidirectional converter has a higher critical inductance requirement for three-phase circuits except at low load voltages. This is due to the different waveforms obtained with multiple systems. As has been seen previously, a unidirectional circuit gives half the pulse numbers compared to the same bi-directional system, so that the choice between the two types of converters would now depend on the output voltage range required. Comparison of Figures 9.23 and 9.24 shows the less stringent inductance requirements of three-phase circuits so that where continuous load current is required, from a low inductance load, the pulse number of the circuit should be made as large as possible.

9.5 The effect of source reactance

In the discussions so far the impedance of the a.c. lines has been omitted for convenience. Generally, this would consist of the reactance of the distribution transformer and any associated cabling, and for reasonably 'stiff' supplies it may be neglected. There are instances, for example, when the converter is operating from its own local generator, or the load current approaches its short-circuit conditions, when this approximation is no longer valid. The effect of source reactance on a converter can be explained with reference to the single-phase bridge circuit of Figure 9.25, in which TH_2 and TH_3 are conducting at time t_0 and feeding power back

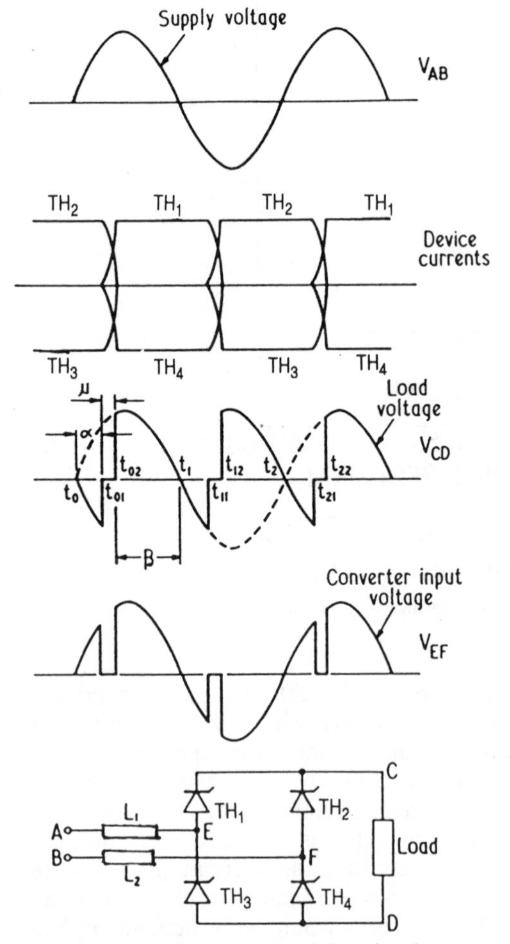

Figure 9.25 Bridge two-pulse bi-directional converter with a.c. source impedance

from the supply to the load. Inductors L_1 and L_2 are line impedances. At time t_{01}, after a delay α, thyristors TH_1 and TH_4 are fired and in the absence of line inductors these thyristors would instantaneously carry the full-load current whilst TH_2 and TH_3 would turn off. This, however, requires the reversal of current flow in any line inductors, which cannot occur instantaneously, so that for a period μ all bridge thyristors will conduct. The instantaneous supply voltage appears across the line inductors in such a direction as to help build up the current in TH_1 and in TH_4, and to decay it in TH_2 and TH_3. The overlap period, for any given load current, is minimum when this voltage is at its peak value, i.e. for α close to 90°. During overlap the load current free-wheels through the four thyristors so that load voltage is zero. Another way of expressing the same thing is to consider the load voltage to be the instantaneous mean of the overlapping phases, which for a single-phase supply is always zero. With opposite arms of the bridge conducting the voltage across EF is that of the input AB, and during overlap this falls to zero.

Comparing the waveforms with and without source reactance, the effect of this reactance can be considered as being fourfold:

(i) The mean d.c. voltage is reduced for any delay angle. This is so since a portion of duration μ has now been removed from the output waveform.

(ii) The harmonics of the output load have been changed since its waveform has been modified. Since overlap does not affect the output pulse number, the spectrum of frequencies present is not altered. It can be shown that, for the same mean d.c. load voltage, the amplitude of the harmonics is lower with overlap in the waveform than when overlap is not present. This is primarily due to the fact that with overlap the firing angle must be advanced (α reduced) to maintain an unchanged d.c. voltage.

(iii) The input voltage to the converter is no longer sinusoidal (as at EF) but can be considerably distorted. This is important, since the timing for control circuits is usually derived from this wave and, unless it is allowed for, faulty operation could result.

(iv) Overlap reduces the safety angle β, where $\beta = 180° - \alpha - \mu$. This is not important for rectifier operation, but it must be borne in mind that β is also the time that is available to a conducting thyristor for turn-off. Since β reduces as α increases, commutation failure could occur at large delay angles. Figure 9.26 shows the circuit waveforms during inversion, the delay angle being increased so that TH_1 and TH_4 are fired at almost the end of the half cycle, at t_{01}. Since the instantaneous supply voltage is low, the overlap angle μ is relatively long. When it is over at t_{02} thyristors TH_2 and TH_3 go off, but they are reverse biased for duration β only and they must turn off during this time. If this does not occur, then at t_1 the devices turn on and provide a full half cycle of power from the supply to the load. It will be seen later that μ varies with load current, so that the maximum value of α chosen must make allowances for this. Circuits operating from severe line impedance sources, therefore, are often limited in voltage control range.

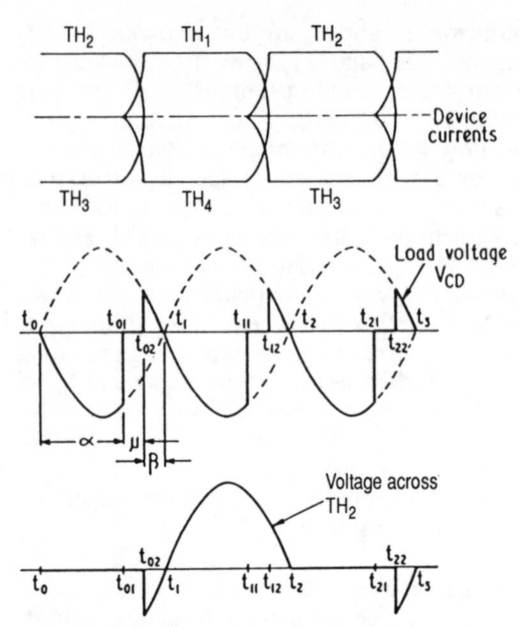

Figure 9.26 Two-pulse bi-directional converter with a.c. source impedance, operating in an inversion mode

Overlap effects are not restricted to bi-directional converters. For instance, Figure 9.27 shows the load voltage and device currents for the circuit of Figure 9.15(a), assuming that the load current is ripple free and the input lines have series reactances. Prior to t_0 thyristors TH_2 and TH_3 were conducting. At t_0 the supply voltage reverses and the load current begins to transfer from the thyristors to the free-wheeling diode D_1. Line reactances prevent this occurring instantaneously, although the load voltage is zero since D_1 is conducting. At t_{01} current has transferred completely to D_1 and TH_2 and TH_3 go off. When TH_1 and TH_4 are fired at t_{02} there is again an overlap angle μ due to the finite time for the current to transfer from D_1 to the thyristors before the load voltage rises to that of the supply, so that the effect of overlap has once again been to remove a portion of duration μ from the load voltage waveform during every half cycle. Since the instantaneous line voltages at the overlap periods are different the two overlap angles μ_1 and μ_2 will also differ, with $\mu_1 > \mu_2$.

Figure 9.27 has illustrated a case where $\alpha > \mu_1$. The situation is complicated when the reverse is true, since now the current in D_1 never reaches the full d.c. value, because before this can happen the opposite bridge thyristor is fired and the current will be caused to decay in D_1. The zero period in the load voltage waveform is equal to the sum of the two overlap periods, during which partial current transfers from thyristor to diode and diode to thyristor, in parallel with thyristor to thyristor.

Overlap in the converter has the effect of increasing the rise and fall times of device currents, this being reflected to the a.c. input current, as shown in Figure 9.28. The fundamental component of the a.c. current is now also seen to be shifted by a lagging angle ϕ_1 from the case for no

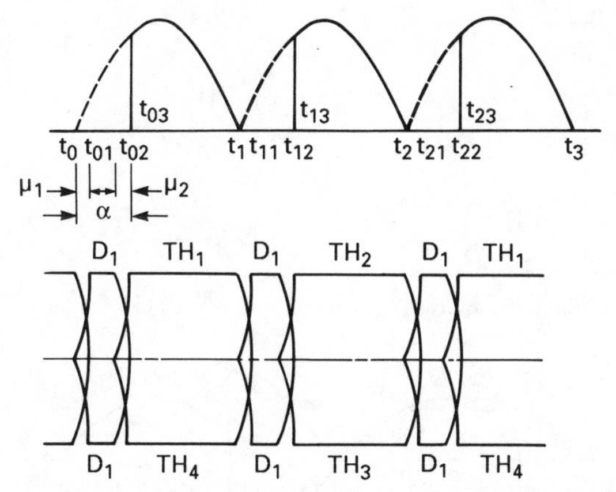

Figure 9.27 Circuit waveforms for the two-pulse unidirectional converter of Figure 9.15(a) operating with a.c. source reactance

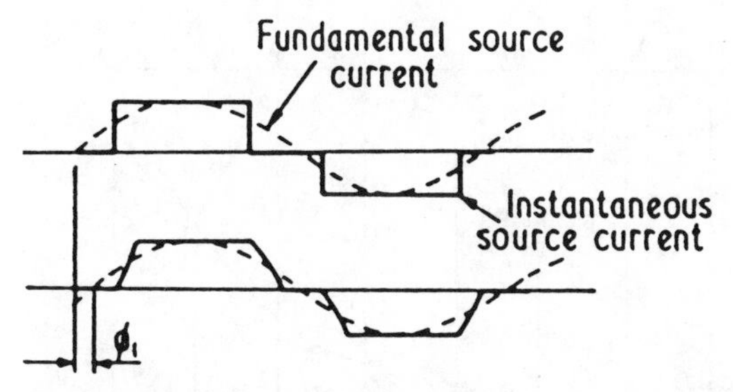

Figure 9.28 Effect of source reactance on the a.c. input current

overlap, so that the power input to the converter in reduced. This is to be expected, since overlap has reduced the mean d.c. voltage, and therefore the d.c. power output, for constant load current. It can also be explained as being due to the introduction of an inductive impedance in the a.c. lines, which would increase the power factor angle.

During the overlap period the two lines between which current is being transferred assume the same potential, which is the mean value of their instantaneous voltages. For a single-phase bridge this has been shown to be zero, but it is not so for three-phase converters. The action of overlap can be illustrated for a three-phase circuit with reference to the diode bridge shown in Figure 9.29, with Figure 9.30 showing its operation for overlap angles of 30°. The diodes will conduct as soon as their anode voltages go positive, i.e. at t_0, t_1, t_2, etc., since they do not possess gate control. Prior to time t_0 diodes D_5 and D_6 were conducting. At t_0 the red phase becomes the most positive one and the load current would normally flow from this to the most negative phase, i.e. the yellow phase, so the load current should commutate instantaneously from D_5 to D_1. However, due to line inductances, there is a finite commutation time during which both D_5 and D_1 pass current to D_6. This overlap period lasts, depending on load current, for up to, say, t_{01}. Diode D_5 now goes off and diodes D_1 and D_6 carry the load current. During overlap of the red and blue phases the top

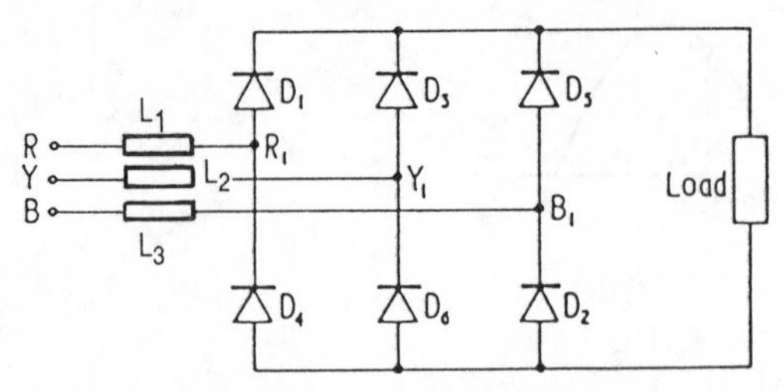

Figure 9.29 Three-phase bridge with source reactance

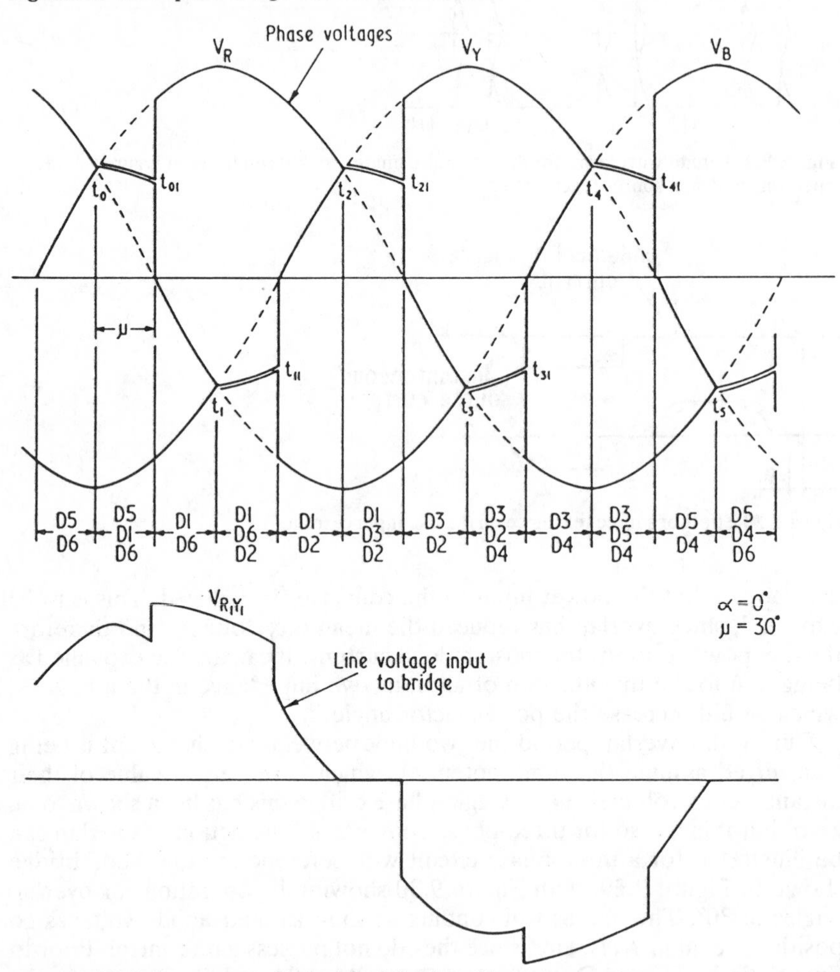

Figure 9.30 Mode-one operation in a diode bridge with source reactance

junction of the load is maintained at a potential which was an instantaneous mean between that of the two phases. The load voltage is the difference between this mean phase voltage and that of the yellow phase, which is itself the potential of the bottom junction of the load. The load voltage can therefore be found in Figure 9.30 as the difference between the heavy lined waveforms.

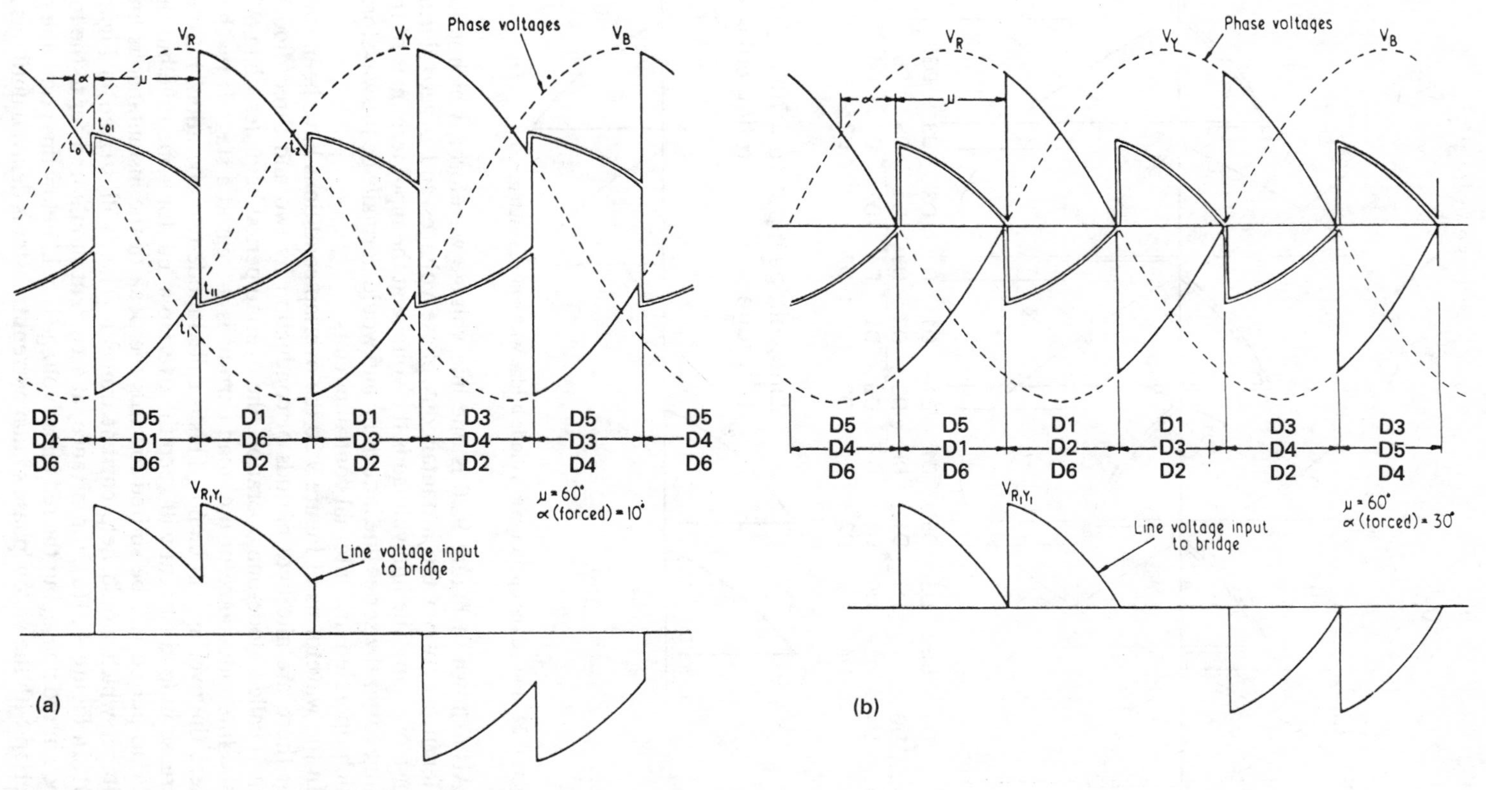

Figure 9.31 Mode-two operation in a diode bridge with source reactance

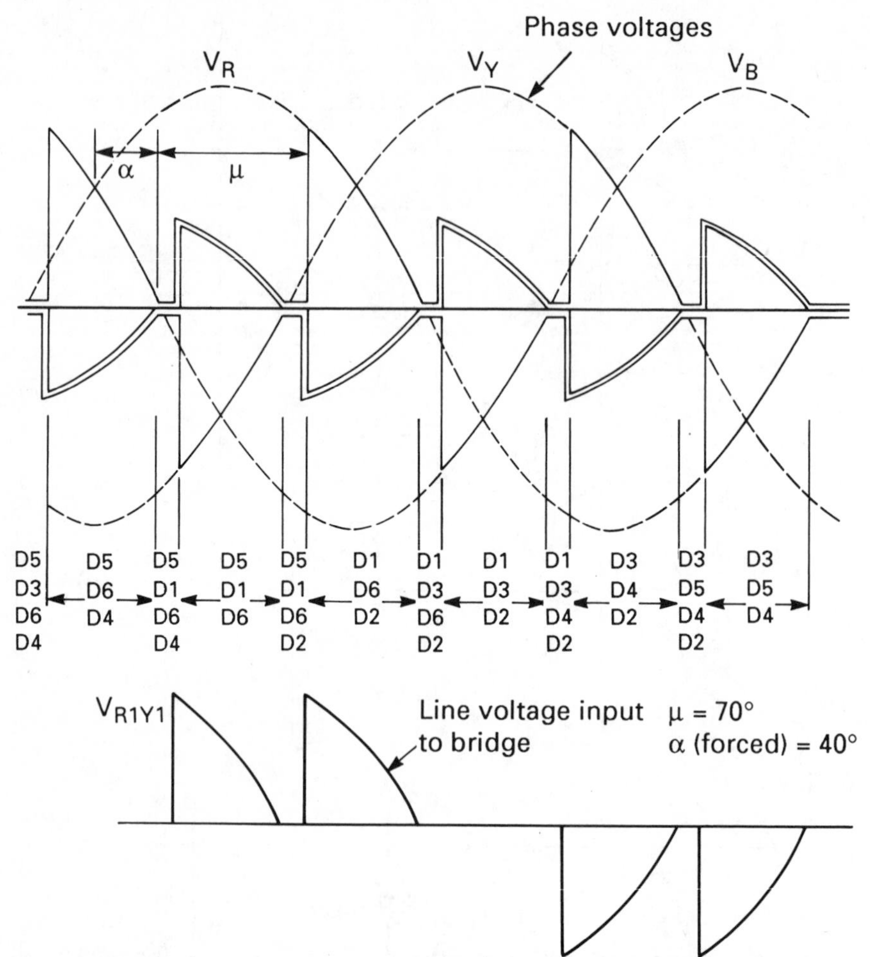

Figure 9.32 Mode-three operation in a diode bridge with source reactance

Also shown in Figure 9.30 is one line voltage waveform, which is the difference between the instantaneous phase voltages and is zero during phase overlap. The line voltage is the voltage at the input terminals of the bridge, after the series line inductors, and would normally be the waveform which operated any internal control circuits.

In the waveforms of Figure 9.30 the overlap conditions have been such that there are alternate periods of conduction by two and three diodes. This is called mode-one operation, the overlap period being less than 60°. If the line inductance or d.c. load current is increased a stage is reached where the overlap will be 60°. This is called mode-two operation. Further increase in load current will keep the overlap at 60° for some time but the overlap period will be shifted towards the peak of the instantaneous line voltage, which would help current transfer. This is illustrated in Figure 9.31. In Figure 9.31(a), for instance, at t_0 current should transfer to the top diode in red phase, but the red phase voltage is held below this point due to overlap with the yellow phase so that current transfer is delayed until this is completed at t_{01}. As soon as the red phase is released by the yellow phase it goes into overlap with the blue phase, as current transfers from D_1 to D_5 and flows to D_6. At t_1 transfer should occur between D_6 and D_2, but since the blue phase voltage is held above that of the yellow phase, owing to its overlap with the red phase, this transfer is delayed until t_{11}. This waveform shows that there has been an artificially created delay angle of about 10° in the changeover point between phases. Similarly, in Figure 9.31(b) the increased load current still gives the same value of 60° overlap, but increases the delay to 30°, which represents the boundary condition between mode-two and mode-three operation. It should be noted that, throughout mode two, three diodes are conducting at any one time.

Figure 9.32 shows mode-three operation, with overlap angles exceeding 60°. The two overlap periods of the top and bottom halves of the bridge now run into each other so that four devices begin to conduct simultaneously over certain parts of the waveform. The load voltage is now

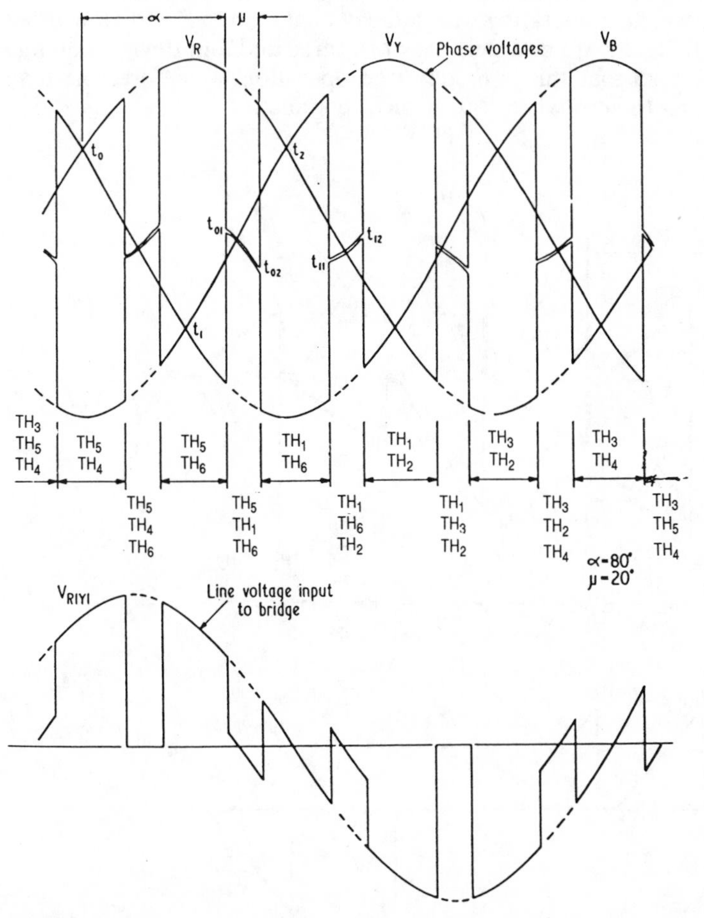

Figure 9.33 Mode-one operation in a thyristor bi-directional bridge with all fully controlled arms and source reactance

zero since the load current free-wheels through devices in the top and bottom half of the bridge, the effect on the input line voltage being seen to be such as to introduce further instances of zero voltage.

Overlap in thyristor converters can result in three operating modes as well, but the occurrence of mode two, which is so important in diode circuits, is rare and is often considered as a special case of mode one. Figure 9.33 gives the waveforms for the bi-directional bridge shown in Figure 9.11, the delay angle α being taken as 80° and the overlap angle μ as 20°. At time t_0 the red phase becomes the most positive line, but the firing of thyristor TH_1 is delayed to time t_{01}. When this is done current transfer occurs between TH_5 and TH_1, this overlap terminating at t_{02} and thyristor TH_5 going off. Similarly, the commutation between the negative half of the yellow and blue phases is delayed from t_1 to t_{11} and so on. The load voltage is given by the difference between the two instantaneous waveforms, as before, and is now seen to have negative (regenerative) periods, the line voltage into the converters also being distorted. As a further example, Figure 9.34 gives the voltage waveforms for the unidirectional converter of Figure 9.18, which is operating with a delay angle α of 30° and an overlap angle μ of 70°. There are now periods when three and four devices conduct simultaneously so that this is mode-three operation, as before, the load voltage dipping to zero when four devices conduct.

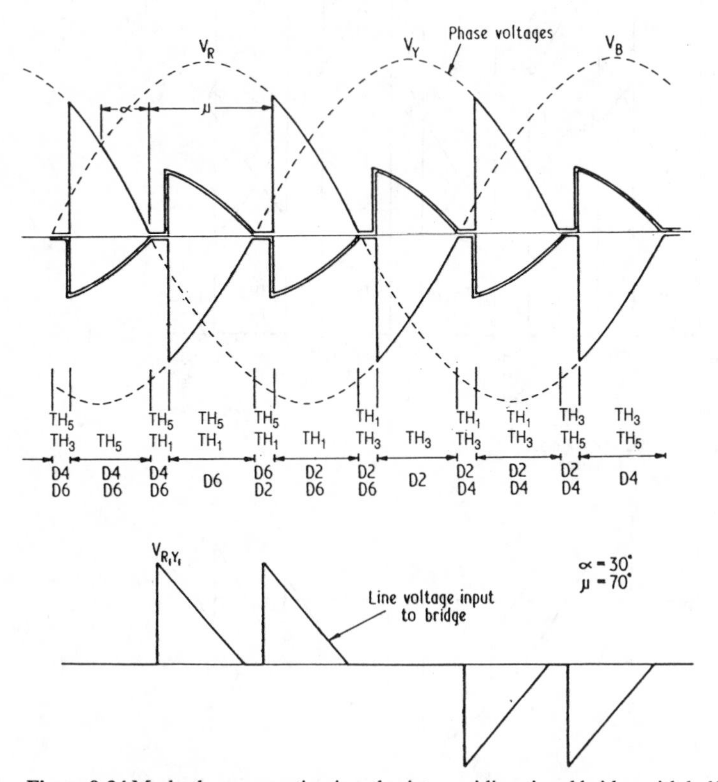

Figure 9.34 Mode-three operation in a thyristor unidirectional bridge with half the arms controlled and source reactance

9.6 Performance factors

The performance of converter circuits can be compared by means of several factors, the commonest in use being the d.c. voltage ratio, the input displacement factor, the input power factor and the input current distortion factor. The d.c. voltage ratio gives the relation between the converter output at any delay angle α and that obtained when the same converter is operating free of delay angles or overlap. The input displacement factor is the cosine of the angle between the fundamental component of a.c. current into the bridge and the supply voltage. It has been shown, when comparing the various circuits, that this angle ϕ increases (lagging) as the phase-control angle α or overlap angle μ increase. For bi-directional converters $\phi = \alpha$. It is not proposed to consider this further here.

The input power factor is the ratio of the total mean input power to the total r.m.s. input volt amperes, and the input current distortion factor is the ratio of the r.m.s. value of the fundamental to the total input current, both these factors defining the relationship between the in-phase and quadrature-phase components of a.c. current. This has been discussed in previous sections and it was shown then that unidirectional converter circuits often have the advantage over bi-directional converters in that they have a lower quadrature, or wattless, current input.

In this section the value of the mean d.c. voltage from a converter will be found and the harmonic current of this waveform discussed.

The output voltage waveform from a converter with zero delay angle or phase overlap is composed of a series of cosine curves which are linked together as in Figure 9.35(a). The duration of each section is equal to $2\pi/p$, where p is the pulse number of the converter. The average value V_{av} of this waveform is given by

$$V_{av} = \frac{p}{2\pi} \int_{-\pi/p}^{+\pi/p} \surd(2)\, V \cos\theta \, d\theta$$

$$= \surd(2)\, V \frac{p}{\pi} \sin\frac{\pi}{p} \tag{9.5}$$

where V is the r.m.s. input voltage.

If a delay angle α is now introduced into the converter, as in Figure 9.35(b), then, since the waveform still follows a cosine wave over its control range, the average output voltage $V_{av(\alpha)}$ is given by

$$V_{av(\alpha)} = \frac{p}{2\pi} \int_{-\pi/p + \alpha}^{+\pi/p + \alpha} \surd(2)\, V \cos\theta \, d\theta$$

$$= \surd(2)\, V \frac{p}{\pi} \sin\frac{\pi}{p} \cos\alpha$$

$$= V_{av} \cos\alpha \tag{9.6}$$

and therefore the d.c. voltage ratio, given by $V_{av(\alpha)}/V_{av}$, is equal to $\cos\alpha$.

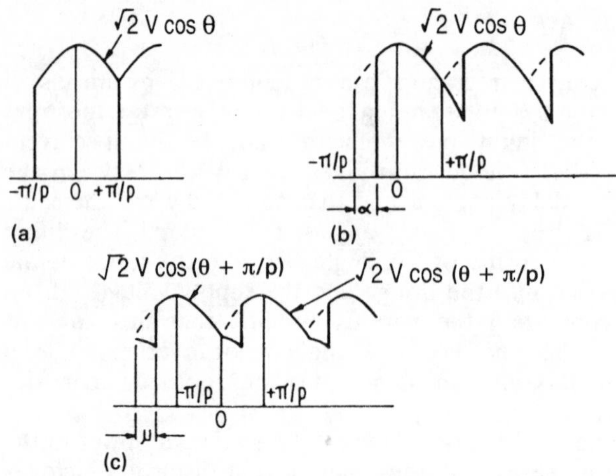

Figure 9.35 Waveforms used in calculating load d.c. voltage for a *p*-pulse converter

To consider the effect of the overlap angle, suppose that the delay angle is zero, as in Figure 9.35(c), so that the datum for the overlapping waveforms has now been moved to the commencement point for overlap, the equations for the two waveforms being given by

$$V_1 = \sqrt{(2)}\, V \cos\left(\theta + \frac{\pi}{p}\right) \tag{9.7}$$

$$V_2 = \sqrt{(2)}\, V \cos\left(\theta - \frac{\pi}{p}\right) \tag{9.8}$$

During the overlap period the load voltage assumes a value intermediate between V_1 and V_2 so that the d.c. voltage is given by

$$V_{\mathrm{av}(\alpha)} = \frac{p}{2\pi}\left[\int_0^{\mu} \frac{V_1 + V_2}{2}\, d\theta + \int_{\mu}^{2\pi/p} V_2\, d\theta\right]$$

$$= \sqrt{(2)}\, V \frac{p}{\pi} \sin\frac{\pi}{p}\,(1 + \cos\mu)$$

$$= \sqrt{(2)}\, V \frac{p}{\pi} \sin\frac{\pi}{p} \cos^2\frac{\mu}{2}$$

$$= V_{\mathrm{av}} \cos^2\frac{\mu}{2} \tag{9.9}$$

Therefore the effect of overlap is similar to that of firing angle delay in that the mean output voltage is reduced, although the extent of variation in the two cases is different. For circuits operating with delay α and overlap angle μ, the value of mean d.c. voltage is given by

$$V_{\mathrm{av}(\alpha)} = V_{\mathrm{av}} \cos\alpha \cos^2\frac{\mu}{2} \tag{9.10}$$

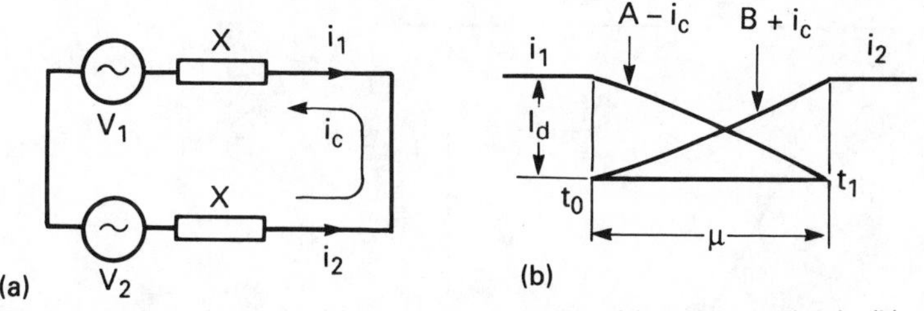

Figure 9.36 Phase currents during converter overlap: (a) equivalent circuit; (b) current waveforms

In order to estimate the overlap angle μ it is necessary to know the value of d.c. current I_d and reactance X in each phase. Figure 9.36 shows the equivalent circuit at the moment of overlap between two phases, V_1 and V_2, where the current in phase 1 is decaying whilst that in phase 2 is increasing, as it takes over the conduction of the load current. The effect of this change can be explained as being due to a circulating current i_c which is caused by the instantaneous phase difference $(V_1 - V_2)/2$ so that the value of i_c is given by

$$
\begin{aligned}
i_c &= \frac{V_2 - V_1}{2X} \\
&= -\frac{\surd(2)\, V \sin \pi/p}{X} \cos\theta \\
&= -I_c \cos\theta
\end{aligned}
\tag{9.11}
$$

where I_c is the peak value of the circulating current.

The current delivered by the phases during overlap can be considered to be composed of a constant part and a variable circulating part, as shown in Figure 9.36(b), and when these two components of current are equal the overlap period is terminated. Considering the instant t_0 at the commencement of overlap, the value of i_1 is given by equation (9.12), and since at t_0 the value of $\theta = 0$ and $i_c = -I_c$, the value of A is given by equation (9.13). Also at this point i_2 is given by equation (9.14), so that $I_c = B$.

$$i_1 = I_d = A - i_c = A + I_c \tag{9.12}$$

$$A = I_d - I_c \tag{9.13}$$

$$i_2 = 0 = B - I_c \tag{9.14}$$

At time t_1 the value of $\theta = \mu$ and i_2 is given by equation (9.15), so that cos μ is given by equation (9.16), and substituting for I_c leads to equation (9.17), where X is the equivalent reactance per phase in the a.c. lines and V is the r.m.s. voltage per phase of the input:

$$i_2 = I_d = B + i_c = I_c(1 - \cos\mu) \tag{9.15}$$

$$\cos\mu = 1 - \frac{I_d}{I_c} \tag{9.16}$$

$$\cos\mu = 1 - \frac{I_d X}{\surd(2)\, V \sin \pi/p} \tag{9.17}$$

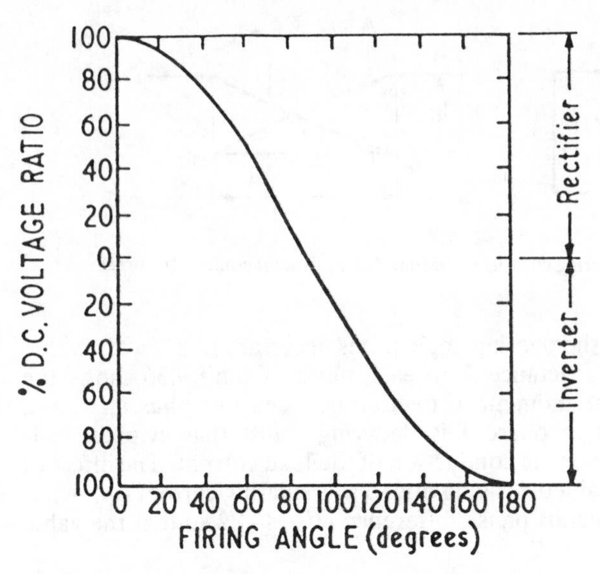

Figure 9.37 Variation of d.c. voltage ratio with firing angle for a bi-directional converter

Equation (9.6) for d.c. voltage ratio is a cosine curve for all bi-directional converters and is shown plotted in Figure 9.37. Unidirectional converters have a different characteristic due to the discontinuity in the cosine curve and the introduction of zero voltage periods during free-wheeling load current. The equation for the d.c. voltage ratio now depends on the circuit used, since the incidence of this point will change. It can be shown for half-controlled converters, in which half the thyristors in the converter are replaced by diodes, that the mean d.c. voltage is given by

$$V_{av(\alpha)} = V_{av}(1 + \cos\alpha) \qquad (9.18)$$

so that the d.c. voltage ratio is $(1+\cos\alpha)$, which is shown plotted in Figure 9.38. Also illustrated in this figure are the ratios for various other systems which use free-wheeling diodes in order to give unidirectional converters.

The harmonics present in the d.c. voltage from the converter have been shown to be related to its pulse number, for a bi-directional system the ratio of the amplitude of the *n*th harmonic to the peak d.c. voltage being given by

$$\frac{V_n}{V_{av}} = \left[\frac{1}{(n-1)^2} + \frac{1}{(n+1)^2} - \frac{2\cos 2\alpha}{(n+1)(n-1)}\right]^{1/2} \qquad (9.19)$$

This is shown plotted in Figure 9.39 up to the 24th harmonic. All the harmonics are not present in every system and generally for a p pulse converter harmonic numbers present are kp, where k is an integer 1, 2, 3, etc. The harmonic plot shows the peaking at 90° firing angles when the mean d.c. voltage is zero, all the output being a.c. ripple.

For unidirectional converters the situation is again complicated due to the variation of the instance of free-wheeling current with the type of

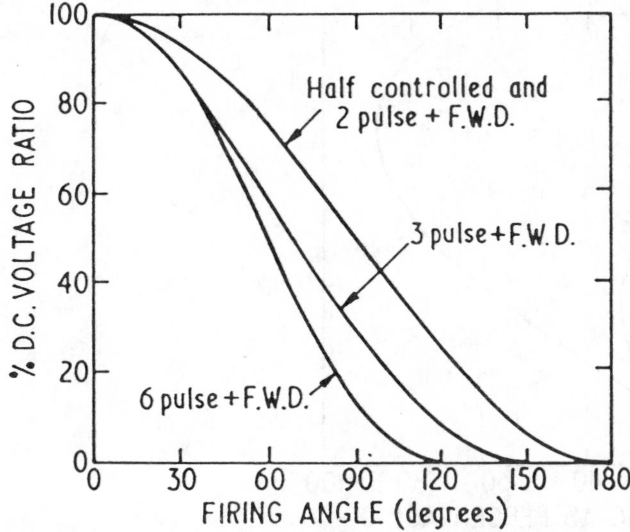

Figure 9.38 Variations of d.c. voltage ratio with firing angle for various unidirectional converters

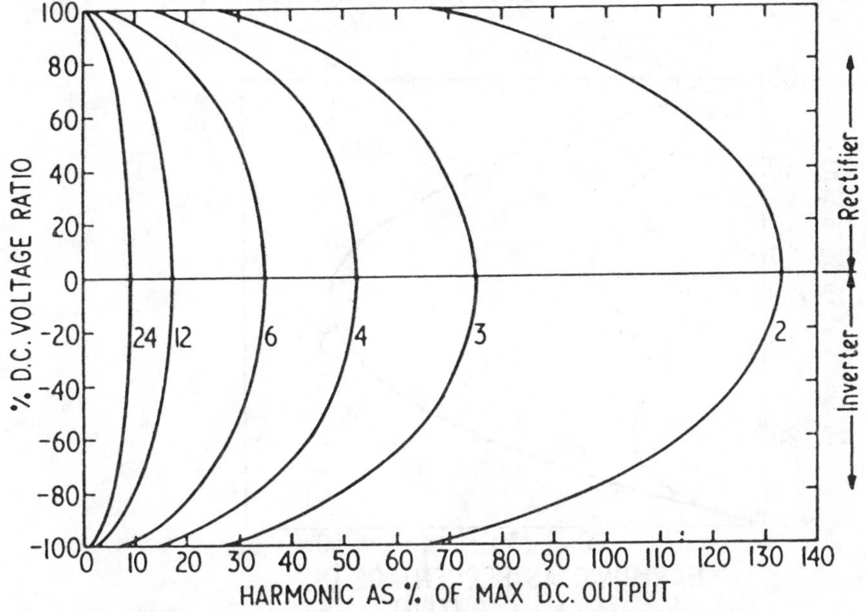

Figure 9.39 Predominant harmonics in the output of bi-directional converters

circuit used. Figures 9.40 and 9.41 give the lower harmonic content for single- and three-phase half-controlled converters, and, comparing this with the bi-directional converter, it is seen that harmonics of the same number are lower in unidirectional circuits.

A useful way of comparing harmonics from several control rectifier systems is to determine their harmonic factors. The r.m.s. voltage output

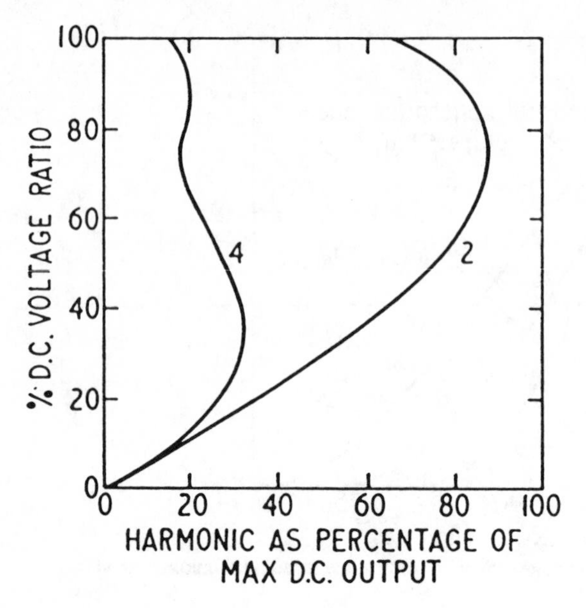

Figure 9.40 Predominant harmonics in the output of a single-phase half-controlled bridge converter

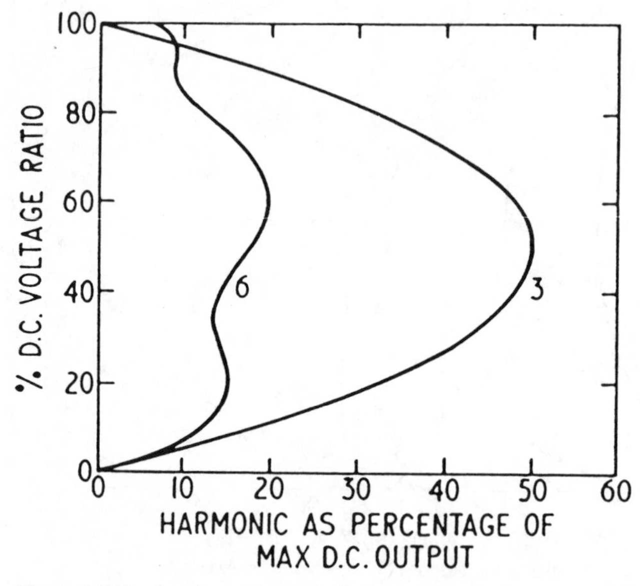

Figure 9.41 Predominant harmonics in the output of a three-phase half-controlled bridge converter

V_{rms} is composed of a d.c. voltage V_{av} plus a spectrum of harmonic voltages of magnitude V_1, V_2, V_3, etc. If V_h is defined by equation (9.20), then the r.m.s. voltage is given by equation (9.21), which gives the value of harmonic voltage as in equation (9.22). Therefore the harmonic factor H_f, which is defined as the ratio of the harmonic to the d.c. voltage, is given by equation (9.23):

$$V_h = \sum_{n=1,2,3\ldots} V_n \tag{9.20}$$

$$V_{rms}^2 = V_{av(\alpha)}^2 + V_h^2 \tag{9.21}$$

$$V_h = \left[V_{rms}^2 - V_{av(\alpha)}^2\right]^{1/2} \tag{9.22}$$

$$\frac{V_h}{V_{av}} = H_f = \left[\frac{V_{rms}^2 - V_{av(\alpha)}^2}{V_{av}^2}\right]^{1/2} \tag{9.23}$$

The harmonic factor can be evaluated for several different circuits, for example considering the bi-directional bridge rectifier circuit of Figure 9.4, where V is the r.m.s. supply voltage, the r.m.s output voltage is given by equation (9.24) and, since the d.c. voltage for zero delay angle is given by equation (9.25) and the d.c. voltage for finite firing angle delay of α is given by equation (9.6), the harmonic factor reduces to that of equation (9.26):

$$V_{rms} = \left[\frac{1}{\pi}\int_{\alpha}^{\pi+\alpha} (\surd(2)\, V \sin\theta)^2\, d\theta\right]^{1/2} \tag{9.24}$$

$$V_{av} = \surd(2)\, V \frac{2}{\pi} \tag{9.25}$$

$$H_f = \left[\frac{\pi^2}{8} - \cos^2\alpha\right]^{1/2} \tag{9.26}$$

Table 9.1 gives the harmonic factors for some of the circuits discussed in the previous sections, which are representative of the various pulse numbers.

Table 9.1 Harmonic factors

Circuit reference	*Harmonic factor H*
Figures 9.2 and 9.4 (two-pulse)	$\left[\frac{\pi^2}{8} - \cos^2\alpha\right]^{1/2}$
Figure 9.5 (three-pulse)	$\left[\frac{2\pi^2}{27} + \frac{\surd(3)\pi}{18}\cos 2\alpha - \cos^2\alpha\right]^{1/2}$
Figure 9.8 (six-pulse)	$\left[\frac{\pi^2}{18} + \frac{\surd(3)\pi}{12}\cos 2\alpha - \cos^2\alpha\right]^{1/2}$

Some of the performance factors and device ratings for typical rectifier circuits are given in Table 9.2. In this table V_{av} is the average or d.c. voltage output; V is the r.m.s. input voltage per phase; V_{rms} is the r.m.s.

Table 9.2 Performance factors

Circuit reference	*Pulses per cycle*	$\frac{V_{av}}{V}$	$\frac{V_{av}}{V_{rms}}$	$\frac{V_{pk}}{V_{av}}$	$\frac{V_{pk}}{V}$	$\frac{I_{av}}{I_d}$	$\frac{I_{rms}}{I_d}$	$\frac{I_{pk}}{I_d}$
Figure 9.1	1	0.45	0.636	3.142	1.414	1.0	1.571	3.142*
Figure 9.2	2	0.90	0.90	3.142	2.82	0.5	0.707	1.0
Figure 9.4	2	0.90	0.92	1.571	1.414	0.5	0.707	1.0
Figure 9.5	3	1.17	0.98	2.09	2.45	0.33	0.577	1.0
Figure 9.7	6	1.35	1.0	2.09	2.83	0.167	0.408	1.0
Figure 9.8	6	1.17	1.0	2.42	2.83	0.167	0.289	0.5
Figure 9.11	6	2.34	1.0	1.05	2.45	0.33	0.577	1.0

output voltage; V_{pk} is the peak or crest working voltage of the devices; I_{av} is the average current per device leg; I_d is the d.c. load current; I_{rms} is the r.m.s. current per leg; and I_{pk} is the peak current per device leg.

9.7 Control circuits

The control or gate-firing circuits used for phase-controlled rectifiers are essentially similar to those used for a.c. phase control, as shown in Figure 8.11. The input detector determines the incidence of zero supply voltage, the delay section provides a variable delay from this point, and this signal is then used to energise the gate-drive circuits which fire the thyristors. There are two further considerations. With source reactance the firing pulses must be of sufficient width to overcome any overlap angle. This was illustrated in Figure 9.27, where it was shown that a pulse width of at least μ_1 is required to ensure that the thyristor reaches its full conduction. This problem of extended firing pulses is not as serious as for a.c. line control, where for inductive loads a pulse duration of 90° is normally required. However, controlled rectifier circuits have an added feature that the input waveform could be distorted if line reactances are severe. Generally, the firing circuits should be able to start the timing pulse when this voltage rises from the zero voltage point.

9.8 Voltage multiplication circuits

The discussions so far have concentrated on power rectification, where the output voltage is controlled to be at a value below that of the input. In this section a special application of power rectification is considered, in which the level of current handled is relatively low, but the output voltage is several orders of magnitude greater than that of the input. Voltage multiplication can vary from a factor of two to several hundred, and there are a variety of circuits which can be used, only a few of these being introduced in this section.

Voltage doublers are first described, since they are the most well known, and these can be categorised as symmetrical, diode pump or bridge. Figure

9.42(a) shows a symmetrical voltage-doubler circuit, which can be considered to be made up of two half-wave rectifier circuits. When input line A is positive capacitor C_1 charges through diode D_1 to the peak of the supply voltage with plate 1 positive, and when the input supply reverses capacitor C_2 charges through diode D_2 to the peak supply voltage with plate 1 positive. Therefore, in the absence of any load, twice the peak input supply voltage would appear across the output lines C and D. With load connected the capacitors are discharged so that the voltage is slightly below this figure; the greater the value of the capacitors, the less they are discharged by the load. The ripple frequency in the output is twice that of the input supply. Each diode must be rated at twice the peak input voltage and the capacitors at the peak supply voltage.

The circuit shown in Figure 9.42(b) is best explained by starting on the negative half cycle of the input, when line B is positive to A. Capacitor C_1

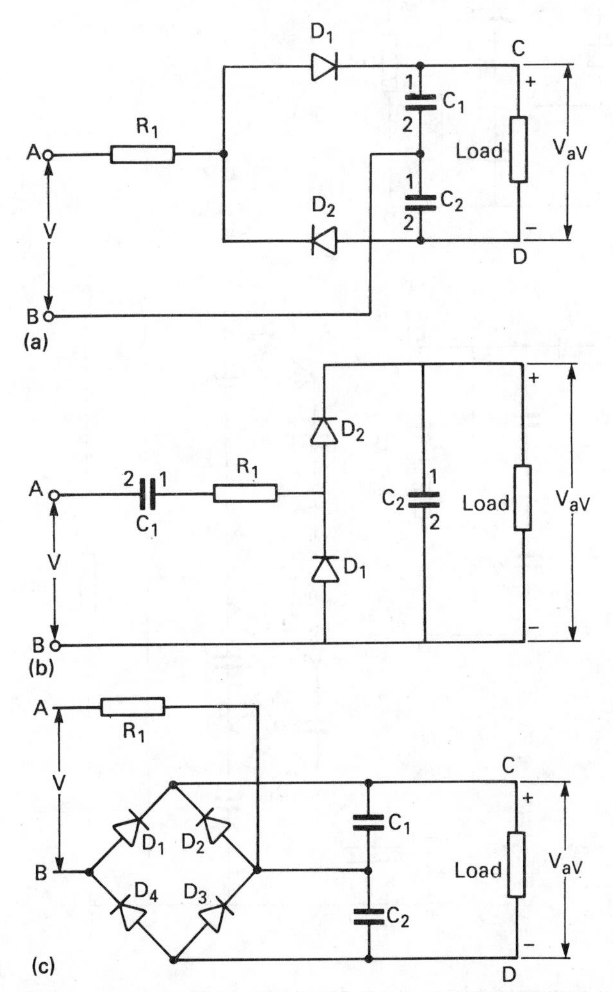

Figure 9.42 Voltage-doubler circuits: (a) symmetrical; (b) diode pump; (c) bridge

charges through diode D_1 to the peak supply voltage with plate 1 positive. On the next half cycle, when line A is positive, the voltage on C_1 adds to the peak of the input supply to pump twice the peak supply voltage onto capacitor C_2, so that the voltage across CD on no load is twice the peak of the input supply, this reducing slightly on load. It is this pumping action which gives this circuit its name, although it is also known as a common terminal voltage doubler, because one side of the load and the input a.c. supply share a common terminal. The output ripple frequency is equal to that of the supply, so that this circuit needs a larger value of capacitor for the same regulation, compared to the symmetrical voltage-doubler circuit.

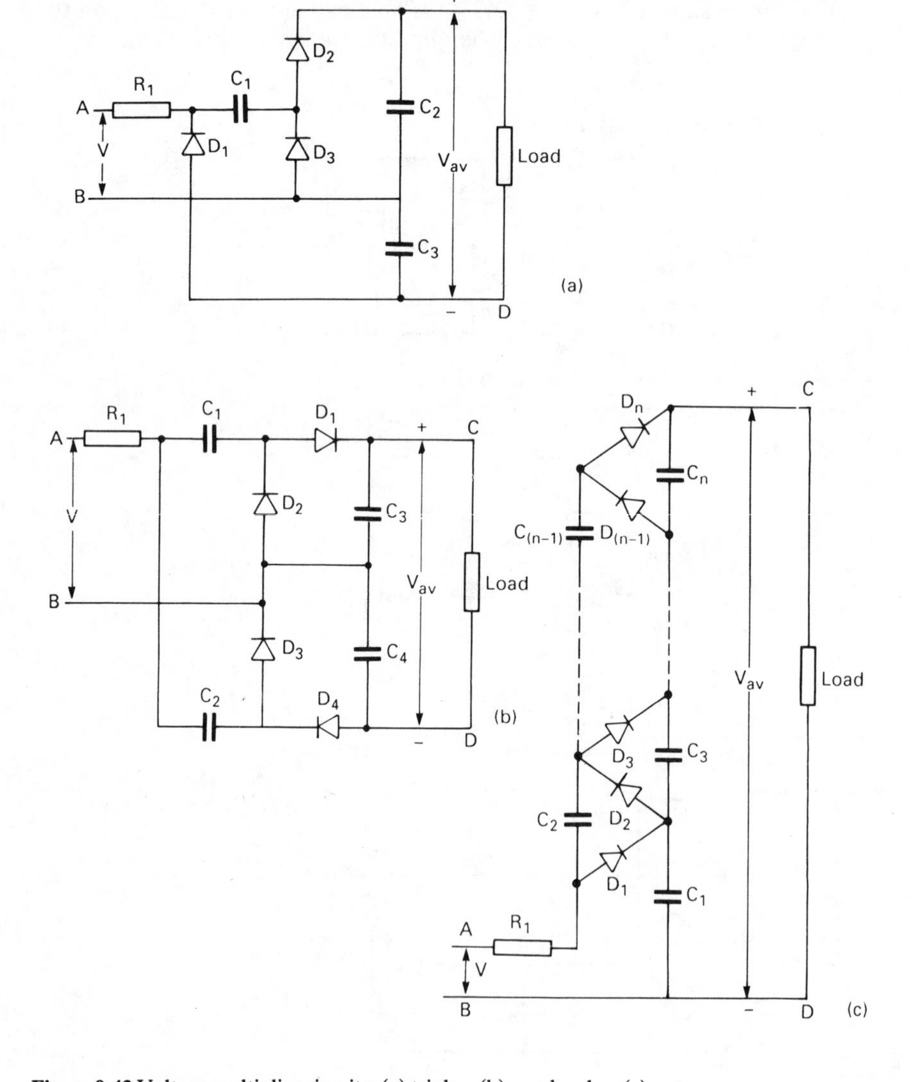

Figure 9.43 Voltage multiplier circuits: (a) tripler; (b) quadrupler; (c) *n* stage

The diodes and capacitor C_2 are again rated at twice the peak of the a.c. supply, whilst capacitor C_1 is rated at the same voltage as the supply, but it must be capable of carrying the r.m.s. load current.

The bridge voltage doubler, shown in Figure 9.42(c), consists of a full wave bridge and a symmetrical voltage-doubler arrangement, so it is capable of good stability and regulation. In this circuit diodes D_2 and D_3 supply the load current in alternate half cycles at the same time that the capacitors are being discharged, so that the output voltage is closer to twice the peak of the input a.c., even on heavy load currents. The ripple frequency is again twice that of the supply, and the diodes each have to be rated at twice that of the peak of the a.c. supply, even though four are used, since the capacitors effectively bypass the a.c. signals.

Voltage-multiplication factors greater than two can again be obtained by a variety of circuits, a few being shown in Figure 9.43. The voltage-tripler circuit of Figure 9.43(a) can be considered to be a combination of the diode pump voltage doubler and a half-wave rectifier circuit. Twice the supply voltage appears across capacitor C_2 due to the action of C_1, D_2 and D_3, whilst the half-wave rectifier D_1 charges capacitor C_3 to the peak of the a.c. supply, so that the voltage across CD is three times that of the supply. The voltage-quadrupler circuit of Figure 9.43(b) is made up of a combination of two diode pump circuits, each giving twice the supply voltage across capacitors C_3 and C_4, so the voltage across CD on no load is equal to four times that of the peak of the a.c. supply.

A more universal voltage multiplier is shown in Figure 9.43(c), this being referred to as the Cockcroft–Walton circuit. Each capacitor is charged to the voltage of the peak of the a.c supply plus that of the capacitor below it, the peak of the output voltage accumulating as one proceeds up the ladder.

Chapter 10

Direct a.c. frequency converters

10.1 Introduction

An a.c. output of a given frequency and voltage can be produced either from a d.c. source, or directly from another a.c. source of a different frequency and voltage. Frequency converters can therefore be classified broadly into two types, d.c. to a.c., called d.c. line frequency converters or inverters, and a.c. to a.c., known as direct a.c. frequency converters or cycloconverters.

Inverters and cycloconverters were introduced in Chapter 6, and in that chapter a simplified operating mode, i.e. single-phase resistive load, was used in order to illustrate the circuit principles. In this chapter cycloconverters will be analysed in greater detail, in particular it will be shown that they form an extension to controlled rectifier circuits, as described in Chapter 9, and can be treated similarly. It is important to keep this similarity in mind since, although controlled rectification forms one of the most frequently used power electronic circuits, the cycloconverter is often regarded as something mysterious and complicated. This reputation has no doubt been gained due to the apparently rather formidable arrangements which cycloconverter systems require, but it will be seen that all such systems can be broken down into basic controlled rectifier circuits, which are interconnected.

Although inverters are described in a later chapter, it is important to recall the main differences between it and the cycloconverter, as introduced in Chapter 6. These are as follows:

(i) An inverter uses fewer power components than a cycloconverter and is usually simpler in construction, even when the commutation components are considered. This usually results in a cheaper system, although in very large current drives the power devices, which are usually thyristors, are much smaller than for a comparable inverter circuit, since they now have a lower duty cycle, so offsetting the cost advantage. In low-power circuits the voltage rating principally determines the cost of power components, so that this no longer applies. It should also be remembered that inverter thyristors need to be specially selected for fast turn-off times and high dv/dt, which results in more expensive components. The thyristor-firing circuits for inverters are simpler than for cycloconverters and this results in a further reduction in system cost.

(ii) An inverter can work from a d.c. or an a.c. source, which has first been rectified and smoothed by relatively cheap components. A cycloconverter, on the other hand, can only work from an a.c. input, so if a d.c. supply is available only it is necessary to first convert it to a.c. before it can feed the cycloconverter. Such a d.c. to a.c. inverter can add appreciably to the original cycloconverter system cost, even if rotating converters are used. Therefore an inverter is much more versatile than a cycloconverter and is the system most frequently used, although, as will be seen later, a cycloconverter is used for special applications, usually when the supply is a.c.

(iii) Both inverters and cycloconverters can regenerate power from the load to the supply, but whereas an inverter requires a fairly complex control system to do this, regeneration occurs as part of the natural process in a cycloconverter. Regeneration is frequently required in high-power motor-drive systems, if frequent stops and starts are necessary, and in this application a cycloconverter has the obvious edge over an inverter.

(iv) Both inverters and cycloconverters (three-phase) can provide a stepless output frequency variation. For an inverter this can vary from any value below to any above the base frequency, since the input is converted to a d.c. line. For cycloconverters, on the other hand, the output frequency is usually limited to about one third that of the input and at any rate to below that of the supply, unless forced commutation is used, which is not usually desirable. The load current waveforms from inverters can be made to resemble those of cycloconverters by using pulse-width voltage-control techniques, as will be seen in Chapter 13.

(v) The inverter has an intermediate d.c. store so that when operating from an a.c. supply the power factor imposed on this source is high, irrespective of that of the load. A cycloconverter does not have any equivalent storage capability so that the load power factor is reflected directly to the supply. By the very nature of the phase control involved in a cycloconverter, the power factor is always lagging, therefore a 60° leading or 60° lagging load angle would both produce an approximately 60° lagging supply power factor angle.

As a generality, an inverter is used where a wide frequency variation is required and a cycloconverter is preferred when most of the output requirements are at low frequency.

10.2 Cycloconverter principles

Cycloconverters can operate in one of two modes, envelope or phase control. In both systems the circuit arrangement is identical, it is the operating mode which determines the type of converter. In Chapter 6 the phase-controlled system was introduced and this is the commonest method in use, being described in this section to illustrate the fundamental properties of cycloconverters. In the following sections the various cycloconverter circuit arrangements, both single- and three-phase, are introduced and this is followed by a description of envelope-type cycloconverters and a performance analysis of phase-controlled converters.

A basic single-phase cycloconverter circuit was shown in Figure 6.4 and it is redrawn in Figure 10.1 in a form which illustrates clearly that the system consists essentially of two bridge-type, two-pulse bi-directional converters, as described in Chapter 9, connected in opposite directions across the load. The converter can be considered to consist of a positive and a negative group, the load voltage and current polarities being as indicated.

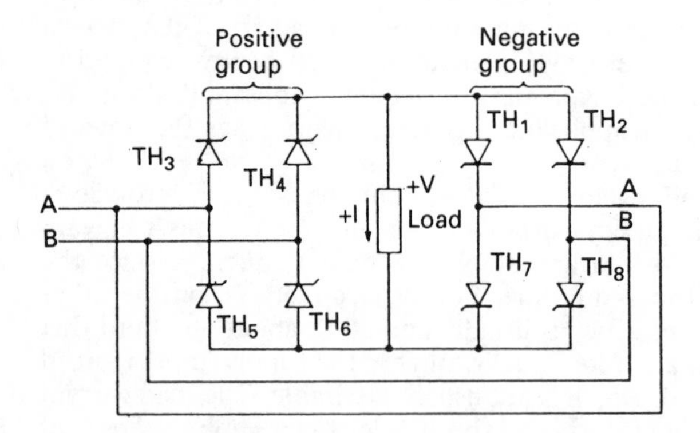

Figure 10.1 Modified arrangement of a single-phase bridge cycloconverter

The operation of the cycloconverter, shown in Figure 10.1, working in a phase-controlled mode, can be explained by reference to the waveforms given in Figure 10.2(a). The load current is assumed here to be filtered and is therefore sinusoidal. At time t_0 line A is positive to B and the load current is negative, i.e. opposite to the direction shown in Figure 10.1, so that thyristors TH_7 and TH_2 are fired. These thyristors are maintained in conduction, even when the voltage across them reverses at t_1, due to the energy stored in the inductive load.

At time t_2 thyristors TH_8 and TH_1 are fired, turning TH_7 and TH_2 off and driving the instantaneous load voltage negative. When load current reverses at t_3, thyristors TH_8 and TH_1 turn off and in order to maintain the load waveform as shown, thyristors TH_6 and TH_3 are fired. These conduct until t_4, when TH_4 and TH_5 are fired turning them off; and so on throughout the cycle.

The instant of firing the thyristors can be varied, as desired. In Figures 10.2(a) and 10.2(b) the angles α_p and α_n represent the minimum and maximum delay angles for the positive and negative group of converters, respectively. When the delay angle is 90° the mean output is zero, the variation of the delay about this point determining the amplitude of the output, as shown in Figure 10.2(b), where the mean load voltage is lower

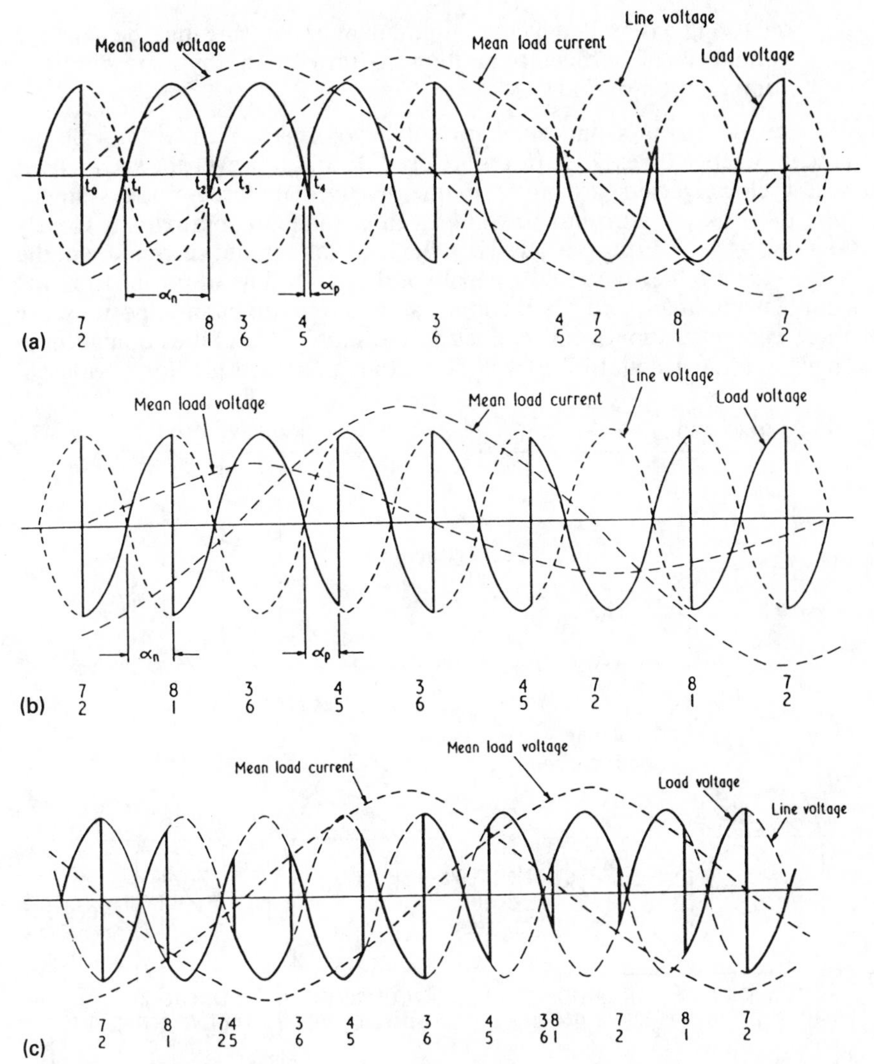

Figure 10.2 Single-phase cycloconverter waveforms: (a) high-voltage output; (b) low-voltage output; (c) regeneration

than that in Figure 10.2(a). Referring to these two figures, the following can be deduced regarding the cycloconverter:

(i) The positive group of thyristors, 3, 6, 4 and 5, conduct in pairs when the load current is positive, the negative group, 7, 2, 8 and 1, conducting when it is negative.

(ii) A firing delay of 90° in any phase corresponds to zero load voltage, the firing points of the thyristors oscillating about this delay angle at a frequency determined by the required output frequency. The greater the swing in the oscillations about 90° (the maximum total excursion being 180°), the greater the mean output voltage. The maximum delay angle is limited by the requirement of ensuring that the negative voltage is of sufficient duration to turn off a thyristor during a commutation period.

Figure 10.2 shows sinusoidal current waveforms, which will not be the case in practice unless load filters are used. It will, however, be shown later that cycloconverters are more frequently used in three-phase systems, where the output current and voltage can be made to approximate closely to a sine wave. During periods when the load voltage and current are in the same sense, either positive or negative, there is a flow of power from the supply to the load, and when voltage and current are in an opposite sense there is regeneration from the load to the supply. Careful examination of Figures 10.2(a) and 10.2(b) will show that there are greater periods of

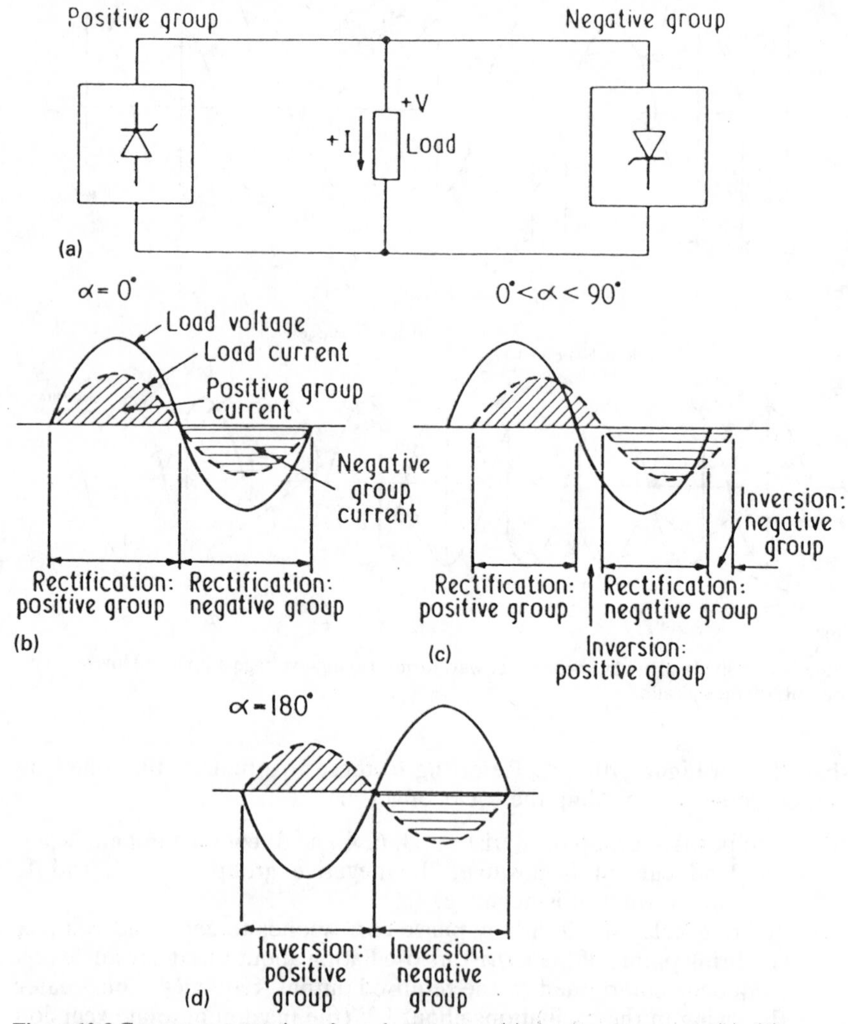

Figure 10.3 Group representation of a cycloconverter: (a) circuit arrangement; (b)–(d) voltage and current for varying firing angles

input power than there are of regeneration. Net regeneration can be obtained by a rearrangement of firing pulses, as shown in Figure 10.2(c). The problem of regeneration is identical to that encountered in usual phase-control circuits, as discussed in Chapter 9, so a cycloconverter can be readily made to control the flow of power in either direction, which is one of its greatest assets.

The operation of the cycloconverter can be explained more clearly with reference to its group diagram, as shown in Figure 10.3(a), the load voltage waveforms being given in Figures 10.3(b) to 10.3(d) for converter delay angles varying from 0° to 180°. The function of the two converters is seen to change from full rectification to full inversion, in varying stages. It is clear that each converter must therefore be able to rectify and invert within a half cycle. Since only one converter carries the load current at any instant, it is possible to fire only this system when required, but it will be seen later that there are often advantages to firing both converters simultaneously, but with their delay angles such that their sum always equals 180°. Figure 10.4 illustrates the operating mode. From this it is seen that at any setting one group is in the rectification mode whilst the second is in inversion, with such a delay angle that the mean output from the two groups are equal, so that it prevents the transfer of mean power between the two converter groups. This system will be referred to again later in this chapter.

10.3 Cycloconverter circuits

Figure 10.3(a) illustrates very clearly that a cycloconverter is basically a combination of various groups of thyristor converters, of the type described in Chapter 9. Figure 10.1 illustrated the use of two-pulse bridge

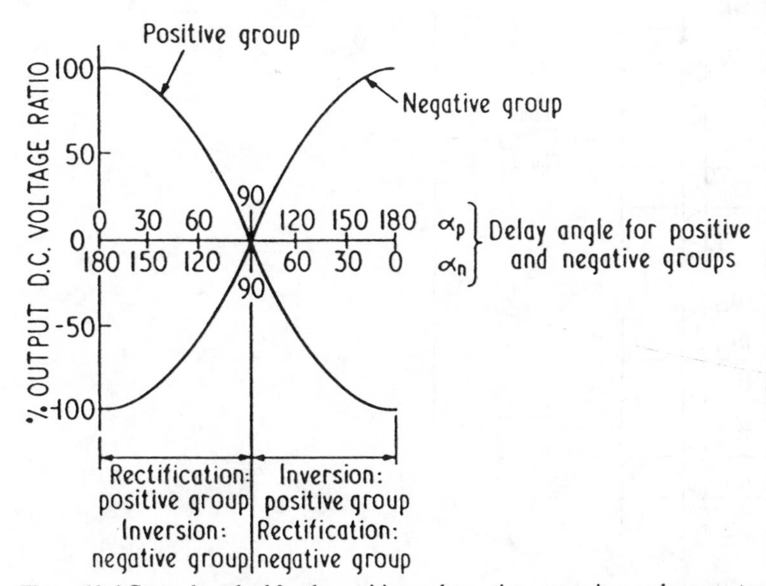

Figure 10.4 Control method for the positive and negative groups in a cycloconverter

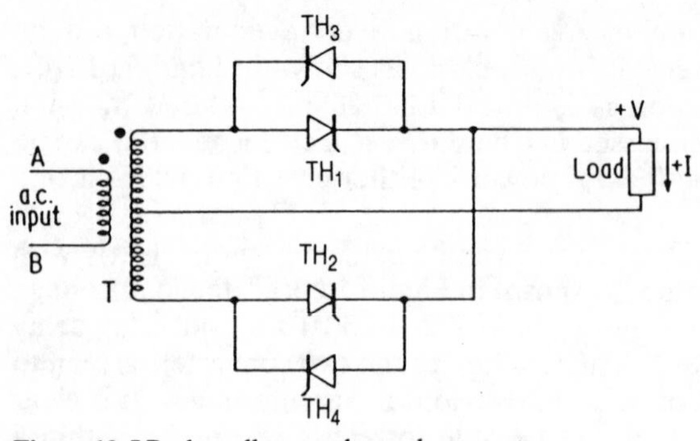

Figure 10.5 Push–pull two-pulse cycloconverter

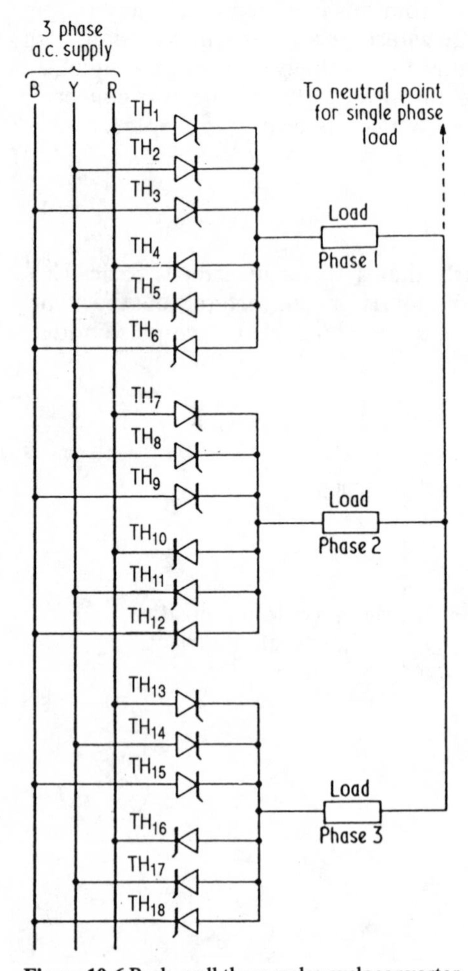

Figure 10.6 Push–pull three-pulse cycloconverter

converters and Figure 10.5 shows a functionally identically circuit made from two push-pull converters. The operation of this system is represented by the previous load waveforms if TH_3 and TH_6 for the bridge are replaced by TH_1 for push–pull, TH_4 and TH_5 for the bridge are replaced by TH_2 for push–pull, TH_7 and TH_2 for bridge are replaced by TH_4 for push–pull, and TH_1 and TH_8 for bridge are replaced by TH_3 for push–pull. TH_1 and TH_2 therefore correspond to the positive group and TH_3, TH_4 to the negative one. In common with rectifier converters, it can be seen that although push–pull systems use half as many thyristors as a bridge circuit, these have now to be rated at twice the load voltage.

The higher the pulse number of the converter, the lower the load ripple voltage, which was found to be the case for a.c. to d.c. converters, so that single-phase cycloconverters are rarely used in practice. Figure 10.6 shows a three-pulse push–pull cycloconverter circuit which is supplying a balanced three-phase load. To obtain a single-phase output only one converter may be used and the load returned to the neutral point of the transformer secondary. The three-pulse circuit can be extended to six pulses either by using a double-star transformer secondary, as in Figure 9.7, or by an interphase transformer connection. The latter system is given

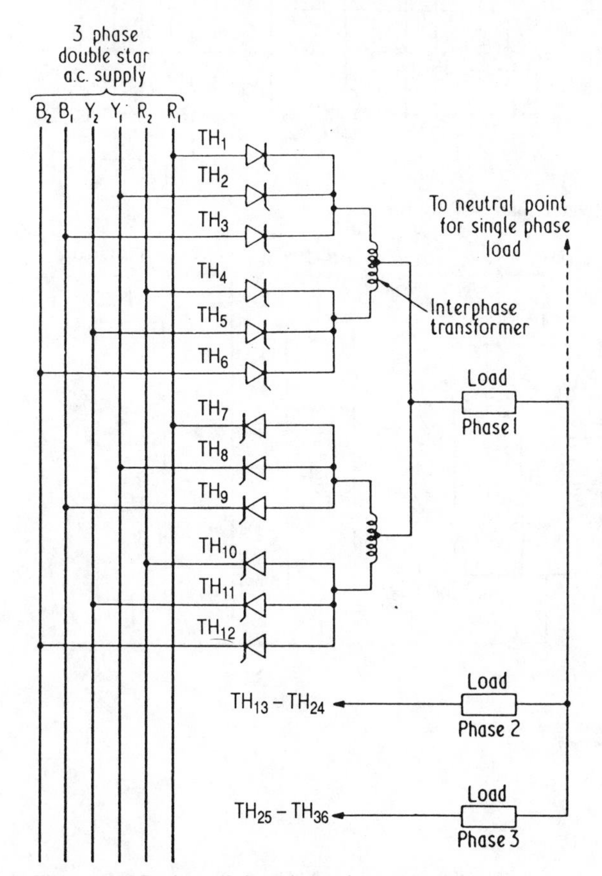

Figure 10.7 Push–pull six-pulse cycloconverter

in Figure 10.7, where only one phase of the converter is shown in detail. If a single-phase output is required, only one converter is used and the load output is returned to the transformer neutral point. Interphase connections are preferable to double star since, as discussed when considering the equivalent a.c. to d.c. converters, the utilisation of the system is much higher. The six-pulse push–pull system can be extended to twelve pulses by using the circuit of Figure 9.10.

Push–pull circuits have a single output point so that there exists a common load terminal to which three-phase systems can be connected. A bridge circuit, on the other hand, requires both ends of the load to be connected to the converter terminals, therefore for three-phase loads either the different load phases, or the input to the bridges, must be isolated. Figure 10.8 shows a six-pulse bridge converter with isolated load

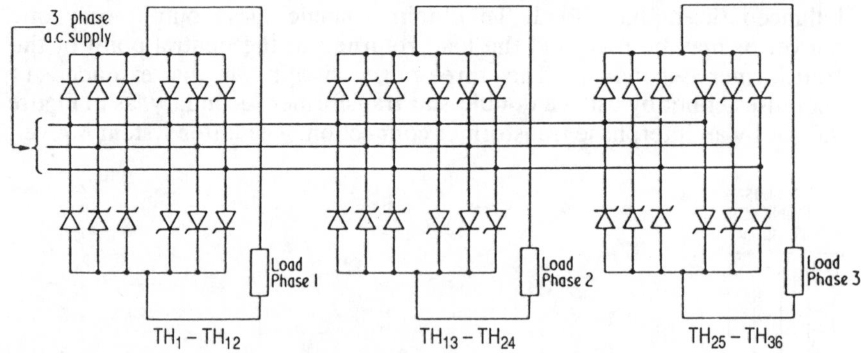

Figure 10.8 Bridge six-pulse isolated load cycloconverter

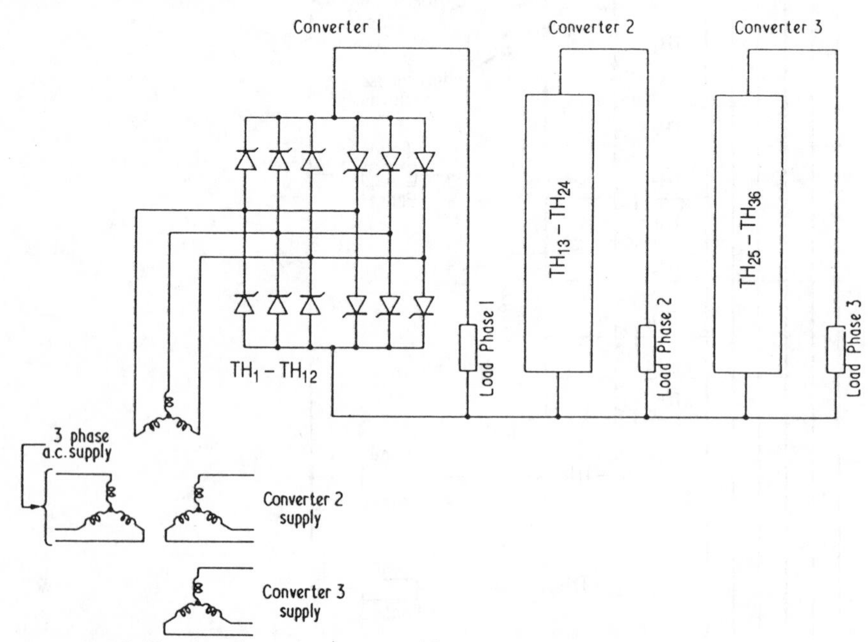

Figure 10.9 Bridge six-pulse cycloconverter with non-isolated load

and Figure 10.9 gives a system in which the load has a common terminal, so that the individual bridge inputs need to be supplied from isolating transformers. Isolated load circuits are preferred owing to the complete absence of any input transformer. In the isolated supply system each secondary winding in effect feeds a separate single-phase load, so that the total secondary rating exceeds that of the primary by over 20%. Extension of the six-pulse bridge to twelve pulses, by the use of series-connected converters, is shown in Figure 10.10.

The circuits described so far can be referred to as symmetrical cycloconverters, since each phase of a three-phase output is made up of a single-phase unit, but this is not always necessary. For instance, Figure 10.11 shows a three-phase load, supplied from an asymmetrical cycloconverter, which has only two single-phase modules. Although the component content of this converter has been reduced by one third compared to symmetrical circuits, the utilisation of the thyristors and

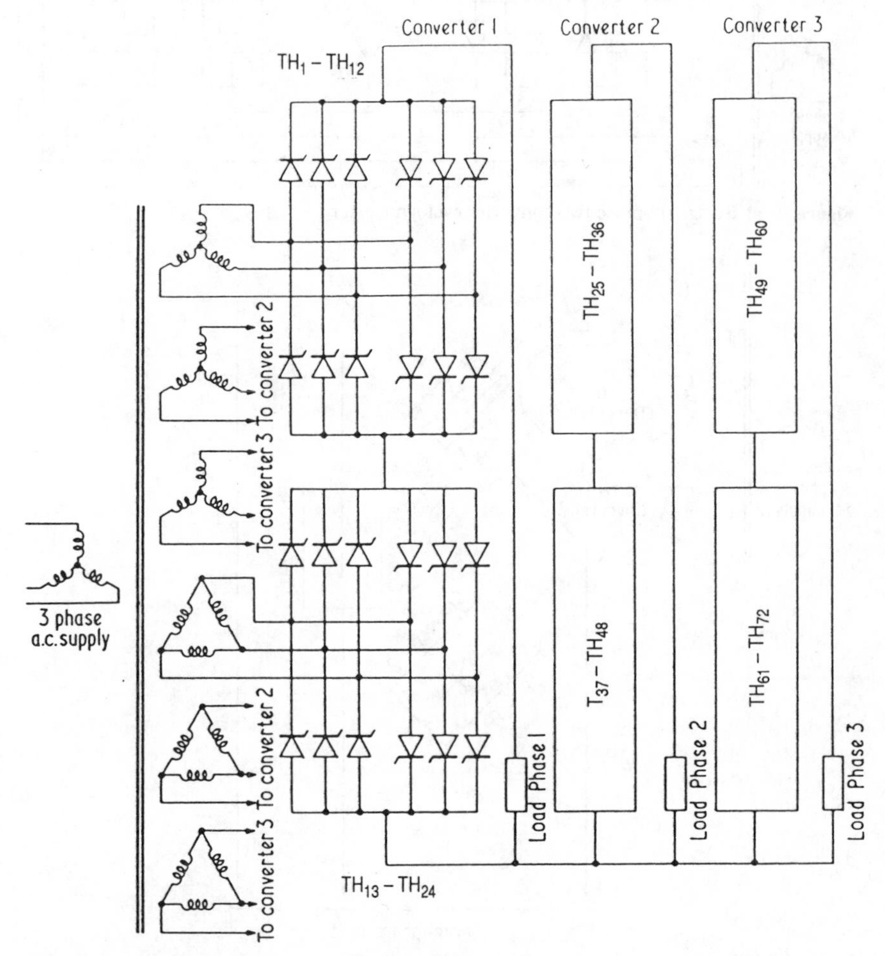

Figure 10.10 Bridge twelve-pulse cycloconverter

transformer has also been appreciably reduced. Furthermore, it is now difficult to maintain a balanced supply to the three-phase load. In spite of this, the system is often economical to use in low-power circuits, where good performance is not a prime criterion. Figure 10.12 shows an alternative form of asymmetrical six-pulse converter, which consists of a ring arrangement of bi-directional a.c. to d.c. converters and is therefore not made from basic single-phase cycloconverter modules. As is seen, the component content has been reduced by 50% over comparable

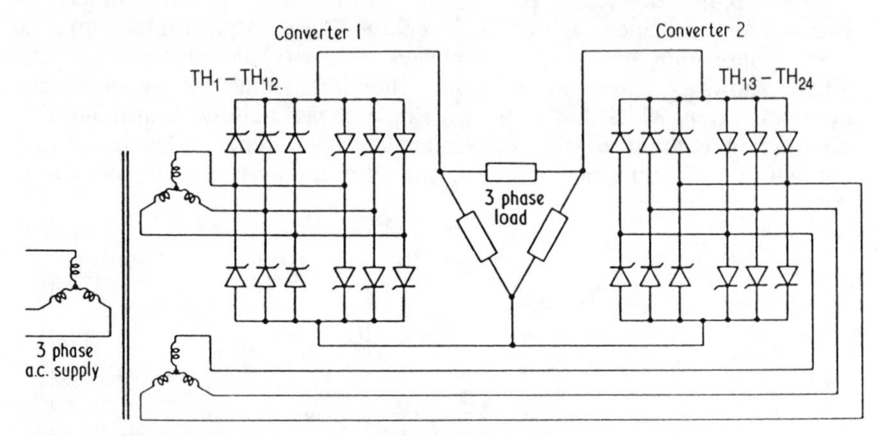

Figure 10.11 Bridge six-pulse two converter cycloconverter

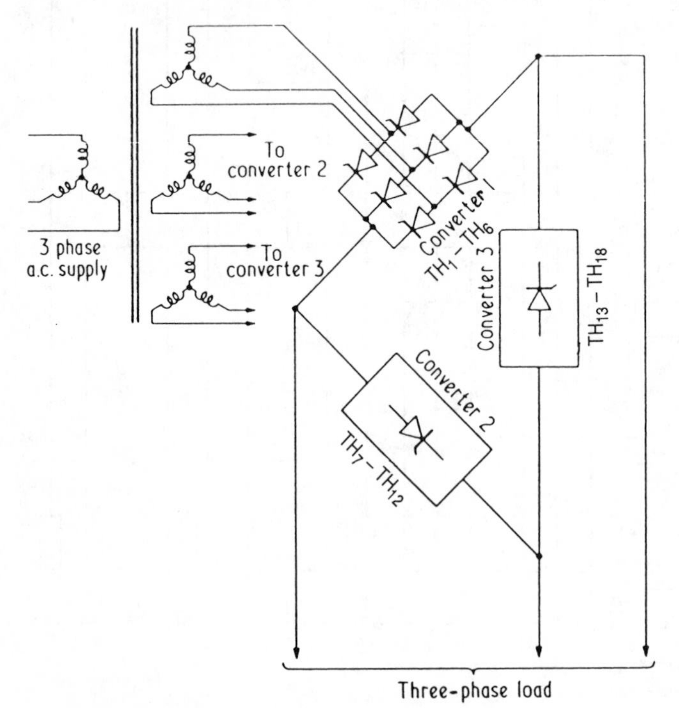

Figure 10.12 Bridge six-pulse ring-type cycloconverter

symmetrical cycloconverters although, as expected, the transformer and thyristor utilisation have now also been considerably reduced. Once again, however, this circuit provides a low-cost medium-performance converter, which is economical in many applications.

10.4 Envelope cycloconverters

In the discussions so far the operation of a cycloconverter has been illustrated with reference to its phase-controlled mode. As mentioned above, this is the most common technique and gives a system in which the load voltage and frequency can be readily controlled. However, in certain applications, especially where the ratio of input to output frequency is fixed, envelope converters can prove simpler and more economical.

Figure 10.13(a) shows the output from a single-phase supply in which the ratio of input to output frequency is three. The converter thyristors are operated at the beginning of every half cycle, so that they follow the envelope of the a.c. supply waveform. Voltage control is now obtained by a correct choice between primary and secondary ratios on the input

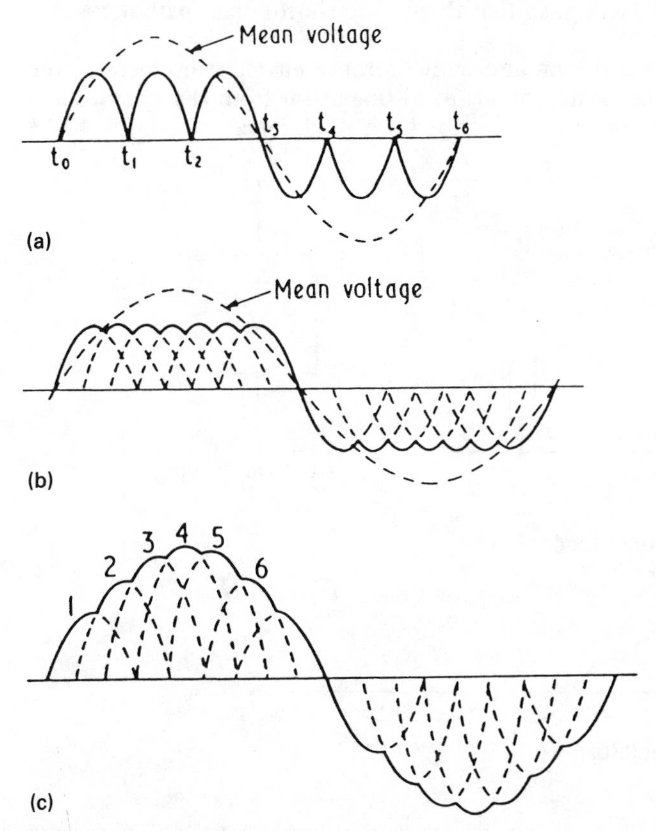

Figure 10.13 Load-voltage waveforms for synchronous envelope converters: (a) single phase; (b) three phase; (c) three-phase stepped output

transformer. From this figure it can be seen that the load voltage will have a very high harmonic content and this can be reduced, as in Figure 10.13(b), by using a six-pulse converter. Once again the load voltage follows the envelope of the various phase voltages and the output voltage is almost rectangular in shape.

Since most cycloconverters have an input transformer, it is possible to vary the ratio of their secondaries with respect to one another. For instance, Figure 10.13(c) shows the output voltage obtained from a system in which phase 1 has 33%, phases 2 and 6 have 73%, and phases 2 and 5 have 97% of the voltage of that of phase 4. The operation is once again of the envelope type, where the load voltage commutates naturally from phase to phase, so that it is always equal to that of the most positive phase. This system gives an output with a good approximation to a sine wave and was widely used in the 1920s and 1930s for traction applications, to provide a stable 16.66 Hz (20 Hz for 60 Hz supplies) voltage.

It can be seen from the above discussions that the half cycle output waveform from envelope cycloconverters is identical to that obtained if the thyristors were all replaced by diodes, so that conduction begins at the commencement of the supply cycle. The function of the thyristor is now purely to ensure that one of the converter groups is switched off when the other group is conducting, so that there is no short-circuit path across the supply.

Figure 10.14 illustrates an alternative form of envelope cycloconverter, often referred to as asynchronous to distinguish it from the synchronous type of envelope converter described with reference to Figure 10.13.

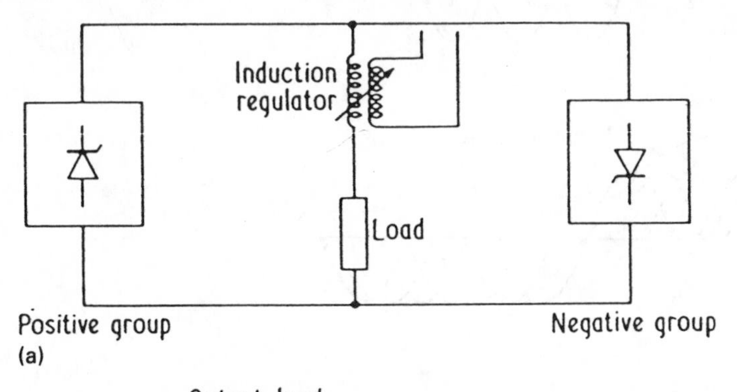

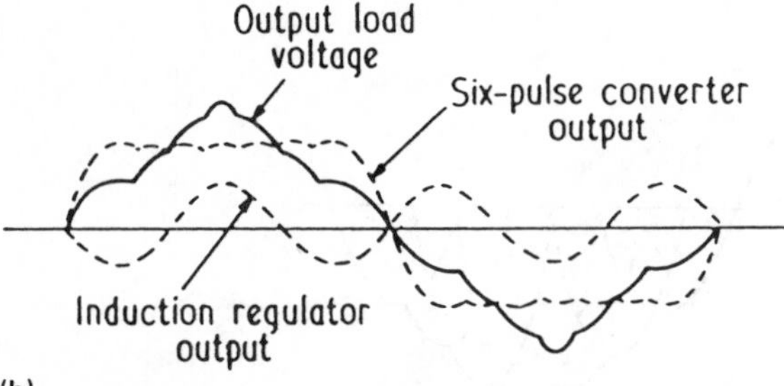

Figure 10.14 One form of asynchronous envelope cycloconverter: (a) circuit arrangement; (b) waveforms

Figure 10.14(a) shows that the circuit is essentially the usual arrangement of a cycloconverter with the addition of an induction regulator, which is fed from the three-phase supply, in series with the load. The output voltage from the six-pulse converter is as illustrated in Figure 10.13(b) and to this is added the induction regulator output, so as to give the resultant load voltage of the form shown in Figure 10.14(b). Once again, the approximation to a sine wave is very close.

Envelope cycloconverters are incapable of variable-frequency operation and, since the converter thyristors operate like diodes during any half cycle, they cannot regenerate load current back to the supply. For this to occur the firing angle on the thyristors would need to be delayed beyond the start of a cycle, as in normal a.c. to d.c. converters. Therefore envelope cycloconverters are incapable of handling inductive loads since they cannot absorb its reactive power. For stable operation it is now necessary to connect a capacitor in parallel across the load, to raise its overall power factor. This disadvantage is not met with in phase-controlled converters, where the firing angle can be shifted readily to meet any required direction of load current flow.

10.5 Phase-controlled cycloconverters

The operation of a phase-controlled cycloconverter has already been described with reference to a single-phase system, as in Figure 10.2. It is seen that the load frequency can be controlled by the oscillation frequency of the firing point about 90°. The load voltage amplitude is governed by the extent of this oscillation about the mean firing point and the converter can be readily switched from rectification to inversion by regulating the firing angle. The same considerations apply for a three-phase converter, and the load waveforms for the three phases of a typical six-pulse system are shown in Figure 10.15.

Each phase of the cycloconverter is made up of basic single-phase converter blocks, as in Figure 10.3(a), and by adjusting the firing angles of positive (α_p) and negative (α_n) systems such that $\alpha_p+\alpha_n$ is always equal to 180°, the mean output voltage from the two groups is equal, so that there is

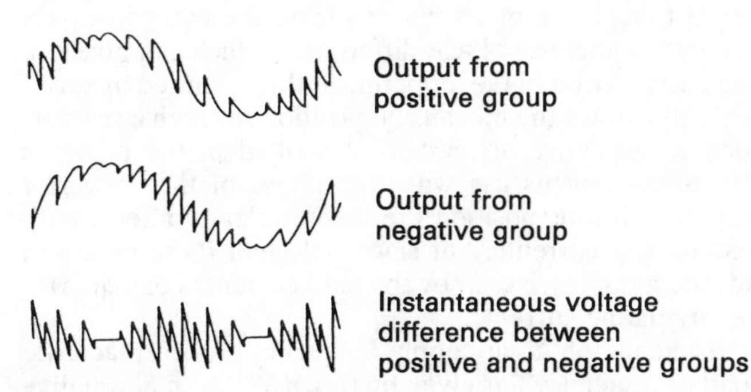

Figure 10.15 Instantaneous voltage difference between positive and negative groups of a cycloconverter

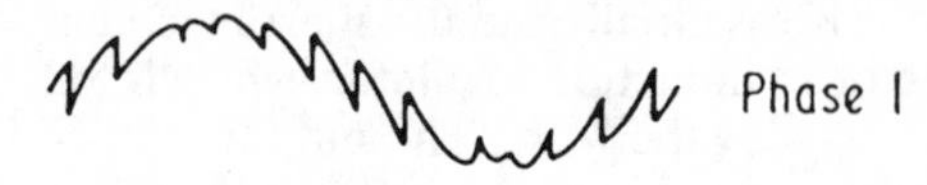

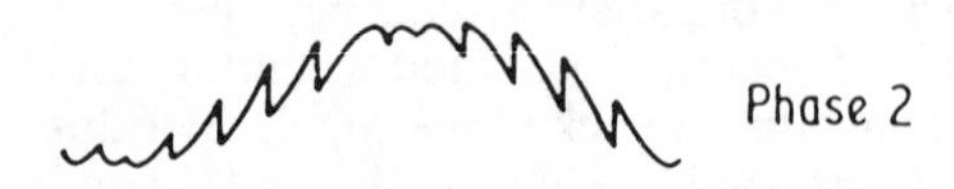

Figure 10.16 Three-phase cycloconverter load voltage for a six-pulse bridge converter

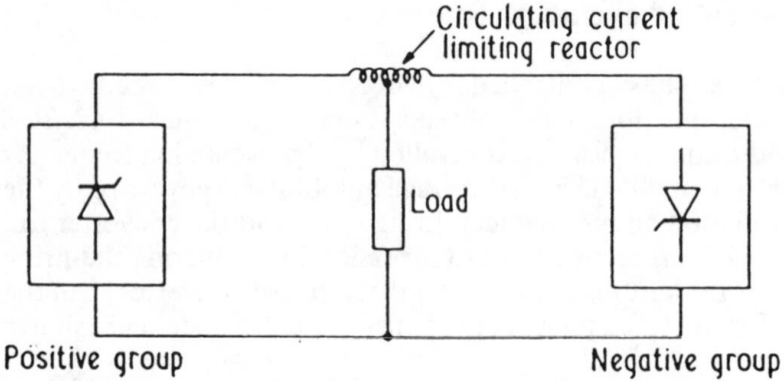

Figure 10.17 Position of a reactor to limit the circulating current between cycloconverter groups

no net transfer of power around the two converters. However, Figure 10.16 shows that even though the mean outputs from the two converters are equal, there are instantaneous voltage differences which can give rise to a large circulating current around the loop, unless this is limited by series reactors. Figure 10.17 illustrates the most usual position for such a reactor, which can be added to all converter circuits described in the previous sections, Figure 10.18 showing its use with one phase of the converter illustrated in Figure 10.7. In this position the reactor clearly affects both the load and the circulating currents, but since only half its turns are in series with the load, the inductance seen by the load current is one quarter of that seen by the circulating current.

Therefore for a reactance of X, at supply frequency f_s, the reactance presented to the load at frequency f_l is given by $(f_l/f_s)(X/4)$. An alternative position for the circulating current reactor is clearly in the supply lines between the two converters, although then it affects the load current to a

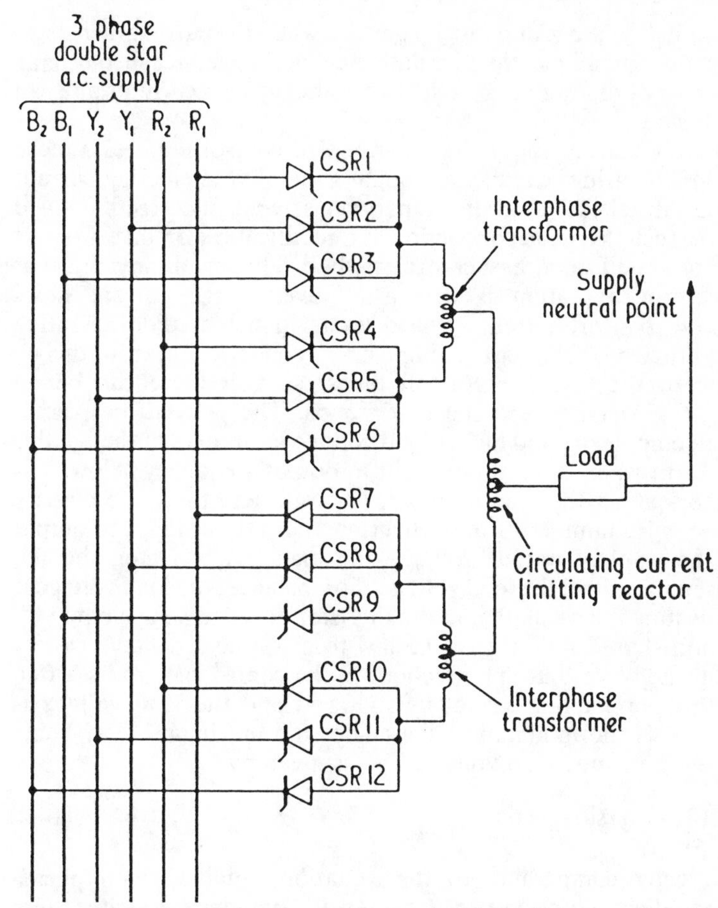

Figure 10.18 Push–pull six-pulse cycloconverter with circulating current limiting reactor

greater extent. If p is the pulse number of the converter system, V the r.m.s. supply voltage to the cycloconverter, and X the effective source reactance, then the circulating current (i_c) is limited to the value given by

$$i_c = \frac{\sqrt{(2)}\,V}{X}\left(1 - \cos\frac{\pi}{p}\right) \tag{10.1}$$

An alternative method for suppressing the flow of circulating current is to ensure that only one converter conducts at any time. This is by far the most satisfactory arrangement, since the circulating currents can often become extremely large, but it does result in a much more complex thyristor control system in which the load current must be sensed and used to block one or other group of converter, depending on the current direction. Another disadvantage of the non-circulating current mode is that the load voltage distortions can become much higher on light loads, when the current becomes discontinuous. Since discontinuous operation is only serious on light loads, it is feasible to run a cycloconverter in a circulating-current mode when the load current falls below a certain critical level and to switch to the group-blocking mode when the current increases. By adopting this technique the circulating-current reactor can be made small, since it is not required to be effective on heavy load currents and can therefore saturate.

The circulating-current suppression method of control requires a more complex thyristor-firing circuit, although generally circulating-current systems are a natural extension of ordinary phase-controlled rectifiers and ensure the correct thyristor conduction for rectification or inversion. It gives the simplest and smoothest control method. The one disadvantage of circulating-current operation is that a relatively large current flows between the two converter groups, which results in increased device rating and reduced efficiency. Non-circulating-current converters prevent this by blocking one of the converters at all times, so that there can be no interchange of power between the two groups. The price to be paid is greater circuit complexity and higher load voltage harmonics, although this is only important when the ratio of input to output frequency is low.

Other factors affect the load harmonics as well, and these are directly related to the pulse number of the converter, the ratio of input to output frequency, the level of output voltage, the load power factor and the technique used to control the load voltage. For instance, the load voltage is varied by adjusting the oscillation of the thyristor delay angle by about 90°. This oscillation should follow a cosine law to give a load voltage closely resembling a sine wave, but imperfections in the control electronics often mean that this perfect law cannot be followed and the load voltage is distorted. If α_m is the minimum delay angle (the maximum delay being $180-\alpha_m$) then the r.m.s. load voltage V_{rms} is given by

$$V_{rms} = \sqrt{(2)}\, V \frac{p}{\pi} \sin\frac{\pi}{p} \cos\alpha_m \qquad (10.2)$$

This is directly comparable to the equation obtained with phase-controlled rectifiers described in Chapter 9. In common with these converters, the load voltage waveform from a cycloconverter is affected by phase overlap caused by source reactance. This is illustrated in Figure 10.19, which shows the shift in the load voltage harmonic current. Source

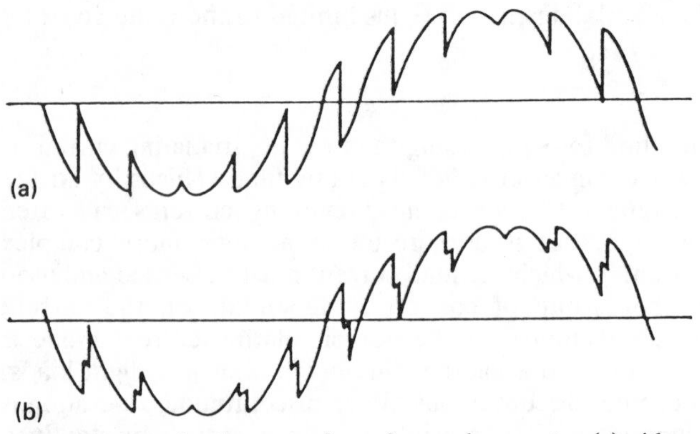

Figure 10.19 Load-voltage waveforms for a cycloconverter: (a) with no source reactance; (b) with source reactance

reactance effects are especially important when the cycloconverter is used in variable-speed constant-frequency (VSCF) systems. These are frequently met with in aircraft, where the cycloconverter is fed from a local alternator which has a high-frequency output. Since the alternator speed is variable so also is its frequency, the cycloconverter now being used to ensure that the load frequency is maintained at a fixed value. The source reactance is that of the alternator, and since the cycloconverter represents its largest load it would create considerable distortion, which would need to be taken into account in any system design. If μ is the overlap angle caused by instantaneous output current I through the source reactance X, α is the firing delay angle and V the r.m.s. supply voltage, then equation (10.3) can be obtained:

$$\cos(\alpha + \mu) - \cos\alpha = \sqrt{\left(\frac{2}{3}\right)}\,\frac{XI}{V} \tag{10.3}$$

The input current conditions to the cycloconverter can be defined by three factors:

(i) The power factor P, which is the ratio of the system watts to the volt amperes;
(ii) The distortion factor D, which is the ratio of the r.m.s. fundamental input current to r.m.s. total input current;
(iii) The displacement factor L, which is the cosine of the angle between the fundamental supply voltage and the fundamental component of the input current.

The relationship between these three factors is given by

$$P = DL \tag{10.4}$$

Therefore for a sinusoidal input current $D = 1$ and the power factor equals the displacement factor, as expected.

The performance criterion for a cycloconverter is dependent on many different conditions. Exact analysis is difficult and is complicated further by the oscillations of the firing angle throughout the output cycle. The equations given above are based on the assumption of a relatively high input-to-output frequency ratio.

10.6 The cycloinverter

The usual form of operation for a cycloconverter is in the step-down mode, where the output frequency is less than that of the input. Generally, this frequency is limited to a maximum of one third that of the supply frequency, since at lower ratios the voltage distortions become appreciable.

When the cycloconverter is running in its step-down mode it is naturally commutated, the leading kVA required to turn off conducting thyristors being derived from the higher-frequency side, which in this case is the a.c. supply. There is no reason why a cycloconverter cannot run with an output frequency greater than that of the input, and such a system is called a step-up cycloconverter, or more commonly a cycloinverter. The important difference between this and the more popular inverter, described in Chapter 13, is that there is no d.c. line in the cycloconverter, the power being converted directly from an input a.c. at one frequency to an output at a higher one. The turn-off energy required by the conducting thyristors must again be derived from the high-frequency side, i.e. the load, in this instance. If the load has a leading power factor then this requirement will be met, and if it has a lagging power factor then a capacitor must be connected across it to artificially reproduce this condition.

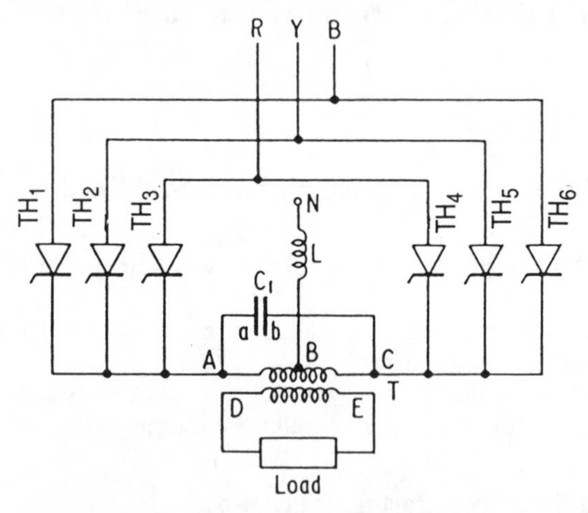

Figure 10.20 Three-pulse push–pull cycloinverter

Figure 10.20 shows one form of commutation which may be used in a cycloinverter. It will be seen in Chapter 11 that this is a parallel-capacitor commutation system, although many of the other techniques described in that chapter may be used instead. The principle of the system is that the thyristors treat the instantaneous value of the a.c. supply voltage as a d.c. base from which the commutation voltage is derived. The overall system is push–pull, where the load is supplied from a centre-tapped transformer T. When TH_1, TH_2 or TH_3 conduct, load current flows from one of the input lines to the neutral, assuming that their instantaneous value is positive, and the supply voltage is impressed across AB. This makes the secondary side D positive to E and when TH_4, TH_5 or TH_6 conduct the voltages across the primary and secondary are reversed. Therefore the load sees an alternating voltage, of a magnitude equal to that of the supply, modified by the turns ratio between half the primary and the secondary of transformer T.

As an example, suppose TH_3 is conducting, the load voltage being proportional to the instantaneous value of the supply phase R, and at the same time side A of the transformer T being raised positive by twice this value, with respect to side C, since B is the centre point. Therefore C_1 charges to twice the instantaneous supply voltage with plate a positive to b. To reverse the load voltage it is necessary to turn off TH_3 and fire, say, TH_4. Thyristor TH_3 will not go off naturally until the end of the half cycle, but if TH_4 is fired then plate b of capacitor C_1 is raised to the same voltage as that of the anode of TH_3. This places a reverse voltage, that on C_1, equal to twice that of the instantaneous a.c. supply at this instant, directly across TH_3, and it turns off. Load current is now supplied by TH_4 and the load voltage has been reversed. Capacitor C_1 need not be connected across the transformer primary, the same effect being obtained with the capacitor on the secondary side.

Cycloinverters are not commonly used, since they do not give any significant advantage over forced commutated inverter systems. They have found limited application for induction heating systems where, for example, the transformer in Figure 10.20 is replaced by a centre-tapped heating coil with a capacitor connected across it. The combination of coil and capacitor forms a resonant circuit which turns the conducting thyristors off at the required times.

10.7 Cycloconverter control circuits

As described in earlier sections, the mean voltage from a cycloconverter must be able to oscillate about zero, moving from a maximum positive to a maximum negative value in each cycle of the output frequency. To obtain this the firing delay is usually biased at an angle of 90° and the firing angle delay is then oscillated by a further 90° about this point, in both the positive and negative directions. This causes the firing point of the positive group of thyristors, making up the cycloconverter circuit, to be advanced and the firing angle of the negative group to be retarded by the same amount, during the positive half cycle, the roles being reversed during the negative half cycle. At all times the sum of the positive and negative delay angles is such that equation (10.5) is satisfied, as described earlier, so that the mean output voltages from the two groups are equal in magnitude but opposite in phase:

$$\alpha_p + \alpha_n = 180° \tag{10.5}$$

The block diagram of Figure 10.21 illustrates functionally the basic system for the thyristor-firing circuit of a cycloconverter. The reference voltage is ideally a steady d.c. level, which produces a firing angle delay of 90°, on which is superimposed a cosine voltage. The frequency of this voltage will determine the load frequency and the amplitude of the cosine curve, relative to the d.c. level, fixes the modulation depth of the load voltage and hence its magnitude. The distributer circuit feeds thyristors in both the positive and negative converter group, but a feedback signal from the load current is shown, in the system illustrated, to be used to blank pulses to the group which is not conducting load current, so that circulating currents are prevented. The reference wave generated is compared to a sample of the low-voltage output and the firing angle is then adjusted such as to minimise the harmonics by causing a delay over successive cycles. The delay is continuously variable from 0° to 180° in response to the control inputs and the reference, the pulses for each group being 120° apart.

The control circuits for cycloconverters can become very complex owing to the relatively involved logic functions which have to be performed and

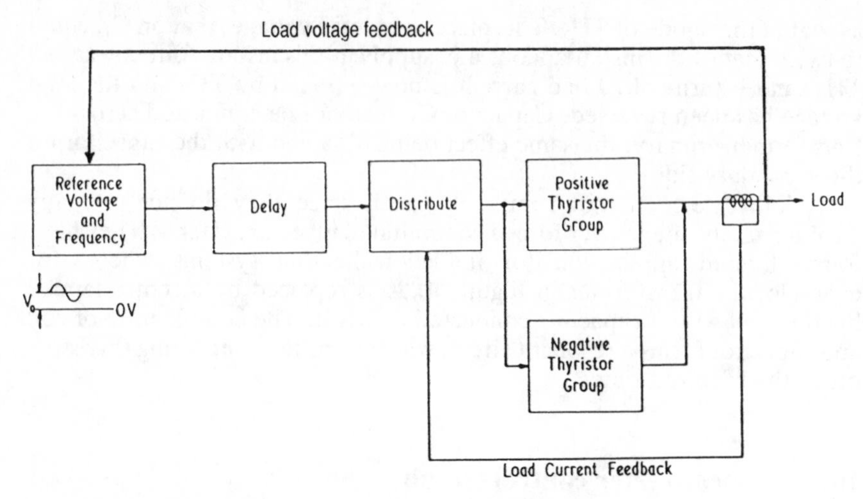

Figure 10.21 Block schematic of a cycloconverter control circuit

the large number of thyristors requiring gate signals. To construct such a drive using discrete components would result in a large and expensive system, so most modern cycloconverter circuits use a considerable amount of integrated circuits, many of them implemented as gate arrays, with complex logic performed in the software of microprocessors.

Chapter 11

Forced commutation techniques

11.1 Introduction

Some power semiconductor switches, such as transistors and gate turn-off switches (GTO), can be turned off by means of a signal on their gate terminal, whilst in others the gate is only able to turn the device on and it will turn off when the current through it has decayed to zero. The turn-off process is known as commutation, and if the power semiconductor is operating from an a.c. supply, then this will occur when the supply reverses, the process being called natural commutation. If the power supply is d.c., or if the conducting semiconductor is to be turned off at the non-zero part of an a.c. cycle, then it must be commutated by forcing the current through the device to zero, and this is called forced commutation.

Choppers and inverters have already been introduced in Chapter 6 and if thyristors are used as the power switches for these circuits, as is usual, then they need to be forced commutated. Figure 11.1 shows the principle

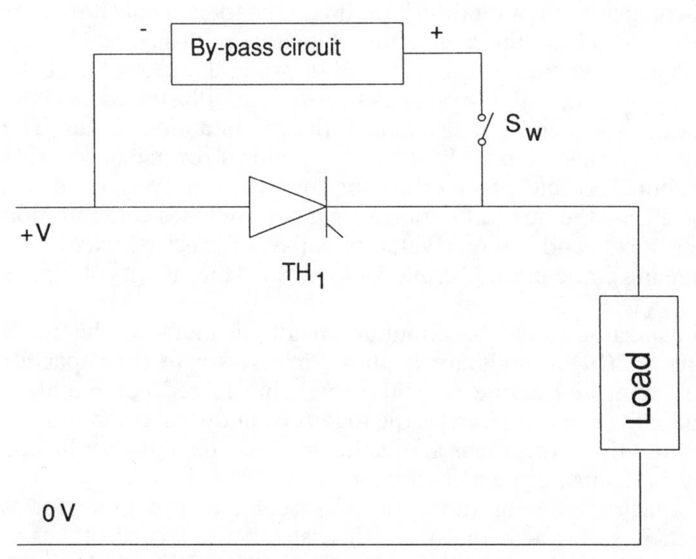

Figure 11.1 A generic forced commutated system

involved in forced commutating a conducting thyristor TH_1 which is operating from a d.c. supply. When switch S_w is closed the load current, flowing via the thyristor, is diverted through the bypass circuit, which also applies a reverse voltage across the device, turning it off. Often this bypass system consists of a capacitor, which has been charged during a previous cycle to the polarity shown.

For successful commutation several conditions must be satisfied:

(i) The time for which the thyristor is reverse biased must exceed its turn-off time.
(ii) The rate at which the forward voltage is re-applied across the device must be less than its dv/dt rating.
(iii) The switch S_w will have to carry a high rate of current increase (di/dt) and this must not exceed its rated value.
(iv) It is probable that the bypass circuit will need to be reset again, so as to be able to apply the required reverse voltage across the thyristor, if it is refired and needs to be turned off.

This chapter examines the different forced commutation techniques used and their application to choppers and inverters are described in Chapters 12 and 13, respectively.

11.2 A classification system for forced commutation

A large number of different circuits are used for forced commutation of power semiconductors, and in order to study them a classification system is required, one system being described here. This is independent of the type of application, e.g. chopper or inverter, so that it will be used again in subsequent chapters when these circuits are described. Four divisions are used in the forced commutation classification, as illustrated in Figure 11.2.

(i) Parallel-capacitor commutation. In this a charged capacitor C is placed directly across the conducting thyristor, as in Figure 11.2(a), turning it off. The circuit which is used to prime the capacitor at the start of every cycle, with the polarity shown, is not illustrated, and the semiconductor switch S_w is also part of the commutation circuit. The capacitor performs a dual role, that of applying a reverse bias across the thyristor TH_1 and of diverting the load current away from this thyristor during the turn-off period. For inductive loads commutation is more difficult and a larger value of capacitor must be used, or a free-wheeling diode placed across the load, as in the circuits of Figure 11.2.
(ii) Parallel capacitor–inductor commutation. In this method, illustrated in Figure 11.2(b), an inductor is placed in series with the capacitor which is connected across the thyristor being forced commutated. Once again the capacitor carries the load current during commutation and provides the reverse bias across the thyristor, the inductor having auxiliary functions, as described later.
(iii) Series capacitor commutation. In this technique the capacitor is connected in series with the power thyristor being turned off, as in Figure 11.2(c), so that it is almost invariably in series with the load.

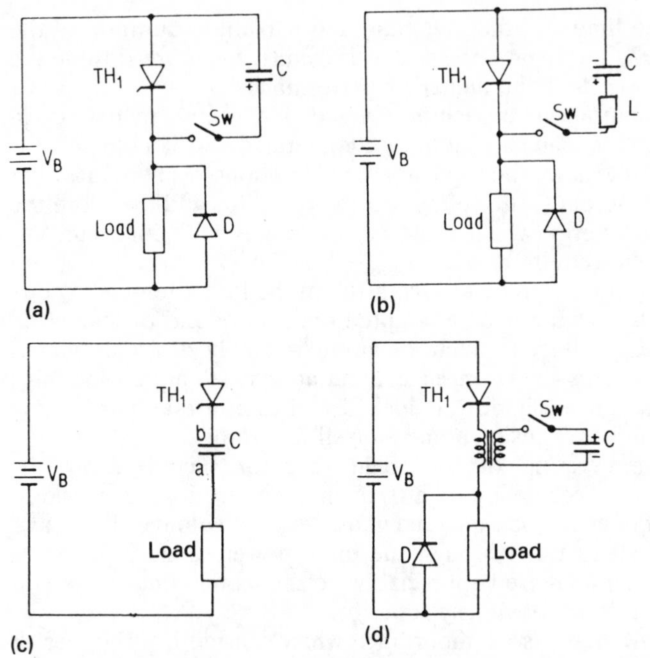

Figure 11.2 Forced commutation techniques: (a) parallel capacitor; (b) parallel capacitor–inductor; (c) series capacitor; (d) coupled pulse

Clearly, once the capacitor is charged to its final value the thyristor is turned off and the capacitor must be reset before the thyristor can be refired.

(iv) Coupled-pulse commutation. For this forced commutation method the turn-off pulse is coupled to the thyristor by means of a transformer or an auto-transformer, a transformer being illustrated in Figure 11.2(d). In this figure the pulse energy is obtained by discharging a capacitor, although many other techniques may be used.

In the circuits of Figure 11.2 only the basic principles of commutation have been illustrated, most of the auxiliary commutation system having been omitted, although these will be included in subsequent sections when each of these four commutation methods are discussed in greater detail.

In common with any engineering design problem, choice of a forced commutation method requires the balancing of technical performance against cost, although for a given price it is true that there is an optimum circuit. To help in drawing up a technical comparison between the various forced commutation systems which are described later, a checklist of six parameters will be considered, as follows:

(i) Does the commutation method enable the power switches within the circuit to be used in either a variable-frequency or a fixed-frequency variable mark-space mode, as desired? Clearly, the greater the flexibility, the better the system.

(ii) What are the limits on the minimum and maximum duration of the output pulse? Once again, the wider the limits, the more flexible the system and therefore the higher its performance.

(iii) Is the commutation capacitor voltage inherently increased in proportion to the load current to be commutated? Such an increase is usually desirable since the capacitor can be optimised to operate over a wide range of loads, the voltage on the capacitor and therefore the commutation energy which it stores, increasing with load current, when it is most required.

(iv) If commutation were to be unsuccessful on the first attempt, say due to an overload, would it be attempted once more and be successful when the load reduced? Such a feature is highly desirable and is superior to systems where once a commutation is attempted and fails, the commutation mechanism is locked and cannot take part in any further operations, causing a once-for-all failure of the system.

(v) Is the current rating of the main thyristor increased by the commutation process? For high-frequency applications a considerable amount of energy is expended in the series of commutations, and if this were all to flow through the main power semiconductor its rating would be increased appreciably, so that commutation systems which avoid this are obviously superior.

(vi) The sixth parameter is of importance when considering chopper or inverter circuits and is related to the inverter or chopper configuration more than to the commutation method. This is whether there is a low-impedance fault current path across the supply, since if such a path does exist then a commutation failure would cause the current to rise rapidly, destroying the semiconductor devices.

In the detailed description of the four commutation methods, given in the following sections, chopper circuits will be used as illustrations, since they are much simpler than inverter circuits and allow attention to be placed on the commutation technique. The same principles apply for inverters and these are described in Chapter 13.

11.3 Parallel-capacitor commutation

The parallel-capacitor commutation method was shown in Figure 11.2(a). To explain its operation, assume that capacitor C is charged to a voltage V_c, from an earlier cycle, with the polarity indicated. The load current just prior to switch S_w closing is $I_{L(pk)}$ and is assumed to remain constant during the short discharge period of capacitor C. If the reverse recovery current through thyristor TH_1 is also neglected, then the time during which the thyristor is reverse biased (t_q) after the switch is closed is given by

$$t_q = \frac{C\,V_c}{I_{L(pk)}} \tag{11.1}$$

and for commutation to be successful this time must exceed the turn-off time of the thyristor.

The rate of re-application of forward voltage across the thyristor is determined by the commutation capacitor discharging through the load and recharging again from the battery, with a polarity opposite to that shown in Figure 11.2. This is given by

$$\frac{dv}{dt} = \frac{I_{L(pk)}}{C} \tag{11.2}$$

and its value must be less than the dv/dt rating of the thyristor for it to remain off after the commutation (reverse bias) period.

Equations (11.1) and (11.2) are applicable to all parallel-capacitor commutated circuits and enable the correct capacitor value to be chosen for a given load and device characteristic. The magnitude of the capacitor voltage (V_c), just prior to the start of commutation, will depend on the auxiliary circuit used to prime the capacitor.

The simplest example of a parallel-capacitor commutated chopper circuit is shown in Figure 11.3. Thyristor TH_1 is fired to initiate the load

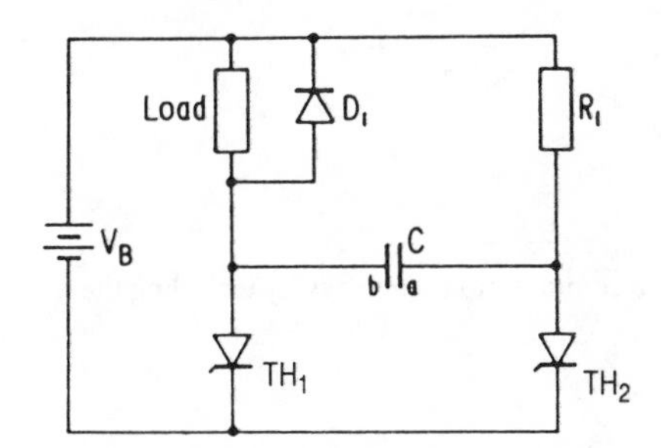

Figure 11.3 An elementary parallel-capacitor commutation system

current and at the same time it charges the commutation capacitor C to the supply voltage V_B, via resistor R_1, with plate a positive. To turn off the main thyristor the auxiliary thyristor TH_2 is turned on, which places the charged capacitor across TH_1 causing it to be reverse biased. The load current now flows through the commutation capacitor, charging it with plate b positive and causing the voltage across TH_1 to change from reverse bias to forward bias. Commutation of this thyristor will be successful if the reverse bias time, given by equation (11.1), exceeds its turn-off time and if the rate of rise of forward voltage, given by equation (11.2), is below its rated value. Thyristor TH_2 would normally remain conducting after commutation of TH_1 has been completed, current flowing through resistor R_1, and it would turn off when thyristor TH_1 is next fired to commence the load cycle and capacitor C is placed across TH_2, causing it to be reverse biased.

The circuit shown in Figure 11.3 is inefficient since current flows via R_1 and TH_2 during the whole of the off period and also since the current needed to prime C ready for the commutation cycle flows through R_1 and TH_1, not adding to the load power. The circuit is also limited in operating frequency and minimum on time of the main thyristor, which determines the minimum output voltage for a chopper. Since R_1 must be made large, so as not to dissipate excessive power during the off period, this then extends the charge time of capacitor C. Commutation of TH_1 cannot begin until this charging has been completed, limiting the minimum time for which the main thyristor can be on.

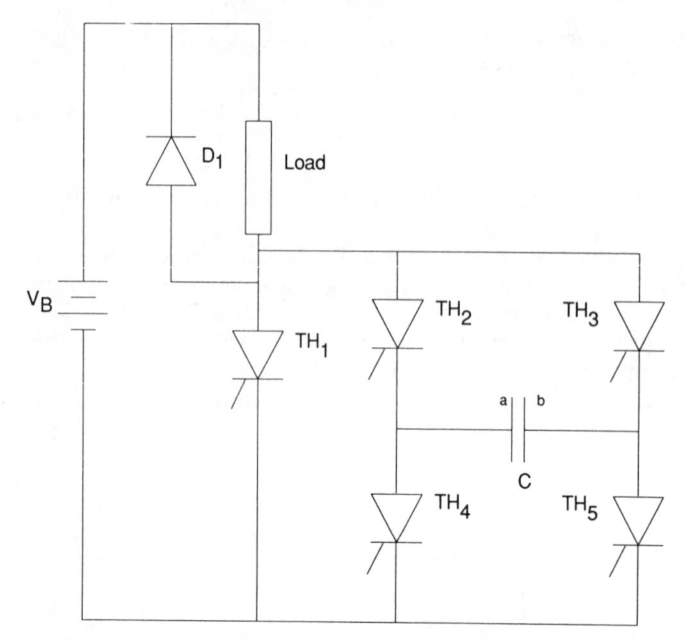

Figure 11.4 Modification to Figure 11.3 to improve commutation efficiency by including the load in the capacitor reset path

Figure 11.4 shows a modification to the elementary circuit of Figure 11.3, which reduces the commutation losses and improves its high-frequency operation by eliminating the need for a separate capacitor charge resistor. Thyristors TH_2, TH_5 and TH_3, TH_4 are fired in pairs to reverse the voltage on commutation capacitor C. At the start of a cycle thyristors TH_3 and TH_4 are fired so that capacitor C is charged to the battery voltage with plate b positive. Since this priming current flows through the load it adds to the load power and is not dissipated in an auxiliary resistor, as in Figure 11.3, so improving circuit efficiency. Also, after the commutation capacitor has been fully charged all the thyristors go off and there is no further power dissipation in the circuit. Thyristor TH_1 is fired to commence the load cycle, and since the commutation capacitor has already been primed there is no requirement for a minimum on time for this thyristor. To turn it off, thyristors TH_2 and TH_5 are fired, applying the reverse voltage of capacitor C across TH_1, which will turn off provided its characteristics satisfy equations (11.1) and (11.2), as before.

Figure 11.5 shows a further modification to the basic circuit of Figure 11.3, which uses a resonant circuit to prime the commutation capacitor. Thyristor TH_1 is fired to commence the load cycle and simultaneously, or some time later, thyristor TH_3 is fired, causing C to charge through inductor L_1 to a voltage V_c with plate a positive. Thyristor TH_3 will go off as soon as this capacitor has reached its full voltage, so that there is no further dissipation, due to the auxiliary commutation circuitry. To turn the main thyristor off, commutation thyristor TH_2 is fired, which places the reverse capacitor voltage across TH_1 causing it to be commutated, as

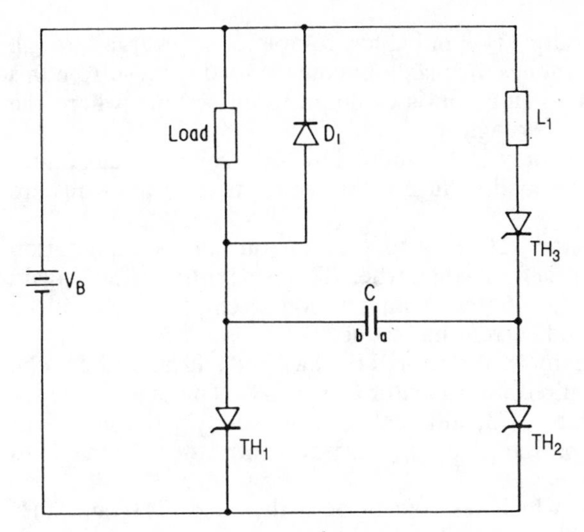

Figure 11.5 Modification to Figure 11.3 by the addition of capacitor resonant charge

before. The capacitor then charges to the supply voltage V_B through the load, and thyristor TH_2 will go off as soon as this has been completed.

Equations (11.1) and (11.2) still determine the conditions for a successful commutation, but since the capacitor is charged via a resonant circuit it will reach a voltage of twice the supply voltage, losses in the resonant path being ignored, so that the size of the commutation capacitor needed is now halved compared to the circuits of Figures 11.3 and 11.4.

The six-point checklist, given in the previous section, can now be used to analyse the circuit shown in Figure 11.5:

(i) The circuit can be operated in either a variable-frequency or a variable mark-space mode, since commutation does not start until an auxiliary commutation thyristor is fired, giving the operator full control over the on and off periods.

(ii) The minimum on time t_c, which determines the minimum load voltage, is given by equation (11.3). The value of capacitor C is fixed by the need for a successful commutation at the peak load current expected, as in equation (11.1), so the only way in which this time can be reduced is by decreasing the inductor L_1. However, the peak current through thyristors TH_1 and TH_3, caused by the charging of capacitor C, is given by equation (11.4), and from this it can be seen that if inductor L_1 is made too small then the charging current through these devices will be very high, adding to their dissipation and rating requirements

$$t_c = \pi \sqrt{(L_1 C)} \tag{11.3}$$

$$I_{TH(pk)} = 2V_B \sqrt{\left(\frac{C}{L_1}\right)} \tag{11.4}$$

The minimum time for which the main thyristor TH_1 must be off (t_o) is dependent on the load current, being longer on light loads, if it is desired not to refire TH_1 until C has completed its charge through TH_2, and this determines the maximum output load voltage from this chopper circuit. This situation is common to all systems where the turn-off pulse flows through the load.

(iii) The commutation voltage is determined by the resonant circuit and is not increased by the load voltage if the inductances of the leads are ignored.

(iv) If the main thyristor TH_1 fails to turn off during a commutation attempt, when TH_2 is fired, then when TH_3 is next fired it will again recharge C ready for another commutation attempt, which will be successful if the load current has reduced.

(v) The rating of the main thyristor TH_1 has been increased by the charging pulse required for capacitor C, via TH_3. This is also true for the circuit of Figure 11.3, although it is avoided in the circuit of Figure 11.4, where the charging current contributes to the load current.

(vi) A fault condition, which resulted in both thyristors TH_3 and TH_4 being gated on simultaneously, would give a low-impedance path across the supply, which would destroy both devices unless they were protected by fuses.

Figure 11.6 shows a parallel-capacitor commutation circuit which has been popularly used for chopper circuits. Auxiliary thyristor TH_2 is fired at the start of any cycle, charging commutation capacitor C to the supply voltage V_B with plate b positive. Main thyristor TH_1 is then turned on,

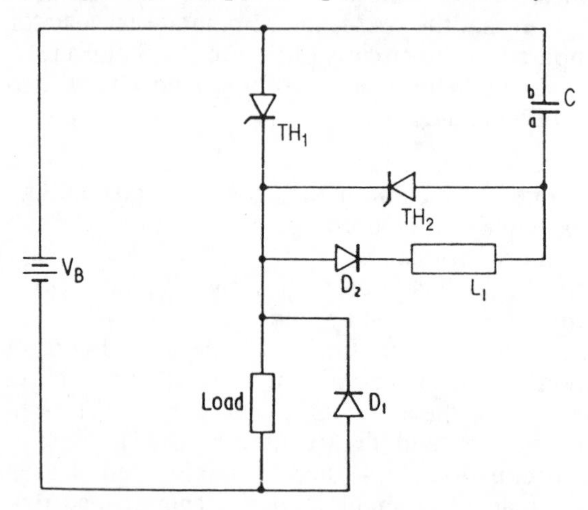

Figure 11.6 An alternative parallel-capacitor commutation system for a chopper

starting the load cycle and causing C to resonate through inductor L_1, via thyristor TH_1 and diode D_2, recharging to V_c, which will be equal to V_B if the resonant losses are ignored, with plate a positive. To turn TH_1 off, thyristor TH_2 is fired, which places the reverse voltage of capacitor C across the main thyristor, as in a normal parallel-capacitor commutation system. The capacitor commences to carry the load current and recharges with plate b positive ready for the next cycle, and when this is complete auxiliary thyristor TH_2 also turns off, so that there is negligible quiescent power loss in the circuit.

The circuit of Figure 11.6 can be analysed using the six points discussed earlier, as follows:

(i) Both variable-frequency and variable mark-space operational modes are possible.
(ii) The same comments regarding minimum on and off times, made when discussing Figure 11.5, apply here.
(iii) The commutation voltage is not boosted by load current.
(iv) If thyristor TH_2 is fired to commence the commutation cycle of the main thyristor TH_1 (plate a of the capacitor being positive) and commutation is unsuccessful, i.e. TH_1 fails to turn off, then the commutation capacitor C will discharge to zero and TH_2 will turn off. Since TH_1 is still conducting, C cannot recharge to V_B so commutation will not be re-attempted. 'Once-for-all' commutation failure has occurred and the chopper can only be restarted by first breaking the supply to TH_1, rendering it non-conducting, and then re-applying the supply.
(v) The rating of the main thyristor TH_1 is increased by the resonant capacitor charge.
(vi) A low-impedance failure path does not exist across the supply voltage, assuming that diode D_1 has not failed to a short circuit.

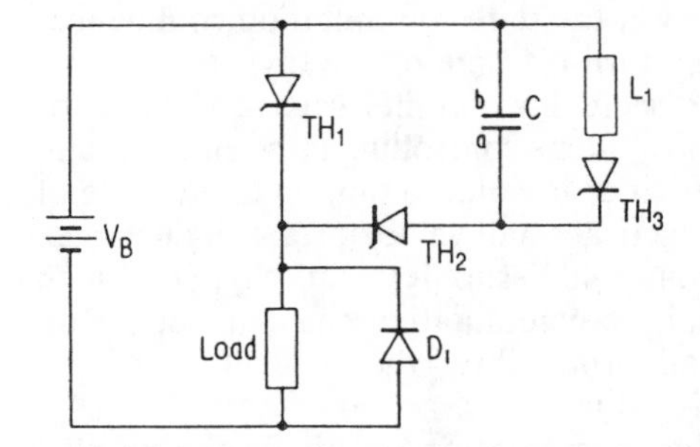

Figure 11.7 Modification to Figure 11.6 to avoid capacitor charge current flowing through the main thyristor

Figure 11.7 shows an alternative circuit to that in Figure 11.6, where the discharge resonant pulse of the commutation capacitor does not flow through the main thyristor TH_1, so that its rating is not increased. Apart from this, the circuit design parameters are the same as for Figure 11.6. At the start of the cycle thyristor TH_2 is fired which charges capacitor C to V_B with plate b positive. Thyristor TH_3 can be fired at any time, but it is usually fired simultaneously with the main thyristor TH_1. This causes commutation capacitor C to resonate with L_1 and recharge with plate a positive, ready for the commutation interval, which commences when TH_2 is next fired.

11.4 Parallel capacitor–inductor commutation

The parallel capacitor–inductor commutation technique was introduced in Figure 11.2(b). Assume that commutation capacitor C is charged with the polarity shown, from a previous cycle, and that the main thyristor TH_1 is on and supplying load current. When switch S_w is closed the current through inductor L starts to increase, due to the reverse polarity on the capacitor. When this current equals that of the load current all the load power is supplied via commutation capacitor C. The reverse voltage of the capacitor is now placed across the main thyristor and it turns off. This thyristor will remain off so long as it is reverse biased for a period in excess of its turn-off time. The commutation capacitor charges via the load to at least the supply voltage. For heavy loads it would charge to a voltage greater than that of the supply since the energy stored in L, whilst it was passing the load current, would transfer to the commutation capacitor, boosting its voltage. The peak voltage on the commutation capacitor, and its rate of rise, must not exceed the rating of the main thyristor or it will break over into conduction.

Assuming that the commutation capacitor is charged to the value of the battery voltage V_B prior to commutation, the value of the commutation capacitor and inductor can be found from

$$C = \frac{2\sqrt{(2)}\, t_{OFF}\, I_{L(pk)}}{\pi\, V_B} \tag{11.5}$$

$$L = \frac{\sqrt{(2)}\, t_{OFF}\, V_B}{\pi\, I_{L(pk)}} \tag{11.6}$$

where $I_{L(pk)}$ is the peak load current at the time commutation commences (switch S_w is closed) and t_{OFF} is the turn-off time of the thyristor.

From these equations it can be seen that parallel capacitor–inductor commutation is unsuitable for use in systems controlling large currents and operating from low supply voltages since it would require a large value of commutation capacitor and an impractical small value of series inductance. For example, to turn off 250 A from a 40 V supply, using thyristors with turn-off times of 50 μs, would require a commutation capacitor of 280 μF and an inductor of 2.5 μH. In all probability, the inductance of the connecting leads would exceed this value.

Figure 11.8 shows the most basic commutation circuit, in the parallel capacitor–inductor group, which is used in practice. When the power is first applied capacitor C charges to the supply voltage with plate b positive and the main thyristor TH_1 cannot be fired until this has been completed. When TH_1 is turned on the load cycle commences and simultaneously the capacitor voltage is placed across inductor L_1, causing it to resonate and recharge with plate a positive. Once this has been completed, the circuit conditions are as in Figure 11.2(b) with the switch closed. The capacitor commences to discharge through the load and as the current builds up through the inductor, to reach the value of the load current, thyristor TH_1 turns off. The capacitor now continues to charge towards the value of the supply voltage, as before.

This circuit can be analysed using the six comparative factors introduced earlier:

(i) It can only operate in a variable-frequency mode, since there is control over the firing time of thyristor TH_1 but no control over when

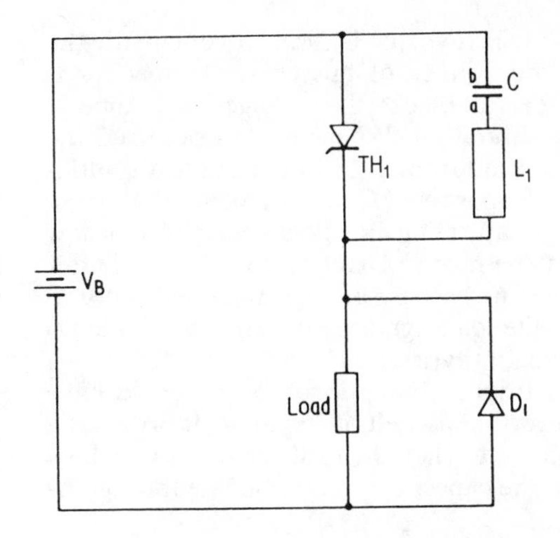

Figure 11.8 A basic parallel capacitor–inductor commutated circuit

it turns off. Turn-off occurs at a fixed time after the thyristor is turned on and C has resonated through L_1.

(ii) Assuming that the current through the inductor L_1 quickly reaches the load current, when the capacitor is placed across it, the on time t_c of this chopper circuit is determined by the time needed for C to resonate through L_1 and is given by equation (11.3). This determines both the minimum and the maximum on times. Once the main thyristor TH_1 has been turned on it cannot be refired until capacitor C has discharged through the load and recharged with plate b positive, which can be a long time on light loads. Therefore the minimum off time can be large, giving a poor ratio of on-to-off time and a low maximum output voltage.

(iii) The capacitor voltage is increased by the load current since it causes energy to be stored in the inductor L_1 which is subsequently transferred to the capacitor. Therefore on heavy loads the commutation energy is increased, which is desirable. If $I_{L(pk)}$ is the peak load current being commutated, then the value of the capacitor voltage is given by

$$V_c = V_B + I_{L(pk)} \uparrow \sqrt{\left(\frac{L_1}{C}\right)} \tag{11.7}$$

(iv) If thyristor TH_1 does not turn off after capacitor C has completed its discharge through L_1 and the load then, since the voltage across the inductor–capacitor commutation is now equal to that of an on thyristor, capacitor C cannot charge with plate b positive and no more commutation attempts will be made. 'Once-for-all' commutation failure has occurred.

(v) The resonant charge current of C flows through the main thyristor TH_1, so its rating is increased.

(vi) A low-impedance path does not exist across the supply in the advent of a commutation failure.

The circuit of Figure 11.9 gives a chopper with greatly improved performance. L_1 is now a saturable reactor. As before, C must be charged with plate b positive before the main thyristor is fired to commence the load cycle. When TH_1 is fired the voltage of this charged capacitor is placed across reactor L_1. The reactor blocks the voltage for a time t_p required to drive it into positive saturation. When this has occurred the reactor presents an after-saturation inductance L_1 to the capacitor and it resonates, recharging with plate a positive. Once resonance has been completed the reactor comes out of saturation and blocks current. It is now driven into negative saturation by the reversed capacitor voltage and after a negative saturation time t_n it again reaches saturation and capacitor C commences its discharge through the load (ignoring the effect of diode D_2 for the present), turning off the main thyristor TH_1.

The diode D_2 across the main thyristor is optional, and it could have been added to the circuit of Figure 11.8 as well, if required. It provides a path for the capacitor discharge current when the load current is very low. Therefore on an open-circuit load the capacitor would resonate through L_1 and D_1.

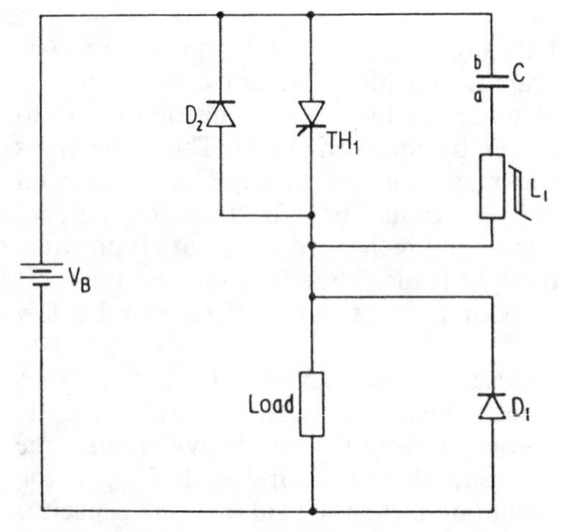

Figure 11.9 Modified parallel capacitor–inductor commutated circuit using a saturable reactor

The performance of the circuit shown in Figure 11.9 can be analysed using the six comparative points, as follows:

(i) Only variable-frequency operation is possible, since the circuit is essentially as in Figure 11.8.

(ii) The on time (t_c) of this circuit has now been increased by the two saturation times of the reactor and is given by equation (11.8). The minimum off time t_o, that is, the time during which the main thyristor must remain off during a cycle, to enable the capacitor to charge with plate b positive, is given by equation (11.9), which is the value on very light loads, when most of the discharge occurs due to the resonance of the capacitor via diode D_2:

$$t_c = \pi\sqrt{(L_1C)} + t_p + t_n \tag{11.8}$$

$$t_o = \pi\sqrt{(L_1C)} \tag{11.9}$$

(iii) The capacitor voltage is increased by the load current, as before, being given by equation (11.7), since on heavy loads most of the capacitor discharge will occur through the load.
(iv) If commutation fails for any reason then it cannot be re-attempted since the main thyristor, being continuously on, prevents the capacitor from charging with plate b positive.
(v) The rating of the main thyristor is increased by the capacitor resonant current.
(vi) No low-impedance fault current path exists across the supply.

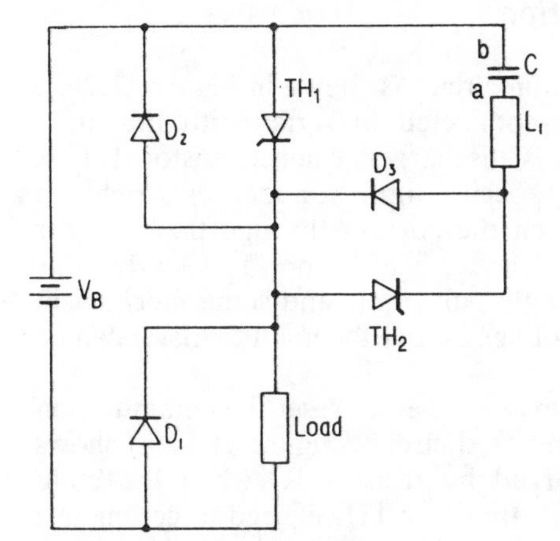

Figure 11.10 A parallel capacitor–inductor commutated circuit capable of variable frequency and variable mark-space operation

Figure 11.10 shows a parallel capacitor–inductor commutation circuit which is frequently used in practice. Although a saturable reactor can be used, as in Figure 11.9, a linear inductor is more usual, as shown. Initially C charges through L_1, D_3 and the load to reach the supply voltage with plate b positive. The load cycle can now commence with thyristor TH_1 being turned on. Capacitor C cannot discharge since D_3 is reverse biased and TH_2 has not yet been turned on. To turn off the main thyristor TH_1 auxiliary thyristor TH_2 is fired, enabling C to resonate with L_1 and recharge with plate a positive. The capacitor now discharges through D_3 and then through both the load and the diode D_2, depending on the magnitude of the load current, as in Figure 11.9. As in that figure, diode D_2 is optional and performs the same role of limiting the discharge time of the capacitor on light loads.

The performance of the circuit in Figure 11.10 is as follows:

(i) Both variable-frequency and variable mark-space operation are possible.
(ii) The minimum on and off times are equal and are given by equation (11.3). However, since the main thyristor is not now commutated until thyristor TH_2 is fired, this can be made long compared to the resonant time given in equation (11.3), so the maximum output voltage can be close to that of the supply.
(iii) The commutation capacitor voltage is increased proportional to the load current, due to the transfer of energy from L_1 to C, as before,

when TH_1 turns off. The capacitor voltage is given by equation (11.7).
(iv) A commutation failure prevents the circuit from re-attempting commutation.
(v) The rating of thyristor TH_1 is increased by the resonance of C through L_1.
(vi) There is no low-impedance short-circuit current path across the supply.

11.5 Series capacitor commutation

The basic series capacitor commutation circuit is shown in Figure 11.2(c), where a commutation capacitor is connected in series with the load. Assume that initially this capacitor is discharged when thyristor TH_1 is fired to start the load cycle. The capacitor now commences to charge towards the supply voltage, and when the current through the thyristor falls below its holding current the device will turn off. Clearly, this elementary circuit is capable of a single pulse only and some mechanism must be used to reset the capacitor voltage before the main thyristor can be refired.

There are several methods which may be used to reset the commutation capacitor in a series capacitor commutated circuit. Figure 11.11(a) shows the capacitor voltage being discharged by resistor R_1 when the main thyristor is turned off. As before, this thyristor TH_1 is fired to commence the load cycle. Generally, the system requires an underdamped resonant circuit, made up of L_1, C and the load, such that the capacitor charges with plate b greater than that of the battery (V_B) so that thyristor TH_1 is reverse biased and turns off more rapidly than when the current is simply reduced to below its holding value.

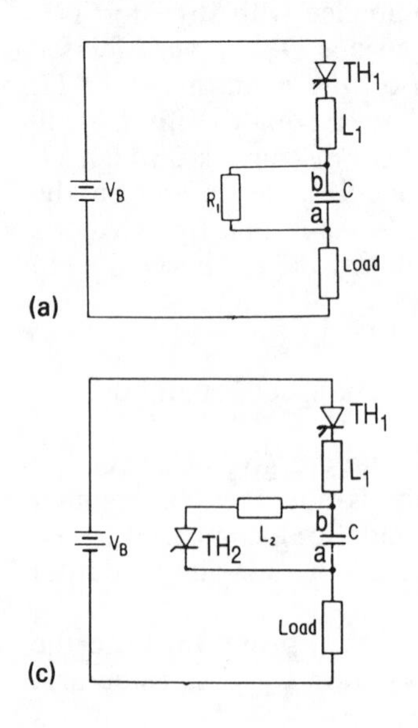

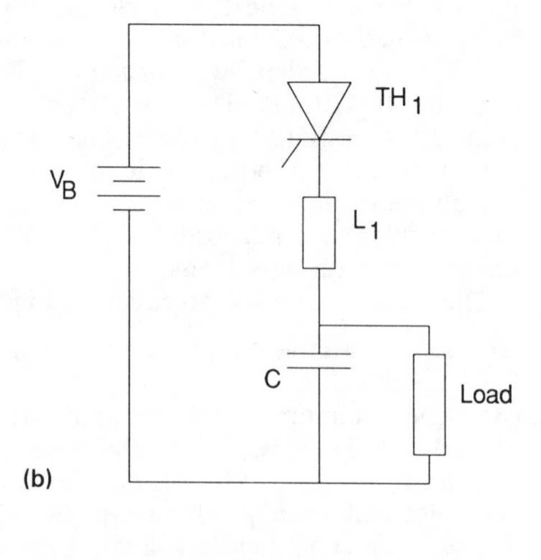

Figure 11.11 Series capacitor commutation: (a) resistor reset; (b) load reset; (c) resonant reset

The circuit of Figure 11.11(a) can be analysed using the six comparison points, as before:

(i) The on time is fixed by the resonant time in the circuit, so that only variable-frequency operation is possible.

(ii) The on time, caused by resonance, is equal to that given by

$$t_o = \pi\sqrt{(L_e C)} \tag{11.10}$$

where L_e is the effective inductance of the series combination of the load and of L_1. This resonant time determines the on period of the chopper circuit shown. After the main thyristor has turned off capacitor C commences to discharge through resistor R_1 and the main thyristor cannot be turned on again, to commence the load cycle, until this voltage has fallen to a sufficiently low value. Otherwise capacitor C will not charge to a high enough voltage for commutation and the magnitude of the load current pulse will be reduced. Therefore the minimum off time of the circuit can be quite large compared to the on time, giving a low maximum output voltage.

(iii) The current flowing in the load and in inductor L_1 both boost the voltage on the commutation capacitor, just prior to the start of a commutation cycle, therefore the commutation voltage is increased in proportion to the load. However, because the voltage on the commutation capacitor is always reset to as close to zero as possible before the start of the next cycle, this voltage boosting can be a nuisance in requiring the capacitor to discharge by a greater amount during the off periods of TH_1.

(iv) A commutation failure would generally occur if R_1 is of a low enough value to provide current in excess of the holding current of TH_1, and TH_1 does not turn off after a half cycle, possibly due to C not being sufficiently discharged from a previous cycle. Under these conditions TH_1 remains on continuously, supplying load current via R_1 and commutation is not re-attempted. An alternative failure mechanism is when R_1 is large and the capacitor is unable to discharge to a low enough voltage between the periods when TH_1 is off, so that when the thyristor is fired it can only deliver very small pulses of load current before its current falls below the holding value, turning it off.

(v) The rating of the main thyristor is not increased by the reset current of the commutation capacitor, which occurs through the resistor R_1.

(vi) There is no low-impedance short-circuit current path across the d.c. supply in the event of a commutation failure.

The circuit of Figure 11.11(a) has a very limited maximum output voltage range, as described in (ii), and is also relatively inefficient since the commutation reset power is dissipated in resistor R_1. Figure 11.11(b) shows a modified circuit in which the load is connected across the commutation capacitor. Once again the capacitor charge circuit must be underdamped to provide an effective reverse voltage to commutate thyristor TH_1, but now when this thyristor turns off, at the end of its resonant half cycle, the capacitor is reset by discharge through the load, rather than an external resistor, so the efficiency of the system is much higher.

Analysis of the circuit given in Figure 11.11(b) shows that it is similar to that in Figure 11.11(a) regarding all the six comparative points, as follows:

(i) Only variable-frequency operation is possible.
(ii) The minimum on time is determined by the resonance through L_1, the load inductance not playing any part. The off time is determined by the load and it should be resistive to prevent resonance during the discharge cycle, which would reverse the voltage across it.
(iii) The commutation capacitor voltage is increased by the load current flowing in inductor L_1.
(iv) Unsuccessful commutation is more likely, with the thyristor remaining permanently on through the relatively low impedance load. When this occurs commutation is not re-attempted.
(v) The current rating of the thyristor is not increased by the commutation capacitor reset action.
(vi) A low-impedance fault current path does not exist across the supply.

A more effective series capacitor commutated circuit is shown in Figure 11.11(c), where a separate resonant circuit is used to reset the commutation capacitor. As before, the main thyristor TH_1 is fired to commence the load cycle and, assuming an underdamped circuit, this thyristor will be turned off at the end of a half cycle of resonance, when C is charged to a voltage in excess of V_B with plate b positive. After a time equal to the turn-off time of TH_1, auxiliary thyristor TH_2 is fired, which enables C to resonate through inductor L_2 and recharge with plate a positive, ready for the next half cycle of load current. This circuit has the following characteristics:

(i) Only variable-frequency operation is possible, since the thyristor is only capable of delivering a half cycle of power during its on period.
(ii) The on time is given by equation (11.10). The off time is also given by this equation where the effective inductance L_e is now equal to L_2. Since this inductance, L_2, can be made much smaller than L_e the minimum off time is much less than the on period, so that the maximum output voltage from the chopper circuit can be made to be close to that of the supply.
(iii) The voltage on the commutation capacitor is boosted by the load current flowing through L_1 and the load. Furthermore, since the voltage on C is built up from two sets of resonant circuits, it will reach a value considerably in excess of the supply voltage V_B, which will result in faster commutation of the main thyristor.
(iv) A commutation failure path exists via L_1, L_2 and the load, and when this occurs both thyristors TH_1 and TH_2 will remain permanently on, supplying the load current, and commutation will not be re-attempted. Since TH_2 needs to carry the resonant reset current during normal operation it is usually a low-current device, so it may be destroyed when called upon to carry full-load current unless protection circuitry is used.
(v) The rating of the main thyristor TH_1 is not increased by the commutation capacitor reset current, which occurs through TH_2.
(vi) There is no low-impedance short-circuit current path across the supply if commutation fails.

Series capacitor commutation circuits have a very limited operational range, due to the capacitor reset time, and are rarely used as choppers. They find more frequent application as sine wave inverters, and this will be discussed further in Chapter 13.

11.6 Coupled-pulse commutation

The elementary coupled circuit shown in Figure 11.2(d) used a transformer to couple the charge from the commutation capacitor onto the conducting thyristor, reverse biasing it and turning it off. The auxiliary circuitry for priming the commutation capacitor is not shown, but since an isolating transformer is employed this can consist of a variety of systems. In addition, the capacitor can be replaced by a separate d.c. source, which is momentarily coupled onto the main thyristor using a power semiconductor. Such a device could be a transistor or gate turn-off switch, which can be turned off by means of signals on its control terminal, the transformer being used to match the voltage and current rating of the device to that of the load.

Figure 11.12 shows a coupled-pulse circuit which couples the charge on commutation capacitor C to the main thyristor TH_1 using transformer T_1. Initially the capacitor is charged to greater than the supply voltage, V_B, due to the resonant action of L_1 and C, with plate b positive. The main

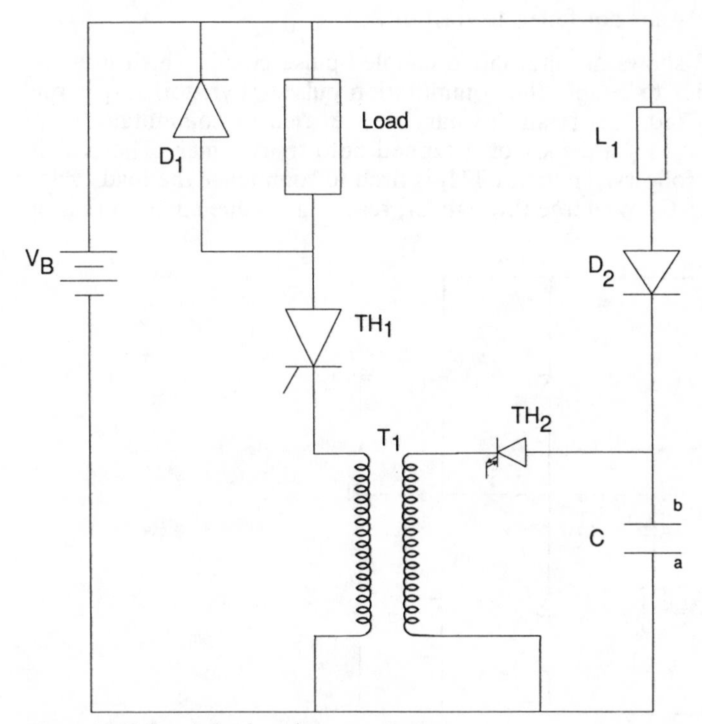

Figure 11.12 Coupled pulse using a transformer

thyristor TH_1 is fired to commence the load cycles and when it is required to commutate its gate turn-off switch, TH_2, is turned on. Capacitor C discharges into transformer T_1, reverse biasing TH_1 and it will turn off providing the reverse voltage is maintained for a time in excess of its turn-off time. C will resonate with the inductance of T_1 and when this has been completed gate turn-off switch TH_2 is turned off by its gate control, enabling C to recharge with its b plate positive, ready for another commutation cycle.

The circuit of Figure 11.12 can be analysed as follows:

(i) It can operate in a variable-frequency or a variable mark-space mode.
(ii) The minimum on time can be made very small, since commutation can start soon after the main thyristor has been fired. The minimum off time is determined by the resonant time for C to charge via L_1 and is given by equation (11.9). Therefore the chopper is capable of a wide output voltage range.
(iii) The commutation voltage is not appreciably increased by the load current, assuming low transformer inductance.
(iv) If commutation is not successful and TH_1 does not turn off when TH_2 is fired, then when TH_2 is turned off the capacitor will recharge so that commutation will be re-attempted on the following cycle.
(v) The current rating of the main thyristor is not increased by the reset action of the commutation circuit.
(vi) A low-impedance current path does not exist across the d.c. supply, in the advent of a commutation failure, assuming that gate turn-off switch TH_2 has not failed to turn off.

Figure 11.13 shows an alternative coupled-pulse circuit which uses an auto-transformer to couple the commutation pulse. Thyristor TH_1 is the main thyristor and TH_2 is an auxiliary device used to commutate TH_1. Inductors L_1 and L_2 are part of a tapped auto-transformer. The circuit operation is as follows. Thyristor TH_1 is fired to commence the load cycle. This also causes C_1 to charge through L_1, reaching a value equal to that of

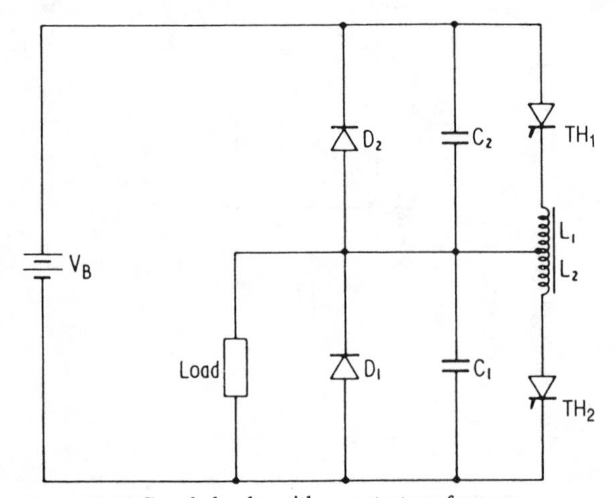

Figure 11.13 Coupled pulse with an auto-transformer

the supply voltage V_B. The clamping action of diode D_2 prevents C_1 charging to a higher voltage. To turn TH_1 off thyristor TH_2 is fired, causing capacitor C_1 to discharge through L_2 and coupling a voltage pulse via L_1 to TH_1, so turning it off. The voltage on C_1 falls to zero, after which D_1 conducts, carrying the current due to energy stored in L_2 and the load. Thyristor TH_2 turns off when the current in L_2 has fallen to below the device-holding value. C_2 is not normally required, but by including it thyristor TH_1 can be fired before TH_2 has turned off, since then C_2 discharges through L_1 coupling a pulse to TH_2 via L_2 and so turning it off. This allows a wider voltage control range.

Assuming a symmetrical system, with $C_1 = C_2 = C$ and $L_1 = L_2 = L$ the values of L and C can be found from equations (11.11) and (11.12). These equations illustrate that this commutation method is not suitable for low-voltage high-current applications, since it would require a large value of commutation capacitor and an impractical small value of inductance:

$$L = \frac{2.76 \times V_B \times t_{OFF}}{I_{L(pk)}} \tag{11.11}$$

$$C = \frac{2.15\, t_{OFF} \times I_{L(pk)}}{V_B} \tag{11.12}$$

The following can be noted about the circuit in Figure 11.13.

(i) The system can be operated in a variable-frequency or a variable mark-space mode.
(ii) The minimum on and off times are approximately equal and are given by $\pi\sqrt{(L_1C_1)}$ and $\pi\sqrt{(L_2C_2)}$ respectively, for low load current.
(iii) Due to the clamping action of D_1 and D_2, the commutation voltage is fixed at V_B, irrespective of the load current.
(iv) An unsuccessful commutation would not allow the circuits to re-attempt commutation.
(v) The charging pulse for C_1 and the discharge pulse for C_2 flow through TH_1 increasing its current rating.
(vi) A low-impedance failure path exists across the d.c. supply.

In comparing the various commutation methods described in this chapter, generally parallel-capacitor commutation circuits are the most flexible and are suitable for operating from a wide range of supply voltages. However, in applications where the supply voltage is high and the load current small, parallel capacitor–inductor commutation should be considered, since these are often simpler than the corresponding parallel-capacitor circuits. Coupled-pulse commutation is more expensive, due to the use of a transformer, and it finds more frequent use in inverter circuits. Similarly, series capacitor commutation is not often used for choppers, due to its limited output voltage range, but is often employed in sine wave inverters.

Chapter 12

D.C. to d.c. converters

12.1 Introduction

To regulate the power between a d.c. source and a d.c. load both linear and switching techniques can be used. Linear regulators are simpler in design and provide a smoother output with less RFI generation, but they dissipate much more power within the regulator. Therefore for high-power applications switching regulators are almost invariably used.

The basic switching d.c. to d.c. regulator was introduced in Chapter 6, and it was shown there that both step-down and step-up controllers are available. The step-down converter is the one most frequently used and this is described in the next section, step-up converters being introduced in section 12.5.

Figure 12.1 shows an elementary transistor step-down converter (chopper) which is similar to the basic circuit illustrated earlier in Figure 6.5 except that a power transistor is used as the switch element and an L-C filter section is included to smooth the load voltage, as would normally be required for a power supply application. Figure 12.1 also shows the circuit waveforms and, as described with reference to Figure 6.6, the output voltage can be controlled by varying the frequency or the mark-space ratio of the conducting transistor switch. The mean output voltage across the load is given by

$$V_L = V_B \frac{t_c}{t_c + t_o} \qquad (12.1)$$

where t_c is the closed time of the switch and t_o is its open time.

When the transistor switch is conducting, the load current builds up through the device. When the transistor turns off the load current switches to the free-wheeling diode D_1 and the current slowly decays, until the transistor is turned on again at the start of the next cycle.

The chopping frequency used must be high, to minimise the load ripple. For example, the filter attenuation factor is given by

$$\tau = \frac{Z_{ca} + Z_{in}}{Z_{ca}} \qquad (12.2)$$

where Z_{in} is the impedance of the inductor L_1 and Z_{ca} is the impedance of

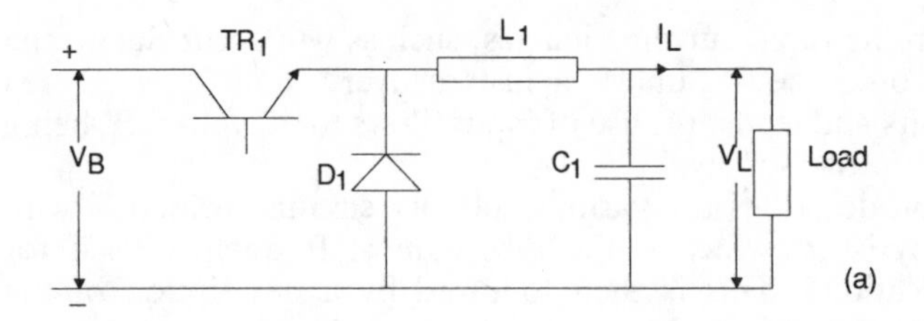

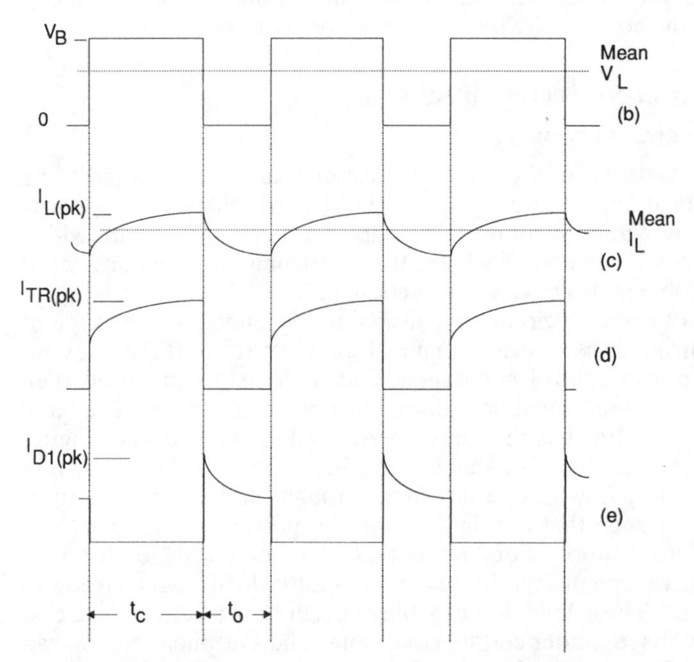

Figure 12.1 Basic switching d.c. to d.c. converter (chopper) operation: (a) circuit arrangement; (b) load voltage; (c) load current; (d) switch (TR_1) current; (e) free-wheeling diode (D_1) current

capacitor C_1. Substituting the values of the inductor and capacitor and the angular frequency (ω), this reduces to

$$\tau = 1 - \omega^2 L_1 C_1 \tag{12.3}$$

For a large value of filter attenuation and ignoring the negative sign, which has no practical significance, equation (12.3) leads to equation (12.4), which illustrates the dependence of the effectiveness of the filter on the value of the chopper-operating frequency, so that a high operating frequency is desirable.

Apart from increasing the power dissipation and complexity of a chopper, a high operating frequency has the disadvantage of giving large RFI generation. This is primarily due to the rapid changes in current in the circuit and can be minimised by the techniques described in Chapter 4. The main precautions are: (1) Keeping wire lengths short to reduce radiation effects; (2) preventing rapid current changes, such as will occur due to the reverse recovery of diode D_1 unless a fast recovery diode is used; (3) screening of circuits and leads; (4) use of input filters to prevent RFI being conducted back into the supply.

The next section describes the various chopper circuits, primarily with reference to a thyristor, which is the device most frequently used for high-power applications. This is then followed by a description of the chopper voltage control methods and the design techniques for chopper circuits. The chapter concludes with a description of the step-up d.c. to d.c. converter and the control circuits which may be used with choppers.

12.2 D.C. to d.c. converter circuits

12.2.1 Chopper arrangements

The two major divisions of d.c. converter circuits can be considered to be the output voltage level relative to the input, i.e. whether the chopper is step-up or step-down, as introduced in Chapter 6. In the discussions which follow, step-down choppers, which are the most common, are considered and step-up choppers are covered in section 12.5.

Three types of chopper circuits are available, depending on the type of load being controlled, as shown in Figure 12.2. Gate turn-off switches are shown as the power-control component and if thyristors are used then commutation components will be required in a practical system. The basic circuit of Figure 12.2(a) has the same operational mode as that of Figure 12.1 and the output is identical, load filtering being ignored. The current in this system can only flow in one direction through the load, from a to b, and the voltage is such that terminal a is always positive.

The regenerative chopper of Figure 12.2(b) is used with loads which have regenerative capability, for example motor loads which require electrical braking. These loads are capable of generating a voltage in excess of the battery voltage, under certain conditions, and this phenomenon can be used to feed energy back from the load to the supply, instead of dissipating the excess energy in resistors. During normal chopper operation gate turn-off switch TH_1 regulates the load power and free-wheeling diode D_2 carries the inductive current of the load when no power is being supplied from the d.c. source. The current in this mode flows from a to b through the load and terminal a is positive, as for the basic chopper of Figure 12.2(a). During the regenerative period the load current reverses and flows out of terminal a, through diode D_1 and into the battery. The amount of regeneration can be controlled by operating gate turn-off switch TH_2, in either a variable-frequency or mark-space mode, since when this device is on the load current will circulate through it rather than being fed back to the supply. By operating TH_2 it is also possible to store energy in the load inductance, which is subsequently fed to the supply when TH_2 is off, so that this system works like a step-up chopper. It should be noted that for regeneration in a chopper the polarity of the current has changed, whereas for regeneration in a.c. to d.c. converters, described in Chapter 9, the polarity of the voltage changes.

The reversing and regenerative chopper of Figure 12.2(c) can be considered to be composed of two regenerative choppers, one set

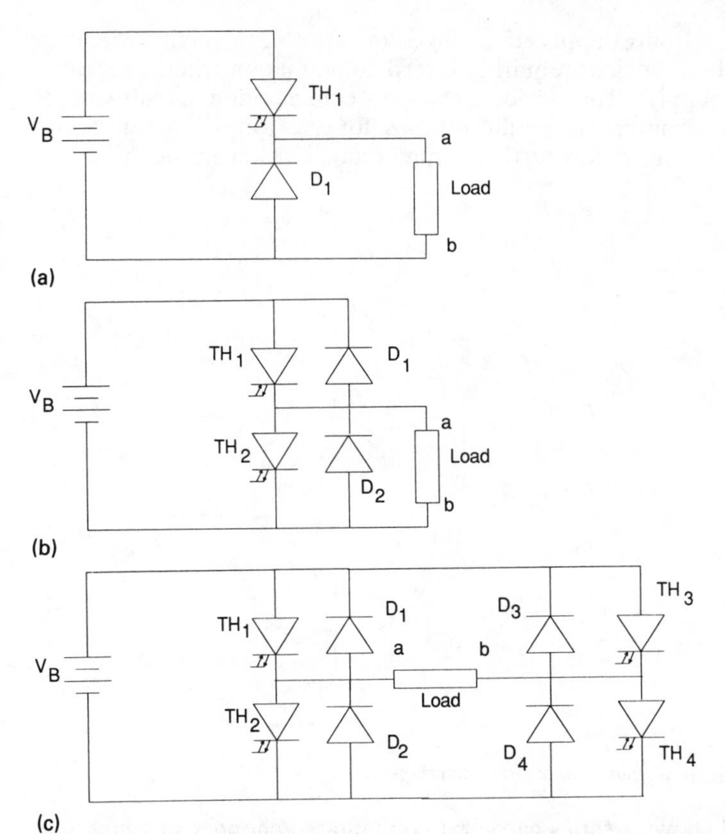

Figure 12.2 Chopper arrangements: (a) basic chopper; (b) regenerative chopper; (c) reversing and regenerative chopper

operating for any given polarity of the load. When the load voltage is to be such that side a is positive switches TH_1 and TH_4 are turned on for normal operation, the load current free-wheeling through TH_4 and D_2 during the off periods of the chopper. For regeneration, in this direction, switch TH_2 and diodes D_1 and D_4 come into operation. To reverse the load voltage, switches TH_2 and TH_3 are now in operation, the free-wheeling current path being through TH_2 and D_4. For regeneration, D_2, D_3 and TH_4 come into play. As seen from Figure 12.2(c), the reversing and regenerative chopper requires many components, especially when thyristors are used as the power switches and commutation components have to be added. In applications where reversing is performed relatively infrequently it is therefore more common to use reversing mechanical contactors rather than power semiconductors.

12.2.2 Commutation methods

Although gate turn-off switches have been shown in the illustrations of the previous section, and power transistors are also widely used in chopper circuits, for high-power applications thyristors are still the main switching component, these devices requiring forced commutation when operating from a d.c. supply. The various chopper commutation circuits were introduced in Chapter 11, as illustrations for the forced commutation classification system. A few further commutation circuits are described in this section.

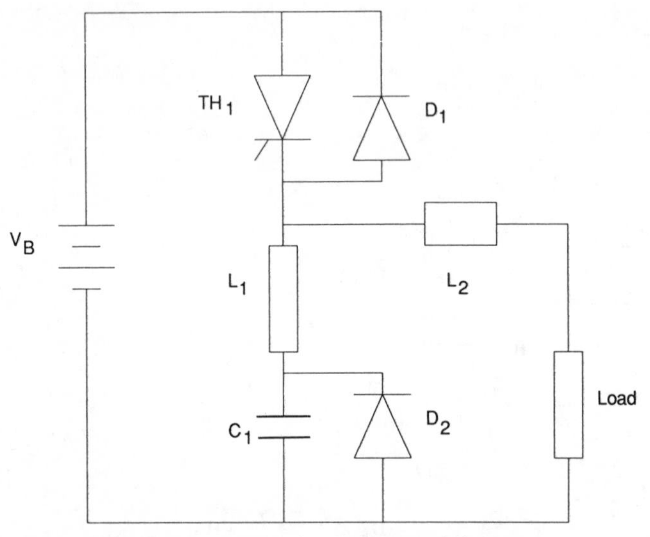

Figure 12.3 A basic series capacitor commutated chopper

Figure 12.3 shows a series capacitor commutated chopper in which the commutation capacitor is reset through the load, as in Figure 11.11(b). Thyristor TH_1 is turned on to commence the load cycle, and when the current through it exceeds that of the load it commences to charge the commutation capacitor C_1 due to resonance. The thyristor turns off when C_1 has resonated through the series inductor L_1. After TH_1 turns off the capacitor discharges through the load, inductors L_1 and L_2 acting to smooth the load current which now flows via the free-wheeling diode D_2. The function of this diode is also to prevent the capacitor voltage from swinging negative, due to resonance with the series inductors and that of the load, so that it will always discharge to zero and the load voltage does not go negative. For effective commutation the capacitor must be discharged to zero volts before the main thyristor can be fired again.

A parallel capacitor commutated basic chopper arrangement, which is similar to those of Figures 11.6 and 11.7, is shown in Figure 12.4. Thyristor TH_2 is initially fired, which charges capacitor C to the supply voltage V_B with plate b positive. Thyristor TH_1 is now fired to commence the load cycle and simultaneously with this thyristor TH_3 is turned on, which allows C to resonate with L_1, the voltage across it reversing and having the same magnitude if the resonant losses are low. To turn TH_1 off thyristor TH_2 is fired, C discharging through the load. Until the capacitor voltage falls to zero TH_1 is reverse biased, and the time from the instant of firing TH_2 is known as the commutation interval, which should exceed the rated turn-off

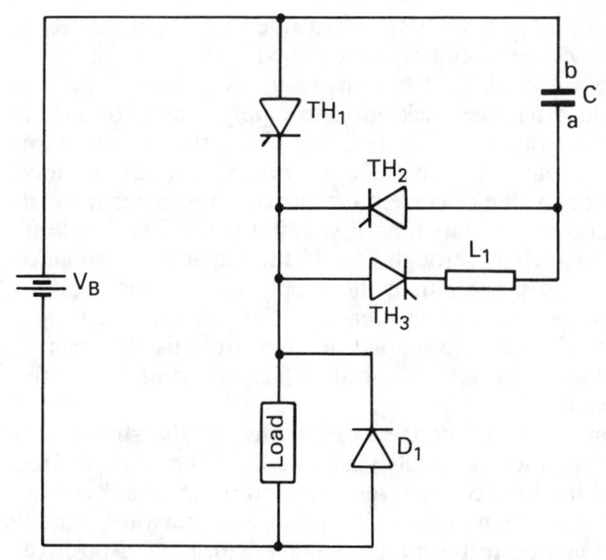

Figure 12.4 A basic parallel capacitor commutated chopper

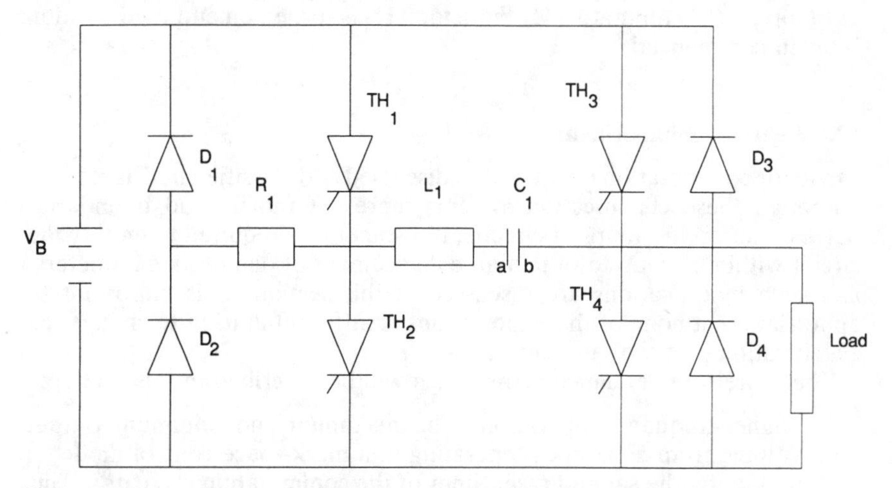

Figure 12.5 A regenerative parallel capacitor–inductor commutated chopper

time for TH_1. Thereafter C continues charging via TH_2 and the load until it reaches voltage V_B with plate b positive. TH_2 now turns off and load current free-wheels in D_1 until TH_1 is again fired to commence the next cycle.

Figure 12.5 illustrates a regenerative chopper arrangement which uses a parallel capacitor–inductor commutation system. During normal chopper operation thyristor TH_3 conducts and provides power to the load. Thyristor TH_2 is turned on simultaneously with TH_3, or a short time later, to charge the commutation capacitor with plate b positive. The capacitor will resonate with its series inductor and initially charge to greater than the supply voltage V_B, but then the capacitor will discharge down to V_B via the current path D_3, the supply, D_2 and R_1. The value of this resistor R_1 is chosen such that the resonant circuit is overdamped.

Thyristor TH_2 turns off once C_1 has charged to its peak voltage and commences its resonant discharge back into the supply. Thyristor TH_1 is fired to turn off the main thyristor TH_3 and this places the commutation L_1-C_1 circuit across the main thyristor. The capacitor provides the load current and after the peak value of current is reached it resonates through TH_1 and D_3 and recharges with plate a positive, after which TH_1 goes off, the load current free-wheeling through D_4. If the capacitor resonance caused it to be charged to greater than the supply voltage with plate a positive, then this voltage will discharge back to the value of the supply voltage through the path R_1, D_1, the supply and D_4. After this, TH_3 can be refired to commence the load cycle, followed by TH_2 to prime C_1 for the next commutation period.

During regeneration the load voltage exceeds that of the supply and current flows from the load to the supply via diode D_3. When TH_4 is fired regeneration stops and the load current free-wheels through this thyristor. Thyristor TH_1 is fired at the same time as TH_4, or soon afterwards, and it causes capacitor C_1 to charge to the supply voltage with plate a positive, any excess voltage being dissipated in the path R_1, D_1, the supply, and D_4, as before. To commutate TH_4 thyristor TH_2 is turned on and regeneration is again commenced.

12.2.3 Circuit enhancements

Thyristor commutation circuits were discussed and classified in Chapter 11. Although these classification systems represent rather rigid boundaries between different groups of circuits, it is sometimes required to modify the circuit within a group to obtain an enhancement of one of its parameters, and such modifications are discussed in this section. It is important to appreciate that none of these alterations is sufficient in itself to change the classification group of a circuit.

There are three enhancements which will be described in this section:

(i) Higher-frequency operation. The maximum and minimum output voltage from a chopper, operating in a mark-space control mode, is limited by the set and reset times of the commutation capacitor. This limits the maximum operating frequency which can be used and still give a useful output voltage range.

(ii) Reduction of commutation losses, which again limits the operating frequency. A chopper or inverter may be designed to give small minimum on and off times and adequate output voltage control range at high frequencies, but excessive commutation losses reduce the efficiency obtained and may make it unsuitable for use at these frequencies.

(iii) Commutation voltage boosting. The higher the commutation capacitor voltage, the smaller its size, for the same commutation energy output, which may lead to cheaper components. If the voltage boost is proportional to load current, optimisation of components at all outputs can be obtained.

12.2.3.1 High-frequency operation

Two problems can arise when a chopper is operated at high frequencies. First, the finite time required to set and reset the commutation capacitor limits the maximum and minimum voltage at any frequency. Second, the necessity of ensuring that a thyristor is reverse biased for the duration of its turn-off time limits the operating frequency unless some form of sequencing is used, as explained later.

In the circuit shown in Figure 12.4 suppose that C is charged to V_B with plate b positive, due to a preceding cycle. Thyristor TH_1 is now fired to supply the load current and simultaneously TH_3 is turned on to cause C to resonate through L_1 and to recharge with plate a positive. The resonant time is given by equation (12.4) and represents the minimum on period for TH_1, since TH_2 cannot be turned on to commutate it until the capacitor has been set:

$$t_c = \pi \sqrt{(L_1 C)} \tag{12.4}$$

When TH_2 is fired thyristor TH_1 turns off and C discharges at constant (assumed) load current I_L, recharging to V_B in a time given by

$$t_o = \frac{2C\,V_B}{I_L} \tag{12.5}$$

Thyristor TH_1 cannot be refired until this is completed, therefore it represents its minimum off period. Capacitor C is chosen to ensure that TH_1 is reverse biased for longer than its turn-off time, under maximum load conditions. If the load current now falls, the reset time can be very large, severely limiting the peak output voltage.

Several techniques exist for resetting the capacitor rapidly, even under light load conditions. Figure 12.6 shows a modification to the chopper of Figure 12.4, where a reset choke L_2 and diode D_2 have been added across the main thyristor to provide a capacitor reset path. Now under open-circuit load conditions the maximum reset time, neglecting resonance losses, is given by equation (12.6), which can be made relatively short.

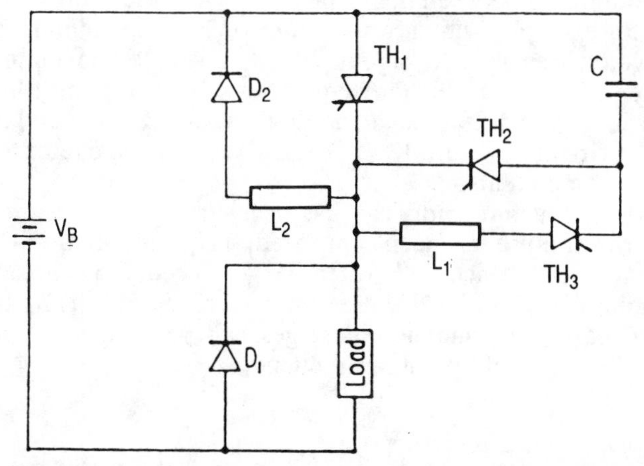

Figure 12.6 Addition of an auxiliary capacitor reset path to Figure 12.4 for higher-frequency operation

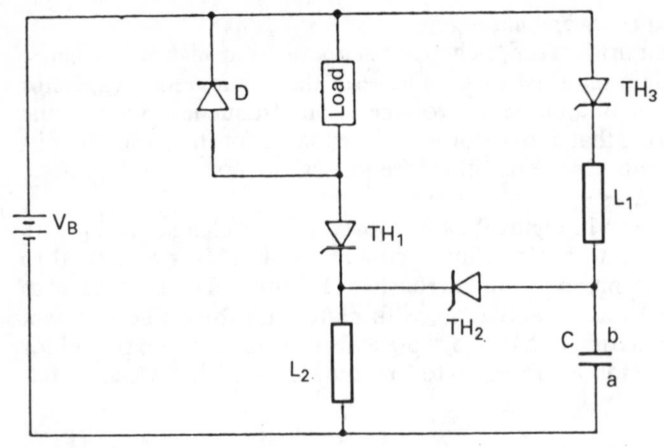

Figure 12.7 A chopper circuit with inherent high-frequency properties

Not all chopper circuits need additions to reduce the commutation capacitor reset time. Figure 12.7 shows a parallel capacitor commutated circuit with inherent high-frequency capabilities. TH_3 is fired simultaneously with the main thyristor TH_1 and it sets the commutation capacitor C within a time given by equation (12.4). When thyristor TH_2 is fired, to turn TH_1 off, the capacitor resets within a time given by

$$t_o = \pi \sqrt{(L_2C)} \tag{12.6}$$

capacitor C resonating and recharging with plate a positive, ready for the next set cycle, when TH_3 is fired.

Apart from the need to set and reset the commutation circuit, which limits the maximum operating frequency, choppers suffer from the limitation imposed by the turn-off time requirements of the thyristors. Returning to Figure 12.6, the values of L_2 and C must be large enough to ensure that TH_1 is reverse biased for longer than its turn-off time. This puts a limit on the minimum time between the firing of TH_2 and the refiring of TH_1. For a device, for example, with a turn-off time of 20 μs the minimum chopping period would be typically about 200 μs, giving a maximum operating frequency of 5 kHz. For higher frequencies sequencing must be used, as illustrated in Figure 12.8, which shows a three-stage sequential chopper based on the circuit of Figure 12.7. The suffixes a, b and c refer to the separate inverter components.

Suppose TH_{1a} was conducting and TH_{2a} has been fired to turn it off. Then, before the turn-off pulse has been completed, TH_{1b} can be fired to recommence the load cycle, which ends when TH_{2b} is fired. To reapply load current, assuming TH_{1a} and TH_{1b} are still reverse biased, TH_{1c} is fired, and so on. Clearly, any number of stages can be connected in sequence to obtain the required operating frequency.

12.2.3.2 Reduction of commutation losses

Losses normally occur in commutation circuitry due to the transfer of energy, from the commutation capacitor to a commutation choke, during

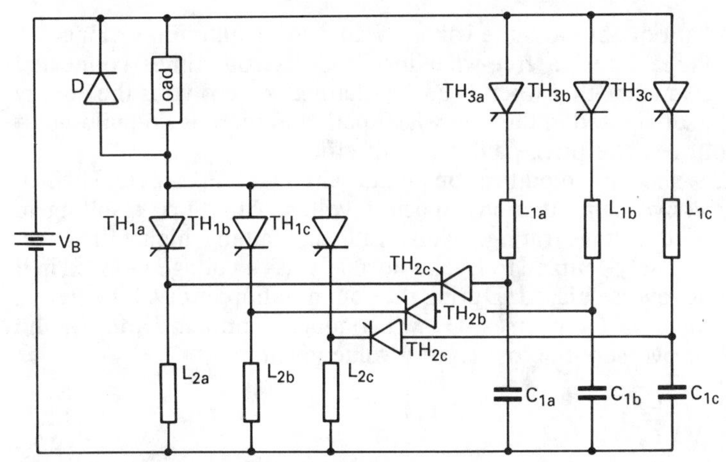

Figure 12.8 Three-stage sequential chopper

the turn-off process and the subsequent dissipation of the energy stored in the choke. The commutation loss is proportional to the load current, which determines the capacitor size for any voltage and the frequency of operation. There is a fixed watts loss per commutation, so that the higher the operating frequency, the larger the energy loss and the lower the system efficiency.

Choppers do not generally have high inherent commutation loss problems. For instance, referring to Figures 12.4 and 12.7, the commutation energy is interchanged between choke and capacitor during the turn-off period, so that there should theoretically be no efficiency loss. In Figure 12.7, assuming C to be charged to V_{pk} with plate b positive, then when TH_2 is fired TH_1 turns off. C now resonates with L_2 and discharges to zero voltage. The commutation energy has transferred from C to L_2, but the commutation interval proceeds beyond this point, with the transfer of energy back from L_2 to C, so that the capacitor recharges to V_{pk} with plate a positive. When TH_3 is next fired C will resonate through the supply and L_1 and be reset, giving zero commutation loss. However, it can be seen that the capacitor voltage builds up over several cycles, to the value V_{pk}, which may be several times the supply V_B depending on the losses in the

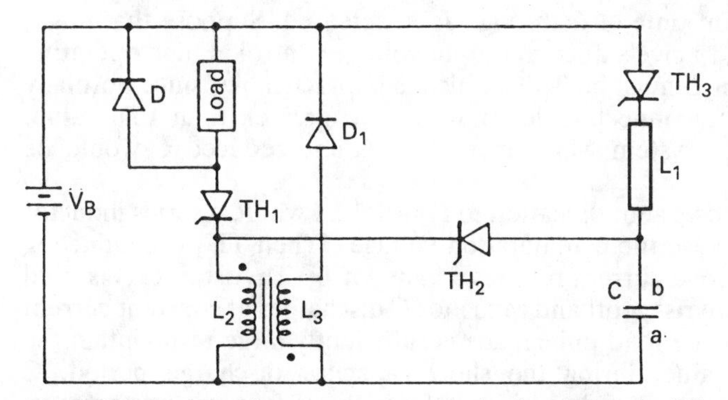

Figure 12.9 Commutation energy recovery in the chopper circuit of Figure 12.7

resonant paths. For zero loss the voltage would reach an infinite value. To limit this voltage boost a free-wheeling diode is sometimes connected across L_2. This means that when C has discharged to zero volts the energy stored in L_2 is dissipated in the free-wheel path and there is no build-up of capacitor voltage. The price paid is loss of efficiency.

Figure 12.9 shows a modification to the circuit of Figure 12.7 which prevents excessive commutation capacitor voltage build-up, resulting in increased device voltage rating, whilst still maintaining high efficiency. Winding L_3 is closely coupled to L_2 and normally has a voltage induced in it such that D_1 is reverse biased. During the commutation interval, however, when the voltage on L_2 reverses, D_1 will conduct, clamping L_3 to V_B and resulting in an overshoot across L_2 of a value given by

$$V_o = V_B \sqrt{\left(\frac{L_2}{L_3}\right)} \qquad (12.7)$$

which can be made small. There is theoretically no commutation loss since the commutation energy stored in L_2 is fed back to the supply by L_3.

12.2.3.3 Voltage boosting

If the commutation voltage is high, the commutation capacitor can have a low value and still provide the same turn-off energy required for the thyristors. For instance, in the chopper circuit of Figure 12.4 if t_{OFF} is the turn-off time of TH_1 then, to commutate a peak current of $I_{L(pk)}$, the capacitor size required is given by

$$C = \frac{I_{L(pk)}\, t_{OFF}}{V_c} \qquad (12.8)$$

In this equation V_c is the commutation capacitor voltage prior to commutation and in the circuit of Figure 12.4 it can have a maximum value of V_B. This equation shows that to commutate a current of 500 A, with a voltage across the commutation capacitor of 20 V and using 20 μs turn-off time thyristors, would require a capacitor of 500 μF, 20 V rating. If the value of V_c is increased by a factor of 10, by some method of voltage boosting, a commutation capacitor of 50 μF, 200 V will be needed, which is smaller and cheaper. Another disadvantage of using high-valued capacitors is that the resonant time of C through L is increased. Suppose that this is limited to 0.1 ms, to give sufficient output voltage control at high operating frequencies. Then L must be 2 μH, which is impracticable, since normally the inductance of connecting leads would be greater than this value. Clearly, a voltage system which increased V_c and reduced C would be advantageous.

Figure 12.10 shows a modification to Figure 12.4 where a series inductor L_S is used to increase the commutation voltage. When TH_1 is conducting suppose that the load current reaches a value of I_L. Thyristor TH_2 is fired to turn the main thyristor off and capacitor C discharges at constant current I_L, assuming that the load inductance is sufficiently large to maintain the current at this value during the short capacitor discharge period. C recharges to V_B with plate b positive. Normally, if L_S were not present, thyristor TH_2 would turn off at this point and the load current would free-wheel in D_1, but due to the presence of this inductor, there is energy (E_{ST}) stored in it, given by

$$E_{ST} = \tfrac{1}{2} L_S I_L^2 \tag{12.9}$$

This energy causes TH_2 to remain on, the voltage on C increasing as the current through it decreases, the difference between this and the load free-wheeling through D_1.

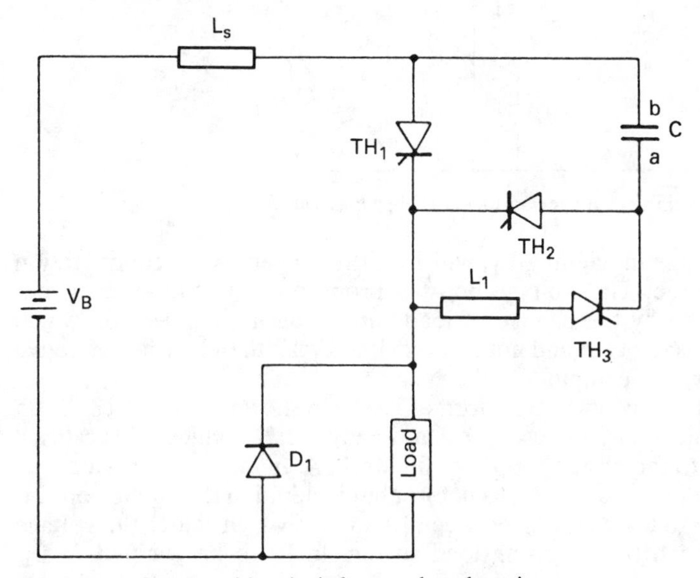

Figure 12.10 Chopper with series inductor voltage boosting

When C has charged to the value given by equation (12.10) the current through thyristor TH_2 has fallen to zero and it goes off, the total load current being carried by D_1. When TH_1 and TH_3 are next fired C resonates with L_1 and recharges to an increased value of commutation voltage:

$$V_c = V_B + I_L \sqrt{\left(\frac{L_S}{C}\right)} \tag{12.10}$$

The value of L_S need not be large. For instance, in the previous example, for $I_L = 500$ A and $C = 50\,\mu$F, a value of L_S equal to only 8 μH gives a boost of 200 V. The principal disadvantage of this system is that voltage boost is not effective during the first cycle, when TH_2 is first fired without any load current flowing. Thereafter the voltage boosting increases proportionally to the load current, which is a very desirable feature.

Figure 12.11 shows a further modification to the basic parallel capacitor commutated chopper, which overcomes the disadvantage of not providing a voltage boost during the first commutation, as experienced in Figure 12.10. Initially, TH_2 is fired to charge C to V_B with plate b positive. When TH_1 is fired C resonates through it and L_1-D_1, recharging with plate a positive. In addition to this discharge, the build-up of load current in L_2

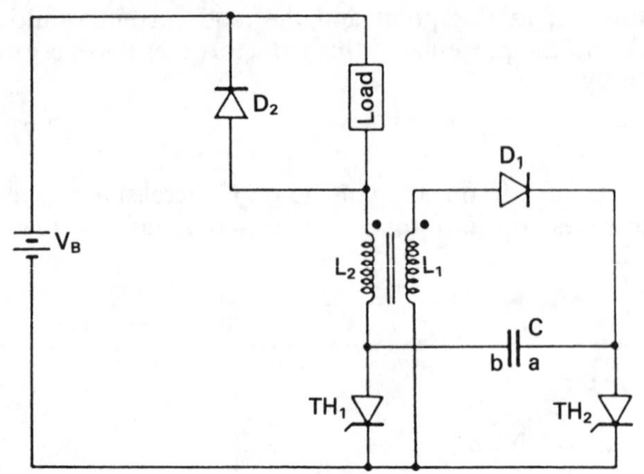

Figure 12.11 Transformer voltage boosting in a chopper circuit

induces a voltage in winding L_1, which further increases the commutation capacitor voltage. This voltage boost is proportional to the load current and it is caused by the change of load current during the cycle in which commutation is to occur and not in a previous cycle, therefore it is effective from the very first commutation.

Although a two-winding transformer has been shown in Figure 12.11, an auto-transformer can be used, as in Figure 12.12, which illustrates a modification to the basic chopper of Figure 12.4. Inductor L_1 provides the resonant path for capacitor C, as before, but in addition the voltage on C is increased due to the coupling between L_2 and L_1, which causes the voltage which is induced by the rising load current in L_2 to be coupled to L_1,

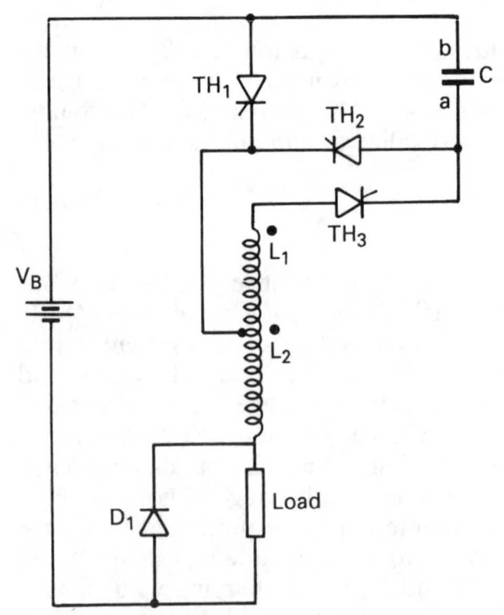

Figure 12.12 Modification to Figure 12.4 for auto-transformer voltage boosting

boosting the voltage on capacitor C by an amount which is proportional to this load current.

12.3 Output voltage control

The method of controlling the output voltage to the load, from a chopper, has already been discussed, and the two methods of variable-frequency and variable mark-space were illustrated in Figure 6.6. The mean output voltage is given by equation (12.1), and the r.m.s. voltage by

$$V_{rms} = V_B \left(\frac{t_c}{t_c + t_o}\right)^{½} \tag{12.11}$$

The voltage waveforms are rich in harmonics and if constant k is defined as the output duty cycle, given by

$$k = \frac{t_c}{t_c + t_o} \tag{12.12}$$

then the nth harmonic in this voltage (V_n) is given by

$$V_n = V_B \frac{\sqrt{(2)} \sin nk\pi}{n\pi} \tag{12.13}$$

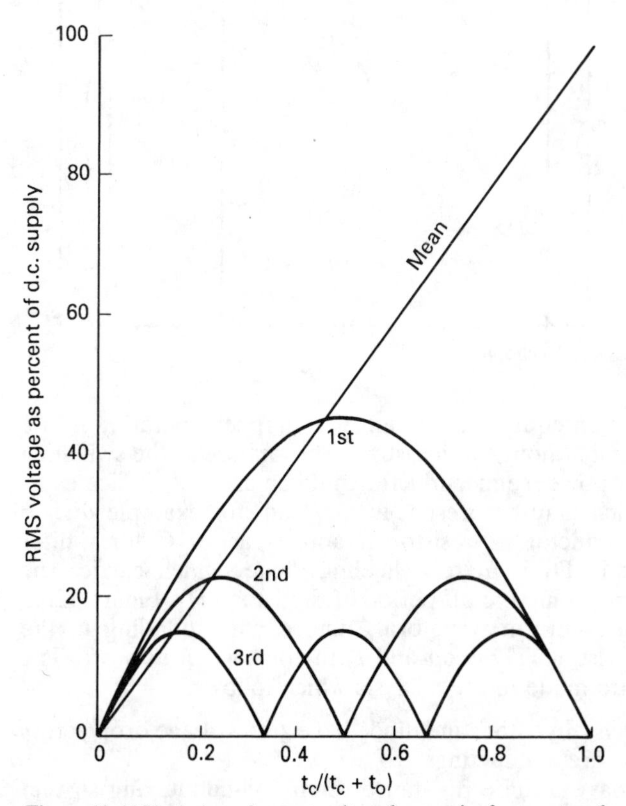

Figure 12.13 Variation of output voltage harmonics from a step-down chopper

The presence of any harmonic, and its magnitude, is therefore determined by the duty cycle of the output pulse, and is shown plotted in Figure 12.13, which also illustrates the variation of the mean voltage, as given by equation (12.1). This plot shows that harmonics are present at all times except when the chopper switch is continuously open ($k = 0$) or continuously closed ($k = 1$), the harmonic with the largest magnitude being that at the chopping frequency, as expected.

12.4 Design of chopper circuits

This section provides an analysis of chopper circuits to enable their design characteristics to be obtained. Initially, the commutation components will be ignored, so that the results are equally applicable to any form of switching control, for example those using transistors or mechanical switches, but later the effects of commutation, as needed for thyristor circuits, on a typical chopper are considered.

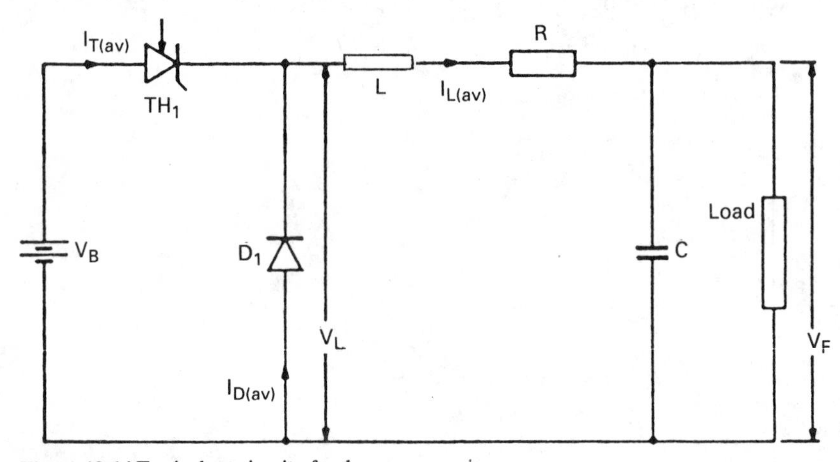

Figure 12.14 Equivalent circuit of a chopper

Figure 12.14 shows an equivalent circuit of a chopper operating into a load of voltage V_L, and although a thyristor is shown here as the switching component any other power semiconductor could be used. V_F indicates an inherent voltage which may be present in the load, for example due to motor back e.m.f. Inductor L, resistor R and capacitor C form filter components and diode D_1 is a free-wheeling diode, which carries the inductive load current during the off period of thyristor TH_1. Figure 12.15 gives this steady state current waveform, time t_c corresponding to the period for which thyristor TH_1 is on and t_o for that when it is off. The assumptions below are made in the analysis which follows:

(i) The power switch (thyristor) and diode have zero voltage drop across them when they are conducting.
(ii) These devices have infinite resistance when non-conducting so that the leakage current through them is negligible.

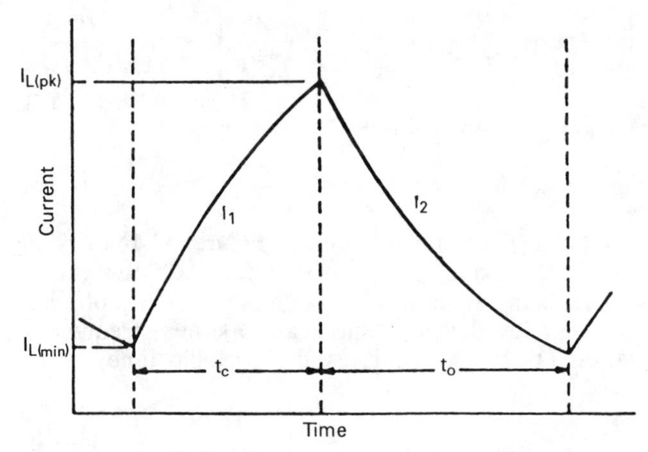

Figure 12.15 Steady state load current waveform of a chopper

(iii) The turn-on time of the thyristors is short compared to the switching period, so the switching losses can be neglected.
(iv) The d.c. source impedance is negligible so that energy can flow in either direction through it without affecting the value of the voltage at its terminals.
(v) The load voltage V_L is constant during a cycle of operation.
(vi) The load current is continuous although it will fluctuate during a duty cycle, as in Figure 12.15.

With these assumptions the values of the minimum ($I_{L(min)}$) and maximum ($I_{L(pk)}$) load current can be written down, from Figure 12.15, as follows:

$$\frac{I_{L(min)}}{V_B/R} + \frac{V}{V_B} = \exp\left(-\frac{R}{L}t_o\right)\left[1 - \exp\left(-\frac{R}{L}t_c\right)\right]\left[1 - \exp\left(-\frac{R}{L}(t_c + t_o)\right)\right]^{-1} \tag{12.14}$$

$$\frac{I_{L(pk)}}{V_B/R} + \frac{V}{V_B} = \left[1 - \exp\left(-\frac{R}{L}t_c\right)\right]\left[1 - \exp\left(-\frac{R}{L}(t_c + t_o)\right)\right]^{-1} \tag{12.15}$$

The mean current ($I_{T(av)}$) through the thyristor TH_1 and the mean current ($I_{D(av)}$) through the diode D_1 can be found from Figure 12.15 as the mean of current i_1 over time t_c and of i_2 over time t_o, respectively. These are as follows:

$$\frac{I_{T(av)}}{V_B/R} = \frac{t_c}{t_c + t_o} - \frac{L/R}{t_c + t_o}\left[1 - \exp\left(-\frac{R}{L}t_c\right)\right]\left[1 - \frac{I_{L(min)}}{V_B/R}\right] - \frac{V_F}{V_B}\left[\frac{t_c}{t_c + t_o} - \frac{L/R}{t_c + t_o}\left(1 - \exp\left(-\frac{R}{L}t_c\right)\right)\right] \tag{12.16}$$

$$\frac{I_{D(av)}}{V_B/R} = \frac{L/R}{t_c + t_o}\left[1 - \exp\left(\frac{R}{L}\,t_o\right)\right]\left[\frac{I_{L(pk)}}{V_B/R} + \frac{V_F}{V_B}\right] - \frac{V_F\,t_o}{V_B\,(t_c + t_o)} \tag{12.17}$$

The mean load current ($I_{L(av)}$) is given by

$$\frac{I_{L(av)}}{V_B/R} + \frac{V_F}{V_B} = \frac{t_c}{t_c + t_o} \tag{12.18}$$

Equations (12.15) to (12.18) are shown plotted in Figures 12.16 to 12.19 and the ripple current in the load, given by the difference between the minimum and peak currents in equations (12.14) and (12.15), is plotted in Figure 12.20. These curves allow device ratings at any operating frequency, determined by equation (12.19) which gives the periodic time, to be obtained:

$$T = \frac{(t_c + t_o)}{L/R} \tag{12.19}$$

All these graphs have the ratio k defined in equation (12.12) as the abscissa since, as seen from equation (12.1), this is the most fundamental ratio in the chopper, determining the magnitude of its output voltage.

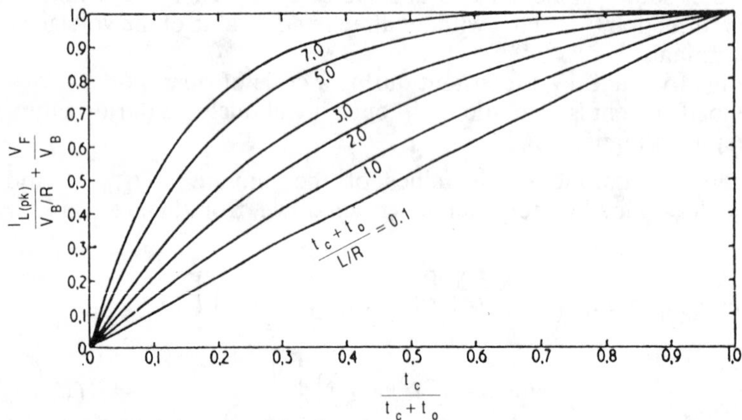

Figure 12.16 Variation of peak load current in a chopper

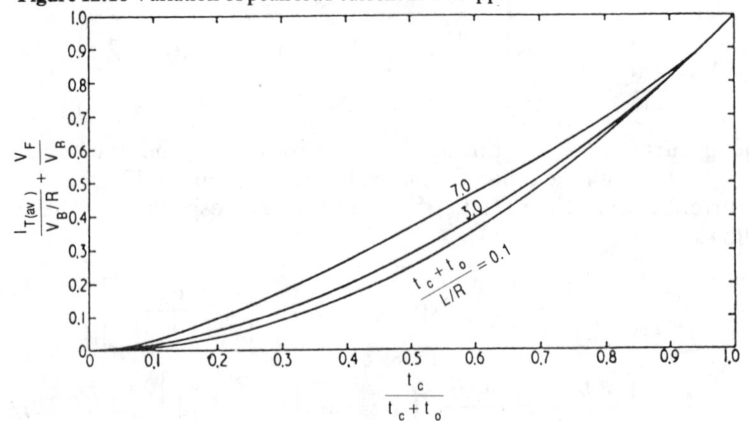

Figure 12.17 Variation of main switch current in a chopper

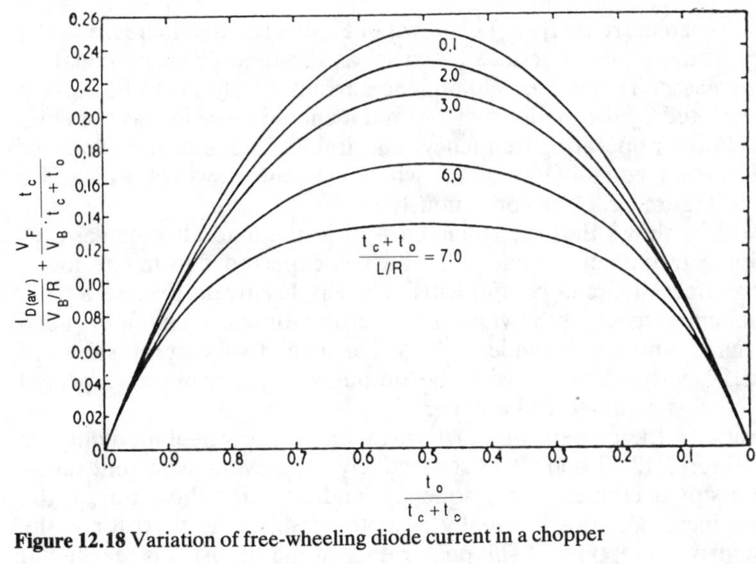

Figure 12.18 Variation of free-wheeling diode current in a chopper

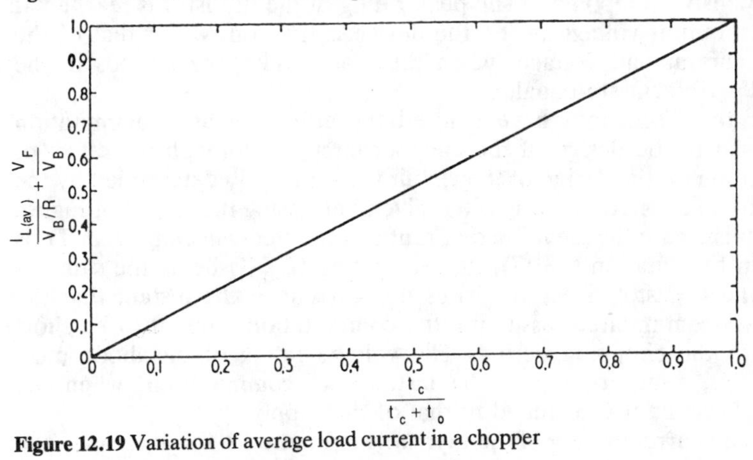

Figure 12.19 Variation of average load current in a chopper

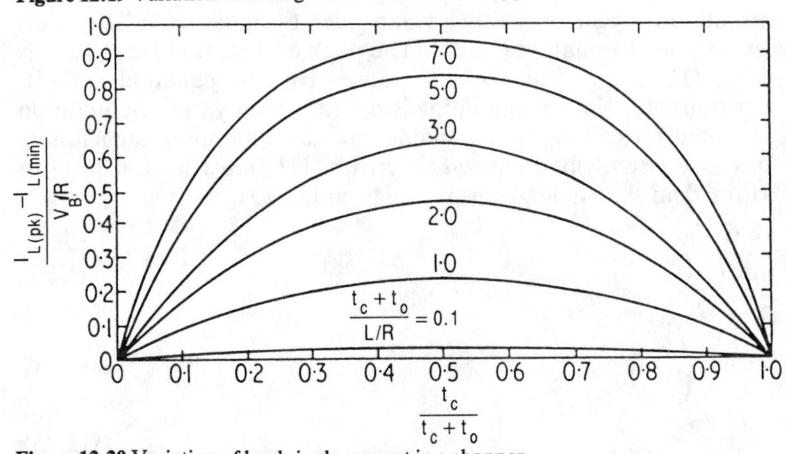

Figure 12.20 Variation of load ripple current in a chopper

The peak load current ($I_{L(pk)}$) is shown in Figure 12.16 to increase as the frequency of the chopper reduces, that is, as the time given by equation (12.19) increases. This is the value of the current which has to be carried and commutated by the main thyristor, and it should be as low as possible, so a high chopper operating frequency is desirable. The maximum value of this load current equals $(V_B - V_F)/R$ when the power switch, which is a thyristor in Figure 12.14, is continuously on.

Figure 12.20 shows that the ripple current in the load also increases as the chopping frequency is reduced, which is expected due to the lower smoothing effect of circuit components. The ripple current reaches a peak at an equal mark-space ratio, which is the setting for which any filters need to be designed, and is independent of V_F. The mean load current, shown in Figure 12.19, varies linearly with the output voltage, again as expected since the load is assumed to be linear.

The rating of the power switch (thyristor) and free-wheeling diode are given by Figures 12.17 and 12.18 respectively. The average current rating of the thyristor decreases with frequency, and although the rating of the diode now increases, this is usually acceptable since the thyristor is the more expensive component. The peak rating of the thyristor is reached at maximum output voltage, when the device is fully on, whilst that of the diode occurs at half voltage, when the mark and space periods in the chopped waveform are equal.

So far the discussions have ignored the effects of any commutation circuit used in the design of the chopper circuit. Although the effect of commutation on the design of the circuit will be largely determined by the commutation method, an example is given here using the circuit of Figure 12.4. In this circuit the mean current rating of the free-wheeling diode D_1 is still given by equation (12.17), its peak rating ($I_{L(pk)}$) being the same as that of the thyristor, since it carries this current at the instant that the thyristor is commutated, assuming the commutation interval to be short relative to the chopping period. The voltage rating of the diode must exceed $2V_B$, which occurs at the instance of commutation, when the voltage of capacitor C is added to that of the supply.

The mean current rating of the thyristor, in Figure 12.4, is increased due to commutation capacitor reset by a value given by equation (12.20). This total current is as in equation (12.21). The peak current is either $I_{L(pk)}$, as in equation (12.15), or the resonant value given by equation (12.22), whichever is greater. Its voltage rating is not affected by the commutation circuit and must exceed V_B. Assuming that commutation capacitor C discharges at constant load current, thyristor TH_1 must have a turn-off time shorter than the value given by equation (12.23).

$$I_c = \frac{2\,C\,V_B}{t_c + t_o} \tag{12.20}$$

$$I_{TH(av)} = I_{T(av)} + I_c \tag{12.21}$$

$$I = V_B \sqrt{\left(\frac{C}{L_1}\right)} \tag{12.22}$$

$$t_{OFF} = \frac{C\,V_B}{I_{L(pk)}} \tag{12.23}$$

Thyristor TH_2 carries a current of $I_{L(pk)}$ during the charge period of C only, so that its mean current is given by equation (12.22), although its size is normally determined by its peak current capability of $I_{L(pk)}$. Similarly, thyristor TH_3 has a mean rating given by equation (12.20) but a peak resonant current of equation (12.22), which can be high. The value of C is fixed by the commutation requirements of equation (12.23), where t_{OFF} is the turn-off time of the main thyristor, and inductor L_1 is chosen such that the resonant time, given by equation (2.4), is small compared to the operating frequency, usually between 5% and 10%.

12.5 The step-up chopper

The step-up chopper was introduced in Chapter 6 and its operation described with reference to the basic circuit of Figure 6.7(a). This circuit is repeated in Figure 12.21(a), the mechanical switch being replaced by

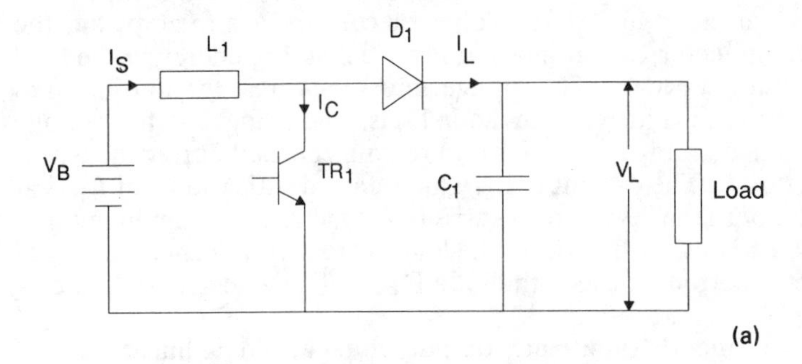

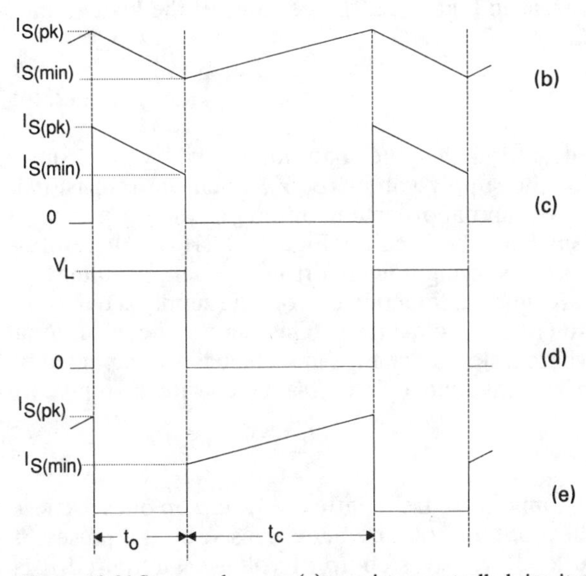

Figure 12.21 Step-up chopper: (a) transistor-controlled circuit; (b) inductor and source current (I_S); (c) load current (I_L); (d) voltage across transistor TR_1; (e) transistor collector current (I_C)

transistor TR_1. Comparison of this circuit with the step-down chopper of Figure 12.1(a) shows that the main difference is that the positions of the main switch (transistor) and the diode have been interchanged, and the inductor is connected to the supply end rather than to the load end.

The step-up chopper works by storing energy in the series inductor L_1 during the period that the transistor is on and transferring this to the load when the transistor is off. Because the load energy is supplied as a series of pulses, a capacitor C is usually added to give some smoothing. The operation of the chopper can be followed by reference to the circuit waveforms of Figure 12.21. When the power transistor is on load current builds up in L_1, reaching a peak of $I_{S(pk)}$ after time t_c, as shown in Figure 12.21(b). The voltage across the transistor is low during this period and has been assumed to be zero in Figure 12.21(d). The current through the transistor builds up to the same peak value of $I_{S(pk)}$, having started from a residual value of $I_{S(min)}$. The current provided from the supply to the load (I_L) during the transistor conduction period is zero, as shown in Figure 12.21(c), all the load current being obtained from the smoothing capacitor.

When the transistor goes off its collector current falls to zero, but the current in the inductor L_1 continues to flow, delivering energy to the load and its smoothing capacitor. The voltage across the transistor now jumps to the value of the load voltage, since diode D_1 is conducting, and this voltage has been assumed, in Figure 12.21(b), to remain constant during the whole of the off period t_o. The current in L_1, which started with a value of $I_{S(pk)}$ at the beginning of the off cycle, now starts to decay, reaching a minimum of $I_{S(min)}$ by the end of the off period. The load current waveform follows that of the inductor current, since with diode D_1 conducting currents I_S and I_L are equal.

Assuming a smooth input and output voltage and a linear current waveform in the inductor, as in Figure 12.21, the value of the load voltage is given by

$$V_L = V_B \frac{t_c + t_o}{t_o} \tag{12.24}$$

From this equation it is seen that the load voltage will vary, from a minimum value equal to the supply voltage of V_B when the transistor is continuously off, to a maximum approaching infinity as the off period of the transistor becomes smaller. The circuit of Figure 12.21(a) is therefore a step-up chopper, the output varying upwards from the supply voltage. If the losses in the circuit are ignored then power from the supply is fed to the load, so that equation (12.25) holds true. This shows the traditional transformer equation where a step-up change in voltage is accompanied by a step-down change in current, and it is the same as with a step-down chopper:

$$V_B I_S = V_L I_L \tag{12.25}$$

Although step-down choppers are frequently used, step-up ones are less popular due to the high smoothing requirements caused by the pulses of energy delivered to the load. Where step-up of voltage is required it is more usual to do this by a inverter feeding into a step-up transformer, as described in Chapter 13.

12.6 Chopper control circuits

The basic control circuit for a chopper consists of a mechanism for turning on the main switching semiconductor at the start of a cycle and of turning off the device when the cycle is to be terminated. If this main switch is a thyristor it can be commutated by firing an auxiliary thyristor. Usually there are two main applications for choppers, to provide a stabilised power supply and to control the speed of a d.c. motor by varying the voltage across it. Both these applications require some form of voltage sensing, in order to keep the load voltage constant under varying current or supply voltage fluctuations. In addition, the current to the load is sensed and current limit applied if this exceeds a predetermined value.

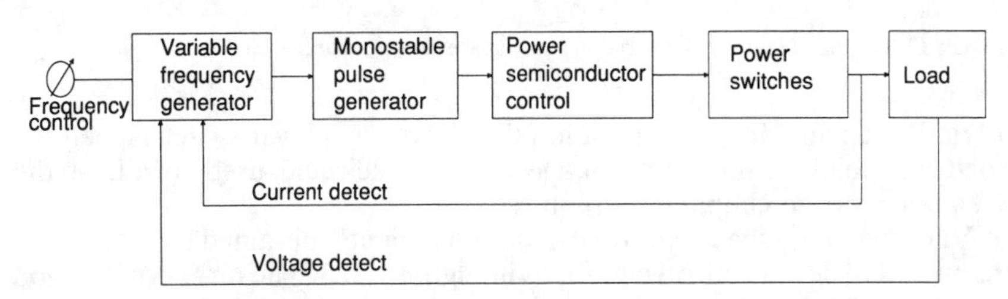

Figure 12.22 Block diagram of a variable-frequency chopper control circuit

Figure 12.22 shows a block schematic of a chopper voltage control arrangement using variable-frequency control. The chopping frequency can be changed within the variable-frequency generator and this feeds a monostable pulse generator, which gives a fixed pulse-width output at a rate determined by the frequency generator. The pulse generator drives the power semiconductor control circuit, the exact nature of which will vary depending on the commutation circuit used. For example, it may consist of a single drive circuit for a power transistor, or a firing pulse for the main switching thyristor, followed by a pulse at the end of the on

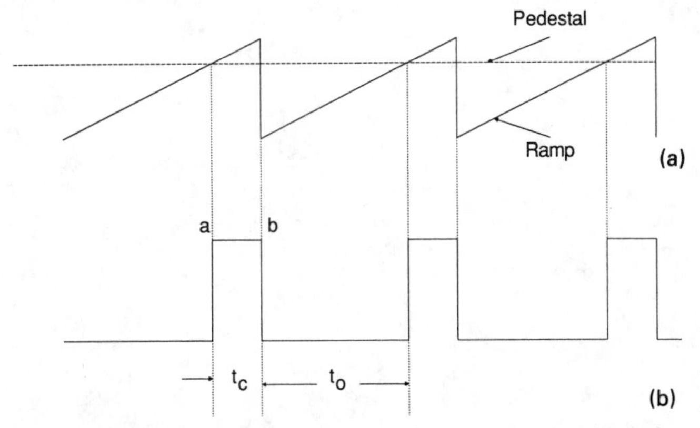

Figure 12.23 Ramp and pedestal control: (a) ramp and pedestal waveforms; (b) output waveform

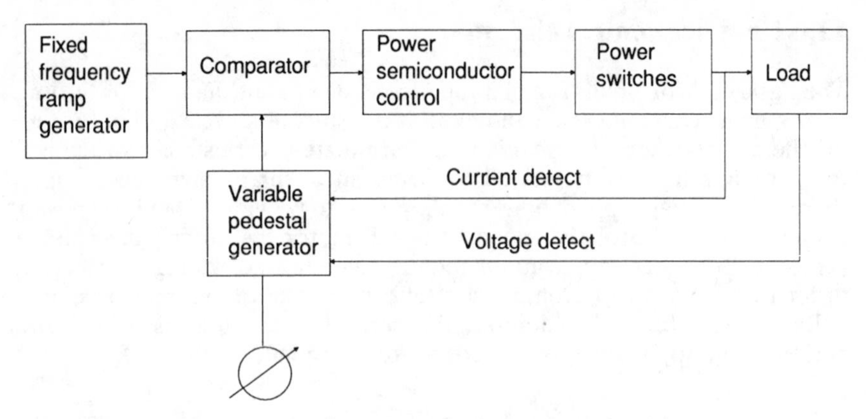

Figure 12.24 Block diagram of a variable mark-space chopper control circuit

period to an auxiliary commutation thyristor. The power switches feed the load and load current and voltage are fed back and used to adjust the frequency of the chopper, as required.

Variable mark-space control can be conveniently obtained by means of a ramp and pedestal control circuit, as in Figure 12.23. The on (or off) period commences when the value of the ramp voltage exceeds that of the pedestal, and it ends when the voltage is again below this value. As seen from Figure 12.23(b), the value of the on and off times, t_c and t_o, can be varied by changing the pedestal voltage level. Figure 12.24 shows a variable mark-space control system for a chopper, using the ramp and pedestal control technique with voltage and current feedback, as in Figure 12.22. The output voltage level is now adjusted by varying the pedestal voltage.

Chapter 13

D.C. link frequency changers

13.1 Introduction

The basic principles of a d.c. link frequency changer, or inverter, were introduced in Chapter 6 and it was seen there that an inverter resembles a chopper in requiring forced commutation if power thyristors are used as the switching semiconductors. The inverter defers in the important aspect that its output is a.c. and not d.c., as was the case for a chopper.

This chapter considers inverters with a view to classifying the multitude of circuits that exist. A classification system based on the commutation method was used in Chapter 12, and a similar technique can be used for inverters. There are, however, several other considerations which did not apply for choppers. First, inverters fall into two major groups:

(i) Push–pull inverters, where the load must be centre tapped, or a separate centre-tapped transformer used to supply the load. Push–pull inverters find frequent application for lower-power inverter circuits, where transistors are used as the switching semiconductors, so avoiding the need for commutation circuitry and sometimes, as will be seen in the next section, for any separate base drive circuitry as well.
(ii) Bridge inverters, in which a centre-tapped load is not essential. There was no such distinction in chopper circuits.

Second, most inverter circuits, whether bridge or push–pull and irrespective of the commutation system used, can be operated in several voltage-control modes. Apart from varying the value of the fundamental a.c. voltage, this also gives a measure of control over harmonics in the waveform. Although choppers were stated as having two control modes, variable frequency and fixed frequency variable mark space, both operate on the principle of giving unidirectional output voltage pulses, the mark-to-space ratio of which is controlled. They cannot be grouped according to a voltage-control method, as is done with inverters.

Several factors were selected in Chapter 11 to enable a comparison to be made between the various commutation systems used, especially in relation to chopper circuits. Seven points may be considered when looking at a goodness factor for inverters:

(i) Is it essential that the load be centre tapped, with the two halves magnetically coupled? Such a push–pull inverter severely limits the type of applications for which it may be used.
(ii) What is the harmonic content of the output waveform? Inverters with high harmonic content, which need to provide a sine wave output, require large and bulky filters. Even if the output is not sinusoidal, a low harmonic content is often advantageous. For instance, in an induction motor it is the fundamental component of the waveform which produces the useful torque, whereas the harmonics result in losses.
(iii) What is the complexity of the inverter and control electronics required to produce the output voltage waveform?
(iv) Can the inverter operate at high frequencies with relatively high efficiency?
(v) What are the maximum and minimum values of the fundamental r.m.s. output voltage?
(vi) Is the commutation voltage increased in proportion to the load being commutated? This is desirable to allow optimisation of commutation components.
(vii) Is the current rating of the main thyristor increased by the commutation capacitor reset pulse?

Two other factors were considered when dealing with choppers, namely:

(i) In the advent of a commutation failure will commutation be re-attempted and will it be successful?
(ii) Does a low-impedance fault current path exist across the supply?

For an inverter a commutation failure almost always results in a low-impedance path across the source, which is protected by fast-acting fuses. Therefore commutation of the thyristors cannot be re-attempted.

The above factors are considered again in the following sections with reference to typical inverter systems.

This chapter follows a format similar to Chapter 12, which described choppers. The various inverter circuits are first introduced, both those using transistors, which do not require commutation, and those with forced commutated thyristors. This is then followed by the techniques used to control the output voltage from an inverter. The design of inverter circuits, both with and without commutation, is then described. The chapter concludes with a description of the current-fed inverter and the control electronics used in inverter circuits.

13.2 Inverter circuits

This section first introduces the various forms of inverter configurations, followed by a description of transistor inverters, popularly used for low- to medium-power applications. For high-power applications thyristor circuits are required and the basic commutation techniques used for these are described. The section concludes with a description of the modifications made to the basic commutated circuits for enhancement of certain performance factors.

13.2.1 Inverter configurations

The two broad classifications for inverter circuits are push–pull and bridge. A push–pull circuit, which uses transistors as the main switching elements, is shown in Figure 13.1(a), the circuit used to drive the base of the transistors not being shown. The operation of this circuit can be followed by reference to the waveforms of Figures 13.1(b) to 13.1(e). Assume that at time t_0 transistor TR_1 is turned on so that end A of the transformer primary goes negative with respect to B and the primary voltage V_{AE} equals that of the supply voltage V_B, the voltage drop across the conducting transistor being neglected. Assuming the load to be inductive and to be flowing due to a previous cycle, this continues to flow via diode D_1 until time t_1, when the current changes direction and flows as collector current through the transistor. At time t_2 transistor TR_1 is turned off and transistor TR_2 is turned on, causing reversal of the load voltage. The load current again continues to flow in the transformer primary via diode D_2 until it reverses at time t_3 and flows as transistor TR_2 current.

It is obviously not essential to use an output transformer, although this is useful for stepping the battery voltage up or down to meet the requirement of the load. It is possible to make the load itself centre tapped, for instance AEB forming the stator winding of a single-phase induction motor. In either case, when TR_1 conducts a voltage equal to the battery, voltage V_B

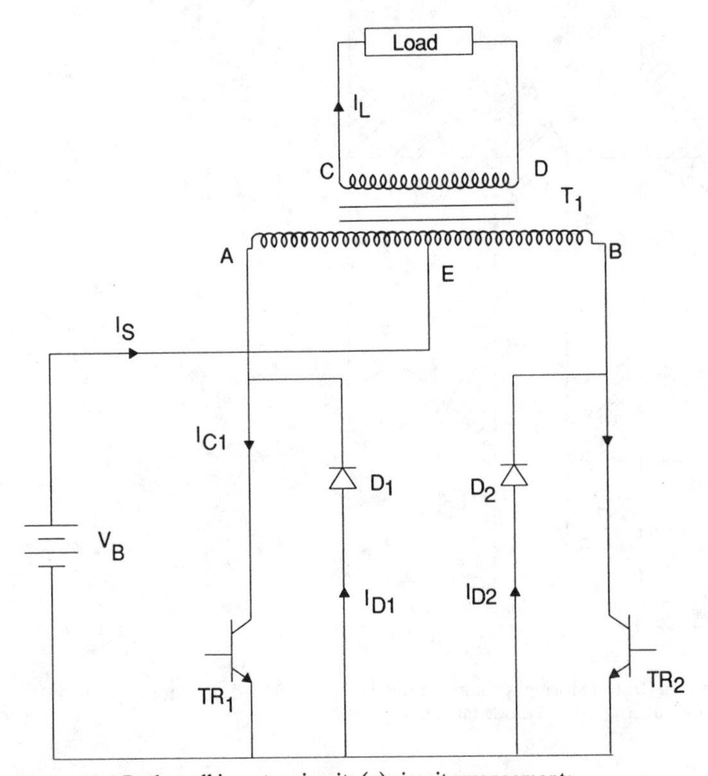

Figure 13.1 Push–pull inverter circuit: (a) circuit arrangement;

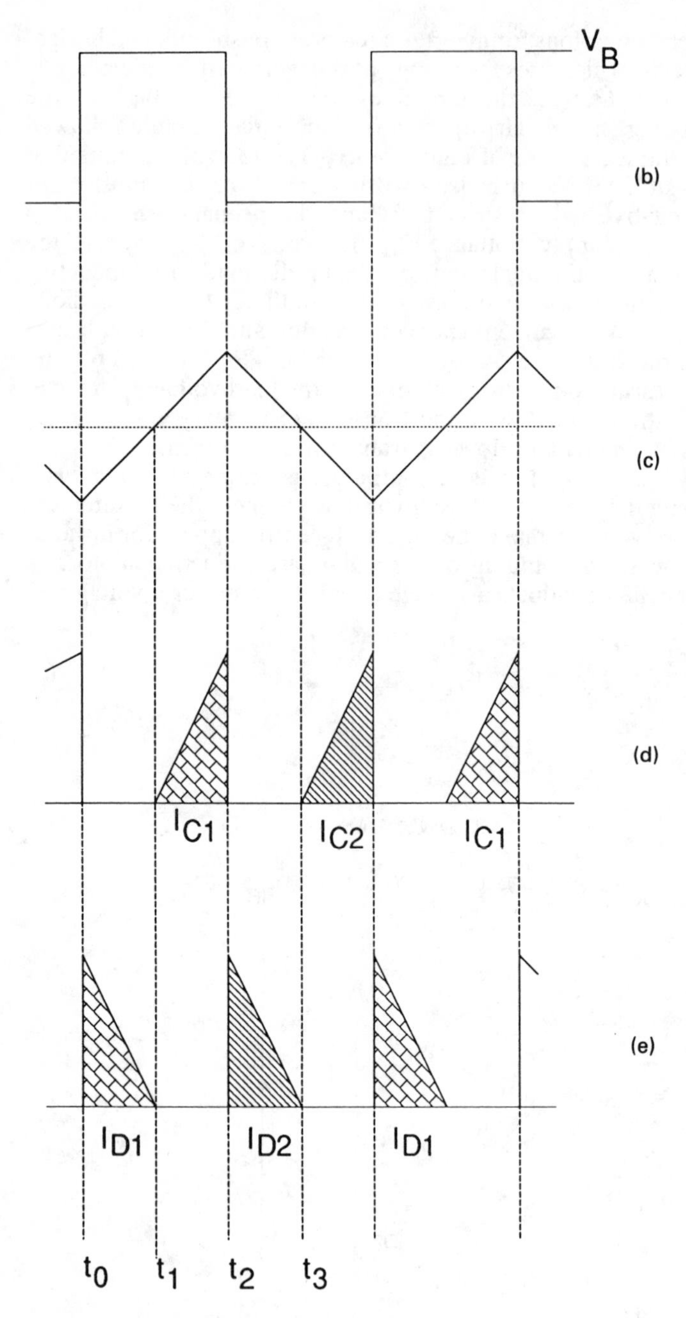

Figure 13.1 continued (b) transformer primary voltage (V_{AE}); (c) load current (I_L); (d) transistor collector current (I_C); (e) diode current (I_D)

is impressed across AE and if the two halves of the load are closely coupled this raises the collector of TR_2 to a potential of $2V_B$ above its emitter. Therefore all the semiconductor devices need to have a voltage rating equal to at least twice that of the supply.

Figure 13. 2 shows a half-bridge inverter in which a centre-tapped d.c. supply is needed. The waveforms for this circuit are identical to those given in Figure 13.1, which is illustrated for an inductive load. Transistor TR_1 is turned on at time t_0 but it does not conduct until t_1, the load current during this period flowing into the supply via diode D_1. When this inductive current reverses the transistor starts to conduct. Similarly, during the following half cycle transistor TR_2 and diode D_2 conduct the load current.

The half-wave bridge circuit can be extended to a full-wave bridge, also known simply as a bridge inverter, as shown in Figure 13.3. Transistors TR_1, TR_4 and TR_2, TR_3 may be turned on in pairs and consequently diodes D_1, D_4 and D_2, D_3 also conduct the inductive current in pairs, the circuit waveforms again being the same as those in Figure 13.1.

Generally, the transistors are not turned on in pairs, the delay being used to vary the magnitude of the output voltage. This is illustrated by the waveforms of Figure 13.4, where the load is assumed to be filtered and is therefore sinusoidal. In Figure 13.4(a) the transistors are turned on in pairs and the load voltage is a maximum value. To reduce this voltage the transistor turn-on sequence shown in Figure 13.4(b) is adopted. At time t_0

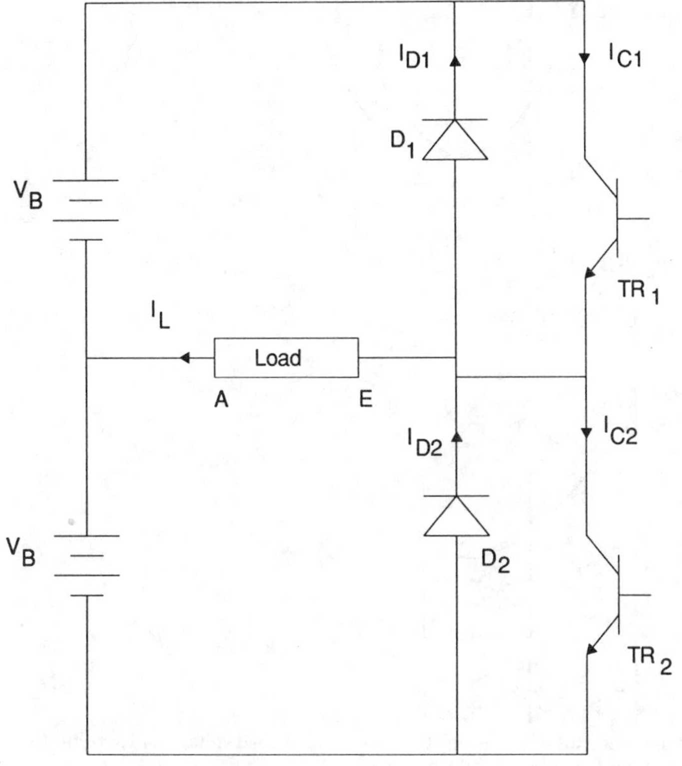

Figure 13.2 Half-bridge inverter

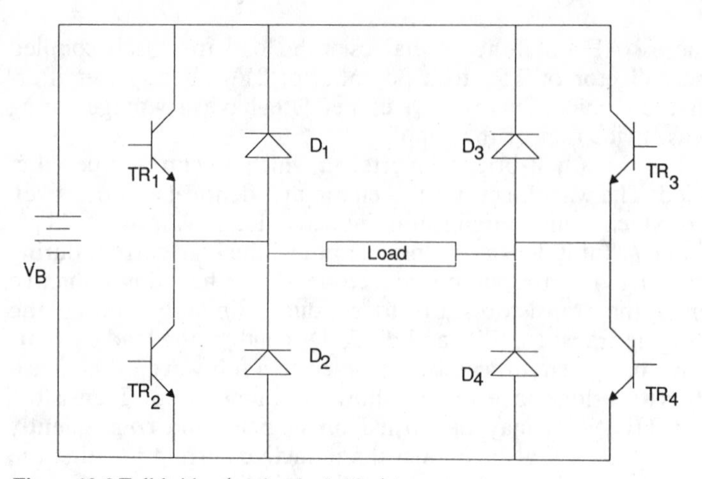

Figure 13.3 Full-bridge (or simply, bridge) inverter

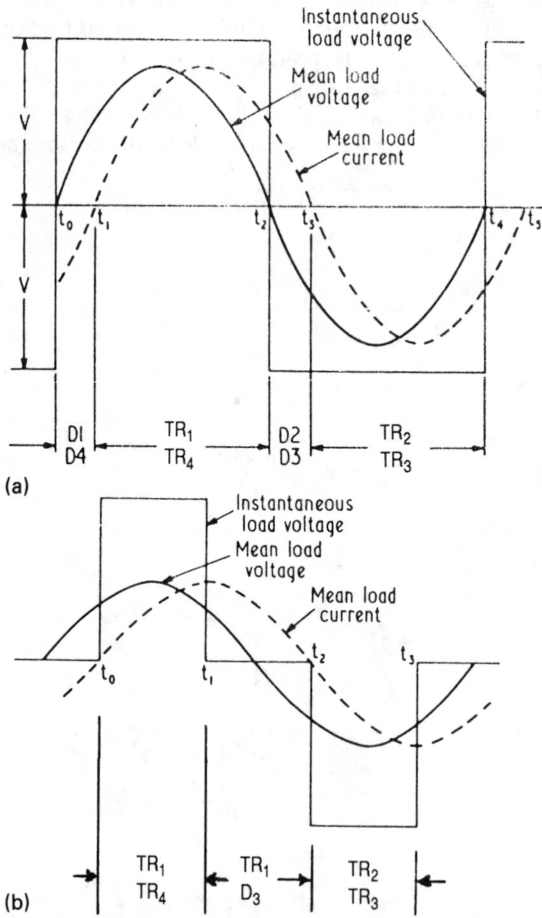

Figure 13.4 Inverter waveforms with filtering to give sinusoidal load voltage and current: (a) maximum voltage output; (b) reduced voltage output

transistors TR_1 and TR_4 are turned on and conduct since the current is starting positive. At time t_1 transistor TR_4 is turned off. The load current continues to circulate through transistor TR_1 and diode D_3, effectively short circuiting the load. At t_2 transistor TR_1 is turned off and TR_2 and TR_3 are turned on, causing the load current to reverse. Finally, at time t_3 transistor TR_3 is turned off and load current circulates through TR_2 and D_4, the load voltage again being zero. As seen from Figure 13.4(b), the mean load voltage has been reduced over that of Figure 13.4(a).

A common requirement in frequency changers is that of regeneration. For instance, a variable-frequency motor control, in, for example, a locomotive, will regenerate when it starts to move downhill and the motor acts as a generator. Regeneration can be obtained in inverters although it is not a natural phenomenon of most inverter systems, as was the case for cycloconverters, and special modifications are necessary to accomplish it successfully. Unless the d.c. supply is 'soft' the regenerated energy will cause excessive voltage increase in the system. For inverters operating from a rectified a.c. source, it is necessary to connect an inverting bridge to feed this energy from the d.c. to the a.c. supply, as shown in Figure 13.5 for a thyristor bridge inverter. Diodes D_1 to D_4 and thyristors TH_1 to TH_4

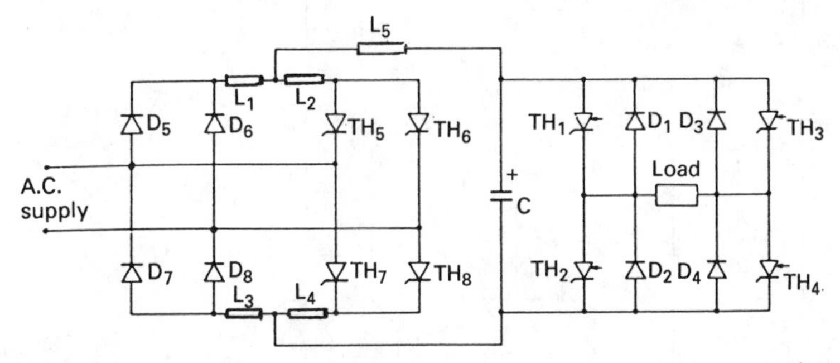

Figure 13.5 Regeneration arrangement for an inverter obtaining its d.c. power from an a.c. source

form the bridge inverter circuit, the d.c. supply for the inverter being derived from D_5 to D_8 and smoothed by L_5 and C. Thyristors TH_5 to TH_8 form the usual arrangement of an a.c. line commutated inverter feeding energy from the d.c. side back to the a.c. supply. It is used during regeneration to prevent excessive voltage build-up across the capacitor. When no regeneration is occurring, inductors L_1 and L_3 are saturated by load current and L_2 and L_4 absorb the instantaneous voltage difference between rectifier and inverter. During regeneration energy flow is from the load to the a.c. supply, and the functions of the inductors are reversed.

The single-phase bridge circuits can be extended to three phases, as shown for transistor bridges in Figure 13.6, the operation of the three-phase variants being very similar to that of the single-phase circuits.

13.2.2 Transistor inverters

Although transistors were the switching semiconductors in the previous section, the base drive circuitry to these devices was not shown, so that any

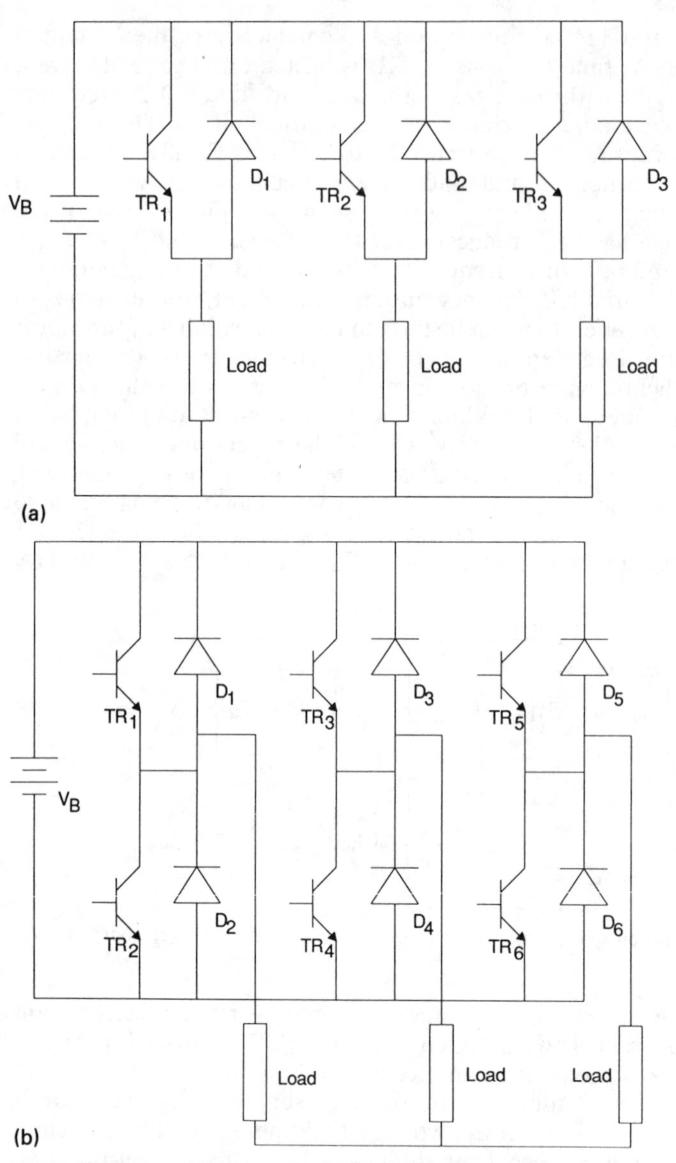

Figure 13.6 Three-phase inverters: (a) half bridge; (b) full bridge

other power semiconductor, such as a thyristor or gate turn-off switch, could have been used in their place. Power transistors are, however, popularly used in inverter circuits, primarily because of the ease with which they can be controlled, and this section will describe inverter configurations which are unique to transistors.

The simplest transistor inverter is the single-transistor circuit which uses feedback from its collector to provide its own base drive. Such a single-ended arrangement is popular for low power blocking oscillator applications, but is not often used in power inverters because of the

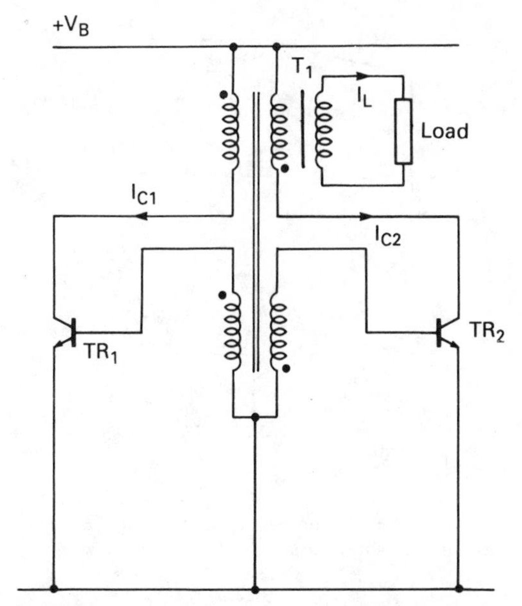

Figure 13.7 Two-transistor, single-transformer inverter

limitation on power output and the asymmetry of the output voltage waveform.

Figure 13.7 shows a relaxation oscillator circuit, with two power transistors and a single transformer, which is frequently used for simple inverter circuits. The operation of this circuit is as follows, the polarity of the transformer windings being indicated by the dots. Assume on switching on that transistor TR_1 starts to turn on. This will cause the current in its collector to slowly increase, which in turn induces a voltage in the transformer winding connected in its collector. This induced voltage is fed back to the base of TR_1 and causes it to turn on further, the whole process being regenerative so that transistor TR_1 will be driven very rapidly into saturation. Because of the polarity of the transformer windings, the base of transistor TR_2 is reverse biased during this period, so it is maintained firmly off.

Figure 13.8(a) shows the collector current waveform in transistor TR_1 during this period. Transistor TR_1 is assumed to turn on at time t_0 and its collector current jumps to the value of the load current I_L at that time. Assuming the load to remain constant, the collector current will now start to increase, due to magnetising current I_{MAG} of the transformer, and this has been assumed to be linear in Figure 13.8. Eventually, at time t_1 the total collector current $I_{C(pk)}$ will reach such a value, given by equation (13.1), that it can no longer be supported by the base current and transistor TR_1 will start to come out of saturation:

$$I_{C(pk)} = I_L + I_{MAG} \tag{13.1}$$

Reduction of collector current will cause a reversal of the voltages induced in the transformer so that the base drive to TR_1 is reduced, forcing

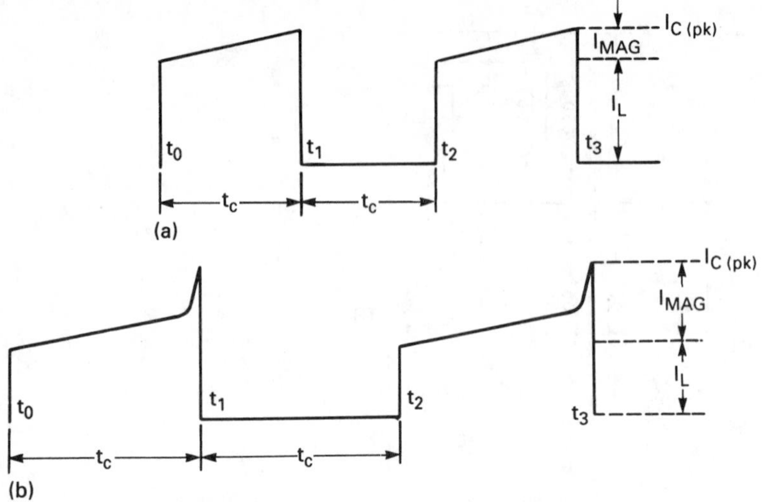

Figure 13.8 Collector current waveform in transistor TR_1 for the inverter of Figure 13.7: (a) with non-saturating transformer; (b) with saturating transformer

it further into the off state, whilst the base current to TR_2 starts to increase, turning it on. The circuit will therefore rapidly flip over into the state where TR_1 is fully off and TR_2 is saturated. The next half cycle will now commence, which ends at time t_2 when the collector current through TR_2 reaches a value where it can no longer be supported by its base current.

The peak collector current which the transistor can support, equal to the sum of the load current and the transformer-magnetising current, is determined by its common emitter gain h_{FE} and the magnitude of the base current I_B. This is given by

$$I_{C(pk)} = h_{FE} I_B \tag{13.2}$$

The value of the transformer magnetising current I_{MAG} depends on the supply voltage V_B, the inductance of the transformer primary winding L_{PRI} and the on time of the transistor t_c, as in

$$I_{MAG} = \frac{V_B t_c}{L_{PRI}} \tag{13.3}$$

Therefore from equations (13.1), (13.2) and (13.3) the on time t_c, which determines the frequency of the inverter, is given by

$$t_c = \frac{L_{PRI}}{V_B} (h_{FE} I_B - I_L) \tag{13.4}$$

This equation shows the major limitation of this inverter circuit in certain applications, since the frequency varies with the magnitude of the load current, being lowest at light loads. This variation in frequency would affect the efficiency of the circuit and also that of any filtering and decoupling components which may be used.

To overcome the problem of frequency variation with load the linear transformer in Figure 13.7 can be replaced by a saturable transformer, the collector current waveform now being as in Figure 13.8(b). The frequency of operation is once again determined by the maximum collector current, but now the magnetising current builds up rapidly, once transformer saturation has commenced, and this swamps the variation in load current, so the operating frequency is less load sensitive.

In the circuit of Figure 13.7 it has been assumed that at switch-on one of the transistors commences to turn on. This may not be the case, especially on heavy load currents when the loop gain of the system is below unity, so that the circuit may fail to oscillate. To prevent this, several starting circuits are used, Figure 13.9 showing two of the simplest techniques. In the resistor-starting circuit the resistor chain R_1 and R_2 provides the initial bias for turn-on, the transistor with the highest gain turning on first. Thereafter the circuit performs as before, the starting resistors having very little effect.

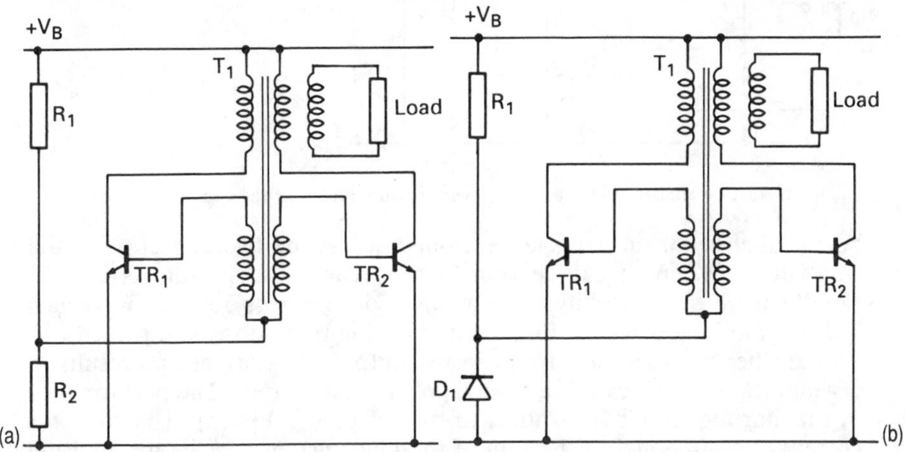

Figure 13.9 Starting circuits for the inverter of Figure 13.7: (a) resistor starting; (b) diode starting

The disadvantage of the resistor-starting circuit is that the resistors need to be of low value in order to provide adequate base current under full starting load conditions, but this results in excessive circuit dissipation during normal running conditions, when the bulk of the base drive is derived via the transformer feedback winding. Diode starting is now preferred, as in Figure 13.9(b), where the reverse biased diode D_1 prevents excessive static dissipation, whilst resistor R_1 can still be made of a value low enough to provide adequate starting current.

In Figure 13.7 a single transformer is used to supply the load current and to feed back the transistor base currents. The disadvantage of this system is that for stable operation a saturable tansformer is required, which is expensive and can lead to distortion of the load supply. A better method is to use a linear output transformer and a second, saturable transformer to feed back the current to the base of the transistors, as shown in Figure 13.10, where use of a starting circuit is optional for starting on heavy load currents. The second saturable transformer can be made much smaller and cheaper since it is only required to carry the base drive currents of the transistors. The circuit performs much the same as before, the base current

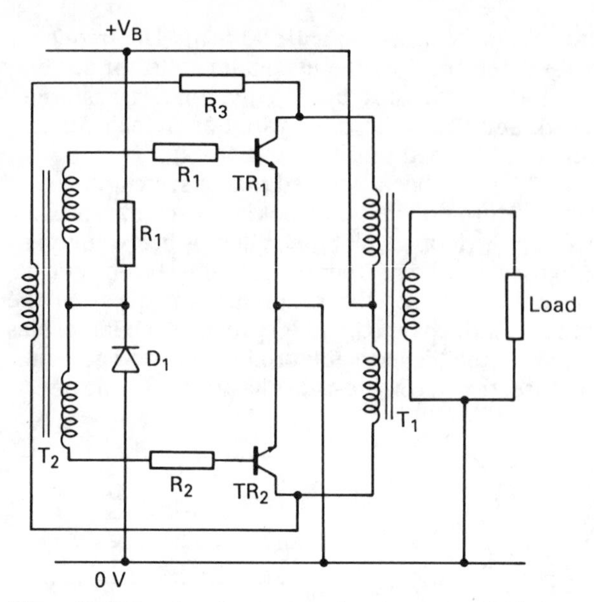

Figure 13.10 Two-transformer, transistor inverter, with diode starting circuit

being taken from the voltage developed across the two collectors to the transistors. When this drive transformer reaches saturation its current rapidly increases, resulting in a voltage drop across resistor R_3 which reduces the base drive causing the conducting transistor to be turned off. This further reduces the current in the output transformer and results in regeneration, which causes a reversal of state, as before. The performance of the starting circuit is identical to that of Figure 13.9(b). The collector current is again equal to the sum of the load current and the magnetising current in the output transformer, but since the output transformer does not saturate and the operation of the circuit is not determined by the magnitude of this current, it can be kept low, reducing the device rating.

13.2.3 Basic thyristor commutation circuits

Irrespective of whether the inverter is push–pull or bridge, and the voltage-control mode adopted, as described in section 13.3, the basic commutation system used for inverter circuits, which use thyristors as the main switch, can be grouped into four classes, as was done for chopper circuits in Chapter 11:

(i) Parallel-capacitor commutation, the commutation capacitor discharging directly into the thyristor being turned off.
(ii) Parallel capacitor–inductor commutation, where a series capacitor and inductor are connected across the thyristors being commutated.
(iii) Series capacitor commutation, the commutation capacitor being placed in series with the load conduction path.
(iv) Coupled-pulse commutation, the commutation capacitor pulse being coupled to the thyristor via a transformer or auto-transformer.

These four methods are examined in this section, with reference to typical bridge and push–pull inverter circuits.

13.2.3.1 Parallel-capacitor commutation

Parallel capacitor commutated circuits are the most popular and the circuit of Figure 13.11(a) is perhaps one of the earliest used thyristor inverter. In fact the push–pull inverter is often called a parallel inverter due to the parallel-capacitor commutation system used, but it will be seen that several other commutation methods can also be used with push–pull inverters.

The circuit of Figure 13.11(a) is similar to Figure 13.1(a) where the transistor switches have been replaced by thyristors, inductor L_1 is added, and diodes D_1 and D_2 have been omitted. Figure 13.11(b) shows a more efficient system. Firing thyristor TH_1 charges capacitor C with plate a positive, to a voltage of $2V_B$. When TH_2 is turned on capacitor C is connected across TH_1 turning it off. The capacitor now discharges to zero voltage, its stored energy then being dissipated in the L_1-D_1-C-TH_2 conduction path. After this, capacitor C charges to $2V_B$ with plate b positive, ready to turn TH_2 off when TH_1 is fired.

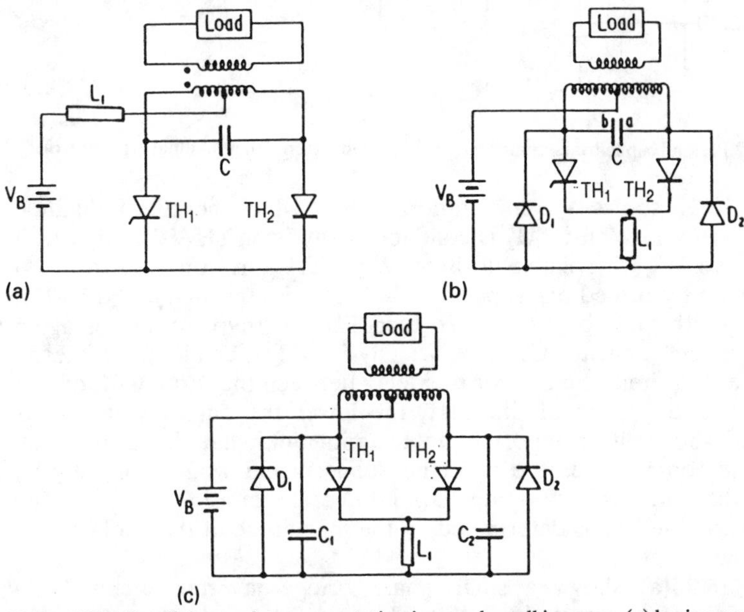

Figure 13.11 Parallel-capacitor commutation in a push–pull inverter: (a) basic arrangement; (b) improved arrangement; (c) use of d.c. rated capacitors

Figure 13.11(c) shows an alternative arrangement of the commutation capacitors, where the capacitors need only have a d.c. rating, which reduces their size, although two capacitors are now necessary.

The simple push–pull parallel capacitor commutated circuits have a disadvantage in that one or another of the main thyristors must be triggered to commutate the conducting device. This means that mark-space voltage-control schemes, as described in section 13.3, cannot be obtained, and even varying the d.c. supply voltage, to control the magnitude of the output voltage, would reduce the commutation energy, which is undesirable. Although h.f. pulse-width modulation systems can be used, either with a rectangular or sine reference, as described in section 13.3, the inverter losses, caused by the circulation of commutation current (for example, via L_1-D_1-C-TH_2) make it inefficient at high frequencies. The most usual voltage-control system consists of two parallel inverters, with a common secondary, whose square wave outputs are phase shifted to give a mark-space controlled voltage output.

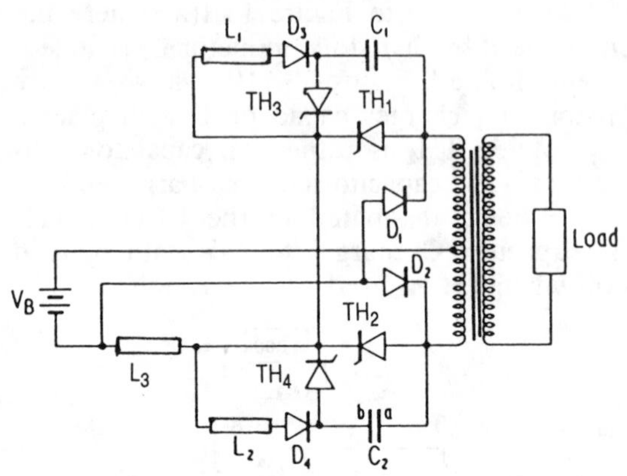

Figure 13.12 Parallel capacitor commutated push–pull inverter with individual arm control

Figure 13.12 shows a commutation system which permits individual thyristor turn-off. Whilst TH_1 is conducting thyristor TH_4 is fired, which charges C_2 to $2V_B$ with plate a positive. When TH_2 turns on, thyristor TH_1 being previously turned off, capacitor C_2 discharges through it and L_2-D_4, recharging with plate b positive. To turn TH_2 off thyristor TH_4 is again fired to connect capacitor C_2 across it. Thyristor TH_1 can be fired later in the cycle as required, the greater the delay between the turning off of one thyristor and the firing of the next, the lower the output voltage. As mentioned above, it is important to bear in mind that this is not true mark-space control since the inductive load current would normally be decaying through feedback diodes D_1 and D_2. Therefore the load voltage at any inverter setting is determined by the magnitude of the load current and its power factor.

Figure 13.13(a) shows a single-phase bridge inverter circuit using parallel-capacitor commutation. With TH_1 and TH_4 conducting, capacitor C is charged to V_B with plate a positive. When TH_2 and TH_3 are fired to commence the next step of the output, capacitor C is connected across TH_1 and TH_4 and turns them off. The circuit is therefore an example of parallel-capacitor commutation in which inductors L_1 prevent the supply from being instantaneously short-circuited during commutation. Figure 13.13(b) shows an extension to a three-phase inverter, where firing any thyristor which has positive anode voltage will commutate all the other conducting thyristors in its row.

An alternative parallel capacitor commutated bridge is illustrated in Figure 13.14. An auxiliary d.c. supply is shown connected to the commutation circuit to enable the main supply to be varied, for example to control the magnitude of the output voltage. Where this is not required a separate supply is not essential and can be shorted out. When TH_7 is fired

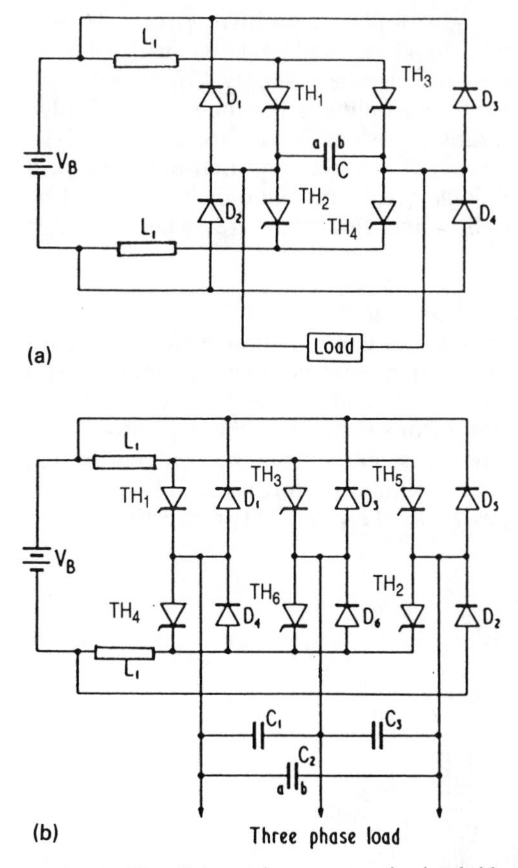

Figure 13.13 Parallel capacitor commutation in a bridge inverter: (a) single-phase; (b) three-phase

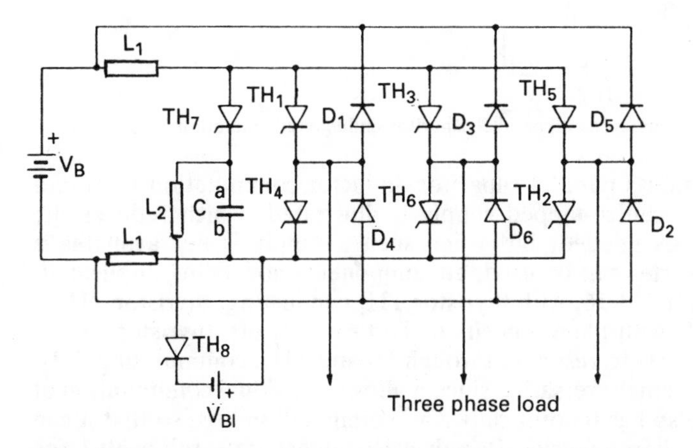

Figure 13.14 Use of an external commutation voltage source in a parallel capacitor commutated bridge inverter

capacitor C is charged to voltage V_B with plate a positive. Firing TH_8 will cause the capacitor to resonate through L_2 and recharge with plate b positive, the voltage being increased if auxiliary supply V_{B1} is included. When TH_7 is next fired capacitor C short-circuits the input to the bridge inverter and causes all the conducting thyristors in both arms of the bridge to turn off. It is important once again to bear in mind that turning off all devices will not make the load voltage zero. This can only be obtained by refiring appropriate devices to allow the current to free-wheel through them.

13.2.3.2 Parallel capacitor–inductor commutation

An example of parallel capacitor–inductor commutation is given in Figure 13.15. This differs from Figure 13.12 in two aspects. First, inductors L_3 and L_4 are in series with capacitors C_1 and C_2, so reducing the rating of the commutation devices. Second, the resonant pulse of the capacitors does not now pass through the main thyristors, so reducing their current rating. Apart from this, the two circuits have similar performance, TH_5 and TH_6 being normally fired simultaneously with TH_1 and TH_2 respectively.

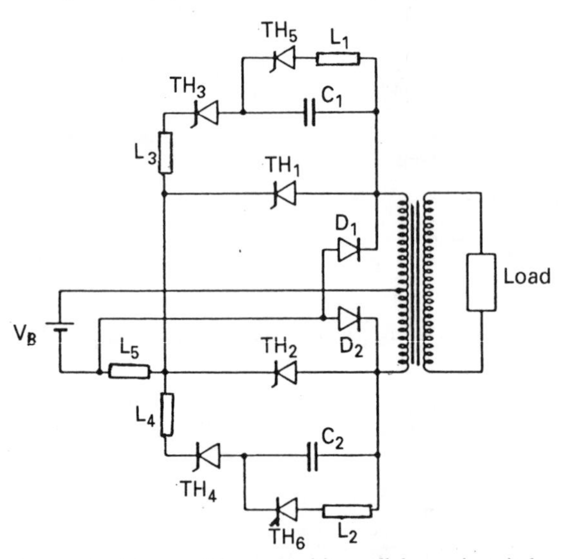

Figure 13.15 Push–pull inverter with parallel capacitor–inductor commutation

Figure 13.16 shows parallel capacitor–inductor commutation used in a bridge circuit. A centre-tapped supply is illustrated, which reduces the number of devices needed, but when such a supply is not available a conventional inverter can be used, all components now being duplicated. Referring to Figure 13.16, with thyristor TH_1 conducting, thyristor TH_4 is fired to charge C with plate a positive. To turn TH_1 off, thyristor TH_3 is turned on, causing C to resonate through D_1 and TH_3, commutating TH_1. This circuit is extremely versatile, since it allows individual commutation of all devices and also has theoretically zero commutation loss, so that it can be operated at high frequencies. Its only disadvantage is the relatively large number of commutation devices required.

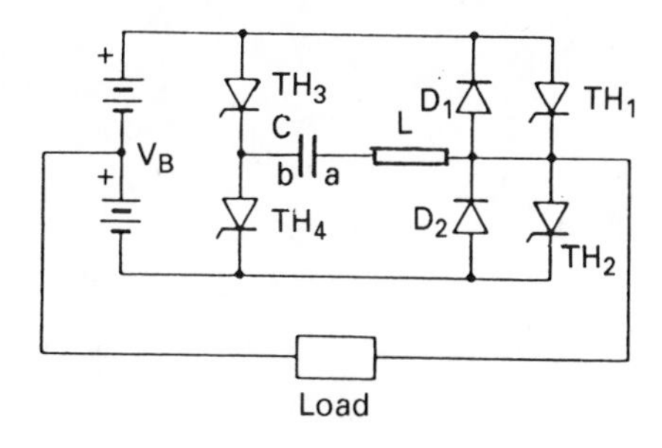

Figure 13.16 Parallel capacitor–inductor commutation in a bridge inverter with tapped supply

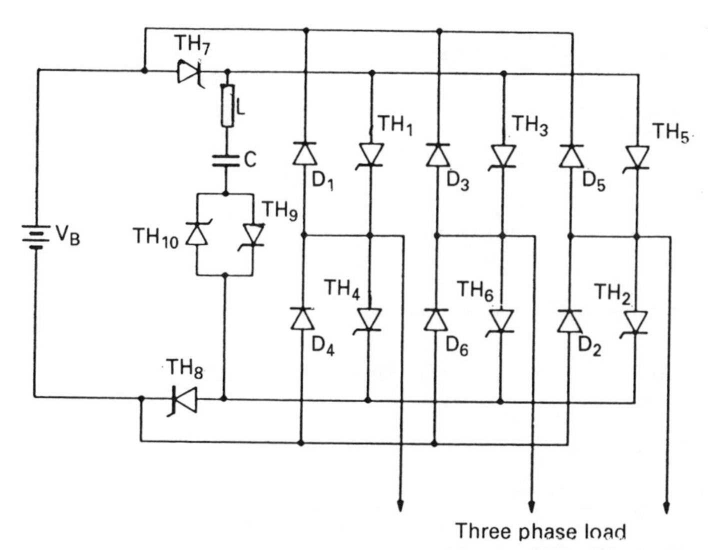

Figure 13.17 Light-weight parallel capacitor–inductor commutated bridge inverter

The inverter of Figure 13.17, although similar in principle to Figure 13.14, has been designed for aerospace applications. Therefore inductors L_1 from Figure 13.14, which can be bulky, have been replaced by thyristors TH_7 and TH_8. With these devices and the appropriate bridge thyristors conducting, TH_9 is fired to prime capacitor C ready for commutation. When TH_{10} is now turned on capacitor C discharges via its inductor L, turning TH_7 and TH_8 off and so eventually commutating the inverter thyristors. This circuit also has high operating efficiencies.

13.2.3.3 Series capacitor commutation

In series capacitor commutated inverters the commutation devices are normally in the main thyristor current path. Figure 13.18 shows a typical circuit, with its waveforms. The inverter is operated at a frequency determined by the oscillatory frequency of the load and C, so that these inverters find most frequent application for producing fixed frequency sine wave output.

Figure 13.19 is another example of a simple series capacitor commutated inverter. Once again the trigger frequency of TH_1 is chosen to coincide with the oscillatory frequency of C_2 and L_1. Capacitor C_1 is large valued

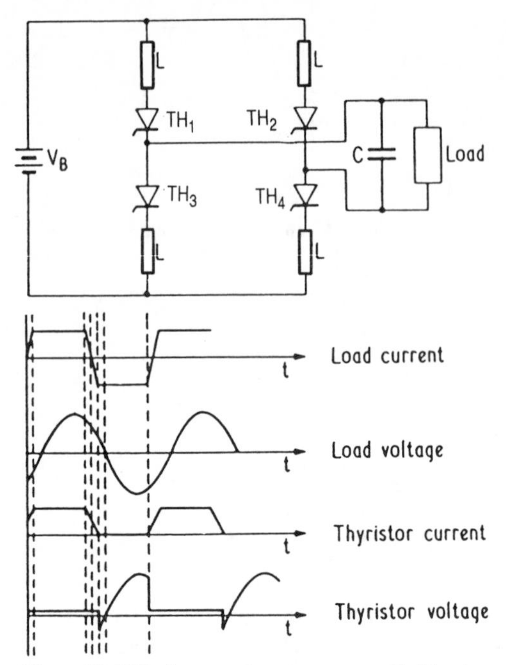

Figure 13.18 Series capacitor commutated bridge inverter

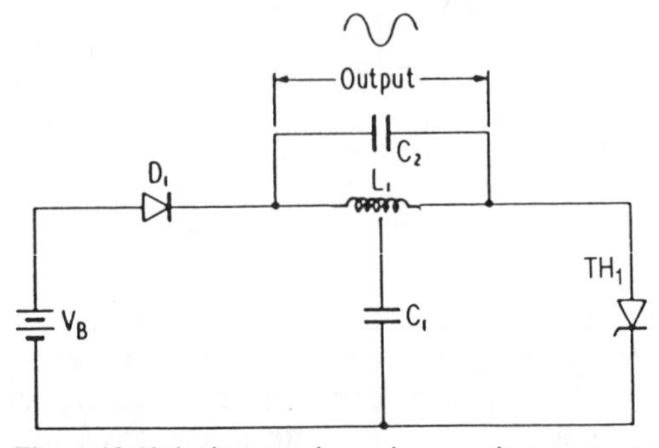

Figure 13.19 An inexpensive series capacitor commutated inverter

and acts as a pivot for the see-saw action of the output. This circuit is very useful as a simple sine wave fixed-frequency generator.

13.2.3.4 Coupled-pulse commutation

A coupled-pulse commutated bridge inverter is shown in Figure 13.20, which will be analysed in section 13.4. This circuit is perhaps the most frequently used and reliable of coupled-pulse inverter circuits. With any device conducting, the commutation capacitor in the corresponding half of the bridge is charged to voltage V_B. When the thyristor in that leg is fired the capacitor discharges through half of the auto-transformer, coupling a turn-off pulse to the conducting thyristor via the other half of the transformer. The circuit is versatile in that firing one device will automatically commutate the other half of the bridge. It can be used to produce outputs with zeros in the waveform with no additional circuitry. It

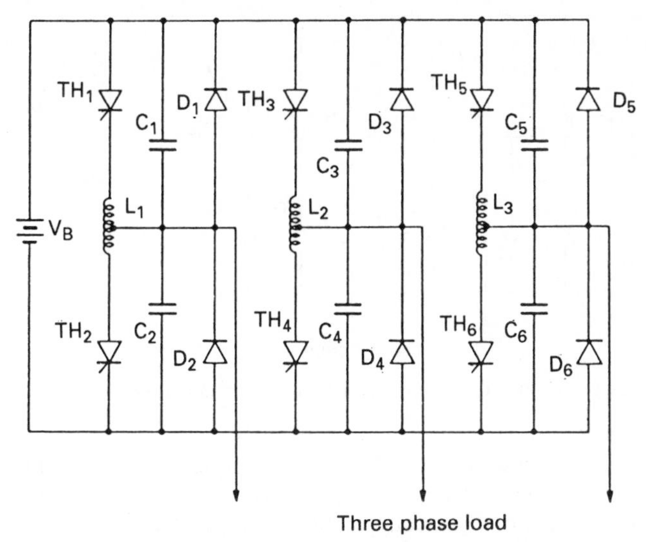

Figure 13.20 Coupled-pulse commutated three-phase bridge inverter

does, however, suffer from low efficiency, due to commutation losses at higher frequencies, and is therefore not suited for voltage-control techniques which require an inherent high-frequency operation, such as pulse-width modulation described in section 13.3. Also it cannot be used in circuits which use d.c. supply variations to change the output, since the commutation voltage is now affected. Figure 13.21 overcomes this disadvantage by using an auxiliary fixed-voltage low-power commutation voltage source V_{B1}, the diodes in series with the main thyristor enabling the commutation capacitors to charge to a voltage which can be greater than V_B.

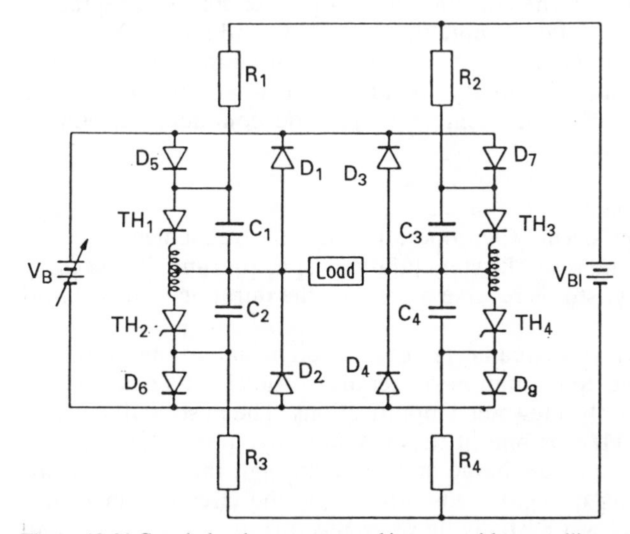

Figure 13.21 Coupled-pulse commutated inverter with an auxiliary commutation voltage source

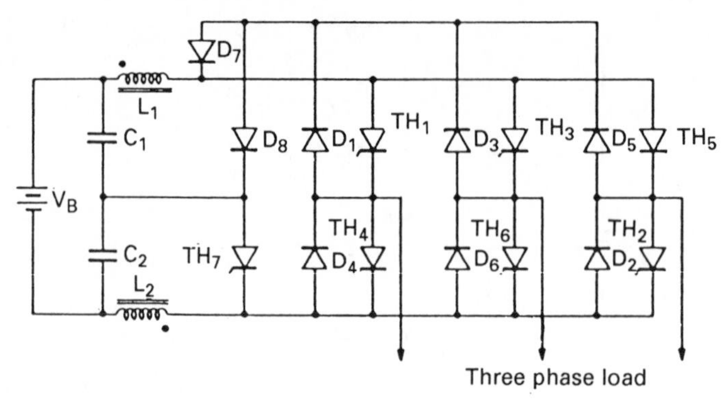

Figure 13.22 Coupled-pulse three-phase bridge inverter with single commutation circuit

Although Figure 13.22 resembles the circuit of Figure 13.14, the inductors L_1 and L_2 are now closely coupled. With thyristors TH_1, TH_5 and TH_6 conducting, (say) capacitor C_2 is charged to a voltage equal to V_B and C_1 is at zero voltage. Firing thyristor TH_7 couples a pulse in L_1 via L_2 which blocks the conducting thyristor. Load current continues to free-wheel via feedback diodes and via D_8, TH_7, therefore true mark-space control is obtained without additional circuitry. Capacitor C_1 is now charged so that during the next part of the cycle, assuming TH_1 is fired, it discharges into L_1 and couples a pulse to L_2 turning TH_7 off.

13.2.4 Modified thyristor commutation circuits

This section describes modifications made to the basic thyristor commutation circuits to obtain enhanced circuit performance in certain parameters. Three types of enhancements are considered, which are the same as those introduced when dealing with chopper circuits in Chapter 12, and, as before, it is important to note that none of the additions change the basic commutation category of the circuits. The three modifications are: (1) enhancement for high-frequency operation; (2) enhancement to reduce commutation losses; (3) enhancement to provide commutation voltage boosting.

13.2.4.1 High-frequency enhancements

Two timing problems occur when operating at high frequencies, i.e. the time required to set and reset the commutation capacitors and the necessity of ensuring that a thyristor is reverse biased for the duration of its turn-off time.

The majority of inverter circuits are not affected by limitations on the set and reset times of the commutation capacitor; of the circuits described in the previous section only a few need modifications. The push–pull inverter shown in Figure 13.11(a) is one of these. When thyristor TH_1 or TH_2 is fired the capacitor has to discharge and reset through the load, and this time can be very long on light loads. Modifying the circuit as in Figure 13.11(b) allows C to reset partially by resonance through inductor L_1 and the feedback diodes, resonance losses only being made up by charging the capacitor through the load, so that for low-loss circuits the reset time is considerably reduced.

Inverters, like choppers, suffer from the maximum frequency limitation imposed by the turn-off time requirements of the thyristors and, as for choppers, sequencing is usually used to overcome this limitation. Figure 13.23 shows a basic two-thyristor inverter which does not use sequencing. Thyristor TH_1 is fired, which charges capacitor C_2 to a voltage equal to V_x, which will be close to $2V_B$ for low resonant losses. The current through the device attempts to reverse and it turns off. Thyristor TH_2 is then fired, which discharges C_2 and charges C_1 to V_x. This thyristor then turns off and thyristor TH_1 is refired. For a low-impedance load the output will be a sine wave provided the device trigger rate corresponds to the resonant time of the inverter. The thyristor waveforms, shown for when the inductors L_1 and L_2 are both coupled and non-coupled, clearly illustrate the larger turn-off pulse obtained if the inductors are coupled, and this is the mode normally used.

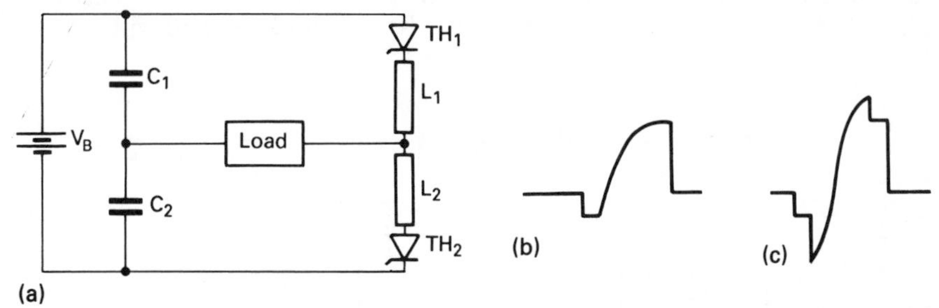

Figure 13.23 Series capacitor commutated inverter: (a) circuit arrangement; (b) thyristor waveforms with inductors non-coupled (c) thyristor waveforms with inductors coupled

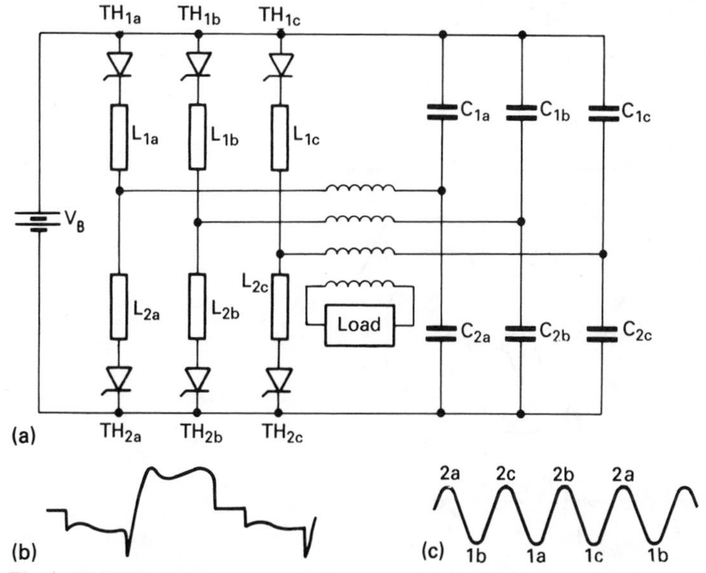

Figure 13.24 Three-stage sequential inverter: (a) circuit arrangement; (b) waveform across TH_{2a}; (c) load waveform

The operating frequency of the inverter of Figure 13.23 is limited by the requirement of ensuring that devices are reverse biased for longer than their turn-off time. Figure 13.24 shows a three-stage sequential inverter which overcomes this limitation, its operation being essentially as for a sequential chopper introduced in Chapter 12, the thyristor conducting periods being indicated in Figure 13.24(c).

It has been seen that series-commutated inverters are most efficient when used with sine wave converters in which the inverter frequency is related to the resonant load frequency. In discussions so far the two frequencies have been maintained equal, but this does not necessarily have to be so. In Figure 13.25 the output load is a tuned circuit consisting of L_2 and C_2 which has a relatively high Q, so that once it is set into oscillation it will produce a sine wave output whose amplitude decreases slowly due to losses. Periodically TH_1 and TH_2 are fired to compensate for this loss. Figure 13.25 also shows various circuit waveforms. Firing TH_1 will charge C_1 with plate a positive. TH_1 will now go off and TH_2 is fired, causing capacitor C_1 to resonate with the load and recharge with plate b positive. Thyristor TH_2 will turn-off when this has been completed and will remain

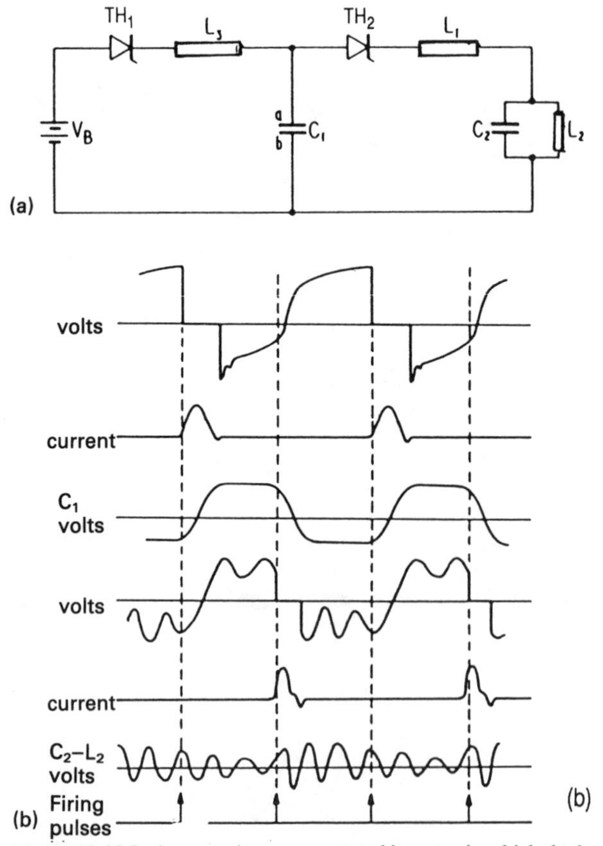

Figure 13.25 Series capacitor commutated inverter in which the load frequency is a multiple of the thyristor operating frequency: (a) circuit arrangement; (b) waveforms

reverse biased so long as the peak amplitude of the load oscillations is arranged to be lower than the capacitor voltage, as shown clearly by the circuit waveforms. Such an inverter is very useful, since the output frequency can be several times the inverter frequency, enabling slower turn-off thyristors to be used, although if the Q of the load circuit is low, the inverter frequency approaches the output frequency and in the limit must be equal to it.

Figure 13.26(a) shows a method of sequencing the basic inverter of Figure 13.25, the operation of the circuit being illustrated by the waveforms of Figure 13.26(b). In the example shown the thyristors are fired at one eighth the tuned output frequency and the bursts of energy are fed to the output at half the output frequency.

As mentioned when discussing choppers in Chapter 12, although there is no fundamental limit to the magnitude of the output frequency which can be attained by sequencing an increasingly greater number of inverter stages, in practice the limit is set by the switching losses which occur in the devices and by the loss per commutation, both of which considerably reduce the efficiency of the inverter.

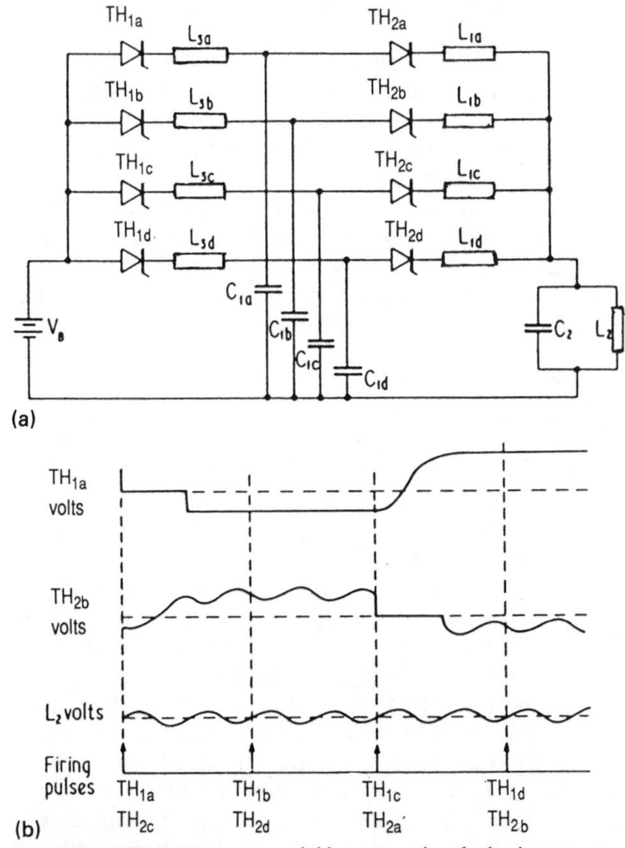

Figure 13.26 Four-stage sequential inverter using the basic arrangement of Figure 13.25: (a) circuit arrangement; (b) waveforms

13.2.4.2 Commutation loss reduction

Commutation losses can be reduced, as was done for choppers, by several techniques. For example, Figure 13.11(b) illustrated a popular push–pull parallel commutated inverter configuration, which is not suitable for high-frequency operation due to its large commutation loss. When TH_1 fires, capacitor C discharges through inductor L_1 and recharges to twice the supply voltage. When this has been completed diode D_1 conducts, feeding back the inductive load current to the supply, and the energy stored in L_1 during the commutation interval is dissipated in the loop L_1-D_1-TH_1.

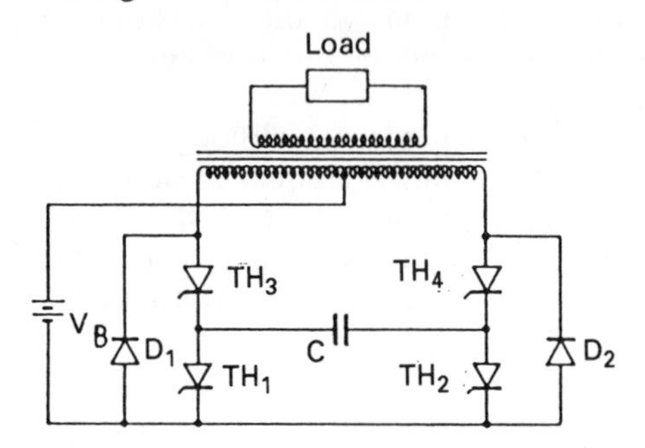

Figure 13.27 'Lossless' push–pull inverter

Figure 13.27 shows a modified push–pull circuit which has zero commutation loss, four thyristors being used to avoid the possibility of short-circuit of the commutation capacitor during the turn-off process and to eliminate the need for a series inductor L_1. Suppose that TH_1 and TH_3 are conducting and TH_4 has been fired to prime the commutation capacitor, after which this thyristor turns off. To reverse the load current thyristor TH_2 is fired, the capacitor discharging and recharging, with reverse polarity, through TH_3, this thyristor also going off when the capacitor has completed its charge. There is no commutation loss in the system since the stored energy in a commutation inductor has been avoided.

Bridge inverter coupled-pulse circuits of the McMurray–Bedford type, as shown in Figure 13.20, also suffer from relatively high commutation losses. For instance, if TH_1 is conducting then capacitor C_1 is discharged and C_2 is charged to V_B. To turn TH_1 off thyristor TH_2 is fired, which couples a pulse to the top device, turning it off. Capacitor C_2 then discharges to zero voltage, after which time the energy stored in inductor L_1 is dissipated in the free-wheeling path TH_2-D_2-L_1.

Several modifications can be made to reduce the commutation loss caused by the circulating current in Figure 13.20, for instance by tapping the load. Figure 13.28 shows the addition of an auxiliary transformer with a feedback winding, which gives a better performance. During the commutation interval, for example when TH_2 is fired as before, the energy stored in L_1 will induce a voltage in winding 1 of the transformer, which will feed back the energy from inductor L_1 to the supply via the step-up winding 4 and bridge rectifier D_7. There is an overshoot voltage on C_2 due to this feedback action, as seen with chopper circuits, but it is small. For example, for a 1:20 turns ratio between windings 1 and 4 it will be 1/20th of the supply voltage.

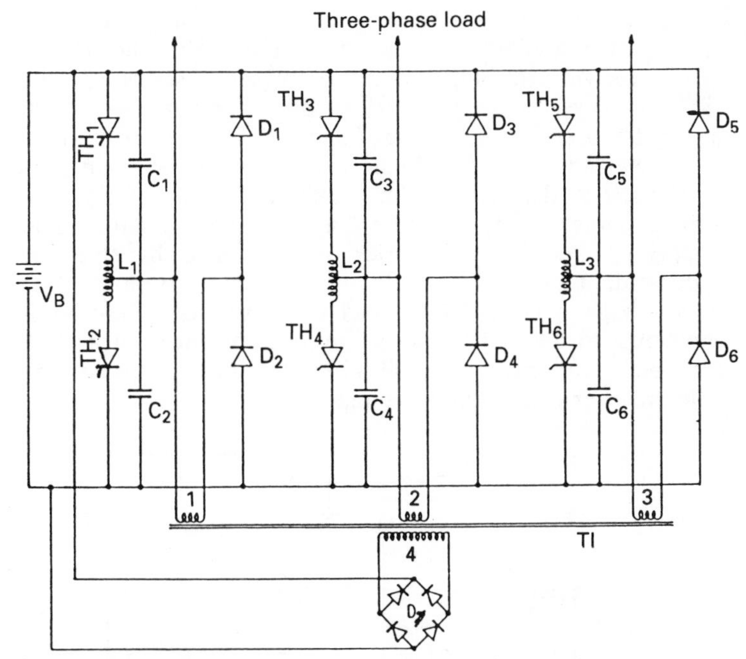

Figure 13.28 Recovery of commutation energy in the inverter of Figure 13.20

13.2.4.3 Voltage boosting

Increasing the available commutation in proportion to the load current improves the efficiency of the circuit, and this occurs in several inverter circuits as part of the circuit function. In these instances it may sometimes be necessary to limit the amount of voltage boost by use of feedback windings to the supply, as discussed in the previous section. In applications where voltage boosting does not occur naturally, auxiliary circuits can be used.

In the coupled-pulse commutated circuit of Figure 13.20 the maximum

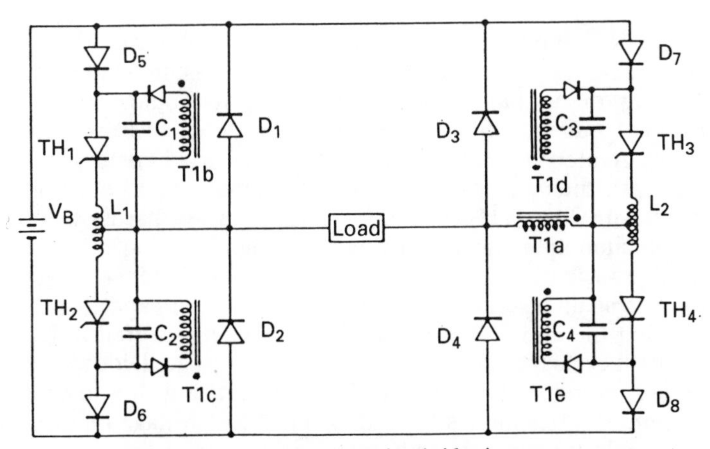

Figure 13.29 Transformer voltage boost in a bridge inverter

commutation capacitor voltage is effectively clamped to the value of the supply voltage V_B due to the action of the free-wheeling diodes. To increase the capacitor voltage, inductors can be placed in series with the supply or in the free-wheeling diode paths. Alternatively, boost transformers can be used, as shown in Figure 13.29, the primary of transformer T_1, referred to as T_{1a}, being connected in the main load current path. Four secondaries feed the commutation capacitors, their polarities being as indicated. Since the voltage on the capacitors is to exceed the supply voltage, diodes in series with the main thyristors are required. Suppose TH_1 and TH_4 are conducting, the load current build-up in T_{1a} induces a voltage in T_{1c} and T_{1d}, which charges C_2 and C_3 to boost voltages proportional to the magnitude of the load current change. This voltage is held on the capacitors until TH_2 and TH_3 are fired, and they form part of the inverter commutation voltage.

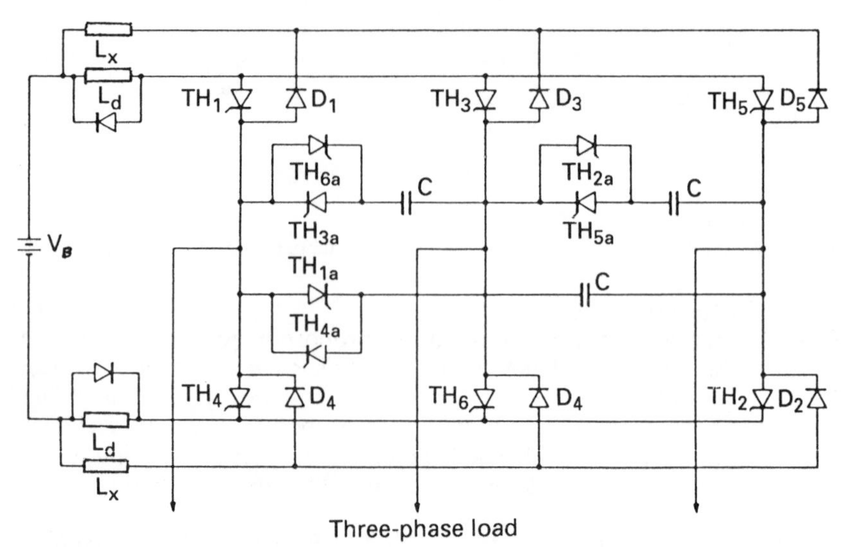

Figure 13.30 Series inductor voltage boost in a bridge inverter

Figure 13.30 shows a modification to the inverter of Figure 13.13. Inductors L_d are required as part of the commutation process, inductors L_x in the free-wheeling diode path being added to give voltage boost proportional to load current. Since the commutation capacitors are charged to a voltage greater than that of the supply, auxiliary thyristors are required in series with them, to prevent a loss of charge when boosting stops. TH_{1a} is fired with TH_1, TH_{2a} with TH_2, and so on, so that they do not perform any function apart from assisting in voltage boosting.

Figure 13.31 shows circuit additions to Figure 13.11(b) for voltage boosting, where once again a series inductor L_2 is used. After capacitor C has charged to $2V_B$ at the end of a commutation, the energy stored in L_2 during a previous load cycle transfers to C, boosting its voltage. Diodes D_1 and D_2 can conduct, carrying the inductive load current, but due to the reverse-biased action of series diodes D_3 and D_4 the boost voltage on C is not lost, being available for the next commutation.

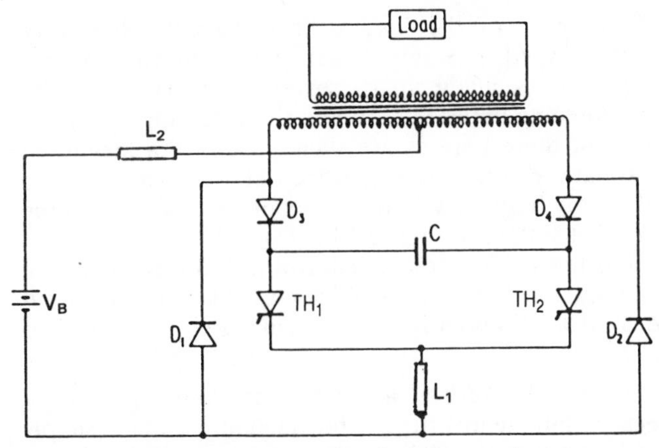

Figure 13.31 Push–pull inverter with series inductor voltage boost

13.3 Output voltage control

One method of output voltage control has already been introduced, with reference to Figure 13.4, for mark-to-space control, or a quasi-square waveform. One of the main considerations in controlling the output voltage is the harmonics which have been introduced, and this was not evident from Figure 13.4 since sine wave filtering had been assumed. In reality this is unlikely to be the case, as for motor-drive applications, and the waveforms will be as illustrated in Figure 13.32(a). The output voltage and current are rich in harmonics, which causes excessive load losses. To

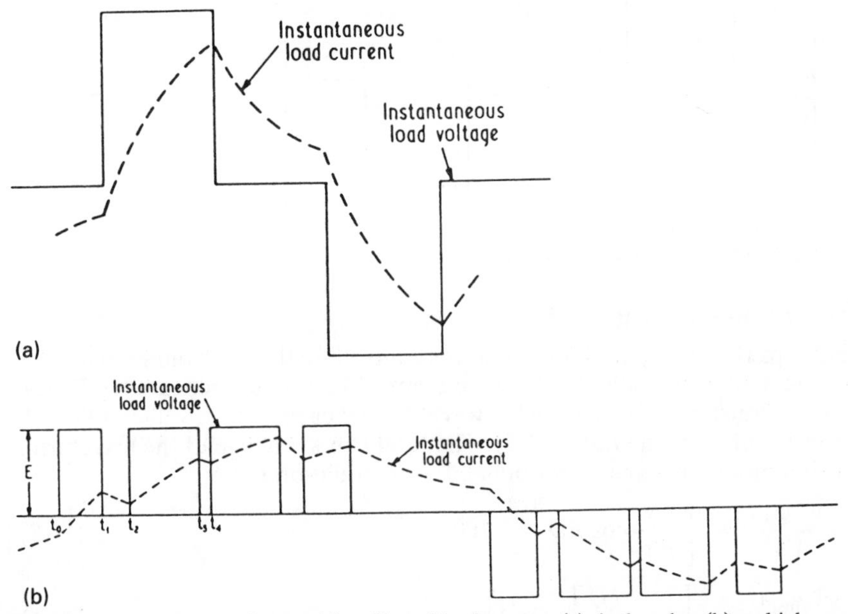

Figure 13.32 Inverter waveforms with unfiltered load current: (a) single pulse; (b) multiple pulse

reduce the harmonic content of the load voltage current, alternative control techniques can be used, one such system being shown in Figure 13.32(b). The load voltage is made up of several sections of varying duration, each block switching between zero and the d.c. supply voltage. The placing and width of these periods are such as to give a mean a.c. voltage which approximates closely to a sine wave, the load current being seen to have a smaller low-frequency harmonic content than that in the traditional voltage-control system of Figure 13.32(a).

This section describes the various voltage-control techniques which can be used for inverter circuits, and the impact which these have on the output harmonics. Voltage control within an inverter is usually required for two applications:

(i) When the output is to be kept at a fixed value, compensating for regulation effects within the inverter, or for fluctuations in the supply voltage or the load current. These requirements usually arise in fixed-frequency inverter supplies.
(ii) When the output is to be varied in a given manner, for example proportional to the frequency to keep the flux within the load constant, such as required for variable-frequency motor drives.

There are several ways in which this voltage control can be achieved, in all these cases the a.c. being composed of a fundamental component and a band of harmonic frequencies. The various control methods all contrive to reduce the harmonic voltages whilst avoiding excessive circuit complexity. In this section these techniques are classified as unidirectional switching, bi-directional switching, and waveform synthesis.

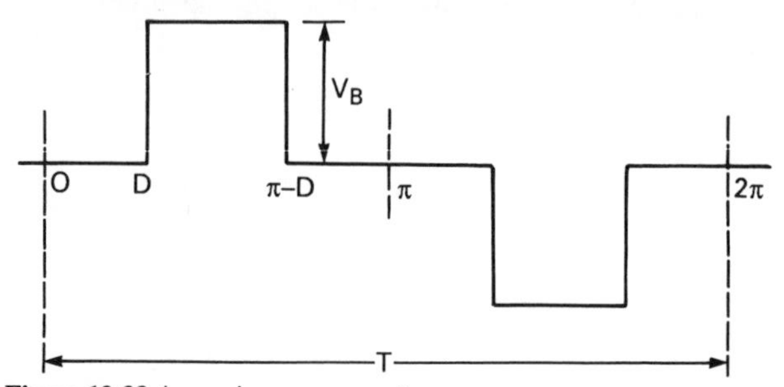

Figure 13.33 A quasi-square waveform

13.3.1 Unidirectional switching

Perhaps the most popular method for controlling the a.c. voltage is to vary its mark-to-space ratio, as shown in Figure 13.32(a) and repeated in Figure 13.33. Fourier analysis of such a waveform gives the r.m.s. value of the *n*th coefficient as in equation (13.5). The total r.m.s. voltage of the waveform, including all harmonics, is obtained as in equation (13.6):

$$\frac{V_{\mathrm{rms(n)}}}{V_{\mathrm{B}}} = \left[\frac{2\sqrt{2}}{n\pi}\cos nD\right] \times 100 \tag{13.5}$$

$$\frac{V_{\mathrm{rms(T)}}}{V_{\mathrm{B}}} = \left[1 - \frac{2D}{\pi}\right]^{1/2} \times 100 \tag{13.6}$$

Table 13.1 Harmonic composition of a quasi-square wave

$2D/T$	*R.M.S. voltage as percentage of d.c. supply*								
	1	3	5	7	9	11	13	15	*Total*
0.00	90.0	30.0	18.0	12.9	10.0	8.18	6.92	6.00	100
0.02	89.8	29.5	17.1	11.6	8.44	6.30	4.74	3.53	98.0
0.04	89.3	27.9	14.6	8.20	4.26	1.53	0.43	1.85	95.9
0.06	88.4	25.3	10.6	3.20	1.25	3.94	5.33	5.71	93.8
0.08	87.2	21.9	5.56	2.41	6.37	7.61	6.87	4.85	91.7
0.10	85.6	17.6	0.00	7.56	9.51	7.78	4.07	0.00	89.4
0.12	83.7	12.8	5.56	11.3	9.69	4.38	1.30	4.85	87.2
0.14	81.4	7.46	10.6	12.8	6.85	1.03	5.85	5.71	84.9
0.16	78.9	1.88	14.6	12.0	1.87	5.96	6.71	1.85	82.5
0.18	76.0	3.76	17.1	8.80	3.68	8.17	3.34	3.53	80.0
0.20	72.8	9.27	18.0	3.97	8.09	6.62	2.14	6.00	77.5
0.22	69.3	14.5	17.1	1.61	9.98	2.03	6.26	3.53	74.8
0.24	65.6	19.1	14.6	6.89	8.76	3.48	6.44	1.85	72.1
0.26	61.6	23.1	10.6	10.9	4.82	7.40	2.55	5.71	69.3
0.28	57.4	26.3	5.56	12.8	0.63	7.92	2.95	4.85	66.3
0.30	52.9	28.5	0.00	12.2	5.88	4.81	6.58	0.00	63.2
0.32	48.2	29.8	5.56	9.37	9.30	0.51	6.07	4.85	60.0
0.34	43.4	29.9	10.6	4.73	9.82	5.60	1.72	5.71	56.6
0.36	38.3	29.1	14.6	0.81	7.29	8.12	3.71	1.85	52.9
0.38	33.1	27.1	17.1	6.19	2.49	6.91	6.80	3.53	49.0
0.40	27.8	24.3	18.0	10.4	3.09	2.53	5.60	6.00	44.7
0.42	22.4	20.5	17.1	12.6	7.71	3.01	0.87	3.53	40.0
0.44	16.9	16.1	14.6	12.5	9.92	7.17	4.41	1.85	34.5
0.46	11.3	11.0	10.6	9.91	9.05	8.04	6.91	5.71	28.3
0.48	5.65	5.62	5.56	5.47	5.36	5.22	5.05	4.85	20.0
0.50	0.00	0.00	0.00	0.00	0.00	0.00	0.00	0.00	0.00

Equations (13.5) and (13.6) are shown evaluated in Table 13.1 up to the 15th harmonic, where the period T equals 2π. From this table it is evident that the harmonic content of the output increases rapidly as the mark-to-space ratio of the waveform is reduced. This is illustrated more clearly in Figure 13.34, which shows the plot of the harmonics, obtained from Table 13.1. Because in reality it is the magnitude of the fundamental which is of interest rather than the mark-to-space ratio, which is a means of varying the fundamental, it is more normal to show harmonic plots against the fundamental, and this is done for the harmonics of Table 13.1 in Figure 13.35.

At low voltages various harmonics are almost equal in value to the fundamental, the total harmonic content being about ten times larger. This represents the greatest disadvantage of the quasi-square voltage-control system, and normally limits the output voltage range to between 30% and 90% of the d.c. supply, i.e. a frequency change of 3:1 in applications where the voltage needs to be varied proportional to the frequency of the load.

Figure 13.33 is also referred to as a single-pulse unidirectional wave. It is unidirectional since in any half cycle the output is either positive or negative, but never both, and there is a single pulse in each half cycle. An obvious extension to this system is shown in Figure 13.36, which illustrates

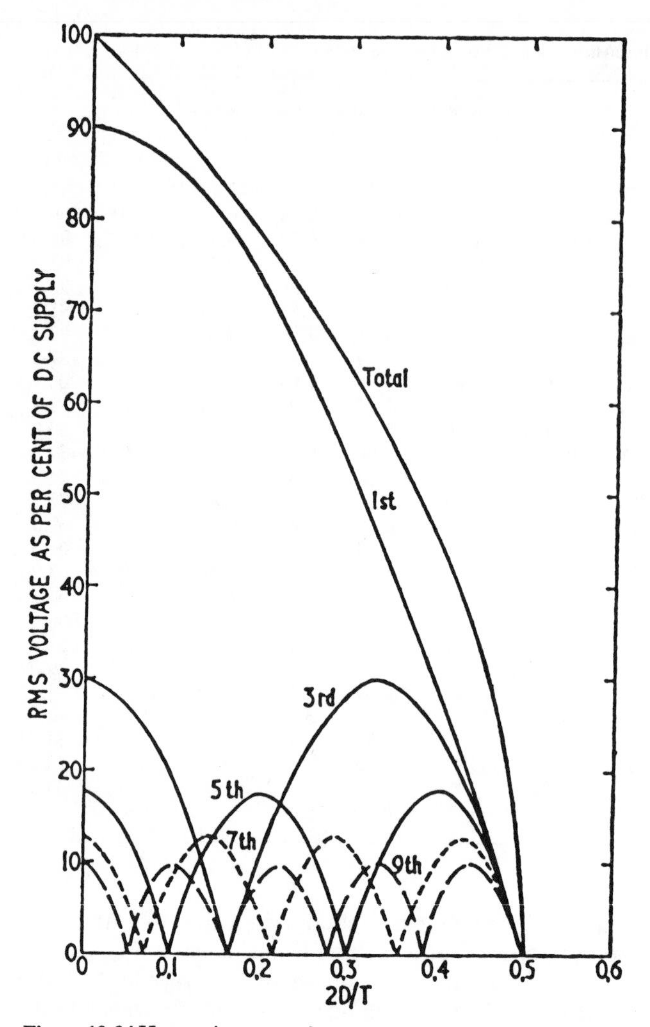

Figure 13.34 Harmonic content for a quasi-square wave plotted against pulse width

a two-pulse unidirectional waveform. Choosing the values of B, A_1, A_2 and ω to satisfy equations (13.7) and (13.8) will give the nth Fourier coefficient, as in equation (13.9):

$$B = \frac{A_1 + A_2}{2} \tag{13.7}$$

$$\omega = \frac{A_2 - A_1}{2} \tag{13.8}$$

$$b_n = \frac{8V_B}{\pi n} \sin nB \sin n\omega \tag{13.9}$$

Obviously, it is possible to arrange the pulses symmetrically in a half cycle, and this is considered later when this technique is extended to a large

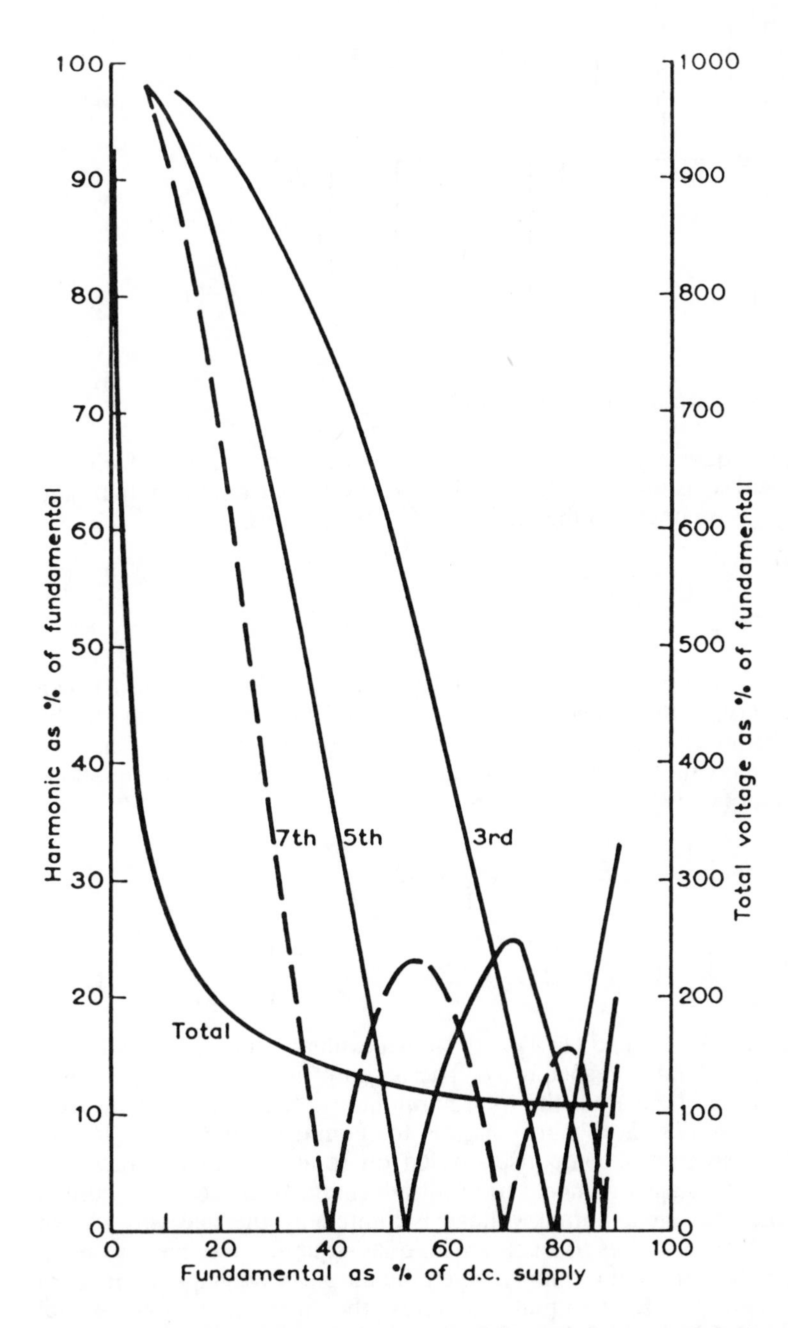

Figure 13.35 Harmonic content for a quasi-square wave plotted against the fundamental

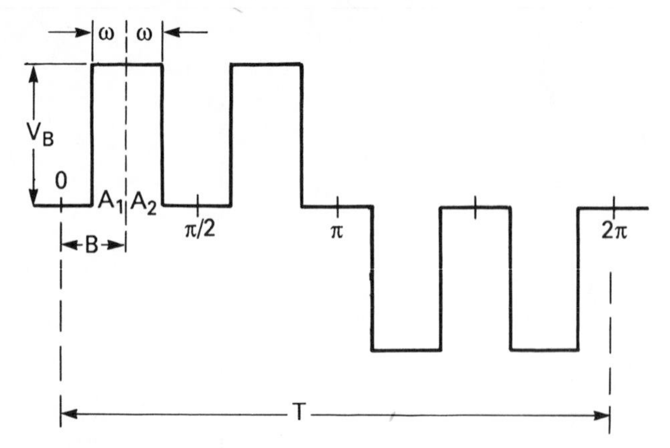

Figure 13.36 A two-pulse unidirectional wave

number of equally spaced pulses. In the present system the waveform will be arranged so as to eliminate, say, the *P*th harmonic from the output, and to do this *B* is chosen so that equation (13.10) is satisfied:

$$\left.\begin{aligned} \sin PB &= 0 \\ \text{or} \quad B &= \frac{\pi}{P} \end{aligned}\right\} \tag{13.10}$$

As long as the pulses are maintained at this value of *B* their width ω can change, varying the effective voltage, but the *P*th harmonic will still be absent. From equation (13.9) the value of the r.m.s. voltage of the *n*th harmonic and of the total waveform can be obtained, as in equations (13.11) and (13.12) respectively:

$$\frac{V_{\mathrm{rms(n)}}}{V_{\mathrm{B}}} = \left[\frac{4\sqrt{2}}{n\pi} \sin n \frac{\pi}{\mathrm{p}} \sin n\omega\right] \times 100 \tag{13.11}$$

$$\frac{V_{\mathrm{rms(T)}}}{V_{\mathrm{B}}} = \left[\frac{2}{\pi} (A_2 - A_1)\right]^{1/2} \times 100 = 200 \left[\frac{\omega}{\pi}\right]^{1/2} \tag{13.12}$$

Equations (13.11) and (13.12) are shown evaluated in Tables 13.2–13.4 for $P = 3$, 5 and 7 respectively. These show that even though lower harmonics are eliminated, the total harmonic content is large. However, higher harmonics are much easier to eliminate than lower-order harmonics, so that a comparison based on total content can often be misleading. Perhaps the most serious disadvantage of this control method is that the maximum output voltage is limited by the necessity of not allowing adjacent pulses to merge. With quasi-square waves the maximum voltage is 90% of the d.c. supply, when the output waveshape has no zero voltage periods. In two-pulse systems the maximum fundamental obtainable is 75.1%, 62.2% or 33.3% of the d.c., according to whether the third, fifth or seventh harmonic is being eliminated. Therefore in a variable-frequency drive, a 240 V, 50 Hz motor could run at a maximum frequency of 340 × 0.333 × 50/20 = 24 Hz and still remain fully fluxed (340

Table 13.2 Harmonics in a two-pulse unidirectional wave with $P = 3$

$2\omega/T$	*R.M.S. voltage as percentage of d.c. supply*								
	1	3	5	7	9	11	13	15	*Total*
0.16	75.1	0.00	18.3	8.20	0.00	9.70	2.98	0.00	80.0
0.15	70.8	0.00	22.0	3.48	0.00	12.6	1.88	0.00	77.5
0.14	66.4	0.00	25.2	1.40	0.00	14.1	6.43	0.00	74.8
0.13	61.9	0.00	27.8	6.21	0.00	13.8	9.92	0.00	72.1
0.12	57.4	0.00	29.7	10.7	0.00	12.0	11.8	0.00	69.3
0.11	52.8	0.00	30.8	14.7	0.00	8.69	11.7	0.00	66.3
0.10	48.2	0.00	31.2	18.0	0.00	4.38	9.70	0.00	63.2
0.09	43.5	0.00	30.8	20.4	0.00	0.45	6.10	0.00	60.0
0.08	38.8	0.00	29.7	21.9	0.00	5.22	1.50	0.00	56.6
0.07	34.0	0.00	27.8	22.3	0.00	9.37	3.35	0.00	52.9
0.06	29.2	0.00	25.2	21.6	0.00	12.4	7.64	0.00	49.0
0.05	24.4	0.00	22.0	19.8	0.00	14.0	10.7	0.00	44.7
0.04	19.5	0.00	18.3	17.2	0.00	13.9	12.0	0.00	40.0
0.03	14.7	0.00	14.2	13.6	0.00	12.2	11.3	0.00	34.6
0.02	9.79	0.00	9.63	9.48	0.00	9.03	8.74	0.00	28.3
0.01	4.90	0.00	4.88	4.86	0.00	4.80	4.76	0.00	20.0
0.00	0.00	0.00	0.00	0.00	0.00	0.00	0.00	0.00	0.00

Table 13.3 Harmonics in a two-pulse unidirectional wave with $P = 5$

$2\omega/T$	*R.M.S. voltage as percentage of d.c. supply*								
	1	3	5	7	9	11	13	15	*Total*
0.20	62.2	54.3	0.00	23.3	6.91	5.65	12.5	0.00	89.4
0.19	59.5	55.7	0.00	21.0	9.29	2.68	13.1	0.00	87.2
0.18	56.7	56.6	0.00	17.8	10.9	0.60	11.5	0.00	84.9
0.17	53.9	57.0	0.00	13.7	11.7	3.82	8.07	0.00	82.5
0.16	51.0	57.0	0.00	9.00	11.5	6.58	3.27	0.00	80.0
0.15	48.0	56.4	0.00	3.83	10.5	8.57	2.06	0.00	77.5
0.14	45.0	55.3	0.00	1.54	8.57	9.54	7.06	0.00	74.8
0.13	42.0	53.7	0.00	6.82	5.98	9.39	10.9	0.00	72.1
0.12	38.9	51.6	0.00	11.8	2.92	8.12	12.9	0.00	69.3
0.11	35.8	49.1	0.00	16.2	0.37	5.90	12.9	0.00	66.3
0.10	32.7	46.2	0.00	19.8	3.63	2.97	10.7	0.00	63.2
0.09	29.5	42.8	0.00	22.4	6.61	0.30	6.70	0.00	60.0
0.08	26.3	39.1	0.00	24.0	9.06	3.54	1.65	0.00	56.6
0.07	23.1	35.0	0.00	24.4	10.8	6.36	3.67	0.00	52.9
0.06	19.8	30.6	0.00	23.7	11.7	8.43	8.39	0.00	49.0
0.05	16.6	25.9	0.00	21.8	11.6	9.50	11.7	0.00	44.7
0.04	13.3	21.0	0.00	18.8	10.6	9.45	13.1	0.00	40.0
0.03	9.96	15.9	0.00	15.0	8.82	8.28	12.4	0.00	34.6
0.02	6.64	10.7	0.00	10.4	6.30	6.13	9.60	0.00	28.3
0.01	3.32	5.37	0.00	5.33	3.28	3.26	5.23	0.00	20.0
0.00	0.00	0.00	0.00	0.00	0.00	0.00	0.00	0.00	0.00

Table 13.4 Harmonics in a two-pulse unidirectional wave with $P = 7$

$2\omega/T$	*R.M.S. voltage as percentage of d.c. supply*								
	1	3	5	7	9	11	13	15	*Total*
0.14	33.3	56.7	22.8	0.00	11.4	15.8	3.22	1.61	74.8
0.13	31.0	55.0	25.1	0.00	7.96	15.6	4.97	0.81	72.1
0.12	28.8	52.9	26.8	0.00	3.89	13.5	5.90	3.06	69.3
0.11	26.5	50.3	27.8	0.00	0.49	9.78	5.86	4.64	66.3
0.10	24.1	47.3	28.1	0.00	4.83	4.93	4.86	5.21	63.2
0.09	21.8	43.9	27.8	0.00	8.79	0.50	3.06	4.64	60.0
0.08	19.4	40.0	26.8	0.00	12.0	5.87	0.75	3.06	56.6
0.07	17.0	35.9	25.1	0.00	14.4	10.6	1.68	0.81	52.9
0.06	14.6	31.3	22.8	0.00	15.5	14.0	3.83	1.61	49.0
0.05	12.2	26.6	19.9	0.00	15.4	15.8	5.35	3.68	44.7
0.04	9.79	21.5	16.5	0.00	14.1	15.7	6.00	4.95	40.0
0.03	7.35	16.3	12.8	0.00	11.7	13.7	5.65	5.14	34.6
0.02	4.90	11.0	8.90	0.00	8.38	10.2	4.38	4.21	28.3
0.01	2.45	5.50	4.40	0.00	4.36	5.40	2.39	2.36	20.0
0.00	0.00	0.00	0.00	0.00	0.00	0.00	0.00	0.00	0.00

$= 240 \times \sqrt{2}$, i.e. the peak rectified mains voltage) when the seventh harmonic is being eliminated.

A second extension to quasi-square wave control is to have several equally spaced unidirectional pulses in a half cycle. This is conveniently obtained by feeding a high-frequency carrier triangular wave and a low-frequency square wave into a comparator, as in Figure 13.37(a). The output voltage swings between the supply voltage V_B and zero volts according to the relative magnitudes of the two waveforms, as in Figure 13.37(b). The width of the output pulses is determined by the ratio of

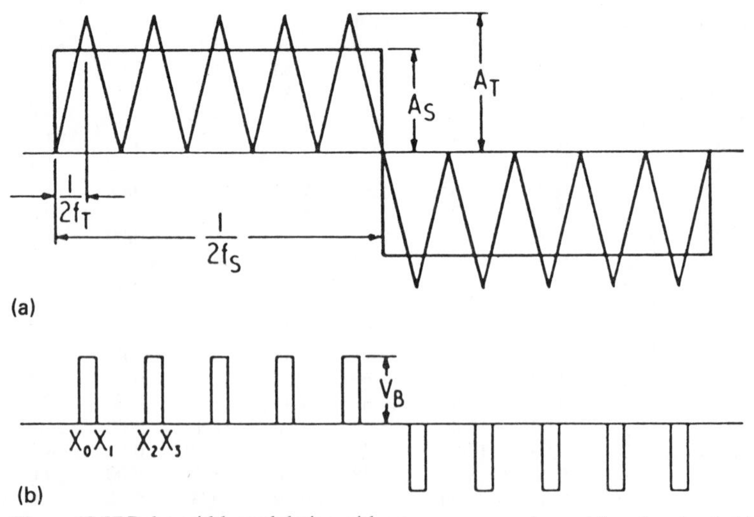

Figure 13.37 Pulse-width modulation with a square wave using unidirectional switching

A_S/A_T and there are as many pulses per cycle as the ratio f_T/f_S. The Fourier coefficient of this waveform is given by

$$\frac{V_{rms(n)}}{V_B} = \sum_{M=0,2,4\ldots} \frac{2\sqrt{2}}{n\pi}[\cos nX(M) - \cos nX(M+1)] \times 100 \tag{13.13}$$

and the total r.m.s. content by

$$\frac{V_{rms(T)}}{V_B} = \left[\frac{f_T}{2\pi f_s}\{X(0) - X(1)\}\right]^{1/2} \times 100 \tag{13.14}$$

Equations (13.13) and (13.14) have been evaluated in Tables 13.5–13.8 for $f_T/f_S = 4$, 6, 10 and 20 respectively. These tables indicate an increase in total harmonic content with pulse numbers where $f_T/f_S = 2$ corresponds to a quasi-square wave. However, the lower harmonics are now considerably attenuated, and this is illustrated more clearly in Figure 13.38, where the

Table 13.5 Harmonic content of a square modulated unidirectional wave with $f_T/f_S = 4$

A_S/A_T	*R.M.S. voltage as percentage of d.c. supply*								
	1	3	5	7	9	11	13	15	*Total*
0	90.0	30.0	18.0	12.9	10.0	8.18	6.92	6.00	100
0.1	82.7	36.2	9.74	17.7	1.11	11.5	2.29	7.84	94.9
0.2	74.8	40.3	0.00	17.3	8.31	6.80	9.31	0.00	89.4
0.3	66.5	42.3	9.74	11.8	13.8	2.70	7.44	7.84	83.7
0.4	57.8	41.9	18.0	2.84	12.6	10.3	1.53	6.00	77.5
0.5	48.7	39.2	23.5	6.96	5.41	10.7	9.05	3.25	70.7
0.6	39.3	34.3	25.5	14.7	4.37	3.58	7.92	8.49	63.2
0.7	29.7	27.6	23.5	18.1	12.1	6.05	0.77	3.25	54.8
0.8	19.9	19.3	18.0	16.2	14.0	11.4	8.72	6.00	44.7
0.9	9.99	9.90	9.74	9.50	9.18	8.80	8.35	7.84	31.6
0.98	2.00	2.00	2.00	2.00	1.99	1.99	1.99	1.98	14.1

Table 13.6 Harmonic content of a square modulated unidirectional wave with $f_T/f_S = 6$

A_S/A_T	*R.M.S. voltage as percentage of d.c. supply*								
	1	3	5	7	9	11	13	15	*Total*
0	90.0	80.0	18.0	12.9	10.0	8.18	6.92	6.0	100
0.1	81.7	29.6	25.5	4.00	8.93	14.6	2.15	4.22	94.9
0.2	73.2	28.5	31.2	5.36	5.90	16.3	10.3	0.02	89.4
0.3	64.5	26.7	34.8	14.0	1.58	12.7	13.8	4.26	83.7
0.4	55.6	24.3	36.0	20.8	3.07	5.04	11.2	6.02	77.5
0.5	46.6	21.2	34.8	24.9	7.05	4.25	3.57	4.26	70.7
0.6	37.4	17.6	31.2	25.6	9.49	12.2	5.65	0.02	63.2
0.7	28.2	13.6	25.5	22.9	9.86	16.2	12.4	4.22	54.8
0.8	18.8	9.25	18.0	17.2	8.07	15.0	13.6	5.98	44.7
0.9	9.44	4.47	9.34	9.23	4.52	8.93	8.73	4.22	31.6
0.98	1.90	0.92	1.90	1.90	0.92	1.90	1.90	0.92	14.1

Table 13.7 Harmonic content of a square modulated unidirectional wave with $f_T/f_S = 10$

A_S/A_T	R.M.S. voltage as percentage of d.c. supply								
	1	3	5	7	9	11	13	15	*Total*
0	90.0	30.0	18.0	12.9	10.0	8.16	6.94	6.00	100
0.1	81.3	27.8	17.8	14.6	18.2	0.85	4.37	5.33	94.9
0.2	72.4	25.4	17.1	15.6	25.0	9.77	1.09	3.51	89.4
0.3	63.6	22.7	16.1	15.9	29.7	17.5	2.37	0.92	83.7
0.4	54.6	19.9	14.6	15.4	32.1	23.2	5.44	1.87	77.5
0.5	45.6	16.8	12.7	14.1	32.0	26.2	7.61	4.26	70.7
0.6	36.5	13.6	10.6	12.2	29.3	26.0	8.52	5.72	63.2
0.7	27.4	10.3	8.19	9.72	24.3	22.8	8.03	5.94	54.8
0.8	18.3	6.93	5.58	6.75	17.4	16.9	6.22	4.87	44.7
0.9	9.17	3.47	2.83	3.45	9.05	8.99	3.38	2.74	31.6
0.98	1.85	0.68	0.58	0.68	1.85	1.85	0.68	0.58	14.1

Table 13.8 Harmonic content of a square modulated unidirectional wave with $f_T/f_S = 20$

A_S/A_T	R.M.S. voltage as percentage of d.c. supply								
	1	3	5	7	9	11	13	15	*Total*
0.1	81.1	27.2	16.5	12.1	9.68	8.28	7.49	7.23	94.9
0.2	72.1	24.3	15.0	11.1	9.16	8.14	7.75	8.07	89.4
0.3	63.1	21.4	13.3	10.0	8.46	7.75	7.69	8.46	83.7
0.4	54.1	18.4	11.6	8.84	7.59	7.13	7.31	8.38	77.5
0.5	45.1	15.4	9.74	7.54	6.58	6.30	6.62	7.84	70.7
0.6	36.1	12.4	7.87	6.14	5.42	5.28	5.66	6.86	63.2
0.7	27.1	9.31	5.94	4.67	4.17	4.10	4.47	5.51	54.8
0.8	18.1	6.22	3.98	3.15	2.82	2.81	3.09	3.85	44.7
0.9	9.04	3.11	2.00	1.58	1.43	1.42	1.58	1.98	31.6
0.98	1.81	0.62	0.40	0.32	0.29	0.29	0.32	0.40	14.1

third harmonic at high pulse numbers varies almost linearly with the fundamental, being 30% of its value. This is clearly highly desirable, since filter designs can be simplified. There is also a very large reduction from the 100% third harmonic content obtained at low voltages with quasi-square wave control methods.

For higher harmonics it is necessary to go to a larger pulse number to obtain appreciable harmonic reduction. For instance, the curves for the seventh harmonic are shown in Figure 13.39, and from these it is seen that with ten pulses per cycle there is still a large and variable harmonic content. With twenty pulses the variation is from 14% to 17% of the fundamental, for the seventh harmonic, at 90% and 10% of the fundamental voltage points. This can be compared with 14% and 90% for the same two points when using quasi-square wave control.

This voltage-control system therefore gives an overall lower harmonic content, compared to quasi-square wave control, provided a sufficiently high ratio of f_T/f_S, say 20, is chosen, whilst the maximum d.c. voltage is still 90% of the d.c. supply. The only disadvantage of this system is that the

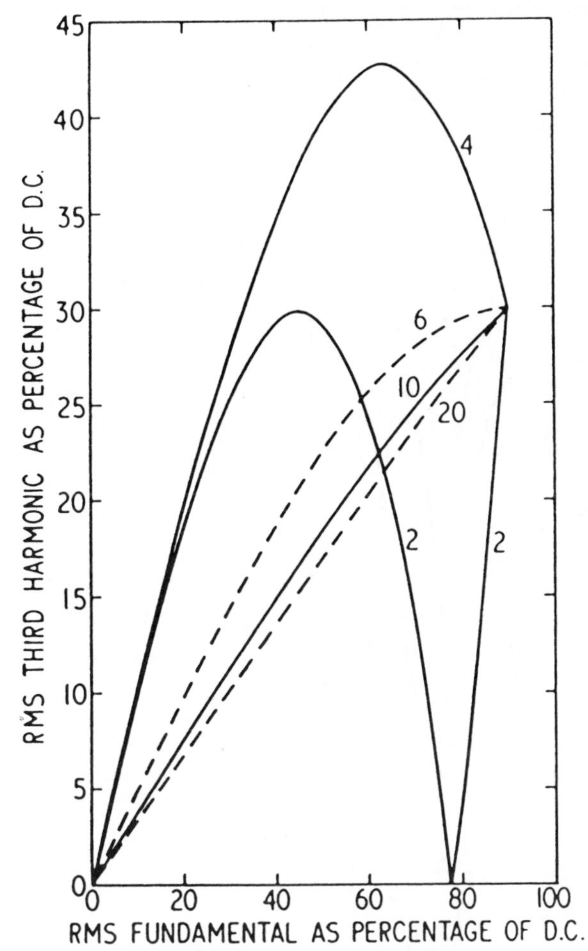

Figure 13.38 Third harmonic content for the square modulated unidirectional wave of Figure 13.37

inverter must now run at twenty times the output frequency required, which can result in extra losses, and special techniques must be used to minimise these losses and to increase high-frequency commutation times, as described in section 13.2.4.

In the voltage-control technique illustrated in Figure 13.37 a square wave is used as the reference source. A greater reduction in harmonic content would be obtained if a sine wave was used as the reference, as in Figure 13.40. The output pulses are now of unequal width, but provided the values of X_1, X_2, X_3, etc. are known the magnitude of the harmonic coefficient can again be found from equation (13.13). Equation (13.14) needs to be modified to take account of the unequal pulse widths, and the total r.m.s. content of the output is now given by

$$\frac{V_{\mathrm{rms(T)}}}{V_{\mathrm{B}}} = \left[\frac{2}{\pi}\sum\{X(M+1) - X(M)\}\right]^{1/2} \times 100 \qquad (13.15)$$

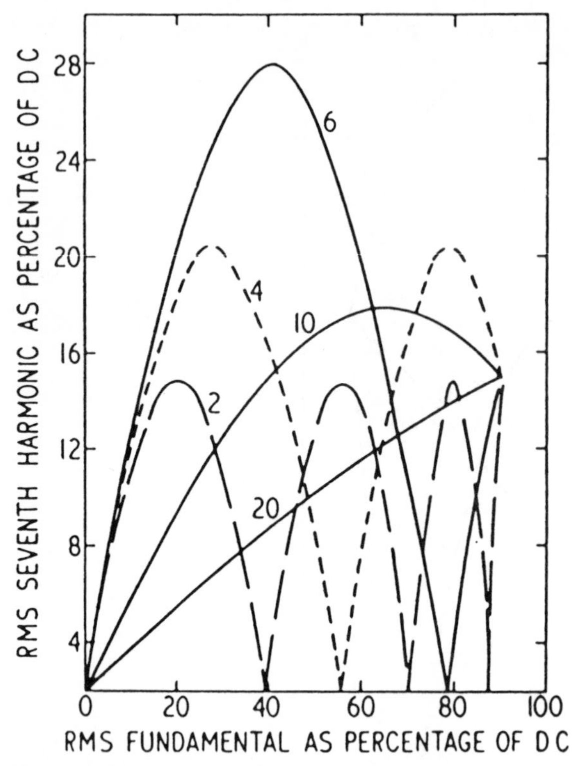

Figure 13.39 Seventh harmonic content for the square modulated unidirectional wave of Figure 13.37

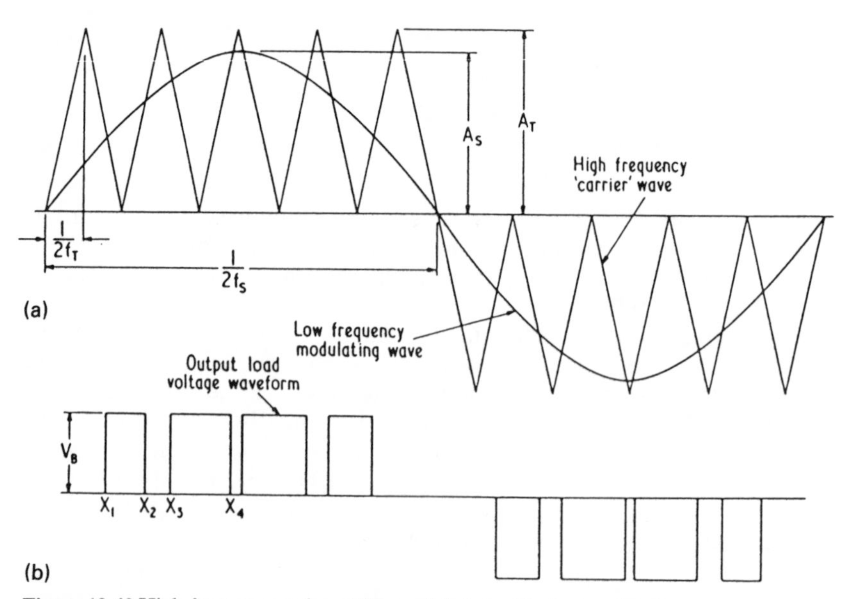

Figure 13.40 High-frequency pulse-width modulation with sine modulating wave and unidirectional switching: (a) modulating waveforms; (b) output waveform

Table 13.9 Harmonic content of a sine modulated unidirectional wave with $f_T/f_S = 4$

A_S/A_T	R.M.S. voltage as percentage of d.c. supply								
	1	3	5	7	9	11	13	15	*Total*
0	0	0	0	0	0	0	0	0	0
0.1	9.93	9.77	9.46	8.99	8.40	7.69	6.88	6.00	26.5
0.2	16.6	15.9	14.4	12.4	9.97	7.32	4.64	2.14	34.4
0.3	23.2	21.1	17.4	12.5	7.18	2.26	1.65	4.13	40.7
0.4	29.6	25.4	17.9	9.19	1.25	4.24	6.48	5.68	46.2
0.5	33.9	27.5	16.9	5.54	3.19	7.27	6.60	2.87	49.5
0.6	40.0	29.5	13.4	1.02	8.41	7.66	2.01	3.52	54.2
0.7	44.0	30.0	10.0	5.36	9.91	5.05	2.41	5.90	57.0
0.8	47.9	29.8	6.01	9.03	9.44	0.95	5.85	5.13	59.7
0.9	51.6	29.0	1.63	11.6	7.08	3.44	6.92	1.61	62.3
0.98	53.4	28.3	0.62	12.4	5.33	5.33	6.38	0.62	63.6

Table 13.10 Harmonic content of a sine modulated unidirectional wave with $f_T/f_S = 6$

A_S/A_T	R.M.S. voltage as percentage of d.c. supply								
	1	3	5	7	9	11	13	15	*Total*
0	0	0	0	0	0	0	0	0	0
0.1	6.98	0.29	7.20	6.57	0.84	7.20	5.98	1.32	24.0
0.2	14.7	0.60	14.7	13.1	1.60	13.1	10.2	2.09	34.8
0.3	21.2	0.25	19.9	18.0	0.59	14.4	11.4	0.55	41.7
0.4	27.8	0.45	23.9	21.5	0.88	12.7	9.08	0.39	47.7
0.5	34.3	1.46	26.7	23.4	2.17	8.47	3.59	0.38	52.9
0.6	40.8	2.74	28.1	23.1	2.72	3.07	3.48	2.69	57.7
0.7	48.8	5.52	27.2	20.4	2.22	2.73	10.9	6.50	63.0
0.8	55.4	7.28	25.7	15.6	0.32	5.67	13.7	6.41	67.1
0.9	60.8	10.4	22.5	10.8	3.59	5.06	12.6	4.51	70.2
0.98	66.3	13.6	18.7	5.12	7.86	2.74	8.82	0.21	73.1

Table 13.11 Harmonic content of a sine modulated unidirectional wave with $f_T/f_S = 10$

A_S/A_T	R.M.S. voltage as percentage of d.c. supply								
	1	3	5	7	9	11	13	15	*Total*
0	0	0	0	0	0	0	0	0	0
0.1	7.26	0.28	0.24	0.88	7.53	6.70	0.43	0.68	25.0
0.2	13.6	0.54	0.12	0.11	13.7	12.3	1.48	0.29	34.6
0.3	20.8	0.95	0.18	0.12	19.9	17.4	2.72	0.27	42.8
0.4	27.5	0.49	0.57	0.02	24.1	21.2	3.63	0.92	49.1
0.5	34.7	0.98	0.69	0.80	27.7	23.0	5.70	0.66	55.2
0.6	42.1	0.60	0.35	3.67	27.8	24.7	6.21	1.17	60.7
0.7	48.9	0.17	0.93	4.90	27.0	23.2	8.51	2.16	65.2
0.8	55.9	1.31	0.91	8.17	24.9	20.8	9.71	2.89	69.9
0.9	62.5	1.43	1.57	11.0	20.7	17.5	10.5	4.11	73.9
0.98	68.5	0.93	1.57	13.4	16.6	12.7	11.9	4.70	77.3

Table 13.12 Harmonic content of a sine modulated unidirectional wave with $f_T/f_S = 20$

A_S/A_T	R.M.S. voltage as percentage of d.c. supply								
	1	3	5	7	9	11	13	15	*Total*
0	0	0	0	0	0	0	0	0	0
0.1	7.58	0.68	1.04	0.22	0.33	0.26	0.38	1.26	25.9
0.2	14.9	0.60	0.09	0.50	0.35	0.30	0.62	0.48	36.2
0.3	21.5	0.15	0.15	0.77	0.53	0.52	0.88	0.24	43.7
0.4	28.2	0.03	0.56	0.89	0.00	0.01	1.02	0.55	50.2
0.5	35.3	0.05	0.00	0.41	0.53	0.15	0.47	0.50	56.2
0.6	42.4	0.16	0.05	0.21	0.22	0.13	0.82	0.53	61.6
0.7	49.2	0.10	0.22	0.39	0.29	0.15	0.85	0.97	66.3
0.8	56.8	0.53	0.59	0.26	0.06	0.36	0.28	0.09	71.2
0.9	64.0	0.55	0.02	0.15	0.49	0.21	0.13	0.76	75.5
0.98	69.0	0.26	0.42	0.19	0.10	0.16	0.32	1.25	78.6

Equations (13.13) and (13.15) are shown evaluated in Tables 13.9–13.12 for various values of the ratio f_T/f_S. Voltage control is again effected by changing the ratio of A_S/A_T. The maximum value of this ratio is limited to 0.98 rather than unity, to prevent adjacent pulses from merging into each other.

Unlike all the previous systems, this method of voltage control results in severe attenuation of frequencies below a certain value. The two-pulse asymmetrical wave technique could be referred to as 'selected harmonic reduction', since in this system the pulse spacing is so chosen as to eliminate any required harmonic. The present system can be called 'lower harmonic reduction', since it works to reduce harmonics below a certain frequency. On examining the tables it will be seen that the harmonic numbers with the largest amplitude occur at $f_T/f_S \pm 1$. Therefore, for example, with $f_T/f_S = 10$ the harmonics are largest at the ninth and eleventh. This is logical, since the tenth harmonic is the 'carrier' wave itself, and no attempt is made to eliminate it. Quite clearly, the higher the ratio of f_T/f_S, the more effective this system becomes. The same statement was made when considering modulation with a square wave, but there the effect of higher carrier frequencies was to keep the proportion of the harmonics constant (and equal to the value for a square wave) as the fundamental was varied, and not to eliminate it.

The system shown in Figure 13.40 has two disadvantages. First, the high inverter frequency required to give effective lower harmonic reduction leads to smaller efficiencies. Second, the maximum output voltage is well below 90% of the d.c. supply, as obtained with a square wave. This would limit the maximum operating frequency when running with some types of loads which need to be fully fluxed, as described above. The system also shows a characteristic increase in total harmonic content with higher operating frequencies. As explained above, this is not normally serious since higher harmonics can be more easily filtered out than lower-order ones.

13.3.2 Bi-directional switching

All the waveforms considered so far have been unidirectional, that is, the instantaneous voltage in any half cycle has been either positive or negative, but never both. For bi-directional switching the output swings positive and negative in a half cycle. The simplest unidirectional system was the quasi-square waveform, shown in Figure 13.33, and it is interesting to note that this can be obtained by combining two square waves phase shifted from each other, as in Figure 13.41, the output being zero when the waveforms are in anti-phase. This single-pulse system was then extended to two-pulse waveforms, as in Figure 13.36. The two-pulse bi-directional

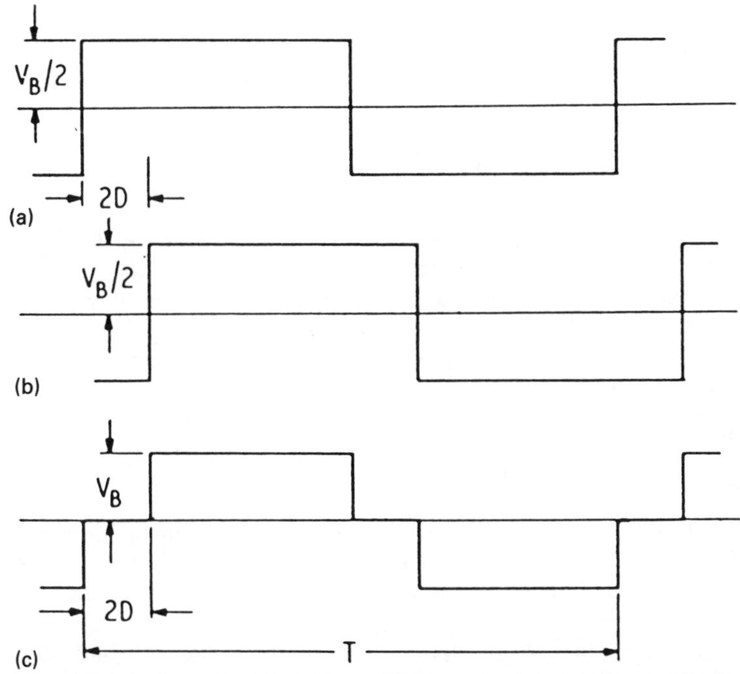

Figure 13.41 Production of a quasi-square wave by combining two-phase shifted square waves: (a) first square wave; (b) phase-shifted second square wave; (c) quasi-square wave

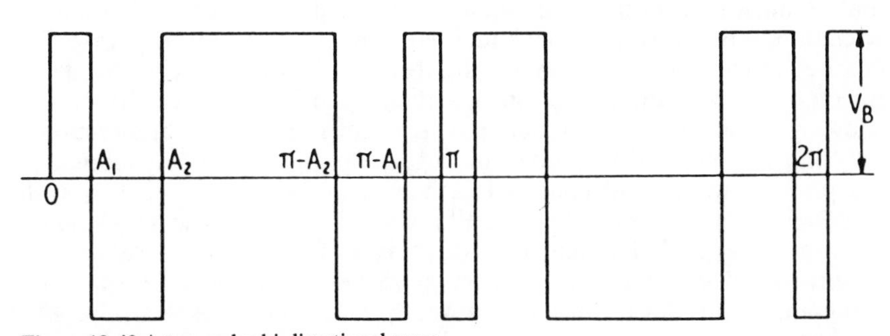

Figure 13.42 A two-pulse bi-directional wave

wave equivalent is shown in Figure 13.42, the Fourier coefficient for this being given by

$$\frac{V_{rms(n)}}{V_B} = \frac{2\sqrt{2}}{n\pi}[1 - 4\sin nB \sin n\omega] \times 100 \qquad (13.16)$$

the values of B and ω being chosen to satisfy equations (13.7) and (13.8), as before.

It is, of course, possible to combine the two-phase shifted waveforms of Figure 13.42 together, as was done in Figure 13.41. This would give a modified result for the nth harmonic content, as in equation (13.17), where $2D$ is the phase shift between the two waveforms. Since the square of the voltage ignores signs, the r.m.s. of the total waveform is given by equation (13.18):

$$\frac{V_{rms(n)}}{V_B} = \frac{2\sqrt{2}}{n\pi}[1 - 4\sin nB \sin n\omega]\cos nD \times 100 \qquad (13.17)$$

$$\frac{V_{rms(T)}}{V_B} = \left[1 - \frac{2D}{\pi}\right]^{1/2} \times 100 \qquad (13.18)$$

In equation (13.17) there are now three variables, two of which, say ω and B, can be used to eliminate any two harmonics, whilst the third, in this case D, is used to control the magnitude of the fundamental voltage. Therefore if B and ω are chosen to satisfy equations (13.19) and (13.20) then harmonics P_1 and P_2 will be absent from the output over the whole range of variation of D:

$$1 - 4\sin(P_1 B)\sin(P_1\omega) = 0 \qquad (13.19)$$

$$1 - 4\sin(P_2 B)\sin(P_2\omega) = 0 \qquad (13.20)$$

Equations (13.17) and (13.18) have been used to obtain a harmonic analysis of waveforms where the third and fifth, the third and seventh, and the fifth and seventh harmonics have been eliminated. The results are given in Tables 13.13–13.15.

The system shown in Figure 13.42 is a method of 'selected harmonic reduction' so comparison will be made with the quasi-square control and two-pulse unidirectional waveforms, i.e. with Tables 13.1–13.4. From these it is clear that the maximum fundamental output voltage is less than that obtainable from the quasi-square wave, and the lower the harmonics eliminated, the less the maximum voltage. This is not the case for unidirectional switching, where the reverse is true, but it must be remembered that the limitation was then fixed by the necessity of not allowing adjacent pulses to merge. This is not required for bi-directional switching since in effect positive and negative pulses are already merged in the primary wave, and control is achieved by phase shifting two such identical waveforms up to the full half cycle. For example, the quasi-square wave gives a peak fundamental which is 90% of the d.c. value. For unidirectional switching this reduces to 75.1%, 62.2% and 33.3% when eliminating the third, fifth and seventh harmonics, respectively. For bi-directional switching comparable figures are 75.5%, 77.5% and 84.0%

Table 13.13 Harmonic content of a two-pulse bi-directional wave with third and fifth harmonic elimination

2D/T	*R.M.S. voltage as percentage of d.c. supply*								
	1	3	5	7	9	11	13	15	*Total*
0	75.5	0	0	22.4	36.8	27.2	2.40	15.2	100
0.02	75.4	0	0	20.3	31.1	21.0	1.65	8.92	98.0
0.04	74.9	0	0	14.3	15.7	5.10	0.15	4.69	95.9
0.06	74.2	0	0	5.58	4.61	13.1	1.85	14.4	93.8
0.08	73.1	0	0	4.21	23.5	25.3	2.39	12.3	91.7
0.10	71.8	0	0	13.2	35.0	25.9	0.41	0.00	89.4
0.12	70.2	0	0	19.7	35.7	14.6	0.45	12.3	87.2
0.14	68.3	0	0	22.4	25.2	3.41	2.03	14.4	84.9
0.16	66.2	0	0	20.9	6.90	19.8	2.33	4.69	82.5
0.18	63.8	0	0	15.4	13.6	27.1	1.16	8.92	80.0
0.20	61.1	0	0	6.91	29.8	22.0	0.74	15.2	77.5
0.22	58.2	0	0	2.81	36.7	6.76	2.18	8.92	74.8
0.24	55.0	0	0	12.0	32.3	11.6	2.24	4.69	72.1
0.26	51.7	0	0	19.0	17.7	24.6	0.89	14.4	69.3
0.28	48.1	0	0	22.3	2.31	26.3	1.02	12.3	66.3
0.30	44.4	0	0	21.3	21.6	16.0	2.29	0.00	63.2
0.32	40.5	0	0	16.4	34.2	1.71	2.11	12.3	60.0
0.34	36.4	0	0	8.26	36.2	18.6	0.60	14.4	56.6
0.36	32.2	0	0	1.41	26.8	27.0	1.29	4.69	52.9
0.38	27.8	0	0	10.8	9.16	23.0	2.36	8.92	49.0
0.40	23.3	0	0	18.2	11.4	8.40	1.95	15.2	44.7
0.42	18.8	0	0	22.1	28.4	10.0	0.20	8.92	40.0
0.44	14.2	0	0	21.7	36.5	23.8	1.53	4.69	34.6
0.50	0	0	0	0	0	0	0	0	0

when the third and fifth, the third and seventh, or the fifth and seventh harmonics are being eliminated, respectively.

A comparison of the harmonic tables shows that, in general, unidirectional switching has a higher total harmonic content than bi-directional switching, the quasi-square wave having the least.

An interesting aspect of harmonic elimination is that the amplitude of the harmonics increases with the harmonic number, which is contrary to that found with quasi-square waves. For bi-directional switching the ninth harmonic is the largest when the third and fifth harmonics are eliminated, whereas when the third and seventh are eliminated the eleventh predominates. The same is true for unidirectional switching, but to a lesser extent and at higher harmonic numbers. It should also be noted that for unidirectional switching every harmonic which is an odd multiple of the harmonic being eliminated is absent from the output waveform. Therefore the third, ninth, fifteenth, etc. are absent, and so on. For bi-directional switching only the two harmonics chosen are eliminated.

Generally, harmonic elimination causes the remaining harmonics to be of much larger magnitude. Perhaps, however, it is worth noting the phenomenal increase in the third harmonic when unidirectional switching is used to eliminate either the fifth or the seventh harmonic. The results given in Table 13.15 also deserve special mention. Although the fifth and

Table 13.14 Harmonic content of a two-pulse bi-directional wave with third and seventh harmonic elimination

2D/T	*R.M.S. voltage as percentage of d.c. supply*								
	1	3	5	7	9	11	13	15	*Total*
0	77.5	0	11.8	0	20.0	33.2	31.3	16.7	100
0.02	77.4	0	11.3	0	16.9	25.6	21.5	9.81	98.0
0.04	76.9	0	9.57	0	8.51	6.22	1.97	5.16	95.9
0.06	76.2	0	6.95	0	2.51	16.0	24.2	15.9	93.8
0.08	75.1	0	3.66	0	12.7	30.8	31.1	13.5	91.7
0.10	73.7	0	0.00	0	19.0	31.5	18.4	0.00	89.4
0.12	72.1	0	3.66	0	19.4	17.8	5.87	13.5	87.2
0.14	70.1	0	6.95	0	13.7	4.16	26.5	15.9	81.9
0.16	67.9	0	9.57	0	3.75	24.2	30.4	5.16	82.5
0.18	65.5	0	11.3	0	7.36	33.1	15.1	9.81	80.0
0.20	62.7	0	11.8	0	16.2	26.8	9.69	16.7	77.5
0.22	59.7	0	11.3	0	20.0	8.25	28.4	9.81	74.8
0.24	56.5	0	9.57	0	17.5	14.1	29.1	5.16	72.1
0.26	53.1	0	6.95	0	9.63	20.0	11.5	15.9	69.3
0.28	49.4	0	3.66	0	1.25	32.1	13.3	13.5	66.3
0.30	45.6	0	0.00	0	11.7	19.5	29.8	0.00	63.2
0.32	41.5	0	3.66	0	18.6	2.08	27.5	13.5	60.0
0.34	37.3	0	6.95	0	19.6	22.7	7.80	15.9	56.6
0.36	33.0	0	9.57	0	14.6	32.9	16.8	5.16	52.9
0.38	28.5	0	11.3	0	4.97	28.0	30.8	9.81	49.0
0.40	24.0	0	11.8	0	6.18	10.3	25.4	16.7	44.7
0.42	19.3	0	11.3	0	15.4	12.2	3.93	9.81	40.0
0.44	14.5	0	9.57	0	19.8	29.1	20.0	5.16	34.6
0.46	9.72	0	6.95	0	18.1	32.6	31.3	15.9	28.3
0.48	4.87	0	3.66	0	10.7	21.1	22.8	13.5	20.0
0.50	0	0	0	0	0	0	0	0	0

seventh harmonics have been completely eliminated from the output, the third and ninth harmonics are also less than those obtained when using the quasi-square method of control. Higher harmonics are considerably greater, but this waveform has special merit when used in three-phase inverters, as interconnections of the output transformer can then be used to eliminate triplen harmonics, making the eleventh harmonic the lowest present.

The two-pulse bi-directional wave can be extended to several equi-spaced bi-directional pulses by mixing a triangular and square wave, as in Figure 13.43. This waveform will contain odd and even harmonics. Knowing the value of the intersection points X_0, X_1, X_2, etc. the r.m.s. of the nth harmonic can readily be obtained, as in equation (13.21), by the arithmetic sum of the individual pulses over the zero to π interval:

$$\frac{V_{\mathrm{rms(n)}}}{V_{\mathrm{B}}} = \frac{\sqrt{2}}{n\pi}\Big\{1 - \cos nX(0) + \Sigma\left[\cos nX(2M+1) - \cos nX(2M+2)\right] - \Sigma\left[\cos nX(2M) - \cos nX(2M+1)\right]\Big\} \times 100 \qquad (13.21)$$

The r.m.s. value of the total harmonic is constant, irrespective of the depth of modulation and the operating frequency, and is equal to that of a square wave since there are no zero periods in the output.

Table 13.15 Harmonic content of a two-pulse bi-directional wave with fifth and seventh harmonic elimination

2D/T	R.M.S. voltage as percentage of d.c. supply								
	1	3	5	7	9	11	13	15	Total
0	84.0	14.5	0	0	8.16	17.6	23.1	21.6	100
0.02	83.8	14.3	0	0	6.89	13.5	15.8	12.7	98.0
0.04	83.3	13.5	0	0	3.48	3.29	1.45	6.67	95.9
0.06	82.5	12.3	0	0	1.02	8.47	17.8	20.5	93.8
0.08	81.3	10.6	0	0	5.20	16.3	22.9	17.5	91.7
0.10	79.8	8.54	0	0	7.76	16.7	13.5	0.00	89.4
0.12	78.1	6.18	0	0	7.91	9.42	4.32	17.5	87.2
0.14	76.0	3.61	0	0	5.59	2.20	19.5	20.5	84.9
0.16	73.6	0.91	0	0	1.53	12.8	22.3	6.67	82.5
0.18	70.9	1.82	0	0	3.01	17.5	11.1	12.7	80.0
0.20	67.9	4.49	0	0	6.60	14.2	7.12	21.6	77.5
0.22	64.7	7.00	0	0	8.15	4.37	20.9	12.7	74.8
0.24	61.2	9.26	0	0	7.15	7.48	21.4	6.67	72.1
0.26	57.5	11.2	0	0	3.93	15.9	8.49	20.5	69.3
0.28	53.5	12.7	0	0	0.51	17.0	9.81	17.5	66.3
0.30	49.3	13.8	0	0	4.80	10.3	21.9	0.00	63.2
0.32	45.0	14.4	0	0	7.59	1.10	20.2	17.5	60.0
0.34	40.4	14.5	0	0	8.02	12.0	5.73	20.5	56.6
0.36	35.7	14.1	0	0	5.95	17.4	12.4	6.67	52.9
0.38	30.9	13.1	0	0	2.03	14.8	22.6	12.7	49.0
0.40	25.9	11.8	0	0	2.52	5.43	18.6	21.6	44.7
0.42	20.9	9.94	0	0	6.29	6.47	2.89	12.7	40.0
0.44	15.7	7.78	0	0	8.10	15.4	14.7	6.67	34.6
0.46	10.5	5.35	0	0	7.39	17.3	23.0	20.5	28.3
0.48	5.27	2.72	0	0	4.37	11.2	16.8	17.5	20.0
0.50	0.00	0.00	0	0	0.00	0.00	00.0	0.00	0.00

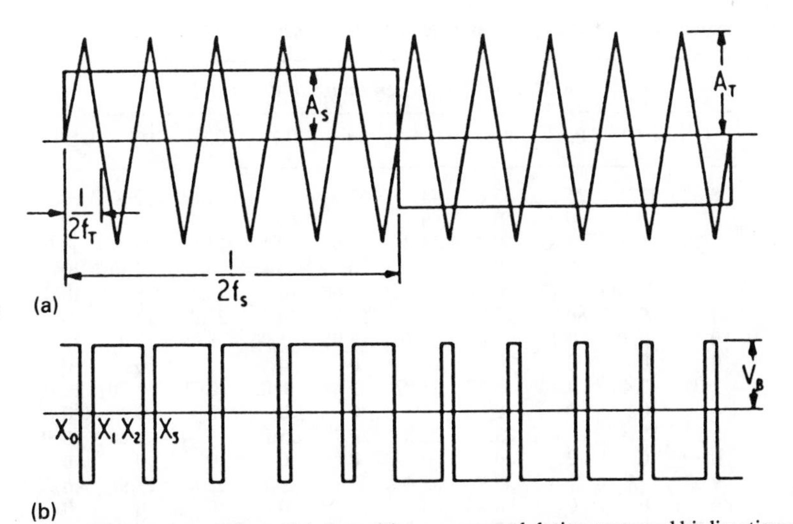

(b)

Figure 13.43 Pulse-width modulation with a square modulating wave and bi-directional switching: (a) modulating waveforms; (b) output waveform

The harmonics, as calculated from equation (13.21), are given in Tables 13.16–13.19 for values of f_T/f_S = 4, 6, 10, 20 respectively.

It should be noted that for bi-directional switching even harmonics are absent from the output, apart from the inverter operating frequency and its triplen harmonics. For unidirectional switching all even harmonics are missing.

The lower harmonic content with bi-directional switching, when the inverter operating frequency is higher than the output frequency, is less than for a quasi-square wave. It is also less than that obtained with

Table 13.16 Harmonic content of a square modulated bi-directional wave with $f_T/f_S = 4$

Harmonic number	*R.M.S. voltage as percentage of d.c. supply*									
	0.1	0.2	0.3	0.4	0.5	0.6	0.7	0.8	0.9	1.0
1	8.6	17.3	26.2	35.1	44.1	53.2	62.4	71.6	80.8	90.0
2	0	0	0	0	0	0	0	0	0	0
3	1.67	3.73	6.16	8.91	12.0	15.3	18.8	22.4	26.2	30.0
4	88.9	85.6	80.2	72.8	63.7	52.9	40.9	27.8	14.1	0
5	1.11	1.48	1.11	0	1.80	4.22	7.18	10.5	14.2	18.0
6	0	0	0	0	0	0	0	0	0	0
7	7.94	14.3	18.7	20.6	20.1	17.1	11.8	4.70	3.74	12.9
8	0	0	0	0	0	0	0	0	0	0
9	8.98	18.1	26.2	32.3	35.6	35.8	32.8	27.0	19.0	10.0
10	0	0	0	0	0	0	0	0	0	0
11	2.17	5.45	9.22	12.8	15.5	16.9	16.7	14.9	11.9	8.18
12	26.7	17.6	4.67	9.27	21.2	28.5	29.6	24.3	13.6	0
13	0.52	0.86	3.79	7.51	11.1	13.6	14.4	13.3	10.6	6.92
14	0	0	0	0	0	0	0	0	0	0
15	7.04	9.68	7.04	0	9.06	17.1	21.4	20.5	14.7	6.00

Table 13.17 Harmonic content of a square modulated bi-directional wave with $f_T/f_S = 6$

Harmonic number	*R.M.S. voltage as percentage of d.c. supply*									
	0.1	0.2	0.3	0.4	0.5	0.6	0.7	0.8	0.9	1.0
1	8.82	17.7	26.6	35.6	44.6	53.7	62.7	71.8	80.9	90.0
2	0	0	0	0	0	0	0	0	0	0
3	2.45	5.06	7.83	10.7	13.8	16.9	20.1	23.4	26.7	30.0
4	0	0	0	0	0	0	0	0	0	0
5	0.78	1.86	3.22	4.82	6.66	8.68	10.9	13.2	15.6	18.0
6	88.9	85.6	80.2	72.8	63.7	52.9	40.9	27.8	14.1	0
7	0.41	0.38	0.10	1.00	2.30	3.95	5.90	8.09	10.4	12.9
8	0	0	0	0	0	0	0	0	0	0
9	2.06	3.45	4.10	3.97	3.07	1.44	0.82	3.58	6.70	10.0
10	0	0	0	0	0	0	0	0	0	0
11	8.34	15.3	20.4	23.0	23.2	20.7	15.9	9.04	0.80	8.18
12	0	0	0	0	0	0	0	0	0	0
13	9.02	17.8	25.3	30.8	33.4	33.1	29.7	23.8	15.9	6.92
14	0	0	0	0	0	0	0	0	0	0
15	2.75	6.00	9.25	12.0	13.8	14.5	13.8	12.0	9.25	6.00

Table 13.18 Harmonic content of a square modulated bi-directional wave with $f_T/f_S = 10$

Harmonic number	*R.M.S. voltage as percentage of d.c. supply*									
	0.1	0.2	0.3	0.4	0.5	0.6	0.7	0.8	0.9	1.0
1	8.94	17.9	26.9	35.9	44.9	53.9	62.9	71.9	81.0	90.0
2	0	0	0	0	0	0	0	0	0	0
3	2.81	5.67	8.60	11.6	14.6	17.6	20.7	23.8	26.9	30.0
4	0	0	0	0	0	0	0	0	0	0
5	1.47	3.04	4.70	6.44	8.26	10.1	12.1	14.0	16.0	18.0
6	0	0	0	0	0	0	0	0	0	0
7	0.80	1.74	2.82	4.01	5.32	6.71	8.18	9.71	11.3	12.9
8	0	0	0	0	0	0	0	0	0	0
9	0.32	0.84	1.54	2.40	3.42	4.57	5.83	7.18	8.57	10.0
10	88.9	85.6	80.2	72.8	63.7	52.9	40.9	27.8	14.1	0
11	0.10	0.05	0.44	1.05	1.8	2.90	4.08	5.38	6.76	8.18
12	0	0	0	0	0	0	0	0	0	0
13	0.57	0.83	0.77	0.39	0.30	1.26	2.46	3.84	5.35	6.92
14	0	0	0	0	0	0	0	0	0	0
15	1.23	2.07	2.46	2.38	1.84	0.86	0.49	2.15	4.02	6.00

Table 13.19 Harmonic content of a square modulated bi-directional wave with $f_T/f_S = 20$

Harmonic number	*R.M.S. voltage as percentage of d.c. supply*									
	0.1	0.2	0.3	0.4	0.5	0.6	0.7	0.8	0.9	1.0
1	8.99	18.0	27.0	36.0	45.0	54.0	63.0	72.0	81.0	90.0
2	0	0	0	0	0	0	0	0	0	0
3	2.95	5.92	8.90	11.9	14.9	17.9	20.9	24.0	27.0	30.0
4	0	0	0	0	0	0	0	0	0	0
5	1.72	3.47	5.23	7.02	8.83	10.6	12.5	14.3	16.2	18.0
6	0	0	0	0	0	0	0	0	0	0
7	1.17	2.38	3.62	4.89	6.18	7.49	8.82	10.2	11.5	12.9
8	0	0	0	0	0	0	0	0	0	0
9	0.85	1.75	2.69	3.66	4.67	5.71	6.76	7.83	8.91	10.0
10	0	0	0	0	0	0	0	0	0	0
11	0.63	1.32	2.06	2.85	3.68	4.54	5.43	6.33	7.25	8.18
12	0	0	0	0	0	0	0	0	0	0
13	0.47	1.00	1.60	2.25	2.96	3.70	4.48	5.28	6.10	6.92
14	0	0	0	0	0	0	0	0	0	0
15	0.33	0.75	1.23	1.78	2.39	3.05	3.75	4.48	5.24	6.00

unidirectional switching and is due to the absence of any 'zero' periods in the waveform. For maximum modulation the output reduces to a square wave and the harmonic content is identical to comparable quasi-square and unidirectional systems.

With bi-directional switching the odd harmonic content generally increases as the operating frequency is raised, whilst the reverse was true for unidirectional switching. However, the presence of the large-valued even harmonics still makes it desirable to operate this system at an inverter frequency of the order of twenty times the desired output frequency.

The square reference wave may be replaced by a sine wave, the output being very similar to Figure 13.43 except that the modulation depth varies linearly along the cycle. This waveform will also contain odd and even sine terms, and since $X_0 = 0$ the r.m.s. voltage of the nth harmonic can be derived from

$$\frac{V_{\text{rms(n)}}}{V_B} = \frac{\sqrt{2}}{n\pi}\Big\{\Sigma\left[\cos nX(2M+1) - \cos nX(2M+2)\right] - \Sigma\left[\cos nX(2M) - \cos nX(2M+1)\right]\Big\} \times 100 \quad (13.22)$$

The solutions for equation (13.22) for four values of f_T/f_S are given in Tables 13.20–13.23. It is seen from these that the harmonic content of the waveform is very similar to unidirectional switching, as in Tables 13.9–13.12. The harmonic with the largest amplitude is that which occurs close to the chopping frequency f_T, both odd and even harmonics being considered. As an example, when operating at $f_T/f_S = 6$ the sixth harmonic is very large. For zero modulation depth the fundamental is zero and the sixth harmonic has a value 90% of the d.c. supply, since the output is a square wave at this frequency. As the modulation depth increases the fundamental also increases in value and the sixth harmonic reduces, whereas adjacent even harmonics, i.e. the fourth and the eighth, increase in value. When operating with $f_T/f_S = 20$ the harmonic content up to the 15th is very similar to unidirectional methods, except that there are now odd and even terms. However, the total harmonic content, which for bi-directional switching is 100% of the d.c. supply irrespective of the modulation depth, is much higher. As in previous methods using 'lower harmonic reduction' it is clear that the inverter frequency should be several orders larger than the output frequency for effective harmonic reduction. This is therefore a disadvantage of the system as it can lead to lower efficiencies.

Table 13.20 Harmonic content of a sine modulated bi-directional wave with $f_T/f_S = 4$

Harmonic number	*R.M.S. voltage as percentage of d.c. supply*									
	0.1	0.2	0.3	0.4	0.5	0.6	0.7	0.8	0.9	0.98
1	8.34	16.0	21.4	29.4	34.9	44.4	49.8	57.2	63.4	70.9
2	0.09	0.61	1.03	6.58	7.28	11.1	12.0	14.9	19.4	22.2
3	2.52	1.23	2.16	0.11	0.64	0.01	0.67	1.22	1.79	1.80
4	89.2	87.2	84.9	80.6	77.0	69.4	64.4	57.0	50.7	42.4
5	2.50	1.25	2.18	3.04	4.38	6.81	8.17	11.2	12.7	15.2
6	0.28	1.79	3.00	2.79	5.18	9.21	12.4	15.8	20.7	25.5
7	8.24	15.3	19.8	22.3	24.7	25.1	25.2	22.6	17.9	13.6
8	0.93	0.87	2.01	1.73	3.34	5.29	7.50	11.6	10.2	13.4
9	8.17	14.8	18.7	24.7	26.2	26.7	24.7	21.1	16.1	8.92
10	0.47	2.88	4.70	10.7	11.9	13.6	13.5	11.0	9.10	5.19
11	2.39	1.41	2.29	0.39	0.57	0.94	2.11	1.44	4.55	5.86
12	27.5	22.0	16.2	7.43	1.14	7.07	10.5	11.8	8.42	4.28
13	2.34	1.47	2.33	6.24	7.08	8.65	7.44	3.31	3.97	1.47
14	0.65	3.82	5.99	10.6	13.6	18.1	18.4	18.9	15.8	11.0
15	7.87	12.9	14.3	8.93	6.41	0.89	3.78	5.53	4.63	1.00

Table 13.21 Harmonic content of a sine modulated bi-directional wave with $f_T/f_S = 6$

Harmonic number	*R.M.S. voltage as percentage of d.c. supply*									
	0.1	0.2	0.3	0.4	0.5	0.6	0.7	0.8	0.9	0.98
1	6.98	14.1	21.0	28.7	35.6	42.7	49.1	56.1	63.5	70.5
2	0.02	0.14	1.29	1.13	0.15	0.11	0.24	0.10	1.86	1.46
3	0.44	0.41	0.82	0.29	0.75	0.88	0.22	0.44	0.95	1.28
4	0.20	0.80	2.72	4.34	6.71	8.52	12.6	14.8	19.6	22.1
5	0.26	2.11	1.46	0.65	0.21	1.66	0.36	1.47	0.04	0.62
6	89.5	87.8	85.1	81.1	76.4	70.8	65.2	58.2	50.4	42.5
7	0.26	2.05	0.87	1.34	0.36	1.58	0.82	1.62	2.97	3.33
8	0.41	1.58	1.71	4.68	6.11	9.46	11.4	15.3	17.1	21.2
9	0.41	0.21	2.22	0.78	3.89	5.49	7.89	10.3	13.0	15.6
10	0.09	0.65	1.08	1.62	0.44	0.12	0.30	0.21	1.14	2.17
11	6.91	13.4	18.7	23.2	25.5	26.8	24.6	22.9	16.8	12.8
12	0.09	1.29	1.07	1.19	0.48	2.87	1.26	3.46	3.85	5.39
13	6.88	13.2	19.0	23.0	26.0	25.9	25.8	22.3	20.0	13.9
14	0.13	0.86	0.70	0.62	2.26	2.42	5.38	6.60	9.29	10.9
15	0.37	0.17	0.76	2.59	3.23	6.16	6.65	10.2	10.7	13.1

Table 13.22 Harmonic content of a sine modulated bi-directional wave with $f_T/f_S = 10$

Harmonic number	*R.M.S. voltage as percentage of d.c. supply*									
	0.1	0.2	0.3	0.4	0.5	0.6	0.7	0.8	0.9	0.98
1	6.94	13.3	21.1	27.4	35.2	43.2	48.9	57.5	63.8	71.6
2	0.01	0.46	0.84	1.25	0.01	0.75	0.04	0.11	0.14	0.72
3	0.42	0.26	0.12	0.05	0.44	0.05	0.16	0.12	0.38	0.26
4	0.04	0.90	0.66	0.41	0.21	0.95	1.45	0.60	0.81	1.08
5	0.26	0.63	0.25	0.62	0.22	0.17	0.43	0.29	0.41	0.23
6	0.02	0.76	0.45	0.26	0.17	0.17	0.20	0.81	1.45	1.01
7	1.31	0.11	1.44	0.32	1.06	0.66	2.09	0.84	0.06	1.38
8	0.24	1.35	2.77	4.55	5.23	9.38	11.4	16.4	19.3	22.4
9	1.89	2.50	1.26	1.64	1.03	1.00	1.07	0.83	0.55	0.11
10	89.4	88.0	85.1	81.8	76.7	70.4	65.3	56.9	50.1	41.4
11	1.88	2.67	1.66	2.41	0.96	1.69	0.99	0.68	0.84	0.63
12	0.35	0.70	2.15	3.55	7.53	9.74	12.8	15.4	18.3	23.2
13	1.31	0.28	0.71	0.14	0.94	0.83	0.68	0.37	0.79	0.59
14	0.04	1.05	0.59	0.54	0.51	0.48	0.60	0.70	0.23	1.99
15	0.25	0.50	0.13	0.53	0.10	0.65	1.27	1.58	1.72	2.99

13.3.3 Waveform synthesis

Several techniques for output voltage control can be considered to fall into the general category of waveform synthesis. One of these is called staggered phase carrier cancellation, and its principle is introduced in Figure 13.44. It essentially consists of combining several high-frequency modulated waves in which the carriers are out of phase, whereas the modulating low-frequency wave is in phase. This results in a strengthening of the low-frequency and a weakening of the high-frequency carrier. This is

Table 13.23 Harmonic content of a sine modulated bi-directional wave with $f_T/f_S = 20$

Harmonic number	*R.M.S. voltage as percentage of d.c. supply*									
	0.1	0.2	0.3	0.4	0.5	0.6	0.7	0.8	0.9	0.98
1	7.23	13.9	21.1	28.1	35.0	42.1	49.9	56.5	63.9	70.7
2	0.01	0.02	0.02	0.10	0.34	0.07	0.79	0.11	0.03	0.27
3	0.05	0.56	0.24	0.19	0.35	0.16	0.14	0.10	0.49	0.22
4	0.01	0	0.06	0.71	0.43	0.03	0.02	0.18	0.86	0.03
5	0.50	0.13	0.05	0.42	0.12	0.13	0.43	0.61	0.75	0.14
6	0	0.04	0.01	0.37	0.11	0.01	0.35	0.28	0.76	0.62
7	0.63	1.02	0.32	0.29	0.80	0.40	0.17	0.17	0.78	0.62
8	0.02	0	0.02	0.76	0.28	0.22	0.70	0.54	0.25	0.01
9	1.03	0.98	0.63	0.86	0.19	0.03	0.19	0.03	0.32	0.22
10	0.02	0.09	0.04	0.31	0.42	0.10	0.12	0.06	0.37	0.22
11	0.35	0.18	0.25	0.13	0.48	0.36	0.12	0.12	0.03	0.14
12	0.03	0.18	0.14	0.07	0.35	0.11	0.11	0.45	0.10	0.10
13	0.22	0.03	0.15	1.05	0.13	0.78	0.14	0.65	0.73	0.33
14	0.06	0.14	0.06	0.05	0.14	0.17	0.34	0.66	0.50	0.30
15	0.13	0.76	0.29	1.14	0.28	0.75	0.07	0.48	0.70	0.56

evident from the waveforms of Figure 13.44 where the carrier in (a) and (b) are 180° out of phase, whereas the low-frequency signal is in phase. Combining the two waveforms will give the lowest carrier harmonic at twice the carrier frequency.

This system can be extended. For instance, combining four waveforms with their low frequencies in phase but their carriers phase shifted by 90°, 180° and 270° would result in the carrier and its first, second and third harmonics being eliminated from the output.

The disadvantage of the staggered phase carrier cancellation system is the extra hardware needed. For example, to combine the two waveforms shown in Figure 13.44 would require two independent inverters, so doubling the power and control circuits. Its prime use is in fixed-frequency

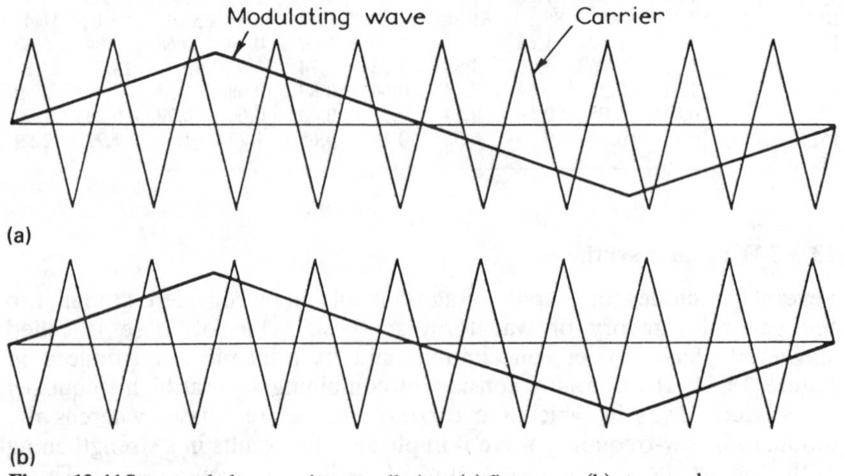

Figure 13.44 Staggered phase carrier cancellation: (a) first wave; (b) staggered wave

sine wave inverters where the extra cost can be justified on account of the reduction in harmonics, and therefore in the cost and size of the output filter.

Another method of waveform synthesis is that using stepped waveforms, which is frequently employed in larger, three-phase, installations, where several inverters are run in parallel but phase shifted, their outputs being summed by a transformer to produce a stepped waveform with reduced harmonic content. The same effect can be obtained by using a tapped supply, as in Figure 13.45(a), the thyristor firing sequence being indicated in Figure 13.45(b), alongside the output waveform. Alternatively, instead of tapping the supply, either the primary or the secondary of the output transformer can be tapped. Taps on the primary of the transformer are shown in Figure 13.46(a), the firing sequence of the thyristors and the output waveform being given in Figure 13.46(b). Figure 13.47(a) shows an inverter arrangement with a tapped secondary, the primary being the normal form of a push–pull inverter. Provided the tappings on the

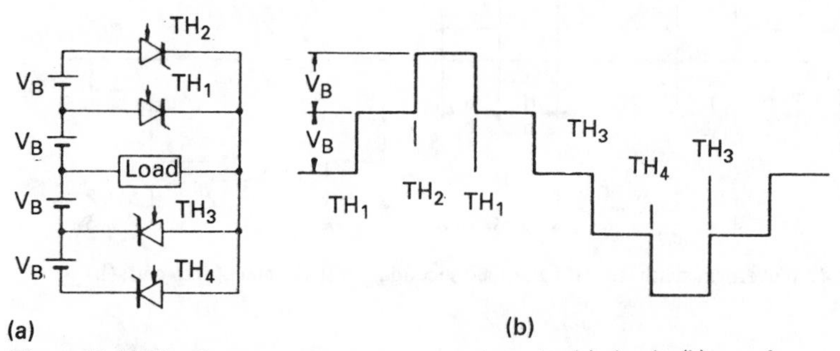

Figure 13.45 Waveform synthesis using a tapped supply: (a) circuit; (b) waveform

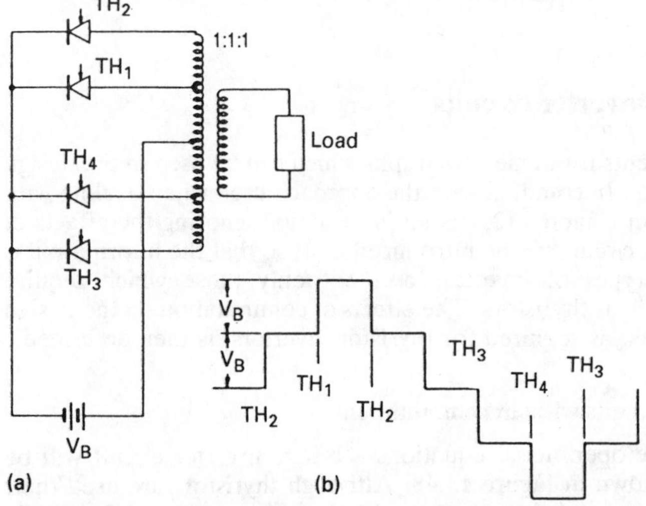

Figure 13.46 Waveform synthesis with a tapped primary transformer: (a) circuit; (b) waveform

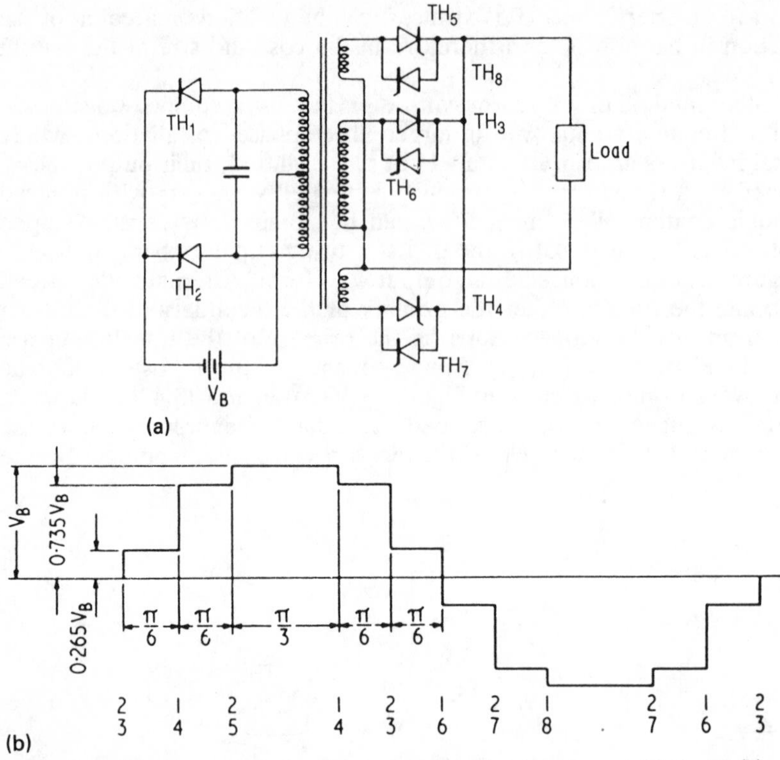

Figure 13.47 Waveform synthesis with a tapped secondary transformer: (a) circuit; (b) waveform

transformer are chosen so as to give the waveform indicated in Figure 13.47(b), it can be calculated that the output contains no harmonics below the eleventh.

13.4 Design of inverter circuits

This section presents formulae and graphs which can be used in the design of inverter circuits. In common with the approach used when dealing with chopper circuits in Chapter 12, design information ignoring the effects of any commutation circuit will be introduced first, so that the information is common to all types of inverters and not only those which require commutation, such as thyristors. The effects of commutation on the design of inverter circuits, as required for thyristor inverters, is then described.

13.4.1 Inverter circuits without commutation

To determine the operational equations a bridge inverter circuit will be considered, as shown in Figure 13.48. Although thyristors are used their commutation components have been omitted, so that the circuit is equally applicable to inverters using transistor or gate turn-off switches as the

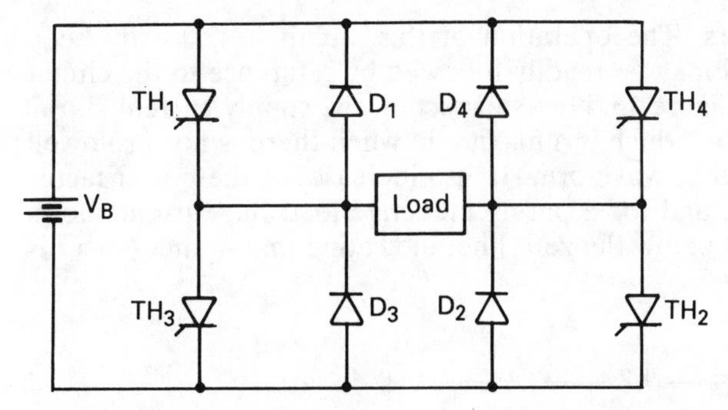

Figure 13.48 Basic thyristor bridge inverter without commutation components

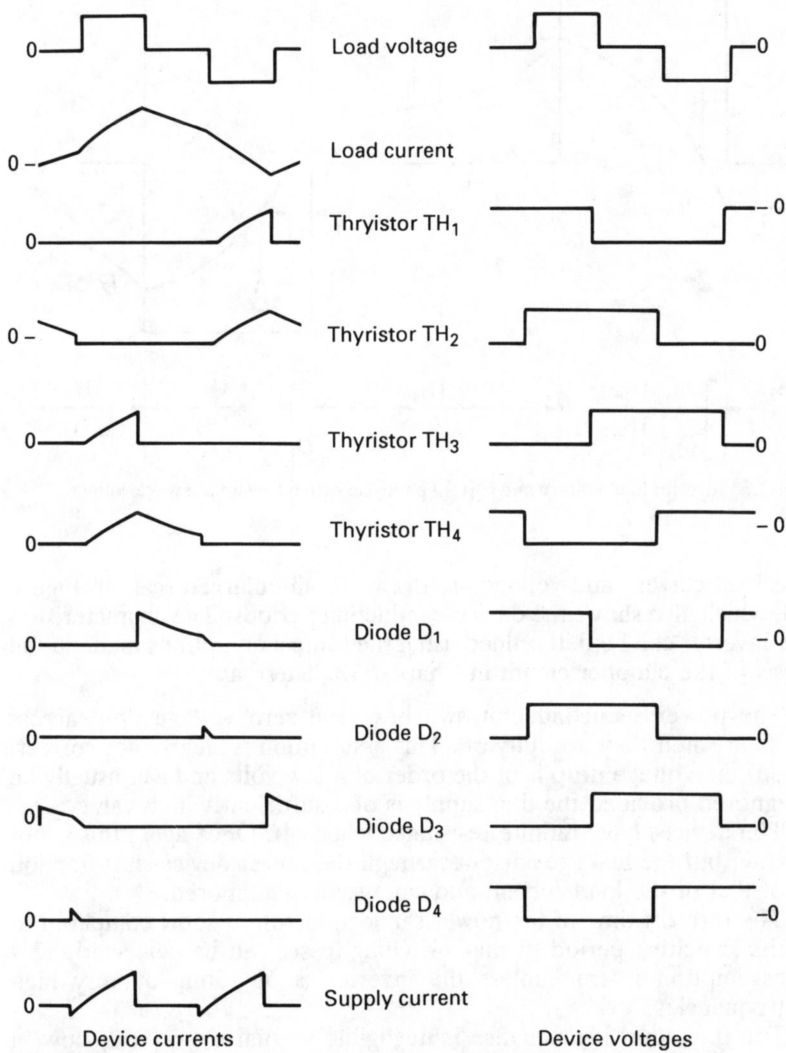

Figure 13.49 Circuit waveforms for the inverter of Figure 13.48

control semiconductors. The operation of this circuit has already been briefly introduced and may be readily followed by reference to the circuit waveforms shown in Figure 13.49. As expected, the supply current shows periods of regeneration, which is a maximum when there is no zero dwell period in the load voltage waveform. It also increases as the power factor of the load decreases, and for a purely inductive load the current shows equal areas above and below the zero line, indicating that no net power is taken from the supply.

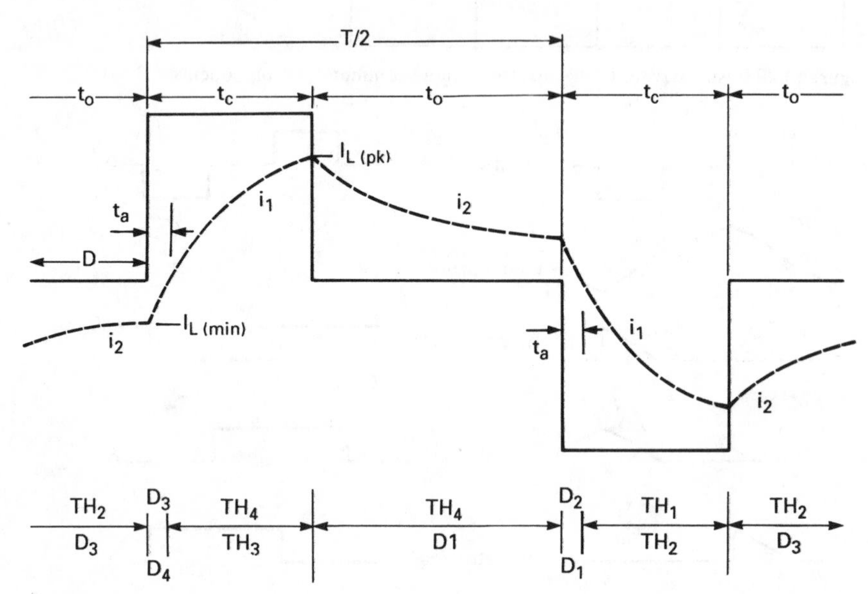

Figure 13.50 Inverter load voltage and current per cycle with quasi-square wave voltage control

The load current and voltage are drawn to an enlarged scale in Figure 13.50, which also shows the device-conducting periods. The characteristics of the inverter can be determined using the same assumptions made in the analysis of the chopper circuit in Chapter 12. These are:

(i) The power semiconductor switches have zero voltage drop across them when they are fully on. This assumption is clearly not correct, but the voltage drop is of the order of a few volts and can usually be ignored provided the d.c. supply is of a sufficiently high value.
(ii) The devices have infinite resistance when off. Once again this is not true, but the leakage current through the power devices is a fraction of that of the load current and can usually be ignored.
(iii) The turn-off time of the power semiconductors is short compared to the switching period so that switching losses can be neglected. This assumption is true unless the inverter is operating at very high frequencies.
(iv) The d.c. source impedance is negligible so that energy can flow in either direction through it without affecting the terminal voltage. This

is usually the case since large reservoir capacitors are connected across the d.c. source, so that any voltage fluctuation is very small.

(v) The load current is assumed to be continuous, which is again usually the case since most inverter loads are inductive, and the operating frequency is high enough to give continuous current.

With the above assumptions the peak load current $I_{L(pk)}$ can be found by an analysis of the waveform of Figure 13.50, as in equation (13.23):

$$\frac{I_{L(pk)}}{V_B/R} = \frac{1 - \exp\{-(R/L)[(T/2) - 2D]\}}{1 + \exp[-(R/L)(T/2)]} \tag{13.23}$$

The r.m.s. current rating of the devices is obtained by an integral of the currents indicated in Figure 13.50, noting that one arm of the bridge carries a larger current than the other, so that the worst case should be considered. Therefore the rating of the thyristors can be taken as i_1 over the time interval $(t_c - t_a)$, or as the sum of i_1 over the period $(t_c - t_a)$ plus i_2 over t_2. The latter gives the worst case and should be the one considered.

The values of the thyristor r.m.s. current, $I_{TH(rms)}$, diode r.m.s. current, $I_{D(rms)}$ and load r.m.s. current $I_{L(rms)}$ are given by equations (13.24), (13.27) and (13.28) respectively, the values of I_{TO} and P being given by equations (13.25) and (13.26):

$$\frac{I_{TH(rms)}}{V_B/R} = \left[\left(\frac{I_{L(pk)}}{V_B/R}\right)^2 \frac{L/R}{2T}\left\{1 - \exp\left[-\frac{T}{L/R}\frac{LD}{T}\right]\right\} + \left(\frac{I_{TO}}{V_B/R}\right)^2\right]^{1/2} \tag{13.24}$$

$$\frac{I_{TO}}{V_B/R} = \left[\frac{L/R}{T}\left\{\frac{T/2 - 2D}{L/R} - \log P + 2P\exp\left(-\frac{T/2 - 2D}{L/R}\right) - \frac{P^2}{2}\exp\left(-\frac{T - 4D}{L/R}\right) - \frac{3}{2}\right\}\right]^{1/2} \tag{13.25}$$

$$P = 1 + \frac{I_{L(pk)}}{V_B/R}\exp - \frac{Rt_o}{L} \tag{13.26}$$

$$\frac{I_{D(rms)}}{V_B/R} = \left[\left(\frac{I_{TH(rms)}}{V_B/R}\right)^2 - \left(\frac{I_{TO}}{V_B/R}\right)^2 + \frac{L/R}{T}\left\{\log P - 2P + \frac{P^2}{2} + \frac{3}{2}\right\}\right]^{1/2} \tag{13.27}$$

$$\frac{I_{L(rms)}}{V_B/R} = \left[2\left\{\left(\frac{I_{D(rms)}}{V_B/R}\right)^2 + \left(\frac{I_{TO}}{V_B/R}\right)^2\right\}\right]^{1/2} \tag{13.28}$$

The above equations can be plotted, as in Figures 13.51, 13.52 and 13.53, to give the maximum load current, and the rating of the thyristor and diode. The curves are shown plotted against the fundamental voltage, the value of the fundamental for any ratio of $2D/T$ being obtained from Figure 13.34. These plots allow the peak load current and the r.m.s. ratings of the diode and thyristor to be determined, at any load voltage and r.m.s. load current, and for a given operating frequency.

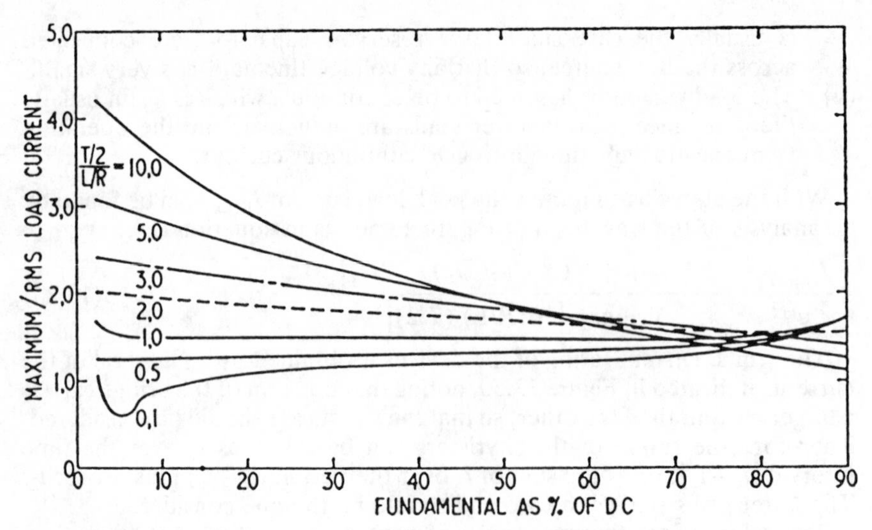

Figure 13.51 Variation of the ratio of peak/r.m.s load current with fundamental load voltage

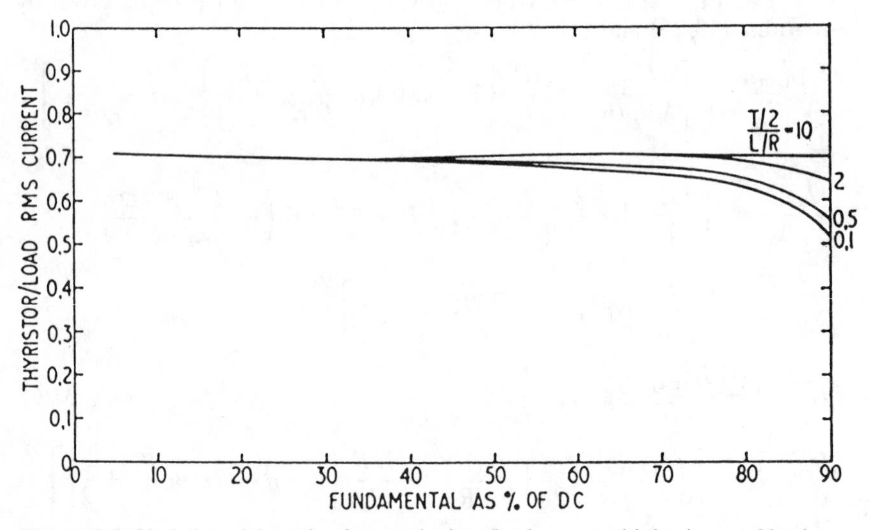

Figure 13.52 Variation of the ratio of r.m.s. thyristor/load current with fundamental load voltage

Figure 13.51 shows the large increase in commutation current requirements and the device peak load currents, at low frequencies and voltages, illustrating the unsuitability of this control method for low mark-space operation. The thyristor r.m.s. current is almost constant at $1/\sqrt{2}$ times the load current, indicating that the devices conduct for approximately half the load current period. The diode ratings also tend to this value at low voltages.

The equations derived so far have been on the assumption of zero inverter loss, so that there is power balance at every stage. If $\cos\phi$ is the

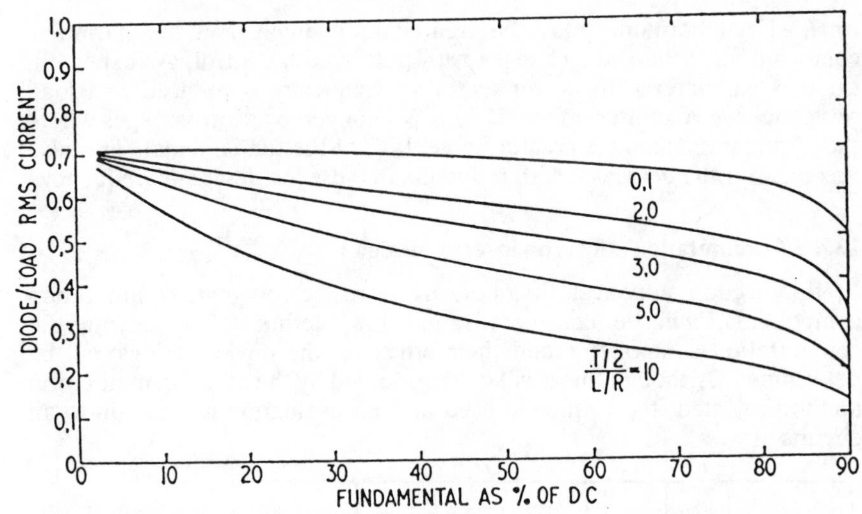

Figure 13.53 Variation of the ratio of r.m.s. diode/load current with fundamental load voltage

power factor of the load, $I_{S(av)}$ is the mean supply current, and $V_{L(rms)}$ is the total r.m.s. load voltage, including harmonics, then equation (13.29) can be obtained, and the power factor is given by equation (13.30):

$$I_{L(rms)}\, V_{L(rms)} \cos \phi \;=\; I_{S(rms)}\, V_B \tag{13.29}$$

$$\cos \phi \;=\; \frac{I_{S(rms)}}{I_{L(rms)}} \; \frac{V_B}{V_{L(rms)}} \tag{13.30}$$

Equation (13.30) is plotted in Figure 13.54. It may seem odd at first to notice from this that the power factor at any frequency is not fixed although both L and R are assumed to be so, but it must be remembered that the applied voltage is not sinusoidal so that the load reactance is not

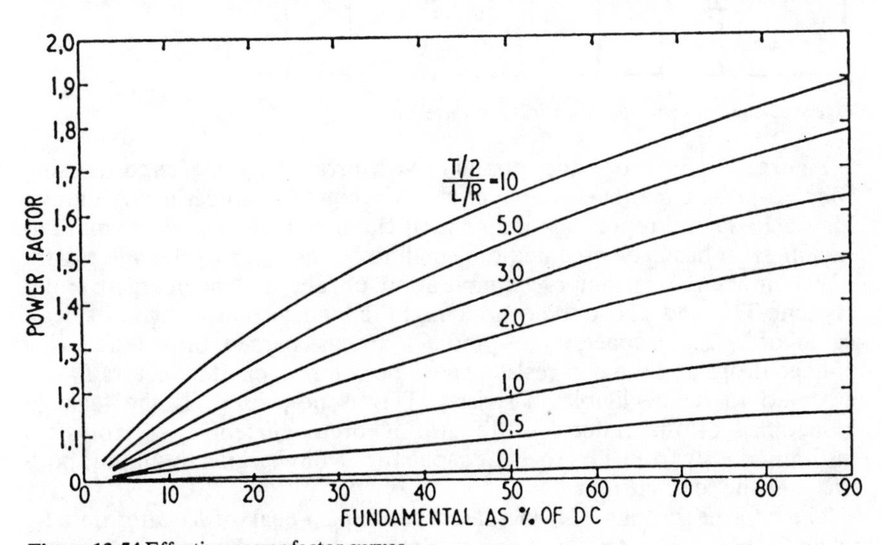

Figure 13.54 Effective power factor curves

$2\pi fL$. Each harmonic must be treated individually and the harmonic composition of the wave changes with pulse-width control. As expected, there is an increase in power factor as frequency is reduced and load resistance has a greater effect. The curves converge at low voltages where the harmonics become a greater percentage of the fundamental. This also accounts for the decrease, with reduction in pulse width, at any frequency.

13.4.2 Commutation effects on inverter design

In this section a typical thyristor inverter, including all commutation components, will be considered, and the method of specifying the commutation components and their effect on the device ratings will be determined. Although these will be determined by the actual commutation method selected, the technique used in their evaluation is common to all circuits.

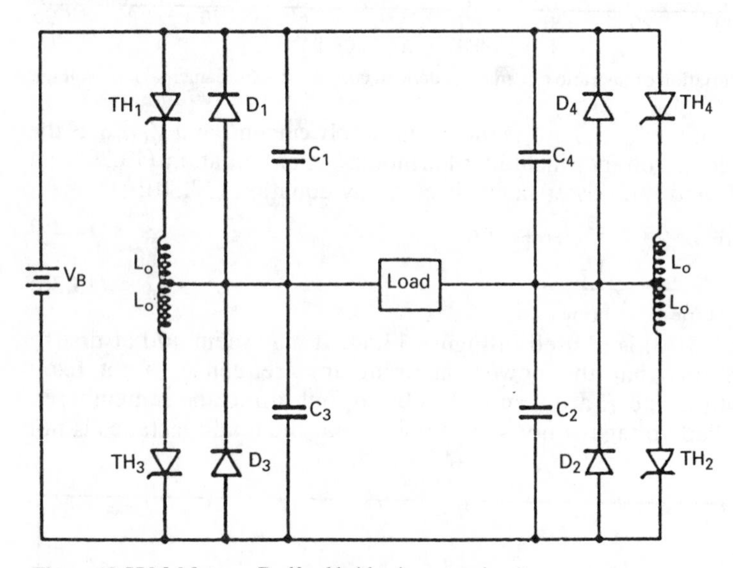

Figure 13.55 McMurray–Bedford bridge inverter circuit

Figure 13.55 shows the popular McMurray–Bedford circuit, using coupled-pulse commutation, which has been introduced earlier. Although Figure 13.49 still represents the general circuit waveforms, the commutation interval has been assumed too small to be distinguishable, and Figure 13.56 shows the instant of commutation of TH_3 to an enlarged scale. Assume TH_3 and TH_4 to be conducting, the load current being at its peak value of $I_{L(pk)}$. Capacitors C_3 and C_4 are discharged provided device voltage drops and the d.c. resistance of the centre-tapped inductors L_0 are assumed to be negligible. Thyristor TH_1 is now fired. If the leakage inductance of the inductors are also ignored, current $I_{L(pk)}$ transfers instantaneously from TH_3 to TH_1, capacitor C_1 discharging to support both this and the load current.

The current through TH_1 increases, reaching a peak of I_{pk} after time t_1, when C_1 has completely discharged. Energy stored in L_0 now free-wheels

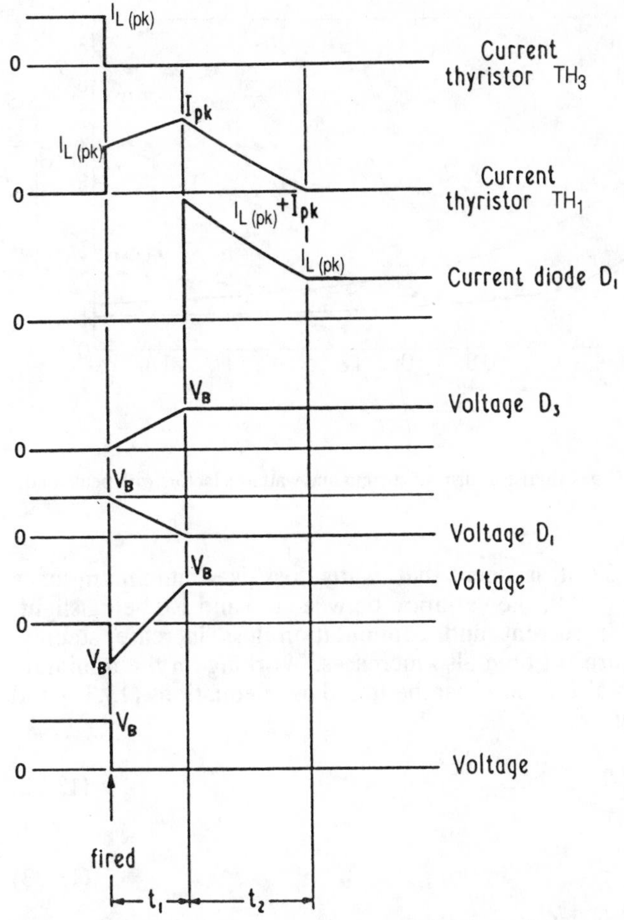

Figure 13.56 Circuit waveforms for Figure 13.55 during the commutation interval of thyristor TH_3

through diode D_1 and is dissipated, falling to zero after a further time t_2. Load current is carried by TH_4 and D_1.

When TH_2 is fired at a later interval, to commutate TH_4, the load current will have decayed to $I_{L(min)}$. Therefore commutation requirements and increased device ratings are not as severe as before, and in the worst case, when there are no zero dwell periods in the output voltage waveform, the ratings of all devices are equal. This will be considered here.

If t_F denotes the turn-off (reverse voltage) time seen by the thyristor being commutated and P_{COM} is the watts loss per commutation, caused by the dissipation of the energy stored in L_0, equal to $1/2L_0(I_{pk})^2$, then Figure 13.57 shows the variation of turn off, peak current and watts loss factors with the commutation factor F_C, given by

$$F_C = \frac{I_{L(pk)}}{V_B}\left[\frac{L_0}{2C}\right]^{1/2} \qquad (13.31)$$

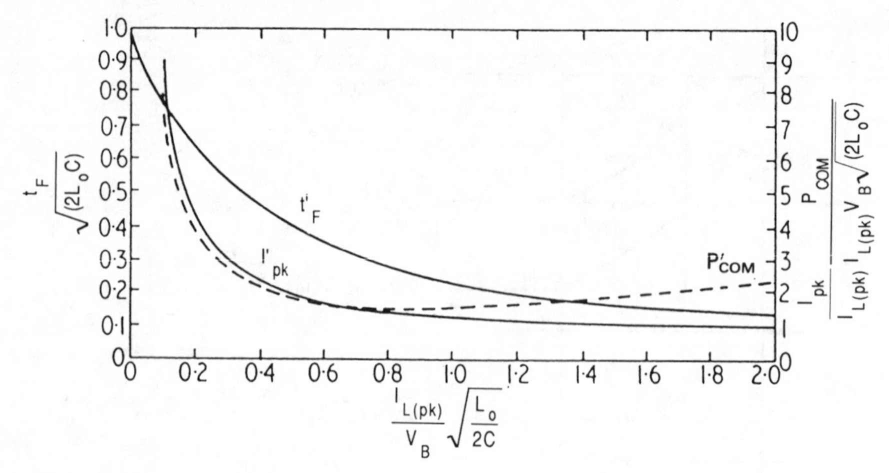

Figure 13.57 Variation of peak thyristor turn-off current and watt loss factors with peak load current

From Figure 13.57 it is seen that watts loss is a minimum at a commutation factor of 0.8, the variation between 0.6 and 1.0 being slight. Below 0.6 the peak current and commutation loss increase steeply, although available turn-off time also increases. Working on the minimum loss point the values of L_0 and C can be found as in equations (13.32) and (13.33), respectively:

$$L_0 = \frac{2.76\, V_B\, t_{OFF}}{I_{L(pk)}} \tag{13.32}$$

$$C = \frac{2.15\, t_{OFF}\, I_{L(pk)}}{V_B} \tag{13.33}$$

The contribution to device ratings by the commutation interval is directly dependent on time t_2 in Figure 13.56. This is the period required for the current in TH_1 to decay from I_{pk} to zero due to losses across TH_1 and D_1 (assumed to be a constant and equal to ΔV) and due to the 'effective loss' resistance (R_e) of the TH_1-L_0-D_1 loop. To reduce this time to a minimum means introducing an external resistance in series with D_1, which gives greater loss during normal operation, with a subsequent reduction in efficiency. There is no one acceptable solution for all cases. Some inverters may be using devices which are overrated, so that higher r.m.s. currents may be acceptable, whilst in others efficiency may not be important so losses across a larger R_e are tolerated. In any case it is always important to ensure that t_2 is less than half a cycle of inverter operation, to prevent commutation failures.

If I_{D1} and I_{T1} represent the commutation current, in r.m.s. terms, through the diode and thyristor respectively, and T is the inverter periodic time, Figures 13.58 and 13.59 show plots of $I_{D1}\sqrt{(T)}/I_{L(pk)})$ and $I_{T1}\sqrt{(T)}/I_{L(pk)})$ against supply voltage. These are calculated for values of L_0 and C given by equations (13.32) and (13.33) and for $\Delta V = 2.5$ V and $R_e = 0.2V_B/I_{L(pk)}$. These figures show the advantage of using fast turn-off

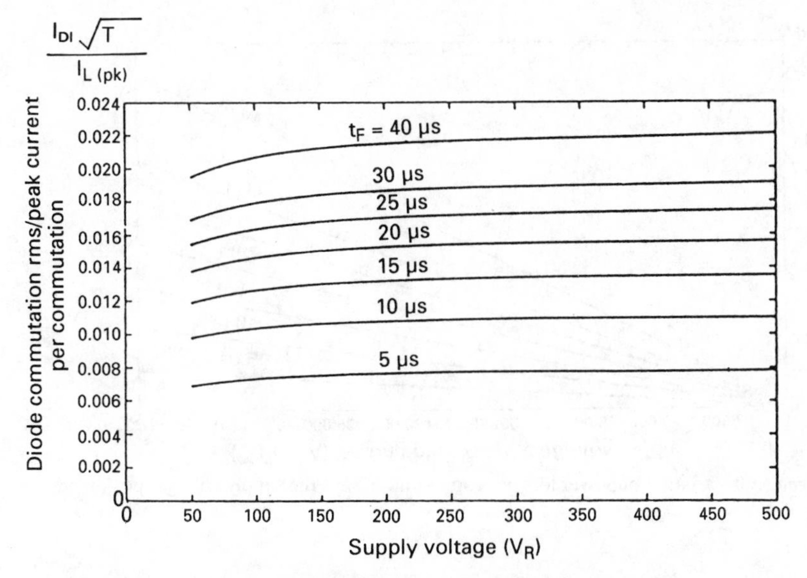

Figure 13.58 Variation of diode current factor with supply voltage

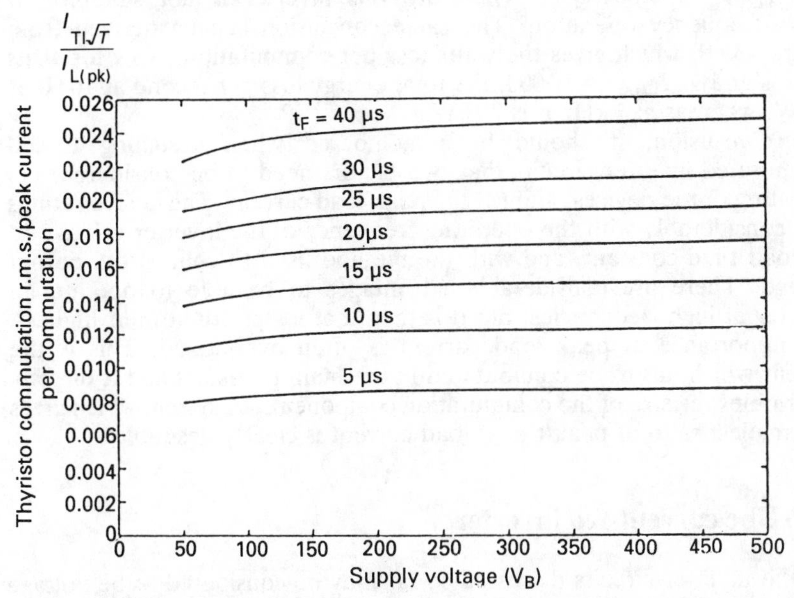

Figure 13.59 Variation of thyristor commutation current factor with supply voltage

time devices. The total thyristor and diode ratings are given by the geometric sum of values read off from Figures 13.52, 13.53, 13.59 and 13.60. Therefore when operating at an inverter frequency of 10 Hz, $I_{T1}/I_{L(pk)}$ is nearly constant at $0.0248/\sqrt{0.1} = 0.079$, whereas Figure 13.52 gives the thyristor current without commutation as $0.35 I_{L(pk)}$. The total current is also approximately this value. For 1 kHz operation, however, $I_{T1} = 0.79 I_{L(pk)}$, and Figure 13.53 still gives the same value as before, so that

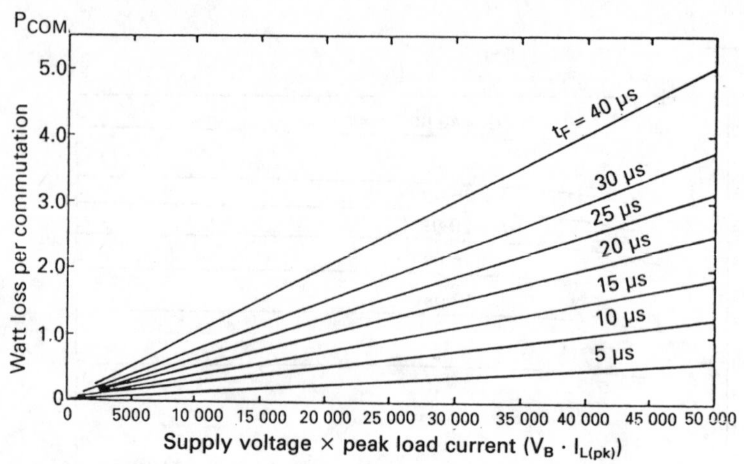

Figure 13.60 Variation of power loss per commutation with the supply voltage–peak load current product

the total thyristor r.m.s. current with commutation is $\sqrt{(0.79^2 + 0.35^2)}I_{L(pk)} = 0.864I_{L(pk)}$. Therefore this inverter is not suitable for higher-frequency operation. The same conclusion can be derived from Figure 13.60, which gives the watts loss per commutation. Now for 40 μs devices and $V_B I_{L(pk)} = 50000$, the total energy loss per second at 10 Hz is 200 W, whereas at 1 kHz it is 20 kW.

In conclusion, it should be remembered when designing forced commutated inverter circuits that two factors need to be considered: (1) the rating of the devices, and (2) the peak load current. The device ratings vary considerably with the operating frequency of the inverter relative to the load time constant, and with the method used to control the output voltage. There are considerable advantages to be able to operate an inverter at high frequencies, but it is then that losses are at their highest. The importance of peak load current is often overlooked. This is the current which has to be commutated in the main thyristors and it directly determines the size of the commutation components. A system which gives the smallest ratio of peak/r.m.s. load current is clearly desirable.

13.5 The current-fed inverter

All the inverter circuits discussed so far may be considered to be voltage fed, since the inverter forces the voltage across the load, whilst the magnitude and waveshape of the current is determined by the power factor of the load. Usually the voltage is fixed, by the d.c. supply being connected directly across the load via the semiconductor switch.

In contrast with this the current-fed inverter maintains a constant current through the inverter switches and the load voltage is determined by the load power factor. Because the current through the inverter is constant no feedback diodes need to be connected across them, for example as required in the inverters of Figures 13.1 and 13.3.

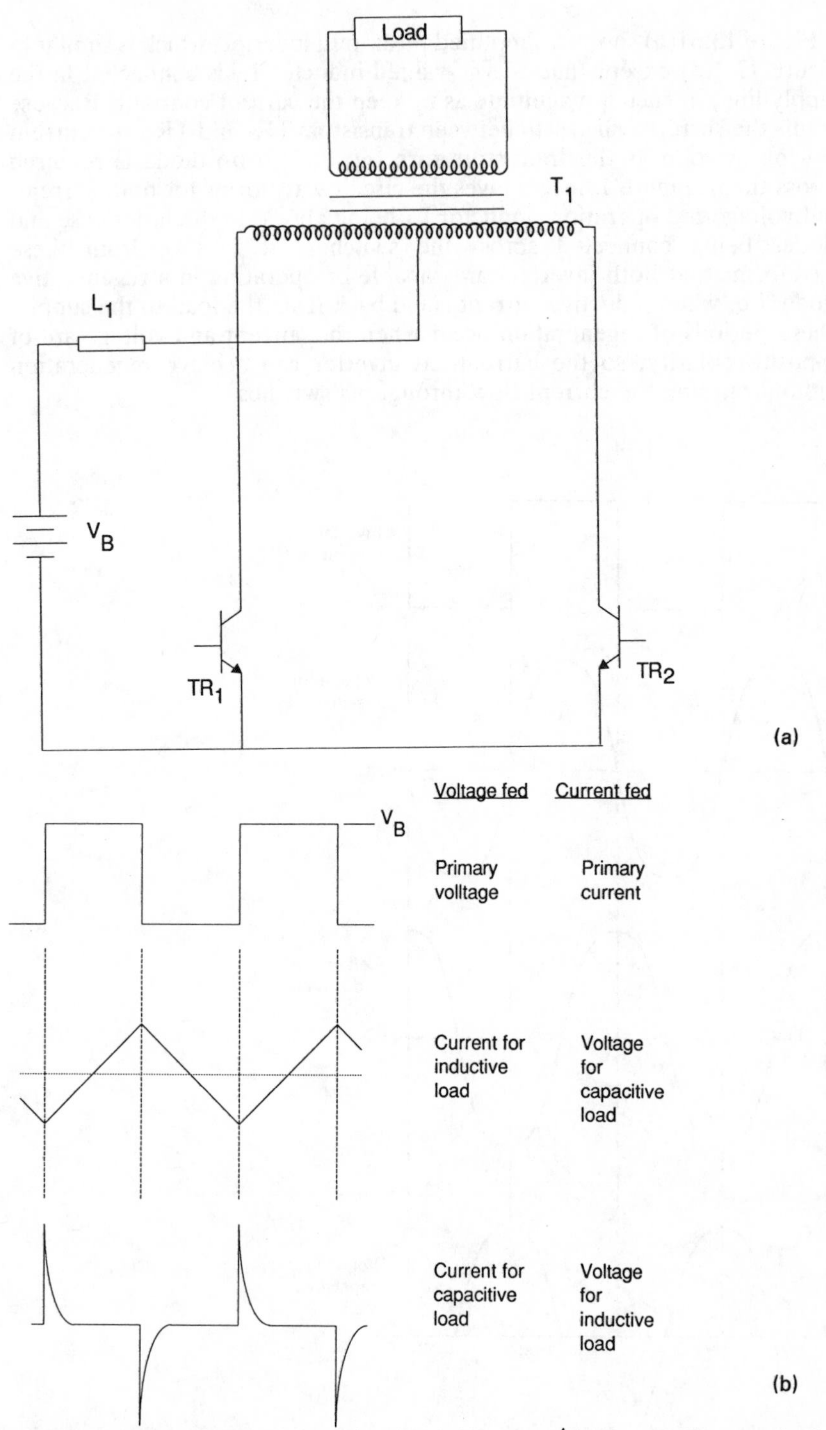

Figure 13.61 Illustration of voltage and current-fed inverters: (a) current-fed inverter circuit arrangement; (b) waveforms

Figure 13.61(a) shows a simplified push–pull inverter, which is similar to Figure 13.1(a) except that a large-valued inductor L_1 is connected in the supply line, of such a magnitude as to keep the current constant. Because of this the current will switch between transistors TR_1 and TR_2, the current flowing as soon as the transistors turn on, and so no diode is required across them. Figure 13.61(b) gives the circuit waveforms for both current- and voltage-fed operation, inductor L_1 being absent in the latter case and diodes being connected across the switches. It is seen from these waveforms that both inverters are capable of operating in a regenerative mode, i.e. when inductive current is fed back from the load to the supply. These periods of regeneration occur when the current and voltage are of opposite polarity, so the current-fed inverter can achieve regeneration without altering the current flow through its switches.

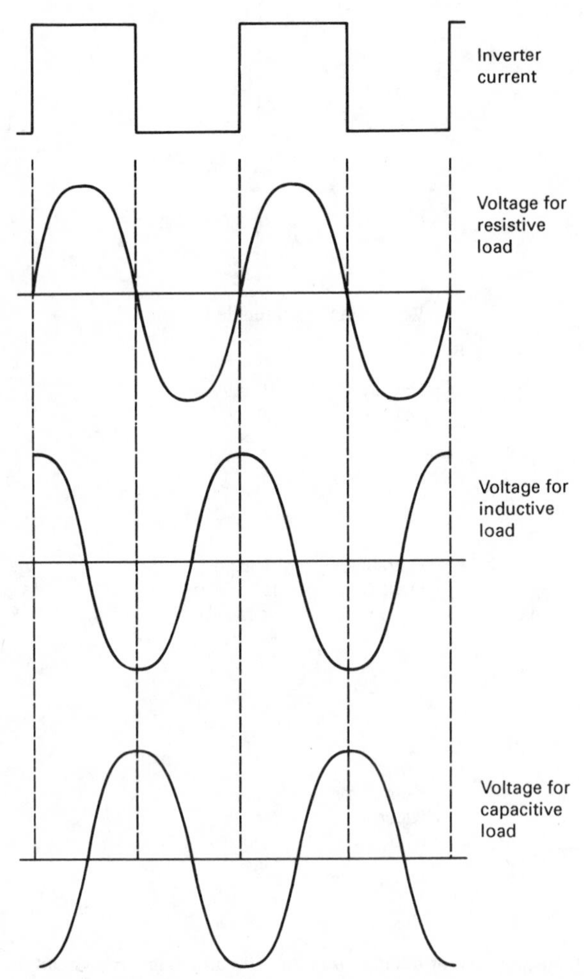

Figure 13.62 The current-fed inverter waveforms when operating into a tuned load

It is seen from the waveforms that the voltage-fed inverter is not suitable for operating into capacitive loads, due to the high spikes of current, whilst the current-fed inverter results in high-voltage transients, when operating into inductive loads or where the load gives a high impedance to harmonic currents. The current-fed inverter is ideal when the load presents a low impedance to harmonic currents and is close to unity power factor, for example a tuned circuit used for induction heating where a large capacitor is connected across the heating coil. In these instances the inverter operates at substantial constant frequency and the load voltage is sinusoidal, as shown in Figure 13.62. The current through the inverter switches is still square and these provide the energy to compensate for resonant losses in the tuned load.

Because a constant current is assumed in a current-fed inverter, this current must flow through one of the inverter switches, so that it is not possible to operate with both devices off, as required for voltage-control systems having unidirectional switching. In these instances a modification is required, as shown in Figure 13.63, where a feedback winding is used on the supply inductor, so that with both transistors off the energy of the choke is fed back to the d.c. supply. Diode D_1 prevents a current from flowing in this winding during normal operation, when one of the transistors is on.

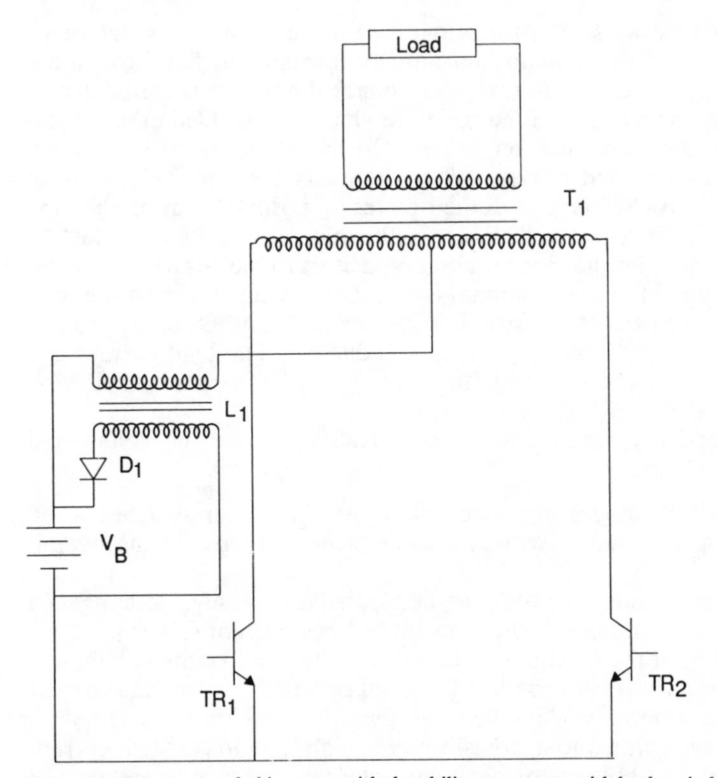

Figure 13.63 A current-fed inverter with the ability to operate with both switches non-conducting

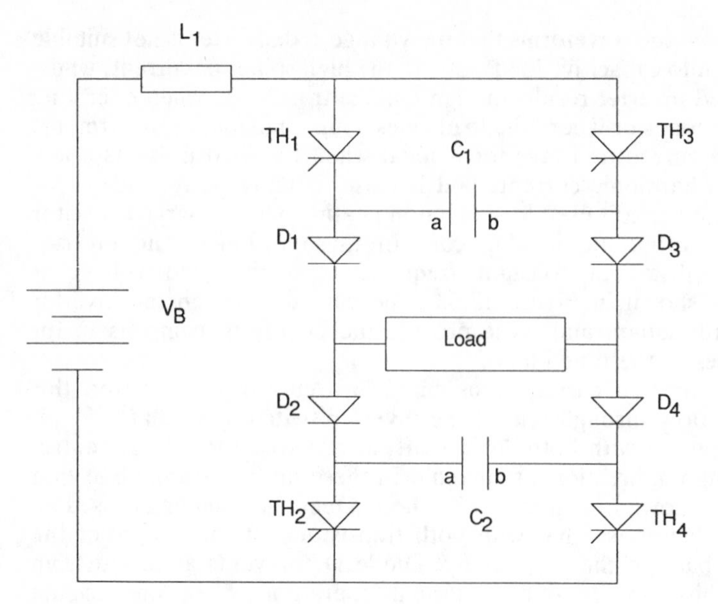

Figure 13.64 A thyristor bridge current-fed inverter

Figure 13.64 shows a thyristor bridge current-fed inverter, inductor L_1 being large enough to maintain substantially constant current through the switches during a cycle. Note that no feedback diodes are required across the main thyristor switches, although diodes have been added in series and take part in the commutation process by isolating the commutation capacitors from the load voltage. When thyristors TH_1 and TH_4 are fired capacitors C_1 and C_2 charge with their plates 'a' positive. During the next half cycle thyristors TH_2 and TH_3 are fired, which turns off the conducting thyristors and enables the commutation capacitors to recharge with reverse voltage, ready for the next commutation interval. Depending on the load power factor, the current will switch from diodes D_1 and D_4 to D_2 and D_3, after a delay, for example if the load is inductive. The load current is a square wave, as before, and the voltage will be sinusoidal with commutation spikes, if the load is tuned.

The current-fed inverter has several advantages over the voltage-fed inverter, as follows:

(i) No feedback diodes are required across the power switches when operating into inductive loads, so reducing the cost of the overall system.
(ii) Capacitive loads can be handled relatively easily, although in voltage-fed inverters they can result in large current spikes.
(iii) Utilisation of the power switches is high since, unlike the voltage-fed inverter, the switches conduct for a full conducting cycle, the current from the supply switching between them.
(iv) Utilisation of the output transformer is high, due to constant current flow with the absence of peak currents, which occur in voltage-fed inverters.

(v) The RFI generated in a current-fed inverter is much lower than that in voltage-fed inverters, since RFI is essentially generated by changing currents and in a current-fed inverter the current is substantially constant.
(vi) The current drawn from the supply in a current-fed inverter is much smoother than that in a voltage-fed inverter, so that the filtering requirements are also much less.

In spite of some of the advantages of current-fed inverters, voltage-fed inverters continue to be used, since they are more versatile in a variety of applications, and they are easier to control.

13.6 Inverter control circuits

A variety of control circuits exist for inverters, dependent on the system used to vary the output voltage and the inverter configuration. All these will consist of the power semiconductor drive circuit and a form of sequencing to turn the devices on at the appropriate instances in the cycle.

Figure 12.23 illustrated a method of mark-to-space control of the output voltage, using a sawtooth waveform and a pedestal reference, and this was incorporated into a control circuit for a chopper in Figure 12.24. The same technique can be used to provide mark-to-space control within an inverter, although Figure 13.65 shows an alternate technique using two shifted square waves to achieve the same effect, as illustrated in Figure 13.41. The

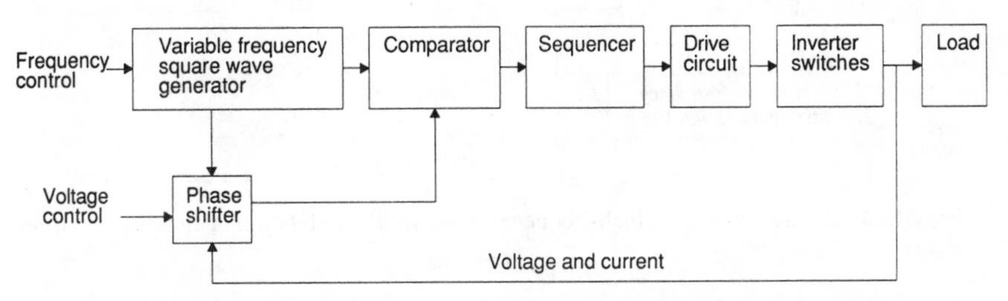

Figure 13.65 System for mark to space control of the output voltage from an inverter

square waveform from the generator is passed through a phase shifter and the two direct and shifted waveforms are compared to provide the input to the sequencer and drive circuits, which turn the inverter semiconductor switches on at the correct instances. The load voltage may be fed back and used to regulate the output, and the current can also be monitored to provide a current limit function, if required. The frequency of the inverter is controlled by varying the frequency of the square wave generator and the output voltage by regulating the phase shift between the two square waves prior to the comparison stage.

Selected harmonic reduction, of the type illustrated in Figure 13.36, can be obtained by the control circuit shown in Figure 13.66. The output from the pulse generator, which determines the frequency of the inverter, is

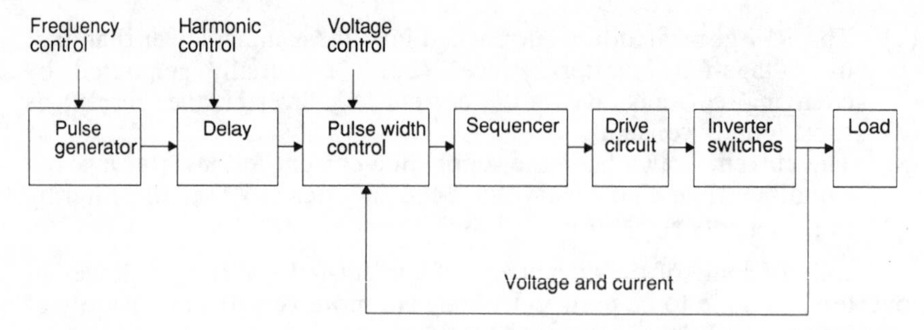

Figure 13.66 Control system for obtaining selected harmonic reduction in the output voltage of an inverter

varied so as to produce a series of pulses spaced at the required distances apart, to eliminate a given harmonic in the output voltage. The magnitude of the fundamental voltage in the output is altered by controlling the width of the pulses. As before, sequencing is needed to drive the correct power switches in the inverter and drive circuits to provide the drive power. Voltage and current feedback can be used to sense and adjust the magnitude of the load voltage and therefore the current.

Figure 13.67 shows a high-frequency pulse-width modulation control system, which gives the waveforms illustrated in Figures 13.37, 13.40 or

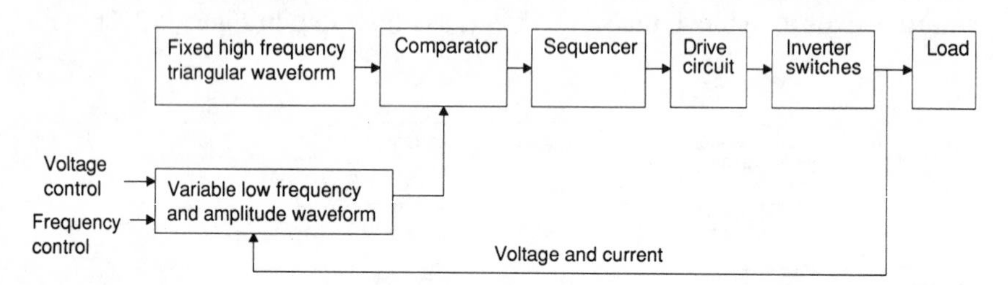

Figure 13.67 Control system for high-frequency pulse-width modulation of an inverter output voltage

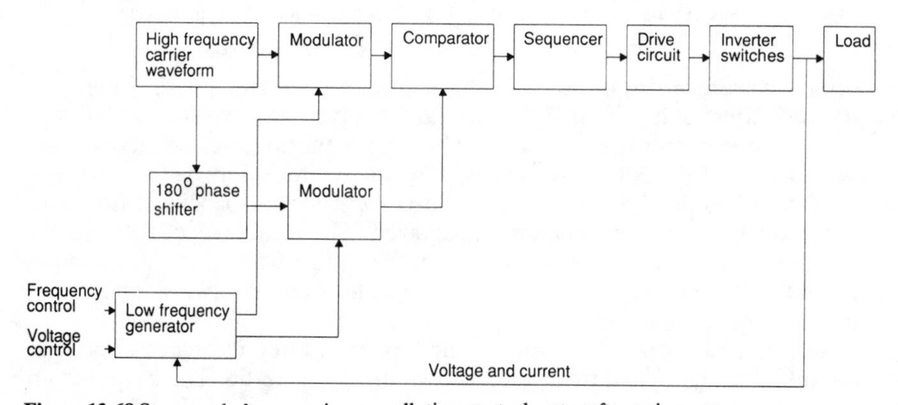

Figure 13.68 Staggered phase carrier cancellation control system for an inverter

13.43. The high-frequency triangular wave is compared with a low-frequency square, triangular or sine waveform, the inverter frequency being determined by the frequency of this low-frequency wave and the magnitude of the output voltage by its amplitude relative to the reference high-frequency wave. The remainder of the circuit, with sequencer, driver and current or voltage feedback, is as before.

Staggered phase carrier cancellation is illustrated in Figure 13.68. The two high-frequency carrier waveforms are phase shifted by 180° before being modulated by the low-frequency waveform and then compared, as in Figure 13.44. The subsequent control stages are then as before.

Part 3

Applications

Chapter 14

Power semiconductor circuit applications

14.1 Introduction

The preceding chapters have already described a variety of applications for power semiconductors, such as static switching, a.c. line control, phase-controlled rectification and inversion, frequency changers like inverters and cycloconverters, and d.c. to d.c. converters or choppers. When describing these applications the main consideration was for analysing the systems from a circuit design point of view. In the present chapter applications will be considered from another aspect, that of usage, although the circuits within each of these groups will already have been described in an earlier chapter. For example, inverters and choppers are used within power supplies, phase-control rectification and inversion in d.c. motor control, and so on.

Although there are a large number of different applications for power semiconductors only a few of these are described here, and they are grouped into four sub-sections: power supplies; electrical machine control; heating; and electrochemical.

14.2 Power supplies

A power supply, or a power source, is anything which is capable of providing energy. With this loose definition almost everything is a power supply, although what is usually referred to by this term is equipment used to provide stabilised and closely regulated d.c. voltage and current output from an a.c. mains input, or from a d.c. source such as a battery. There are many ways in which this can be achieved, a few of these being shown in Figure 14.1. In these forms of power supplies it is important to isolate the output from the input, and this is usually done by a transformer.

Figure 14.1(a) shows a power supply arrangement in which voltage regulation is applied to the a.c. input. This regulation technique can use any of the methods described for a.c. line control in Chapter 8, although phase control of thyristors or a triac is commonly used. The regulated voltage is then rectified and filtered to give the d.c. output. Voltage regulation is used to enable the operator to vary the magnitude of the output voltage, but it also provides a control mechanism for keeping the

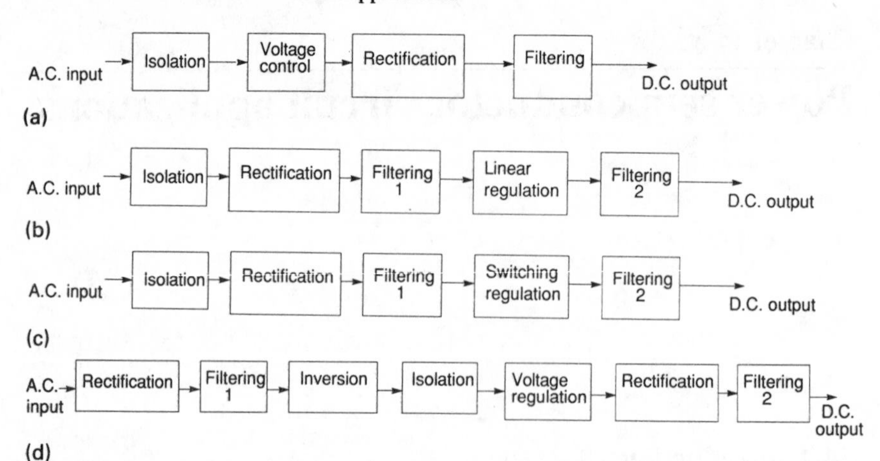

Figure 14.1 Common arrangement for a.c. to d.c. power supplies: (a) a.c. voltage control; (b) linear d.c. regulation; (c) switching d.c. regulation; (d) switching a.c. regulation

voltage constant under changing load current or input voltage and a method for limiting the current under fault conditions.

Regulating the d.c. voltage can also be used, as in Figures 14.1(b) and 14.1(c). After isolation through the transformer the a.c. voltage is rectified and filtered to give a raw d.c. voltage. This can then be adjusted to the required d.c. output value by a transistor operating in a linear mode, or by switching chopper regulation, as described in Chapter 12. A second stage of filtering is required to provide the final smoothed d.c. output. Switching regulation contains a higher ripple content than linear regulation, so it requires a greater amount of smoothing, but it is much more efficient since it does not dissipate the voltage difference between the input and output across its regulating power semiconductor, so it is used more often for high-power supplies.

Figure 14.1(d) shows an arrangement in which the a.c. input is first rectified and filtered to give a d.c. supply, which is not isolated from the a.c. input. This is then inverted, the inverter frequency being several orders higher than that of the a.c. input, and the high frequency is isolated in a transformer. Usually voltage regulation occurs within the inversion stage, using one of the many techniques described in Chapter 13, although a separate a.c. line control voltage regulator may be used, as shown in Figure 14.1(d). Finally the a.c. voltage is rectified and filtered to provide the d.c. output voltage. This arrangement gives a power supply which is light and small, since the high-frequency isolation enables a small transformer size to be used, but it also generates a higher level of noise and higher ripple compared to non-switching forms of power supplies. If the primary input source is d.c. then the main technique used is that of Figure 14.1(d), with the initial rectification stage omitted, since it provides isolation following d.c. at the first filtering stage.

Performance factors used in the measurement of power supplies include:

(i) Line regulation, the variation in the load voltage as the input voltage changes.

(ii) Load regulation, where the load voltage fluctuates as the load current changes.
(iii) Temperature regulation, resulting in changes in output voltage due to variations in temperature, primarily due to temperature effects on components used within the power supply.
(iv) Ripple and noise, caused by switching effects within the power supply and insufficient filtering at the output.

Power supplies usually provide two forms of protection for the load, overvoltage and overcurrent. Overvoltage protection is normally achieved by sensing the power supply output voltage and applying a short-circuit across the power supply output lines when this voltage exceeds a preset value. Thyristor crowbar circuits are used for this since the semiconductor switch can be made to operate within a fraction of a second, so preventing damage to the load. Once the crowbar has operated, overcurrent circuitry within the power supply comes into play, as described below.

Overcurrent protection is required not only to guard the load from excessive currents under certain fault conditions, but also to protect the power supply from damage. It is achieved by sensing the load current and feeding this back to the voltage control section within the supply, so that the voltage reduces to limit the current. Three techniques are used for this, as illustrated in Figure 14.2. In the normal current limit mode the voltage begins to drop rapidly as the load current increases, so the maximum fault current is only slightly greater than that of the current limit setting. This is the simplest and most commonly used method of overcurrent protection. For constant current operation the current limit point is never reached since the circuit is designed to provide a constant current to the load under all conditions. In re-entrant or foldback protection the voltage rapidly decreases once the current limit point is reached, so that the load current falls below the full-load value. This method of current limit is usually used with linear regulators, to limit the dissipation across the voltage controlling element when a fault occurs.

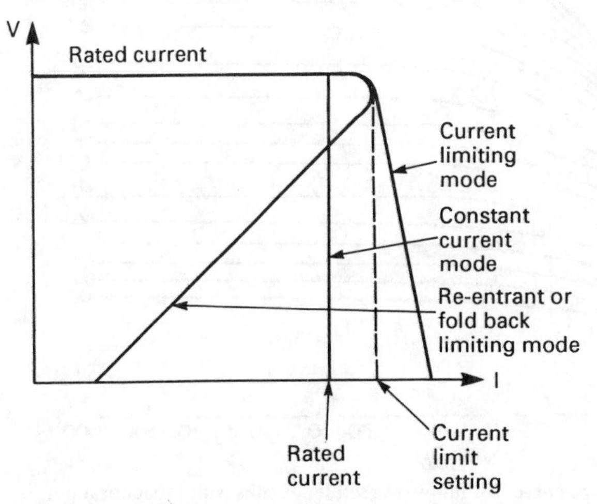

Figure 14.2 Current-limiting techniques in a.c. to d.c. power supplies

The a.c. supply and rectification shown in Figure 14.1 can be single phase or three phase, and the filtering used can consist of several types, capacitor input filters being illustrated in Figure 14.3. This circuit, as applied to a mercury arc rectifier, has been extensively analysed by O.H. Schade (*Proc. IRE,* July 1943) and his design curves, one of which is illustrated in Figure 14.4, are still used today. The series resistor R_S is

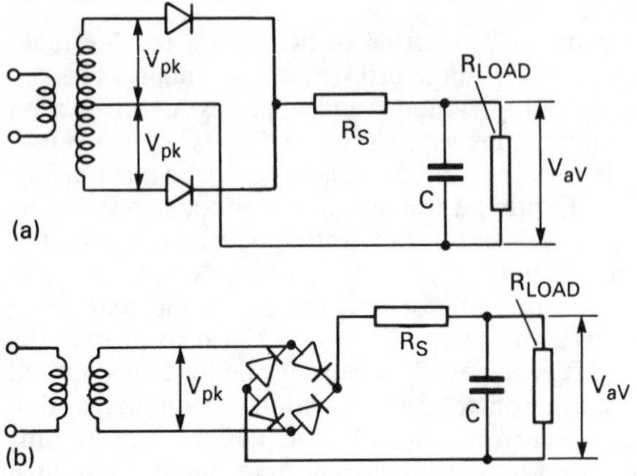

Figure 14.3 Full-wave rectifier circuits with capacitor input filters: (a) push–pull; (b) bridge

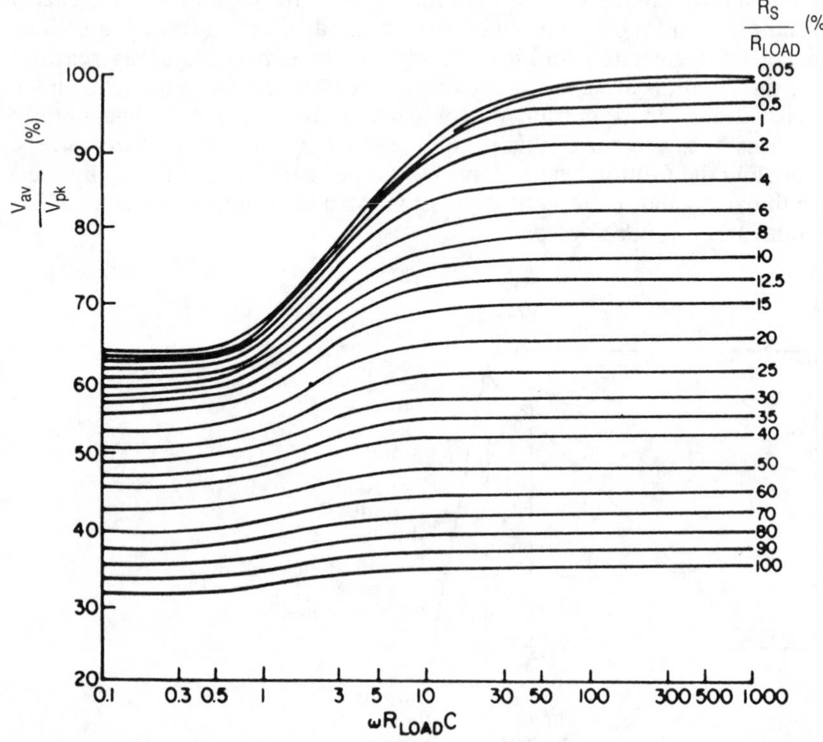

Figure 14.4 D.C. output voltage curves for full-wave rectifier circuits with capacitor input filters

usually that of the transformer winding, the rectifier resistance and any external resistance added to limit the surge current caused by the charging effect of the capacitors in the filter.

Figure 14.4 gives the magnitude of the d.c. output and Figure 14.5 the load ripple voltage. Generally, the value of $\omega R_{LOAD}C$ is chosen to operate on the flat portion of the curves of Figure 14.4 and to reduce the ripple to the value desired.

At high-load currents a choke input filter is usually preferred, since it gives better smoothing without needing a large value of capacitance

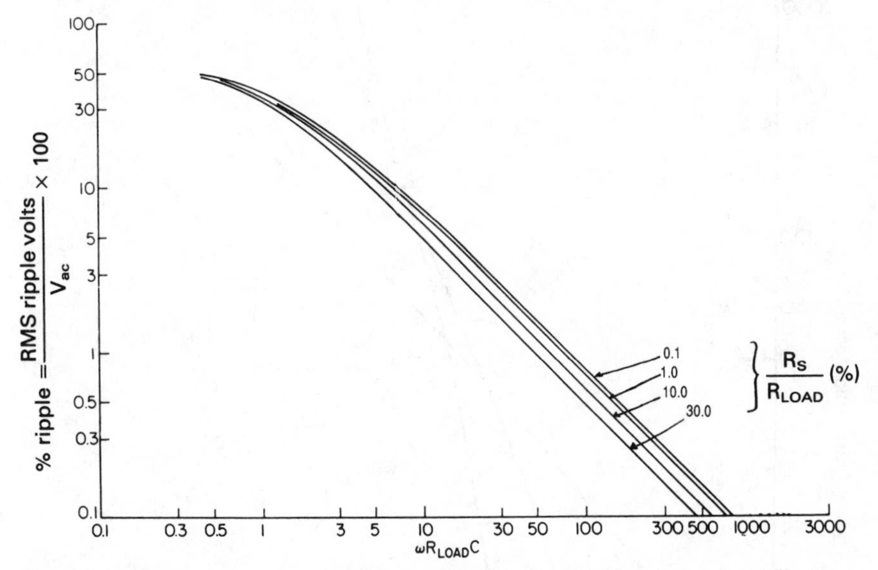

Figure 14.5 Ripple voltage curves for full-wave rectifier circuits with capacitor input filters

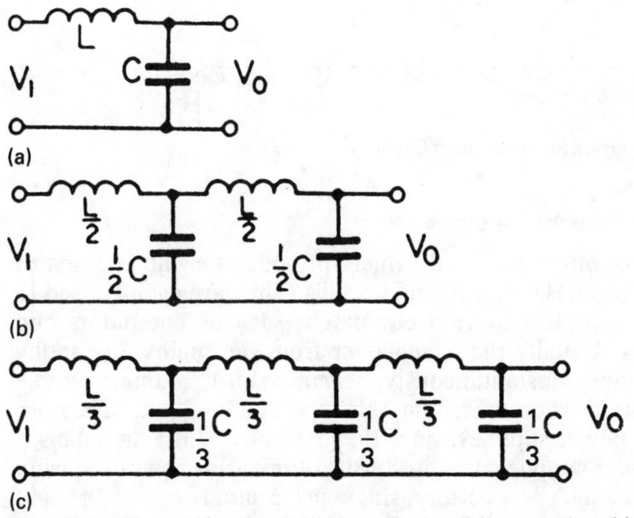

Figure 14.6 Choke input filters: (a) single-section; (b) two-section; (c) three-section

carrying high ripple current. This filter can be obtained by replacing the series resistor R_S in Figure 14.3 by a choke. For applications which require very smooth supplies even this single-stage filter is inadequate, resulting in large-valued capacitors and chokes, and multi-section filtering is then preferred, as shown in Figure 14.6. Figure 14.7 illustrates the attenuation curves for these filters, where f_r is the fundamental ripple frequency, and from this it is seen that high attenuation factors are obtained more effectively by multistage filters.

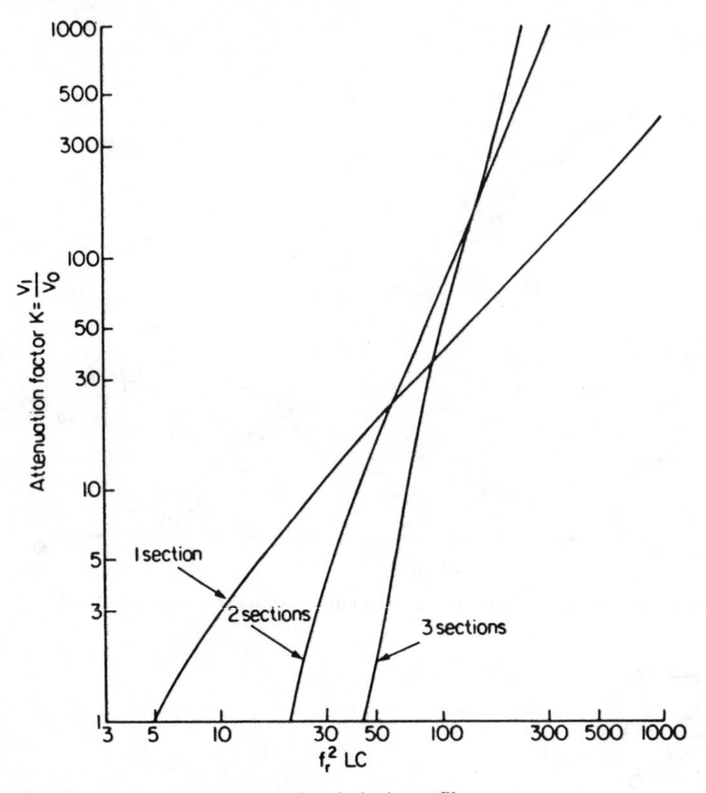

Figure 14.7 Attenuation curves for choke input filters

14.2.1 Uninterruptable power supplies

Power supplies are often required which provide a.c. at the mains frequency of 50 Hz or 60 Hz once the mains fails. They are usually used to provide emergency supplies to vital equipment such as computers and life-support systems. Usually the changeover from the mains to standby supply must occur almost instantaneously, hence the term 'uninterruptable power supplies' is used when describing this type of equipment. They are also called standby power supplies, since they stand by whilst the mains is working. These power supplies usually consist of inverters operating from a d.c. storage source, such as a battery, since such a supply can be brought into service quickly, and it is switched off once a motor generator is able to

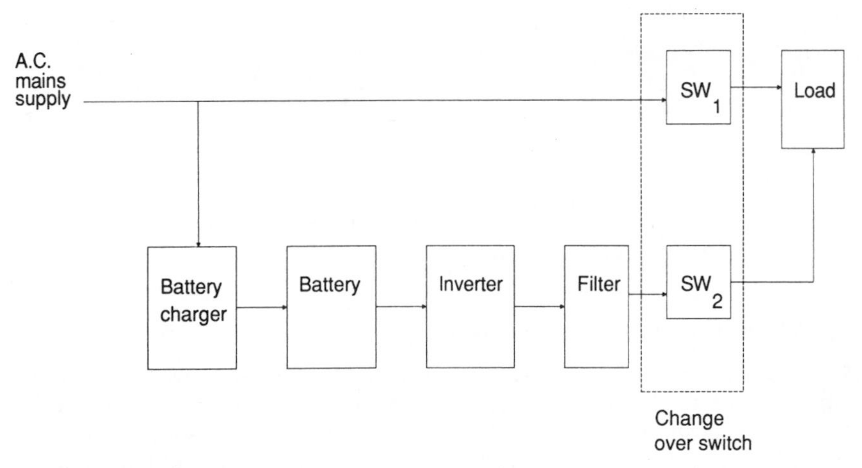

Figure 14.8 An uninterruptable or standby power supply

run up to speed to supply the power. Furthermore, not all emergency loads require uninterruptable supplies: emergency lighting, for example, can usually operate effectively with a break of about 200 ms.

Figure 14.8 shows the schematic of a typical uninterruptable power supply. The load is normally supplied direct from the main a.c. supply, switch SW_1 being closed and SW_2 being open. During this period the battery is trickle charged from the mains supply. The inverter is operated at a fixed frequency and is therefore optimised for a sine wave output, by using techniques such as pulse-width modulation or waveform synthesis, the output waveform being further enhanced by the filter. The inverter is also synchronised to that of the mains so that when the main supply fails, switch SW_2 closing to apply the inverter supply to the load, there is no waveform distortion.

14.2.2 Variable-speed constant-frequency supply

A variable-speed constant-frequency (VSCF) supply provides an accurate fixed-frequency output voltage from a variable-frequency source. Such a system is widely used in aircraft where the main a.c. power is generated by an alternator driven from the aircraft's engines, and since these operate at variable speeds the output from the alternator has a variable frequency.

Traditionally, variable-slip couplings have been used between the engine and the alternator to keep its speed constant, but such mechanical systems need maintenance and have limited life. A better solution, shown in Figure 14.9, uses a cycloconverter, operating in reverse mode to those illustrated in Chapter 10, to provide a fixed-frequency output, which is then filtered to give a highly stable sine wave supply. Such a system requires no maintenance, is reliable and lightweight, and gives better performance by use of feedback since the response time is much shorter than in mechanical systems. Closer control of frequency is also possible, with a better match of the amplitude and phase between different phases of a three-phase system, even when the load is unbalanced.

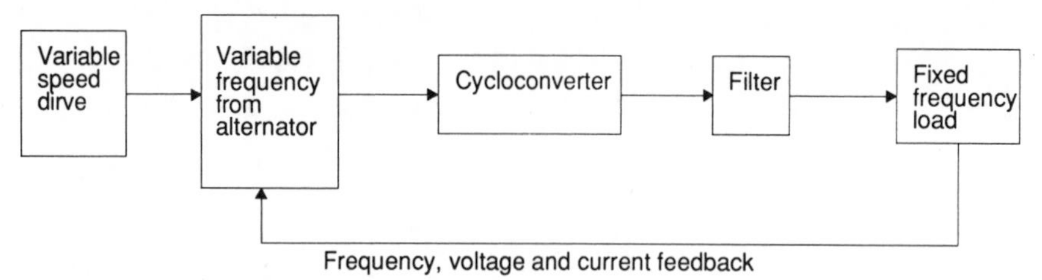

Figure 14.9 A variable-speed constant-frequency (VSCF) supply

14.2.3 High-voltage d.c. transmission (HVDC)

A.C. power is a convenient source to generate and to convert from one voltage level to another, for example by using transformers, and it is therefore the supply most often used both as a power source and for transmission. However, for very long distances, both over land and under the sea, a.c. can result in relatively high transmission-line losses, and in these instances d.c. is preferred. In these applications the a.c. supply is first transformed to a high-voltage and then converted to a d.c. voltage for transmission, the high-voltage d.c. then being converted back to a.c. at the other end, before being transformed down to a low voltage. This is illustrated in Figure 14.10.

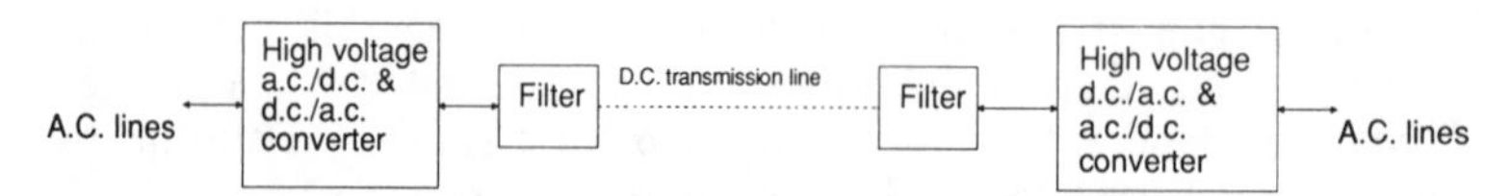

Figure 14.10 A high-voltage d.c. transmission system

The power source used for high-voltage d.c. transmission is usually a thyristor converter, made from series-connected devices to obtain the high-voltage capability. Commonly a twelve-pulse fully controlled bridge is used, providing both the rectification and inversion operation, the ripple being low enough to enable fairly simple filters to be used. These filters minimise ripple voltage, which would result in additional transmission losses.

14.3 Electrical machine control

Electrical machine control marks one of the most important applications for power semiconductors, especially thyristors and triacs. Due to their small size these power devices have given practical reality to systems which previously were only an experimenter's dream.

Control electronics bridges the gap between two widely differing sciences. On the one hand, the power semiconductor, with its associated control circuitry, may be regarded as a tool of the electronic engineer, whereas the machine, which differs only slightly from conventional designs, is the instrument of the electrical engineer. It is not possible to treat power semiconductors and machines separately if they are to work together as a system. The characteristics of the machine affect the performance of the power semiconductor and vice versa, but the rapid growth of electronic control systems has resulted in instances where electronic engineers have developed systems whilst having no knowledge of machine fundamentals, or electrical engineers have blundered into electronic control schemes whilst being unfamiliar with transistor circuitry.

A control engineer is a unique animal who must have a good working knowledge of both electronics and electrical machines. The previous chapters have dealt with power electronic circuitry at some length, and the present section will consider the principles of electrical machines and how they can be controlled by power electronic devices.

Although power semiconductors are in widespread use for machine control, sometimes because they represent the only practicable method for obtaining a control system, often they are in direct competition with traditional controllers and must then prove themselves to be either technically or economically superior. The advantages which power semiconductor drives have over conventional methods, such as Ward–Leonard systems, are:

(i) Quicker response. This is specially advantageous when fast acceleration or high-speed accuracy is required, as in reversing-mill tables.
(ii) High operating efficiency due to the small voltage drop across a power semiconductor, even when it is handling large amounts of power.
(iii) Less weight and lower installation cost. A power semiconductor control cubicle occupies a small floor area, is light and does not require any special floor surface or mounting. This compares very favourably with the aligning and mounting requirements for motor-generator systems.
(iv) Ease of maintenance. A power semiconductor system can be designed to be easily maintained by incorporating diagnostic circuits during the design stage and providing plug-in card replacements for electronics. Servicing of the controller is, of course, not required due to the absence of any moving parts.

The disadvantages of power semiconductor systems are:

(i) The low overload capability, which is caused primarily by the low time constant of the power semiconductor devices. This means that in most cases the control system must be rated to withstand the peak overload expected instead of the usual full-load value.
(ii) Inability of a power semiconductor system to regenerate unless special circuit modifications are included, such as double-bridge controllers. A motor-generator set, on the other hand, can readily pass energy in either direction.
(iii) Low power factors in drives which use phase angle control. As has been illustrated in previous chapters the load power factor is approximately proportional to the delay in the turn-on of the power semiconductor within an a.c. half cycle, and the power factor will therefore be poor at low speeds, when the voltage is low and the delay angle is therefore large.

(iv) Generated harmonics in systems where large voltages are switched, due to the very fast operating time of the power semiconductor. These harmonics cause radio frequency interference which must be suppressed.

14.3.1 Elements of electrical machines

Machines fall into two groups, motors and generators. To explain the difference between them it is necessary to discuss the operating principles of an elementary machine, as in Figure 14.11. The essential requirement for production of magnetic force is interaction of two magnetic fields. The force of attraction and repulsion between two bar magnets is well known, a similar force resulting if one or both magnets are replaced by a current-carrying conductor.

In Figure 14.11 magnet N-S is fixed (stator) whereas coil a-b is mounted on a drum and is free to rotate (rotor). If current flows into the rotor at a and out at b, as indicated, then this will produce a magnetic field so that side Y is a north pole and X a south pole. The drum will rotate, the north pole at Y aligning itself with the stator south pole. The machine is an elementary motor, the electrical energy in the coil being converted into mechanical work on the drum.

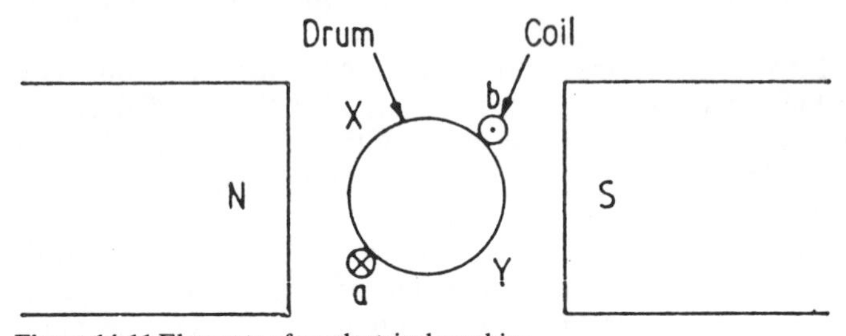

Figure 14.11 Elements of an electrical machine

Now assume that no external voltage is applied to coil a-b, but the drum is rotated clockwise. Work done in rotating the drum must be opposed by current flowing in the coil, since if it were to aid the external force this would create the well-known paradox of a perpetual-motion machine. The direction of coil current will be as indicated in Figure 14.11. Mechanical work is done in overcoming the repulsion between similar poles on stator and rotor, and this is converted to electrical energy in the coil, so that the machine is a generator.

It is clear that the construction of a motor and generator for a given type are very similar, it is only the terms of reference which differ. For a motor energy input is electrical and output is mechanical, whilst for a generator mechanical input energy is converted to an electrical output.

Motors and generators can be further divided into various groups as shown in Figure 14.12, and these will be discussed in the following sections.

Figure 14.13 shows the construction of an elementary d.c. machine, where the rotor is a single coil and the commutator consists of two half cylinders. During a complete coil rotation the commutators are in contact

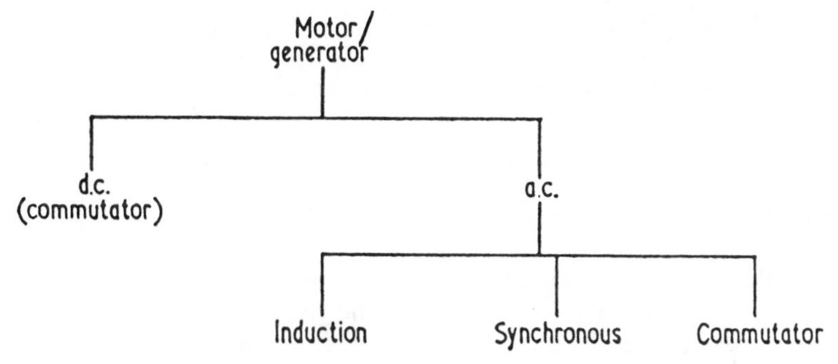

Figure 14.12 Types of electrical machines

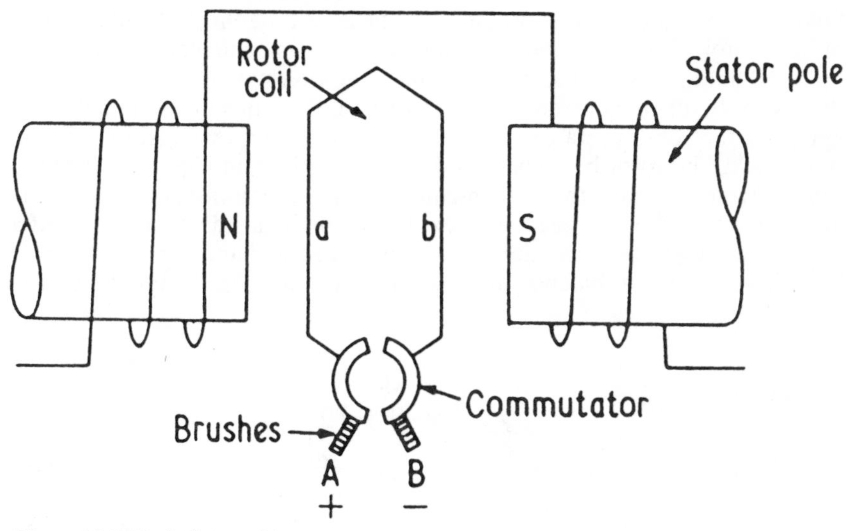

Figure 14.13 Basic d.c. machine

with brushes A and B for half the period. During motor operation a d.c. voltage is applied to brushes A and B, with the polarity shown. Current will flow up coil side a and down side b, this current forming rotor poles which interact with the stator flux to produce a torque on the coil in an anticlockwise direction. The function of the commutator is to switch the rotor poles, by reversing coil current, as shown in Figure 14.14, and so maintain unidirectional motion.

In Figure 14.13 if brushes A and B were connected to a load and the armature rotated in a clockwise direction, by an external mechanical force, an e.m.f. would be induced in coil sides a and b, forcing current down side b and up side a. The commutator ensures that this current is unidirectional in the load and the machine is now a d.c. generator.

During motor action, when the rotor revolves, the changing flux induces in it a voltage which will produce poles to oppose the stator flux, this being referred to as the motor back e.m.f. This voltage is a function of the motor speed and the strength of the field flux, being in effect secondary generator action in a motor.

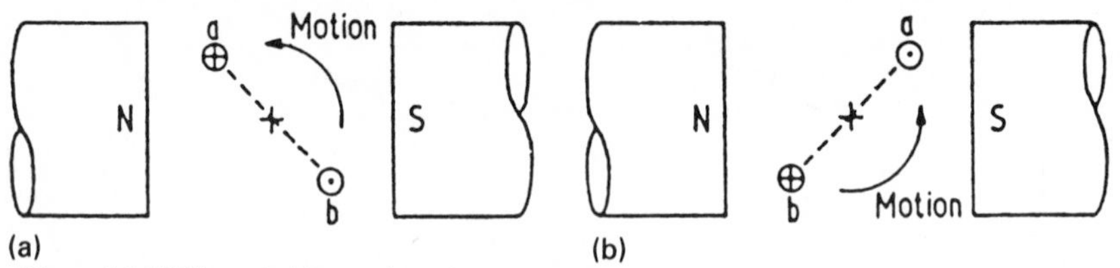

Figure 14.14 The switching action of a commutator

The function of the commutator must be clearly noted. It senses the rotor position, by virtue of its construction, and switches the rotor current, at the appropriate instant, to ensure that torque is unidirectional in a motor and that the generator output is d.c.

Figure 14.15 illustrates an elementary a.c. motor, which differs from Figure 14.13 in that the commutator and brushes are removed and the ends of the rotor short-circuited. Assume magnetising current to flow and produce field poles as illustrated. Now if the poles were caused to rotate they would induce a current in the rotor coil, this current, by Lenz's law, opposing the rotating stator field. Figure 14.15 shows two positions of the stator field, the rotor being assumed at a standstill, and illustrates that the rotor current is such that the torque is always unidirectional. A.C. is induced in the rotor by the stator flux, the action being identical to that of a transformer with short-circuited secondary. An equal and opposite current flows in the stator to balance the rotor ampere turns, this being the power current.

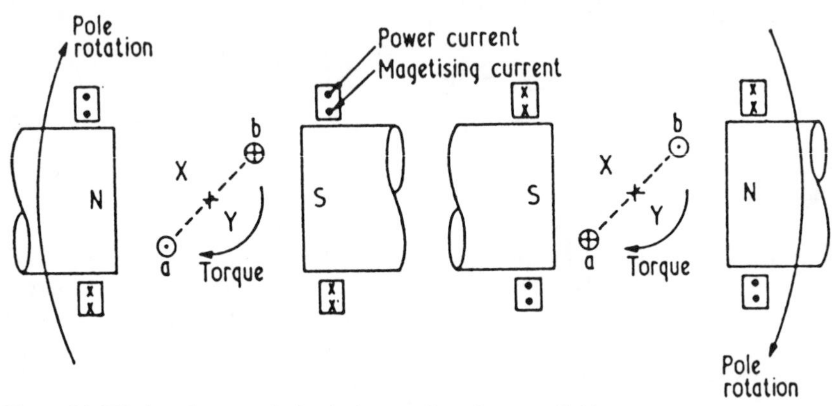

Figure 14.15 Induced rotor polarity during rotation of a stator field

This machine is called an induction motor. In an actual machine the stator poles are not physically rotated, but instead they are supplied with magnetising currents displaced in amplitude and time to produce a rotating stator flux. The stator will carry this input a.c. magnetising current, with an input a.c. power current which produces a mechanical motor output.

The rotor of an induction machine may be driven by an external force at a speed greater than that of the stator field. This is shown in Figure 14.16 where the stator poles are assumed almost stationary. Applying the principle that the poles induced in the rotor by the stator field will oppose

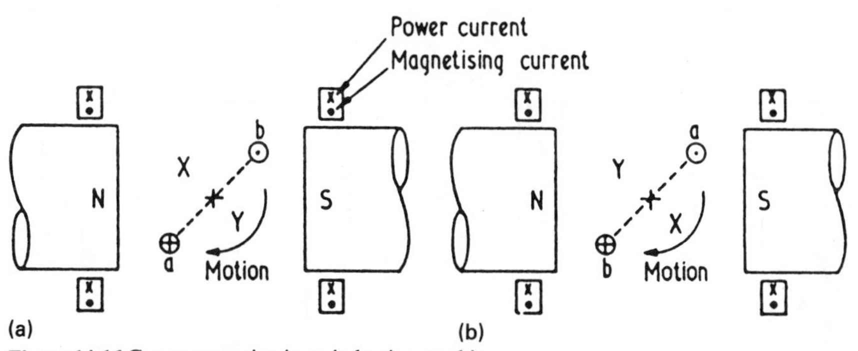

Figure 14.16 Generator action in an induction machine

this stator field and that the rotor ampere turns are balanced by flow of stator power current, the current directions will be as shown. It is seen that the power current is reversed, the machine converting mechanical input into electrical output and is now a generator.

Since current is induced in the rotor due to a changing flux linkage with the stator field, it is evident that an induction machine must always run below (motor) or above (generator) the speed of the rotating stator field, never at the same speed. It was also seen that in a d.c. machine power and magnetising currents are carried by two separate windings, whereas for an induction machine they both flow in the same winding, this being the fundamental difference between the two. It is essential in both systems that stator poles be maintained during both motor and generator action.

If the rotor of Figure 14.13 is replaced by a magnet, the result is a synchronous machine. Like the induction machine, the stator field is rotating which enables the rotor to lock on and follow at the same speed. Any other speed would cause alternate attraction and repulsion, making the machine unstable and causing it to stop. For motor action the input at the stator provides excitation for the revolving field and supplies the power component necessary to overcome the load torque and cause the rotor to follow the stator field. In this respect the machine is similar to an induction motor. However, when generating, no input is required to the stator, since excitation is provided independently to the rotor. The action is now more like that in a d.c. generator, in fact the only difference between the synchronous and d.c. generator is that the commutator (rectifier) is not present in a synchronous machine, resulting in an a.c. output. In many synchronous machines the rotor poles are produced by a electromagnet. The commutator in Figure 14.13 is replaced by slip rings so that the current in coil sides a and b is unidirectional and independent of rotation.

The a.c. commutator machine is very similar to an induction machine, the stator field revolves, the rotor being connected to a commutator as in a d.c. machine. This is illustrated in Figure 14.17, the rotating field inducing alternating currents in the rotor. The frequency of these currents depends on the frequency with which the stator field cuts the rotor conductors and on the speed difference between stator field and rotor. This is as for normal induction machines, but the rotor currents flow through the brushes and since the brushes are stationary the stator field always cuts them at a fixed

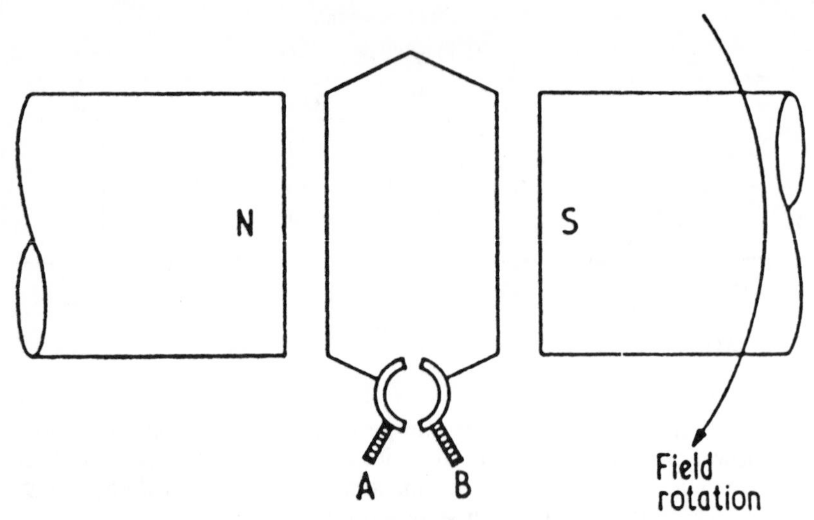

Figure 14.17 Elements of an a.c. commutator machine

speed, that of the rotating field, so that the frequency at the brushes is constant. The commutator acts as a frequency converter, between the speed-dependent frequency in the rotor conductors and the fixed frequency at the brushes. This introduces several advantages during machine control, as seen later.

Having looked at basic types of machines, the methods by which they can be controlled are now examined in the following sections.

14.3.2 D.C. motors

The armature and field coils of a d.c. motor can be arranged in several ways, as shown in Figure 14.18, the system used determining the overall

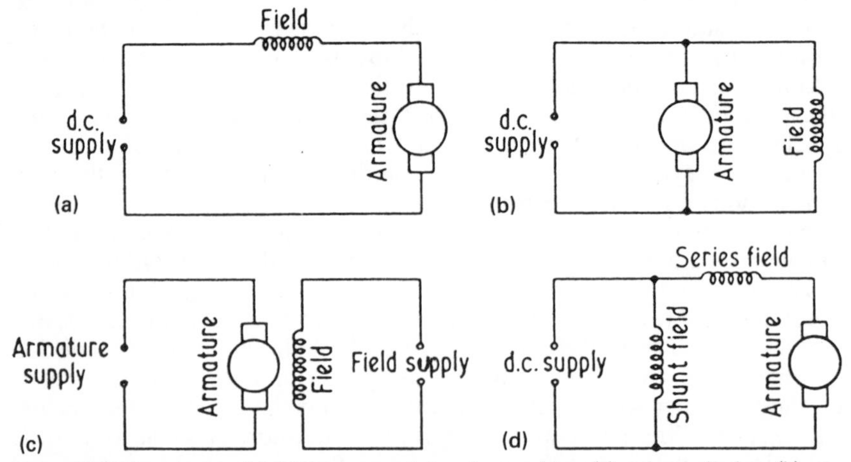

Figure 14.18 Armature and field arrangements in a d.c. machine: (a) series excitation; (b) shunt excitation; (c) separate excitation; (d) compound excitation

performance of the motor. In all cases equations (14.1) to (14.4) hold, where V is the applied d.c. voltage, E is the motor back e.m.f. at speed N and field flux ϕ, I_a is the value of the armature current which gives motor torque T, I_f is the field current and K_f, K_n and K_t are called the flux, speed and torque constants respectively. R is the series resistance of the motor and is equal to either the armature resistance alone or to the sum of armature (R_a) and field (R_f) resistances, depending on the type of motor connection used:

$$E = V - I_a R \tag{14.1}$$

$$N = \frac{E}{K_n \phi} \tag{14.2}$$

$$T = K_t \phi I_a \tag{14.3}$$

$$\phi = K_f I_f \tag{14.4}$$

The performance of the motor can be readily derived from the above equations. For example, equations (14.1) and (14.2) show that the motor speed is roughly proportional to supply voltage V, ignoring secondary effects due to voltage drop caused by armature current, provided that motor flux does not also vary in the same manner with voltage. This can be satisfied with series and separately excited machines, but not with shunt field windings in which the current is determined by the supply voltage and field resistance. Therefore shunt motors cannot be used for speed control using supply-voltage variation.

The methods by which a d.c. motor can be started, controlled and stopped will now be examined with reference to some typical power semiconductor control circuits. These circuits are used for illustration purposes only, and any of the other circuits described in this book may also be used, provided that their output corresponds to that required by the motor.

14.3.2.1 Starting

Equation (14.2) shows that when a motor is starting its back e.m.f. will be very small at low speeds. Therefore, from equation (14.1), the armature current will be large since the armature resistance is small. The high starting current would give a good motor-starting torque, and this is one of the advantages of a d.c. motor, although excessive currents can cause machine damage over a period of time and should be avoided. For large motors the peak current drawn at starting is also limited by the supply authorities, in order to prevent reductions in the voltage of the supply to other consumers. Since the armature current is proportional to the difference between supply voltage and motor back e.m.f. and inversely proportional to the armature resistance, there are two methods which may be used to limit motor-starting current. In one technique an external resistance can be connected in series with the armature to increase the effective armature resistance. As the motor speeds up the resistance is progressively reduced, either manually or by automatic methods such as servo-driven potentiometers. This is the system most commonly used with conventional d.c. motor starters. With power electronic drives it is much easier to arrange for the supply voltage to the motor to be progressively increased as the machine speeds up, so that the peak armature current is held at a predetermined value which would give good starting torque and long motor life. Depending on whether the motor is supplied from an a.c. or d.c. source, it is now possible to use a variety of power electronic circuits to give controlled rectification (Chapter 9) or d.c. line control (Chapter 12) respectively. Static contactors have also been discussed in Chapter 8.

14.3.2.2 Control

The principal parameter of interest in d.c. motor control systems is the speed of the machine, this parameter generally being linked in a secondary loop to the motor current. Such an arrangement is shown in Figure 14.19, the d.c. motor being mechanically connected to a tachogenerator which provides a feedback signal proportional to the motor speed. This feedback

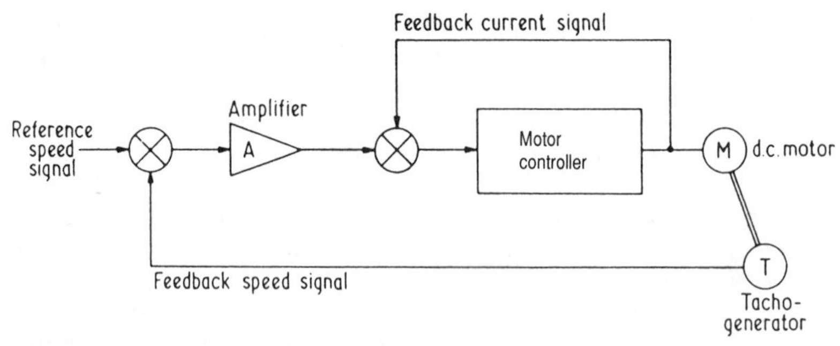

Figure 14.19 Closed-loop d.c. motor speed-control system

signal is compared with the reference, or demand, speed signal and the error, after amplification if required, can be used to adjust the motor controller so that the motor speed equals that of the reference. Figure 14.19 shows a system in which the speed error signal is first passed into another servo point where it is controlled by a current feedback signal. This inner loop has overriding command such that, even though the reference speed may be abruptly changed, the error signal driving the motor controller is modified to ensure that motor current is always below a predetermined maximum value.

Although a tachogenerator has been shown in Figure 14.19 as providing the feedback speed signal, it is not the only system that can be used. For instance, equations (14.1) and (14.2) indicate that the motor speed is proportional to the supply voltage provided allowance is made for the drop in the motor armature resistance. Since this resistance can be measured, all that is required is a knowledge of the armature current at any supply voltage. Therefore the feedback speed signal can be made proportional to the supply voltage on which is superimposed an I_aR drop compensation signal. Since this signal is proportional to armature current it may also be used in the current limit loop if required.

For a certain load torque, motor speed will be such that the resultant current interacting with the field will give a machine torque just sufficient to overcome the load, plus internal losses. Speed control is therefore a form of torque control. In a d.c. motor, if the armature voltage is reduced, but the load torque kept constant, speed will fall so that the back e.m.f. reduces sufficiently to increase the current to almost its original value (flux being kept constant) and give an unchanged machine torque. Similarly, a reduction in field flux would cause a reduction in speed so that the increased current can give the same torque, the flux-current product tending to be maintained constant. If the load torque was allowed to fall as flux decreased, motor speed would increase to compensate for lost torque and so keep the power constant.

Depending on the type of supply, either a form of controlled rectification or d.c. line control may be used. Figure 14.20 shows a system which uses a unidirectional thyristor converter to control the voltage applied to the motor armature. The field may be supplied either from a separate bridge rectifier, or as in Figure 14.20, where diodes D_1 and D_2 are used to pass current to both the armature and the field. The field is supplied by the diode bridge D_1-D_2-D_4-D_5. Although this circuit economises on the number of devices used, the ratings of diodes D_1 and D_2 have been increased, and for large machines a separate field rectifier system is to be preferred.

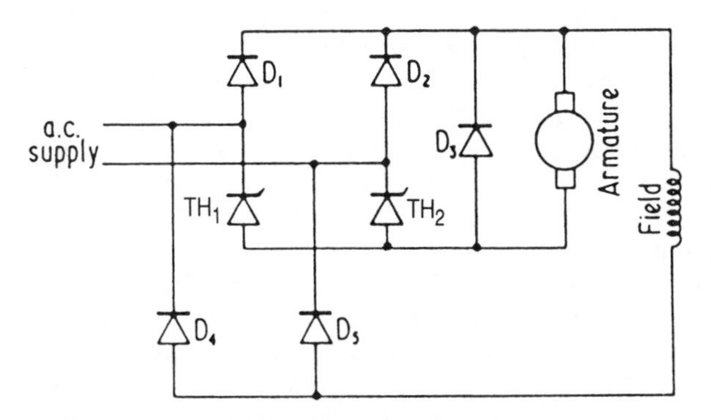

Figure 14.20 Separately excited motor speed-control system

Figure 14.21 shows a simple half-wave series motor drive which has found use in small domestic control systems such as food mixers. The circuit also has an element of feedback speed control, as will be evident from the description of its operation. When line A of the input goes positive to B, capacitor C begins to charge via R_1 and the motor. As soon as the capacitor voltage reaches the breakdown value of trigger diode D_4 it breaks over and fires thyristor TH_1. Load current now flows from the a.c. supply to the motor for the remaining half cycle. When the supply polarity reverses TH_1 turns off and motor load current free-wheels in diode D_1. If the motor is heavily loaded then diode D_1 is forward biased for the whole time that TH_1 is off. This means that when C commences to charge its time constant is determined primarily by the value of R_1, the motor resistance being small. Since the speed of a series motor varies substantially with load this would mean that when the load decreased the motor would speed up, even though the value of R_1 remained unchanged. The load current would

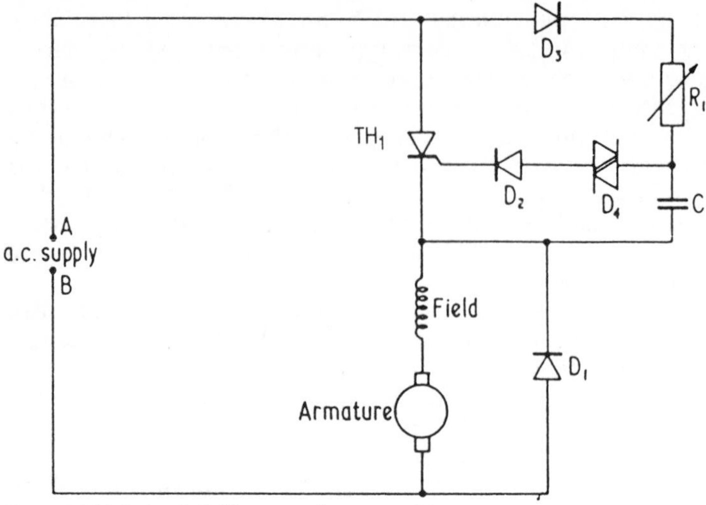

Figure 14.21 A simple half-wave series motor drive

now tend to go discontinuous during the half cycle, so that D_1 no longer clamps the motor voltage. This indicates that the charging time of C will be affected by the motor back e.m.f., which is developed on residual motor flux. Since this e.m.f. increases with speed the capacitor would take longer to reach the trigger voltage of D_4. This effectively increases the firing delay of the half-wave cycle and reduces the motor voltage. The motor speed will therefore fall and remain relatively stable about the value determined by the setting of R_1. Therefore R_1 represents the demand speed and motor back e.m.f. is the feedback speed signal.

Although it is usual to control the armature voltage on a d.c. motor, equation (14.2) shows that speed variations can also be obtained by keeping the armature voltage fixed but changing the field voltage, and hence flux. Figure 14.22 shows such a system where a bi-directional converter is used to supply the field. This has the advantage of being capable of regenerating the highly inductive current stored in the field back to the supply. This facility is often required in fast response drives, such as in Ward–Leonard schemes with power electronic control of the generator field current. Generally, a field control is used only where a narrow speed range, say 4:1, is required. This is because at high flux densities the motor

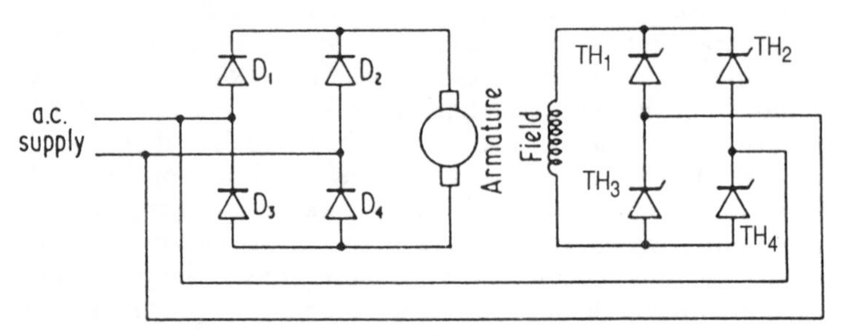

Figure 14.22 Field control system for limited motor speed variation

poles would saturate, whereas at low densities problems would be experienced with motor commutation due to the predominant effect of the armature reactance. No such restriction is placed on armature voltage-control, and with this method drives having speed ranges of 100:1 are quite common.

In the discussion so far it has been assumed that the motor is being supplied from an a.c. source, which is the usual industrial drive since an a.c. supply is almost universally available. For many applications, however, especially in battery vehicles, only a d.c. supply is present, and the motor must now be driven by some form of chopper arrangement. Most of the circuits described in Chapter 12 can be used for chopper drive of a d.c. motor, although a few, such as the circuit shown in Figure 14.23, are not suitable. Referring to the load voltage waveform, suppose that at

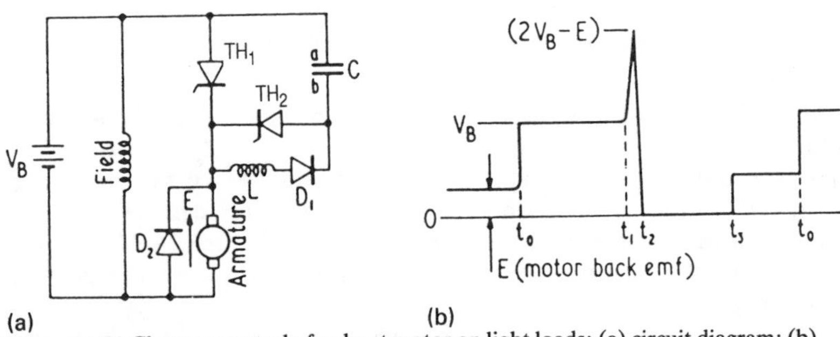

Figure 14.23 Chopper control of a shunt motor on light loads: (a) circuit diagram; (b) load-voltage waveform

time t_0 thyristor TH_1 is fired. The load voltage rises to V_B and power is supplied to the motor. From a previous cycle C was charged to $V_B - E$ with plate a positive, where E is the motor back e.m.f. at this speed. When TH_1 is fired the capacitor resonates with L and recharges with plate b positive, so that when TH_2 is fired, thyristor TH_1 is reverse biased by $V_B - E$ and turns off. Motor current flows in D_2 and C charges to V_B with plate a positive, the voltage drop in D_2 being ignored, the load voltage being zero as at t_2. On light loads the inductive load current will decay to zero, the current flowing against the motor back e.m.f., before TH_1 is refired. When this occurs, say at t_3, the capacitor will discharge through D_1 and its voltage will fall to $V_B - E$. The motor back e.m.f. has therefore had two effects. First, it has distorted the load voltage waveform between heavy loads and light loads. Since the magnitude of the load voltage is determined by its waveform this means a change in the motor voltage and hence its speed, even though the firing period of the thyristors has been unaltered. This is not serious and can be compensated for by using closed-loop speed control, as in Figure 14.19. A much more serious effect of the back e.m.f. has been a reduction in the available thyristor commutation voltage, and although this voltage is reduced most on light loads, when commutation is least demanding, it is a disadvantage of this circuit. A much better solution is to replace D_1 by a thyristor, which turns off as soon as capacitor resonance

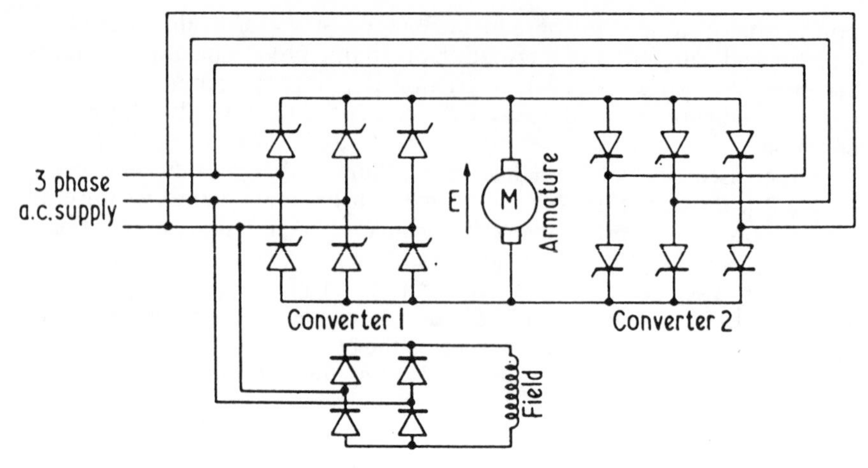

Figure 14.24 Reversible three-phase thyristor controller for a d.c. motor

has been completed, and so prevents the motor back e.m.f. from affecting the capacitor voltage.

Apart from controlling the speed, it is often necessary to be able to reverse the direction of rotation of a d.c. motor. This can be done by electromechanical contactors, if required, although Figure 14.24 shows a reversible converter arrangement in which the motor direction can be readily changed by operating one or other of the converters. It is now necessary to have all elements of the converters controllable, that is, to use bi-directional converters only, so as to avoid short-circuits between supply lines. This converter arrangement was introduced in Chapter 10, in relation to cycloconverters, and it was shown that there are two ways in which it may be operated. In the first method both converters are fired simultaneously but with delay angles $\alpha_1 + \alpha_2 = 180°$. This gives equal mean output voltages from both converters and they can readily be changed from one direction of operation to another by adjusting the delay

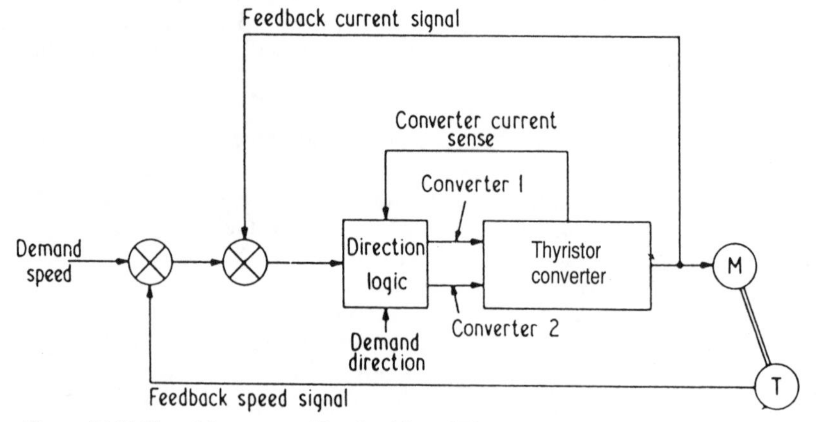

Figure 14.25 Closed-loop control for the drive of Figure 14.24

angles. However, due to instantaneous voltage differences between the converters it is necessary to introduce a reactor between them in order to limit the circulating current flow. A better control method, in the present application, is to sense the converter current direction and to ensure that only one converter is fired at any time.

Figure 14.25 shows the closed-loop speed-control arrangement for such a system. In addition to demand speed there is also now a demand direction, which is used to operate one of the two converters to change the polarity of the armature voltage and hence the motor direction. Feeding into the same direction logic circuitry is a signal which senses the direction of current flow in the converters, and ensures that the changeover from one converter to another does not occur unless the current in the previously conducting converter has fallen to zero. In addition, the magnitude of the motor current can be sensed to provide current limiting as before.

14.3.2.3 Braking

In a purely mechanical method of braking the kinetic energy of a revolving motor may be dissipated as heat by friction brakes. Electrical methods often provide considerable advantages over mechanical methods and three such techniques are discussed in this section.

Plugging

The direction of rotation of the motor is rapidly changed by reversing the supply voltage polarity, the kinetic energy of the motor being dissipated as heat in the motor itself. Figure 14.26 illustrates the principles of plugging using field reversal. When the motor field current is suddenly reversed

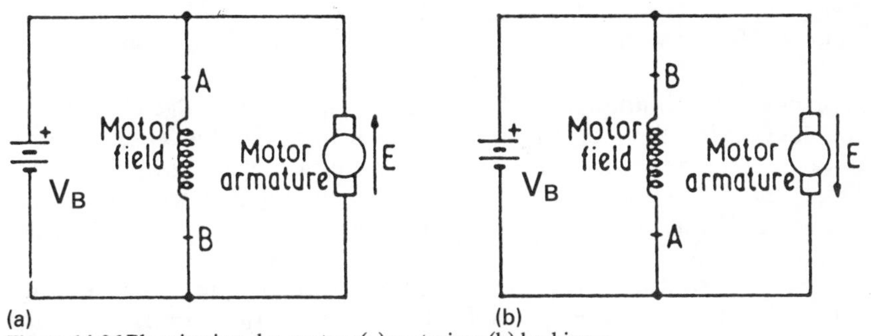

Figure 14.26 Plugging in a d.c. motor: (a) motoring; (b) braking

while the machine is still revolving in the same direction, the d.c. source will oppose the motor rotation and try to turn it in the reverse direction. Since the motor back e.m.f. has reversed due to the reversal of field current, it will help the d.c. source in circulating a large current through the motor. This current exceeds the stalled motor current and gives rise to severe heat losses in the motor, bringing it to a rapid halt. If the supply is still maintained the motor will then commence to rotate in the reverse direction.

The dual converter system shown in Figure 14.24 can readily be extended to plugging. Suppose converter 1 was on and the motor was rotating in the forward direction, the motor back e.m.f. being as shown. To plug the machine converter 1 is switched off and converter 2 operated such that the converter voltage aids the motor back e.m.f. in circulating a large current through it. The converter firing angle delay should not exceed 90° since regeneration would then occur, as explained below. Controlled plugging is clearly possible by varying the delay angle, it being greatest at $\alpha = 0°$.

Dynamic braking
The kinetic energy of the motor is dissipated as heat in the motor itself, or in external resistors. Dynamic braking differs from plugging since the motor now acts as a generator and feeds current into an external resistor or circulates it around its own windings, such a system being shown in Figure 14.27. During motoring thyristor TH_4 is off and the three-phase bridge is

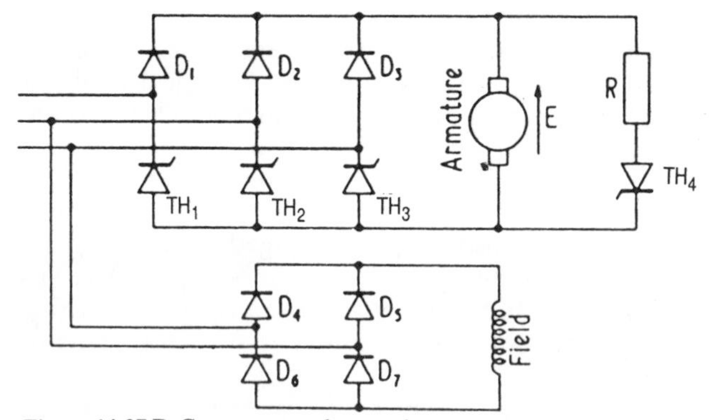

Figure 14.27 D.C. motor speed-control system with provision for dynamic braking

used to control motor speed. To brake the motor the bridge converter is switched off but the motor field current is maintained. Since the motor is still revolving in the same direction its back e.m.f. is as shown. Thyristor TH_4 is now fired. The motor acts as a generator and feeds a current through R, the motor kinetic energy being dissipated as heat in this current path.

A chopper-controlled d.c. motor drive is shown in Figure 14.28 in which battery V_B supplies power to the load, which is a series motor. TH_1 is the main thyristor controlling this motor whilst C, L, TH_2 and TH_3 are its associated commutation circuitry. The inductive energy of the motor is carried by free-wheeling diode D_1. Across the armature is connected diode D_2 and resistor R_{DB} which constitute the dynamic braking circuit. R_{CL} is a low-valued current-sensing resistor which is used primarily to feed back the value of the load current to current limiting circuitry.

The field of the motor is connected between two reversing contactors. When both forward F and reverse R contactors are de-energised, contacts 1 and 3 are closed and 2 and 4 are opened, as illustrated in Figure 14.28. No current can flow through the motor even if TH_1 is triggered.

If contactor F or R is energised the state of contacts 1 and 2 or 3 and 4 will be changed so that current can flow through the motor. The current

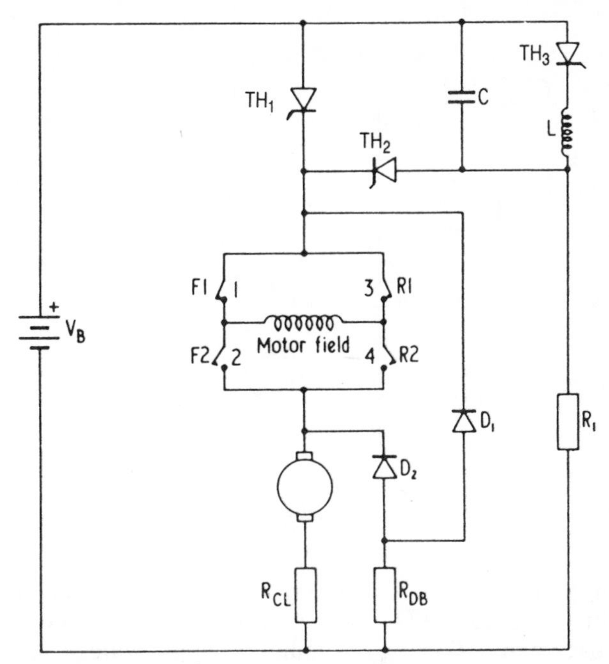

Figure 14.28 Reversing and dynamic braking in a chopper-controlled d.c. motor drive

through the field is in the reverse direction when F or R is energised. It is therefore possible to run the motor in either direction.

Assuming the motor is running in the forward direction, contactor F is energised and R de-energised so that contacts 3 and 2 are closed. The back e.m.f. of the motor opposes the current and diode D_2 is reverse biased. If now the motor is thrown into reverse by de-energising contactor F and energising R, so as to reverse the field current, the back e.m.f. will reverse since the load inertia is still keeping the motor armature rotating in the forward direction. Diode D_2 is forward biased and the kinetic energy of the load is dissipated as heat in the motor armature and the dynamic braking resistor. During braking thyristor TH_1 can still be switched, as in normal motoring operation, to control the field current. The greater this current, the larger the back e.m.f. of the motor and the more severe the braking. After the load has been brought to a standstill, if the power is still applied, the motor will move in the reverse direction in a normal motoring mode.

Regeneration

During regenerative braking the motor acts as a generator and the kinetic energy of the motor and its load is recovered and may be used again. Referring to Figure 14.24, suppose the motor is running in the forward direction, converter 1 being on. To generate, converter 1 is turned off and 2 is turned on. If the delay angle of this converter exceeds 90° there is net regeneration, this being a maximum for $\alpha = 180°$.

Figure 14.29 shows a regenerative system for a chopper controller. During motoring thyristor TH_2 is off and TH_1 operates in the usual on–off

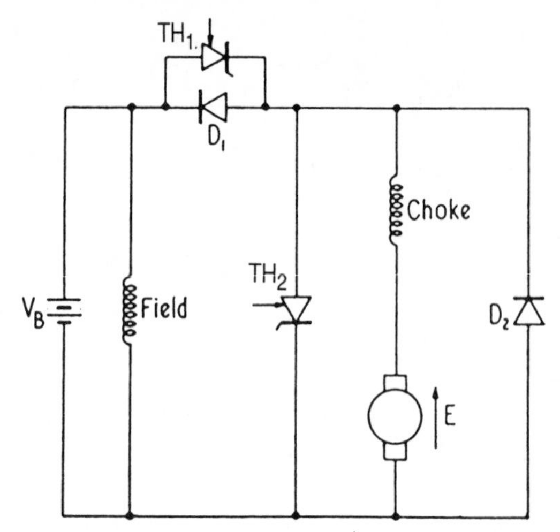

Figure 14.29 Motoring and regenerative braking of a d.c. motor

mode. Arrows on both thyristors indicate that they are forced commutated by circuitry which is not shown. Diode D_1 is also reverse biased throughout the motoring period, back e.m.f. E being less than the supply voltage V_B. Although regeneration could be obtained by switching off TH_1 and strengthening the motor field to increase the value of E, such a system has a limited control range since soon the field would run into saturation. A better system is to maintain the field constant but to operate TH_2 as a chopper during regeneration. Thyristor TH_1 and diode D_2 now no longer play a part in the system, which is essentially a step-up chopper, as described in Chapter 12.

Although regeneration is the best system of braking, due to its high efficiency, it is much more complex and expensive to obtain. Dynamic braking is inefficient, but by the use of external resistors the motor losses are minimised. It is important to appreciate that since both these systems rely on a generated output from the motor they are not effective at low speeds, where the motor back e.m.f. is small. Field strengthening can now be used to reduce this problem but it will not overcome it altogether. In plugging, on the other hand, the motor direction can be reversed if necessary so that it is effective right down to standstill. For quick-response systems it is therefore clearly the preferred method. A system such as that shown in Figure 14.24 is advantageous since regeneration can be used at higher motor speeds and then the firing angle advanced to below 90° in order to plug the motor to rest.

14.3.3 Electronic commutator d.c. motors

The d.c. motor is by far the most popular variable-speed machine, due to its excellent speed–torque characteristics and variable-speed capability. Its one major drawback is its mechanical commutator which places design and environmental limitations on its operation. It is feasible to use an a.c. motor drive system when a mechanical commutator is undesirable but, as is shown later, a d.c. motor has many advantages over a.c. motors in some applications. It is in order to maintain these characteristics, whilst overcoming the limitations of a mechanical commutator, that electronic commutator, or brushless, motors are used.

In this section the construction of electronic commutator d.c. motors is described in more detail. These motors are not commonly used in large sizes but have gained wide acceptance in high-performance small motor drive systems. Transistors are now the switching device most commonly used.

14.3.3.1 The electronic commutator

In a d.c. motor torque is produced by the interaction of the stator and rotor fields. Usually the stator contains salient poles and interpoles, if provided, which are energised by a field coil, and the rotor carries armature current supplied by an external d.c source. To maintain unidirectional torque between the rotor and stator it is necessary to switch the rotor current periodically so as to keep the two fields as closely perpendicular to each other as possible. A conventional d.c. machine uses a mechanical commutator and brushes to achieve this. The commutator switches the current, this occurring at the correct instant in the rotor position by virtue of the commutator location, and the greater the number of commutator segments, the smoother the torque produced.

A two-segment motor would have a torque–position characteristic as shown in Figure 14.30(a), the peak torque occurring when rotor and stator

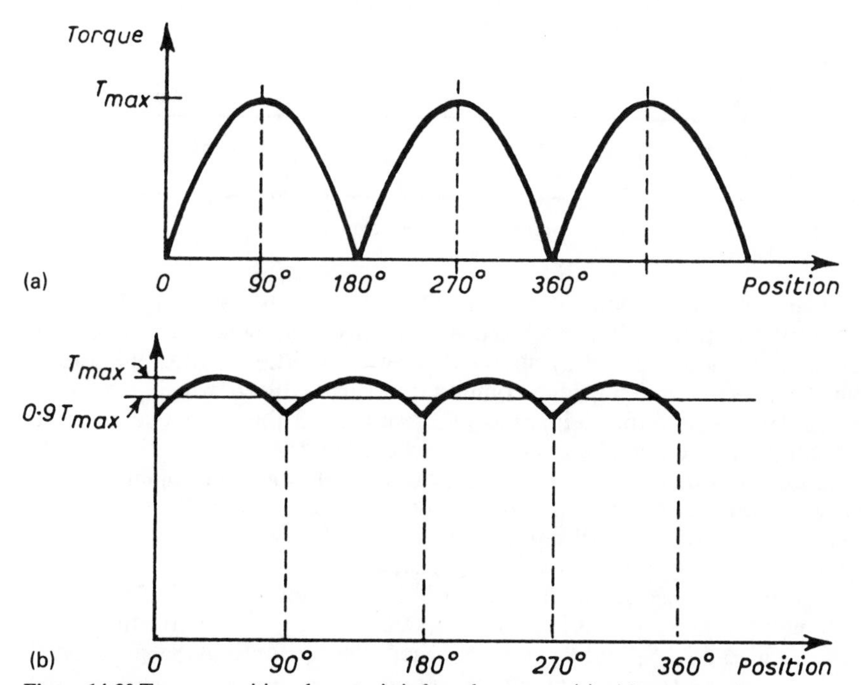

Figure 14.30 Torque–position characteristic for a d.c. motor: (a) with two-segment commutator; (b) with four-segment commutator

fields are at 90°E and falling to zero when this angle is reduced to zero. If four segments are provided, it is possible to switch rotor current four times per revolution so that the stator field is always maintained between 45°E and 135°E ahead of the rotor field, the torque curve being as in Figure 14.30(b). The mean torque is now 90% of the peak value and gives a much smoother operation.

It is possible to design a d.c. motor in which the rotor has permanent magnet poles and the stator current is switched. This would require the brushes to rotate at the same speed as the rotor to maintain unidirectional torque. Although such a system is undesirable in conventional machines it can be used with advantage in electronic commutator motors, since the stationary commutator makes it easier to operate the electronic devices. Figure 14.31 shows such an inverted machine which contains a multi-segment commutator. With rail A or rail B positive the polarity of the stator flux can be readily controlled by closing the appropriate switches. As with most electronic commutator machines, the rotor is salient pole in construction and the field contribution to the air gap flux predominates over the armature contribution, so that the effect of armature reaction becomes negligible.

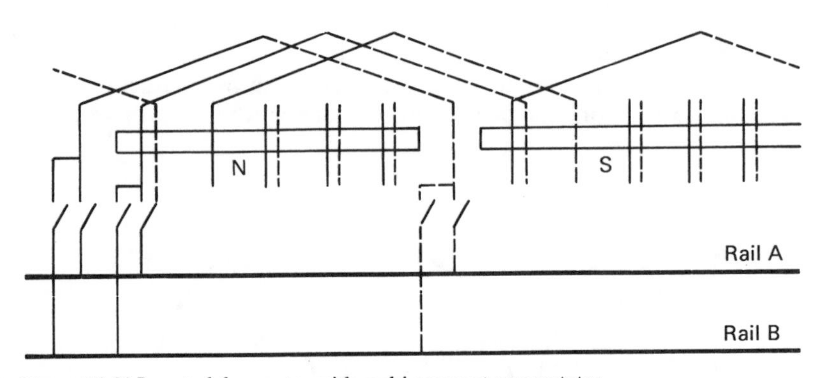

Figure 14.31 Inverted d.c. motor with multi-segment commutator

Although the circuit of Figure 14.31 has been used in electronic commutator motors, it uses a large number of switching devices and can be expensive. Figure 14.30(b) shows that, with a little sacrifice in torque smoothness, a four-segment commutator can be used which would be much more economical. Figures 14.32 and 14.33 illustrate two possible winding arrangements for such a machine, and also show the switch-operating periods. It is seen from these that although the open-winding arrangement uses only half as many switches as the closed-winding machine, its winding utilisation is only one quarter. Therefore it would generally be physically larger and less efficient.

Figure 14.32 also shows that the switching action is very similar to that obtained by a two-phase inverter where S_4, S_7, S_3 and S_8 constitute one phase and S_2, S_5, S_1 and S_6 give the other. Thus essentially, as mentioned earlier, there are two functions required to be performed by the commutator in a d.c. machine. First, it must sense the relative position of

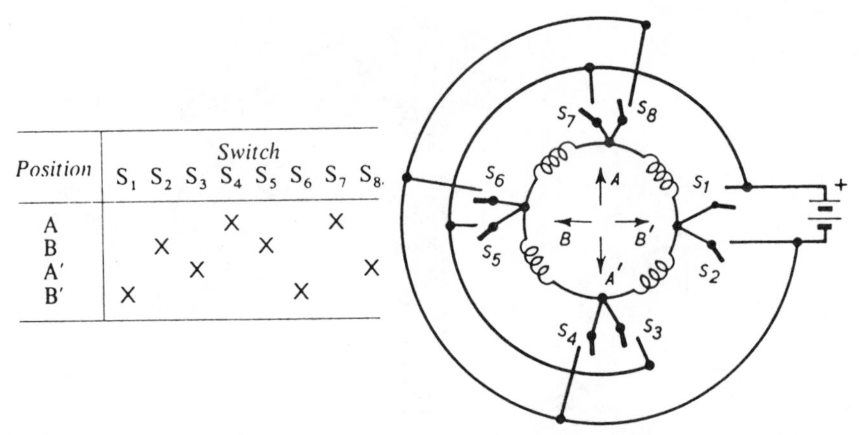

Position	S_1	S_2	S_3	S_4	S_5	S_6	S_7	S_8
	Switch							
A				X			X	
B		X			X			
A'			X					X
B'	X					X		

Figure 14.32 Closed-winding arrangement for a four-segment electronic commutator. The table gives the switching sequence

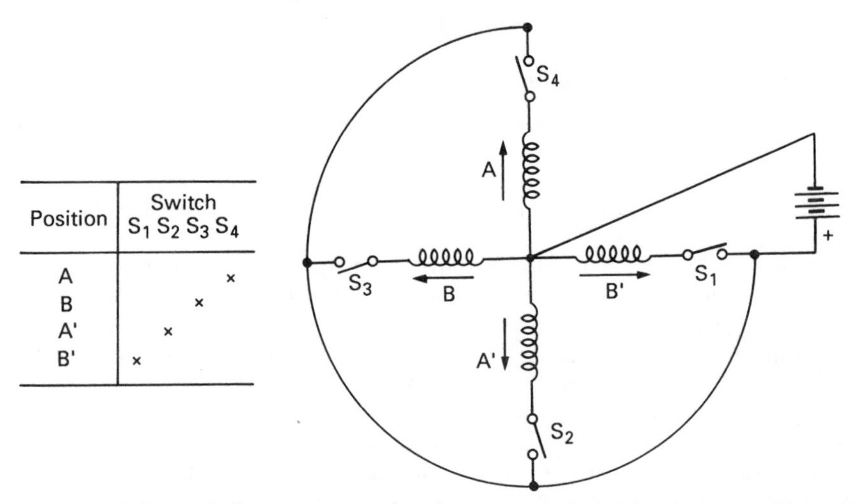

Position	S_1	S_2	S_3	S_4
	Switch			
A				×
B			×	
A'		×		
B'	×			

Figure 14.33 Open-winding arrangement for a four-segment electronic commutator. The table gives the switching sequence

the stator and rotor fields, and second, it must switch the stator field current at the appropriate instance. We will now examine these functions in detail and see how they are performed by an electronic commutator.

14.3.3.2 Position sensors

Rotor-position sensors are required to detect the instant at which the stator current is to be switched. There should therefore be as many position signals as there are commutator segments. A four-segment motor requires switching signals at 0°*E*, 90°*E*, 180°*E* and 270°*E*. Many different types of devices may be used, those most commonly employed in electronic motors being magnetic, optical or based on the Hall effect, so only these three are considered here. In all cases it is important that the detector gives a signal

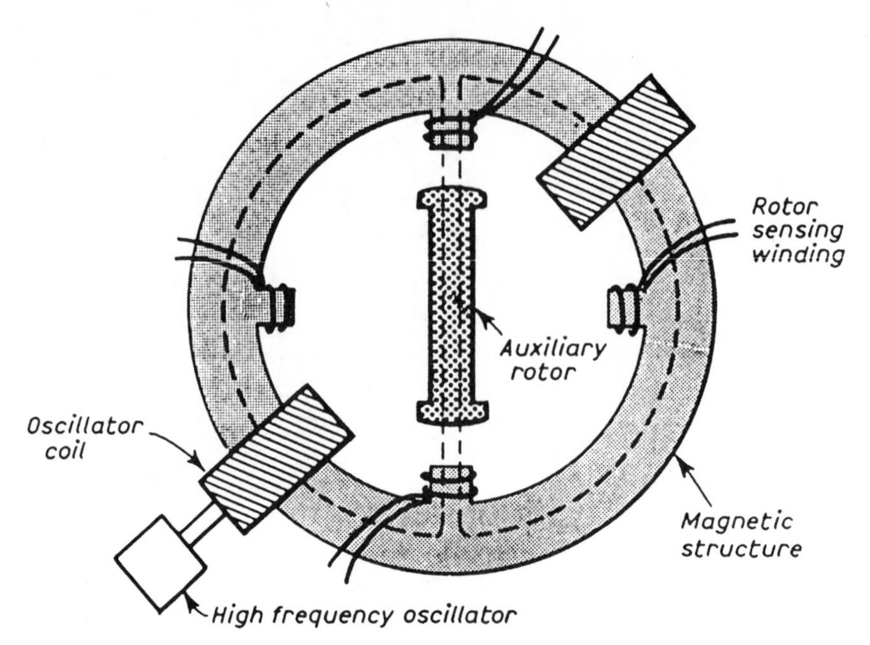

Figure 14.34 Magnetic position sensor

regarding the position of the rotor and not its speed or direction of rotation.

Figure 14.34 shows one form of rotor-position sensor which has been used. Attached to the main rotor is an auxiliary rotor which revolves in a four-pole stator yoke. On each of the poles, which correspond to the four switching points of a four-segment motor, is wound a sensing coil which feeds the respective stator winding switch. A high-frequency oscillator supplies two coils placed at diametrically opposite ends of the yoke and arranged to produce antiphase signals. Clearly, a voltage will be induced in only those coils which are under the rotor poles, so that the system produces rotor position signals.

There are several possible modifications to this arrangement. One system uses four sense coils embedded in the main frame of the stator, the coils being supplied in parallel by a high-frequency oscillator. A series-resonant capacitor gives a sense voltage across the coils unless it is saturated by the close proximity of a rotor pole. The absence of a pick-up signal at any coil therefore indicates that the corresponding stator switch is to be operated.

Magnetic sensing devices are very robust and are not affected by dirt or dust. They require minimal auxiliary components when the main rotor poles are utilised as part of the sense system, and the signal output can normally be used to operate the stator switches directly without further amplification. However, the induced signals in the coils build up gradually, this giving slower turn-on and greater dissipation in the switches, as well as possible uncertainty in the exact switching point. The presence of leakage flux also means that there is always some induced signal in the coils. Since all signals are a.c. they have usually to be rectified before being used to operate the switches.

Optical sensors are generally less robust than magnetic systems but are also much smaller and lighter. They have therefore tended to be used in motors for special applications such as aerospace. Figure 14.35 shows a schematic of an optical system where a light shield with an aperture is connected to the rotor and revolves around a stationary light source. Photocells placed at the four switch points detect the commutation positions for the motor, as before. Generally, the output signals from the light detectors are weak and need further amplification before they can be used to operate the stator switches. However, the signals can be made to rise sharply and since they are generally d.c. no rectification is required.

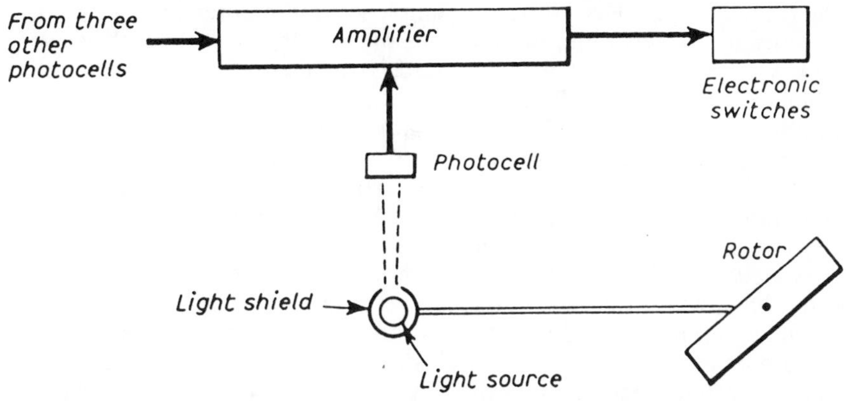

Figure 14.35 Block schematic of an optical position sensor

A sensor which has been extensively used in many small electronic commutator motors is the Hall effect device. It combines the robustness of magnetic sensors with the lightness of optical devices. The Hall effect is well known. A current I_c passed between the two faces of a thin conductor or a semiconductor placed in a transverse magnetic field B would result in a redistribution of charge carriers within the device and induce a voltage across it in a direction perpendicular to both the current flow and the magnetic field. This voltage, known as the Hall voltage, is proportional to I_c and B.

Hall generators placed at the commutator switching positions would therefore react with the rotor field to produce the required position signals. A further advantage of the Hall detector is that the voltage reverses with the direction of the magnetic field so that only two sensors at 90°E to each other need be used for a four-segment commutator.

In addition to the advantages of small physical size, ruggedness and direction sensitivity, the output from a Hall generator is d.c. and so does not need rectification. The disadvantage of the system is that an auxiliary source is required to provide I_c and that, like a magnetic sensor, the output voltage is always present due to leakage flux, and it builds up gradually.

It is possible to use the Hall voltage to feed the stator coils directly, instead of via a switch, but now the generator needs to handle much higher power levels. The Hall generator is predominantly resistive, its efficiency being relatively low, so that motors built on this principle would also be very inefficient and uneconomic for all but the smallest machines.

The choice between the various rotor position sensors is usually determined by the motor application. Large machines use magnetic sensors since efficiency and switching power requirements are important, although optical sensors may be used where robustness is not essential. For smaller motors Hall generators give a small and relatively cheap system. For higher efficiencies magnetic systems have been used and optical sensors are chosen where lightness is equally important.

14.3.3.3 Electronic switches

The switches used to change the direction of stator current, at a signal from the rotor-position sensors, need to satisfy several requirements. First, they must be adequately rated to handle the full winding power, which can vary widely with the size of machine used. It is also important to establish whether the current through the switches will be unidirectional or bi-directional. A mechanical commutator can readily pass current in either direction whereas electronic switches are usually restricted in this aspect, therefore special provisions must be made if bi-directional operation is required. This often arises when the same machine is to act as a motor and a generator and also in some closed stator winding arrangements.

A mechanical switch can only be off or on, but an electronic device can often have a controlled turn-on, as in a transistor. This means that, for instance, in Figure 14.36 instead of controlling the applied voltage (V_B) to the motor to change its speed, the value of this voltage can be kept fixed but the voltage across the switch (V_S) varied. Therefore an electronic switch can act in a switching mode, as for a mechanical commutator, or in an amplified mode, where the extent to which it is on or off is controlled, this mode having several advantages.

For instance, Figure 14.37 shows a four-segment commutator operated from Hall effect sensors. The currents through the sensors are constant, but if the rotor flux is assumed to be sinusoidally distributed, then the Hall

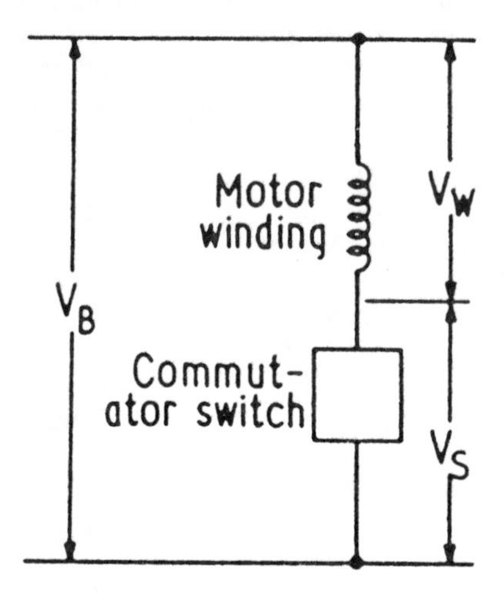

Figure 14.36 Distribution of voltage across the electronic commutator switch and the motor winding

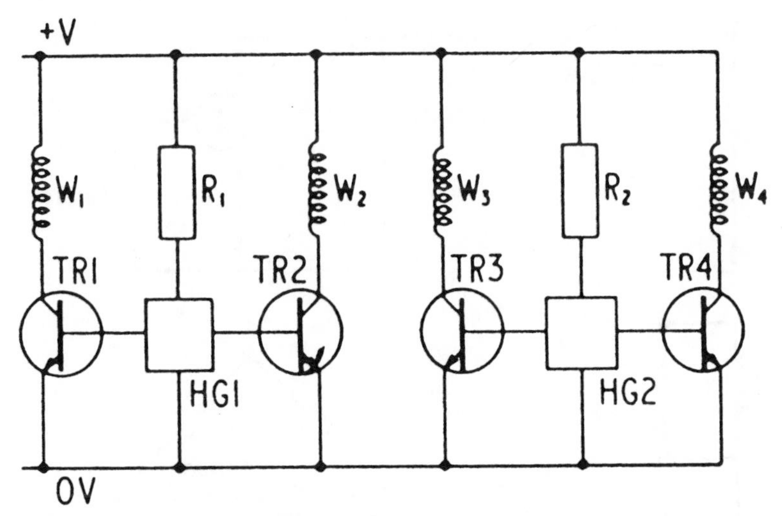

Figure 14.37 Winding arrangement on a four-segment Hall sensor electronic motor

voltage, being proportional to this flux, would cause the transistor to operate in an amplification mode. The stator current distribution in the four windings would then be such as to maintain the stator flux constant, irrespective of rotor position. Since motor torque is proportional to the product of stator and rotor flux, and these are now constant, such a motor is capable of extremely smooth operation. The amplification mode has one obvious disadvantage; the dissipation across the switches is much greater than if they were always either off or saturated, so its use is limited to relatively small motors.

There are two devices which have been predominantly used as switches in electronic commutator motors, namely thyristors and transistors. Where the power to be handled is large, thyristors are the obvious choice. The thyristor is not capable of amplification-mode operation, but at the high-power levels the low efficiency of such a system would be unacceptable. Another characteristic of a thyristor is that it is unidirectional in operation and for bi-directional switching two devices can be connected in a parallel back-to-back arrangement or a bi-directional device such as a triac can be used.

Transistors are much more versatile since they can be operated in amplification or switching mode, Figure 14.38 showing several arrangements that may be used. Figure 14.38(a) is intended for unidirectional operation, whereas Figures 14.38(b) to 14.38(d) give bi-directional capability. Since only one sensor is used in (c) and (d) these are generally the preferred arrangement.

The motor winding shown in Figure 14.37 is inefficient due to poor winding utilisation. Figure 14.39 shows an alternative arrangement which overcomes this disadvantage and also illustrates another form of closed winding. The table shows that the switches have purposely been operated so as to produce a stepped output voltage waveform, which reduces the motor voltage waveform harmonic content and gives greater efficiency.

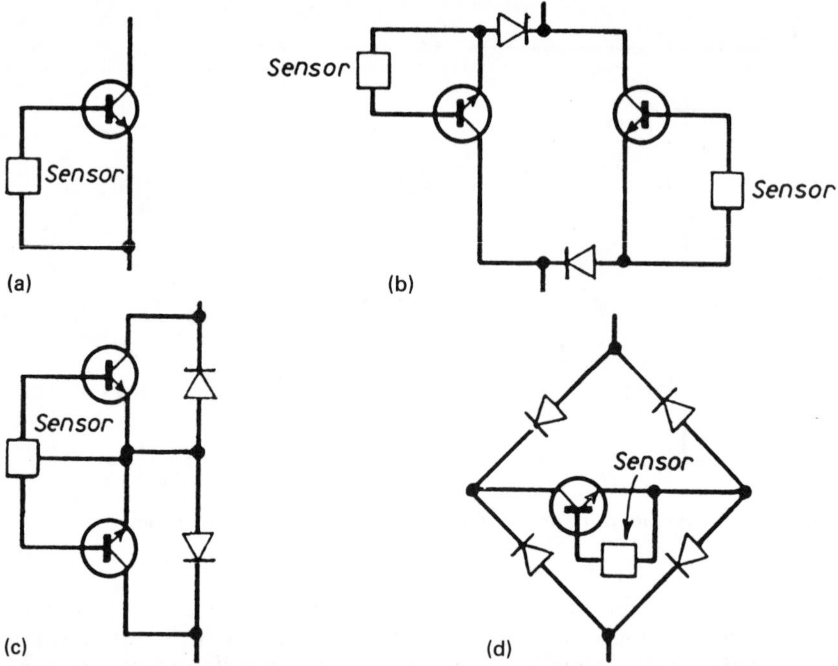

Figure 14.38 Electronic switching using transistors: (a) unidirectional operation; (b)–(d) bi-directional operation

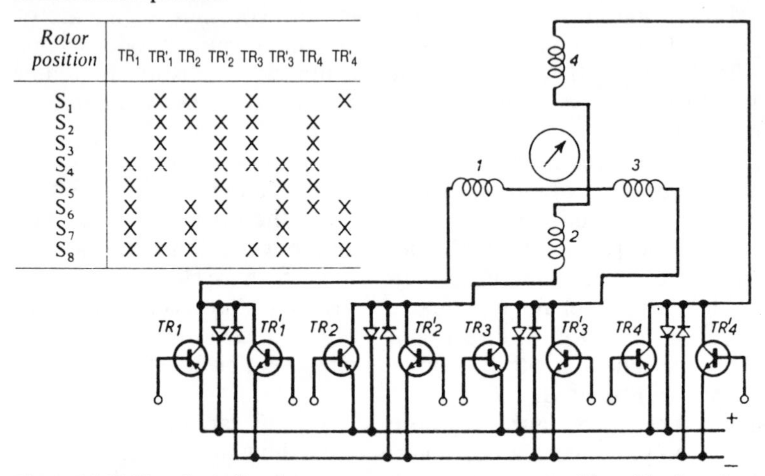

Rotor position	TR_1	TR'_1	TR_2	TR'_2	TR_3	TR'_3	TR_4	TR'_4
S_1		X	X		X			X
S_2		X	X	X	X		X	
S_3		X		X	X		X	
S_4	X	X		X	X	X	X	
S_5	X			X		X	X	
S_6	X		X	X		X	X	X
S_7	X		X			X		X
S_8	X	X	X		X	X		X

Figure 14.39 Closed-winding four-segment commutator motor. The table gives the switching sequence

14.3.3.4 Advantages of an electronic commutator

As mentioned earlier, a d.c. motor has very desirable characteristics for a wide range of applications, its major disadvantage being its mechanical commutator, which leads to the following limitations:

(i) The commutator and brushes are prone to sparking with a consequent production of radiated interference. This can be a serious hazard when operated in certain environments, such as in airborne equipment.

(ii) Sparking reduces the life of the commutator so that conventional d.c. machines need periodic servicing and brush replacement and cannot be mounted in inaccessible positions.

(iii) Usually the commutator is mounted on the rotating member, since an inverted motor would require the brushes to rotate. This limits the maximum speed at which it can run and therefore the power output from a given frame size.

(iv) The commutator also limits the machine design since it requires a large ratio of diameter to length of armature iron.

(v) The commutator is made from good-quality copper which softens at high temperatures and limits the machine-operating temperature. Dirt and moisture also attack the commutator and the machine needs to be specially protected in hazardous atmospheres.

(vi) Rubbing of the brushes on the commutator and clatter of the brushes in their holder give rise to a considerable amount of audible noise.

An electronic commutator motor maintains all the advantages of a traditional d.c. machine, such as high starting torque and excellent efficiency at all speeds, whilst avoiding the disadvantages of a mechanical commutator. In addition it can also produce smoother torques in small motors, due to the fact that the number of commutator segments in a small conventional motor is usually limited by physical size to six or eight, giving a relatively poor torque–position characteristic. An electronic commutator using four segments, on the other hand, can produce a much smoother torque by using its switches in an amplification mode.

The d.c. commutatorless motor gives a performance similar to that obtained from an induction motor driven by an inverter. From the hardware aspect both these systems are also very similar, the induction motor drive being slightly simpler due to the absence of the rotor-position sensors. There is, however, one important difference. The frequency at which the stator switches operate in an electronic commutator motor is determined by the rotor speed, as sensed by the position sensors, this speed being in turn fixed by the winding current. For an inverter drive, on the other hand, the switch frequency is fixed by an internal oscillator in the inverter. The motor merely follows this rotating field and speed control is achieved by controlling the inverter frequency and not the supply voltage. The following further differences can be noted between the two systems:

(i) The inverter of an induction motor drive could be located at some distance from the motor. This would require only two wires connecting the driver to the motor. An electronic commutator motor would need the sensing leads to be brought out as well, these leads then being exposed to pick-up effects.

(ii) Permanent magnet motors take a larger starting current than corresponding induction machines, which could be a serious problem since a current surge can cause demagnetisation of the permanent magnet. Smaller motors, with higher-impedance windings, are less prone to this effect.

(iii) Since the inverter can operate independently of the motor in an inverter drive, several motors can be run in parallel off the same inverter.

In general, the electronic commutator motor is best suited to applications requiring high starting torque, good efficiency at low speeds and continuous speed variation from standstill to full speed. Due to their poor efficiency at low speeds, induction motors driven from inverters are more suited for use in applications where the output is a fixed high speed, and where, by using crystal-controlled oscillators in the inverter and a synchronous motor, the speed regulation can be made exceptionally good.

14.3.4 A.C. motors

As outlined in Figure 14.12, there are three basic types of a.c. motors popularly used with power semiconductor drives, induction, synchronous and commutator. There is yet another type of motor, the reluctance machine, which has not been mentioned so far since it can be considered as a special form of synchronous motor.

Reluctance motors were not popular, since they suffered from several disadvantages, chiefly connected with the low power factor, but developments in new materials has resulted in machines with greatly improved performance. All reluctance motors have a certain salience in their rotors. The stator produces a rotating field similar to an induction motor, which causes the motor to run up close to synchronism by induction motor principles. Thereafter the salient poles lock in onto the stator field and it runs at synchronous speed, as in a synchronous motor, even though the rotor poles are not externally excited. It is this synchronous-induction

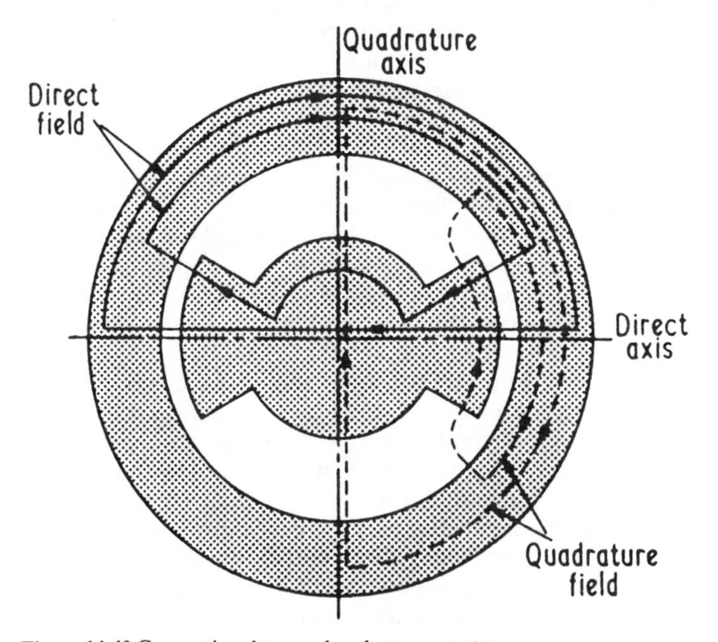

Figure 14.40 Conventional two-pole reluctance motor

action of the reluctance motor that has resulted in its use in applications such as multi-motor converter drives in the textile, steel and plastics industries. When used to drive an alternator it can also provide highly stable computer supplies.

To obtain a high power factor with good pull-in and pull-out torques it is essential to increase the ratio of the direct axis to the quadrature axis magnetising reactance of the motor. Figure 14.40 shows a conventional reluctance motor and indicates the paths of the direct and quadrature fluxes. To achieve a large value of direct axis reactance and a small value of quadrature axis reactance it is necessary to reduce the salient pole arc to as low a value as possible. Although this now increases the power factor it leads to greatly increased magnetising currents and reduces torques. A recently developed form of segmental motor which overcomes these disadvantages is shown in Figure 14.41 for a two-pole motor. Since all the

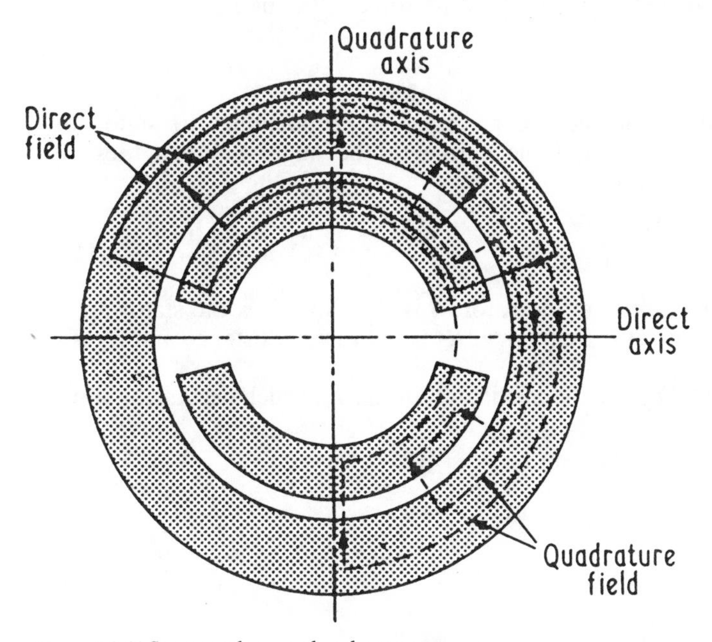

Figure 14.41 Segmental two-pole reluctance motor

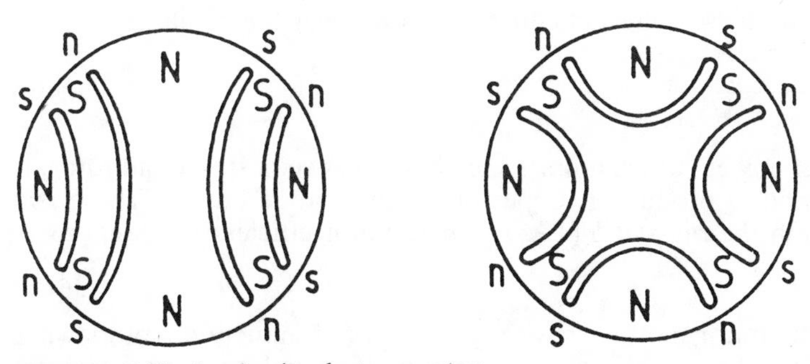

Figure 14.42 Flux barriers in reluctance motors

torque is experienced by the salient poles and none by the central rotor cylinder, this has been eliminated to give a motor with reduced inertia. In Figure 14.41 only the quadrature axis flux crosses the relatively large interpolar space so that its reactance is greatly reduced.

Instead of segmenting the rotor it is possible to attenuate the quadrature axis flux by using flux guides or barriers, as in Figure 14.42. Two forms are illustrated here, in both cases it being seen that the quadrature axis flux is decreased to a greater extent than the direct axis flux, due to the position of the guides.

It has been seen by the above discussion that a reluctance motor is in effect a synchronous machine with unexcited rotor poles. Both machines, on the other hand, behave like an induction motor during run-up periods, and can be considered as special cases of the induction machine. The torque on the rotor of an induction motor is caused by induced currents produced by a rotating stator field. If N_1 and N_2 are the speeds of this field and of the rotor respectively, then the slip S is defined by

$$S = \frac{N_1 - N_2}{N_1} \tag{14.5}$$

The magnitude of the rotor torque is proportional to stator flux, rotor-induced current and the cosine of the angle between these two phases, being given by

$$T \propto \frac{SE_1^2 R_2}{f_1(R_2^2 + S^2X_2^2)} \tag{14.6}$$

where T is the torque on the rotor, S is the slip, E_1 is the stator applied voltage, f_1 is the stator frequency, R_2 is the rotor resistance and X_2 is the rotor reactance.

Equation (14.6) gives the maximum torque T_m which occurs at a slip S_m, derived from

$$T \propto \left(\frac{E_1}{f_1}\right)^2 \tag{14.7}$$

$$S_m \propto \frac{R_2}{f_1} \tag{14.8}$$

$$X_2 \propto f_1 \tag{14.9}$$

Substituting these values in equation (14.6) gives the result as in

$$T = \frac{2\,T_m}{S_m/S + S/S_m} \tag{14.10}$$

The input power, which crosses the air gap between stator and rotor, is primarily used in producing copper losses and torque output. These two are divided in the ratio of $S/(1-S)$ so that motor efficiency is equal to

$$\eta_m = 1 - S \tag{14.11}$$

In a commutator motor, as with some types of induction motors when the motor winding is brought out to slip rings and brushes, a resistor connected in series with the rotor would result in power loss. This would mean that the output power of the machine is reduced and it must slow down to supply the same torque. A further advantage of commutator motors is that the frequency at the brushes is the supply frequency. There can therefore be a direct interchange of power between the a.c. lines and the rotor, as shown in Figure 14.43. If power is fed from the rotor to the

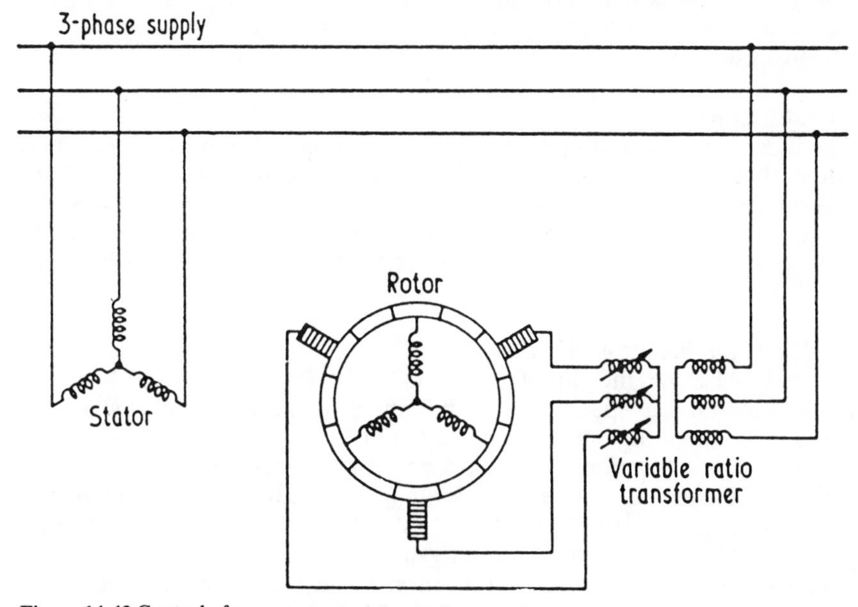

Figure 14.43 Control of an a.c. commutator motor

a.c. supply the motor will slow down, and if power is supplied into the rotor its speed will increase. The a.c. commutator is similar to an induction machine with the addition of a frequency converter in the rotor to allow interchange of power between rotor and a.c. line. These machines are larger, costlier and more difficult to design than other a.c. drive systems. They have mainly been replaced by d.c. motor and power electronic controlled rectifier drivers, and an a.c. commutator machine with a power controller is very rarely used.

14.3.4.1 Starting

Induction motors, and other a.c. machines which run up as induction motors, draw a large starting current when connected directly to a mains supply. This is due to the fact that there is considerable slip between the stator field and the rotor speed, so that at full stator flux there is a large induced rotor current, which is reflected to the stator windings by transformer action. To decrease the starting current one must reduce either the stator flux, by lowering the stator voltage, or the stator frequency, by a reduction of the supply frequency. The stator voltage must now also be changed in order to keep the machine flux constant. When flux reduction is required the stator voltage can be decreased by a.c. line control methods, as described in Chapter 8. Frequency control is only possible when the motor is running from an inverter or cycloconverter, the supply frequency being increased gradually as the motor runs up so that the peak current is limited. It will be seen later that there is now no loss of starting torque, as occurs with voltage-control systems, since the motor flux is kept constant by automatic adjustment of the supply voltage. When a variable-frequency supply is available for starting, synchronous and reluctance motors need not run up as induction machines since they can lock onto the stator field at low speeds, and then run up to full speed in synchronism.

14.3.4.2 Control

The most usual operating mode for synchronous and reluctance motors is when they are rotating at the synchronous speed of the stator field. Therefore the speed of these machines can only be varied by changing the frequency of the stator supply. The same considerations apply to an induction motor, but due to its construction there are four methods which may be used to control it, as follows.

(1) Excitation field strength control

Figure 14.44 shows the plot of equation (14.10), from which it can be noted that at synchronous speed, when $S = 0$, the torque is zero, and at standstill $S = 1$, the operating point being located at $1/S_m$.

Since S_m is a function of rotor resistance and reactance, the amount by which the torque–slip curve extends along the X axis is determined by the rotor construction. If $S_m = 0.2$ then at standstill $S/S_m = 5$ and the starting torque is 0.385 times the maximum value.

From equation (14.7), T_m is proportional to $E_1{}^2$, other parameters being fixed. Therefore, if stator voltage is halved the maximum torque is reduced by a factor of four, although Figure 14.44 remains unchanged since it is normalised to T_m. The starting torque will still be $0.385T_m$ but since T_m is one quarter of its original value the torque is also reduced to a quarter.

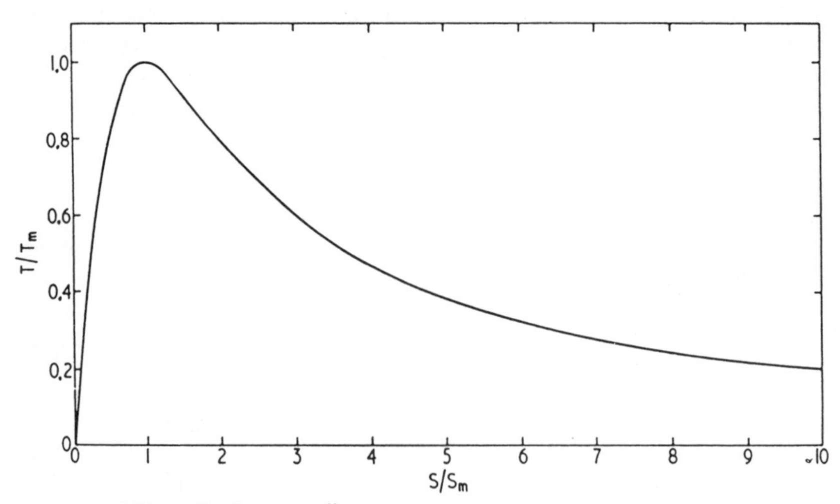

Figure 14.44 Normalised torque–slip curve

If the torque–slip (not normalised to maximum values) curve is plotted for variable voltage control, the characteristics, derived from Figure 14.44, would be as in Figure 14.45. For operation with a constant load (torque line AB) variation of the stator voltage from V_4 to V_3 will reduce the speed from E to F. Further reduction in voltage will reduce the available motor torque to below load torque and it will stop. Therefore the range of speed control available is limited. Operation on points I and J is not possible since an increase in speed results in an increase in available motor torque so that the motor will rapidly run up to the positive slope part of the characteristic.

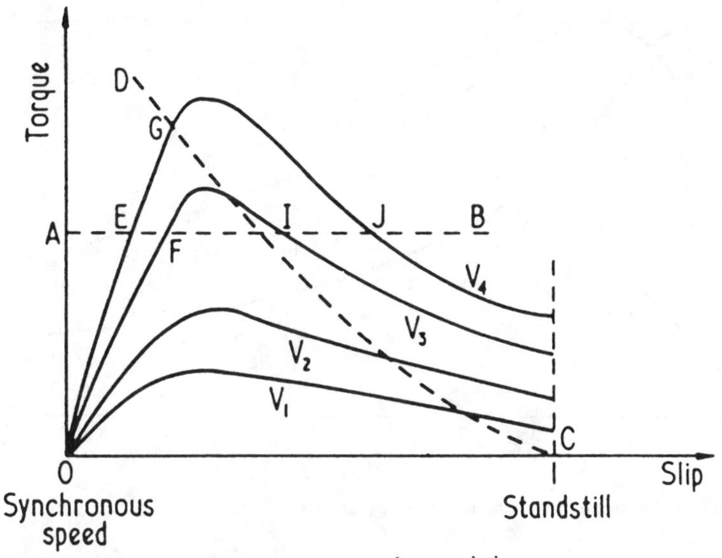

Figure 14.45 Torque–slip curves for stator voltage variation

A load having a characteristic such as DC (fan loads) is more suited to stator voltage control. On any part of the curve beyond the operating point, an increase in motor speed would increase the difference between the load torque and available motor torque and so cause the motor to slow down again. Therefore, the system is inherently stable and the speed can be controlled over a very wide range.

The main disadvantage of this system of speed control is the large slip at low speeds. From equation (14.11) efficiency is equal to $(1-\text{slip})$ so that for low speeds the efficiency is poor. At half speed the maximum efficiency is only 50%. Another consequence of this slip is that heat is generated in the motor, rotor loss being equal to S times the rotor input, so that at low speeds the heat can be excessive. Both these factors limit this method of speed control to small motors where efficiency is relatively unimportant and the heat easier to dissipate.

Voltage applied to the induction motor can be controlled by some form of line impedance such as saturable reactors, although these are bulky and relatively slow in response. A better technique uses thyristors in a.c. line control systems, as described in Chapter 8.

(2) Excitation field speed control

An alternative control method is to vary the speed of the stator field by controlling the supply frequency, but now the supply voltage must be changed in proportion to frequency in order to maintain motor flux constant at its most economical value.

If supply voltage E_1 is changed in proportion to the supply frequency f_1 then from equation (14.7) T_m is constant. However, from equation (14.8) S_m will vary inversely as f_1 so that as frequency is reduced the starting torque approaches the peak torque.

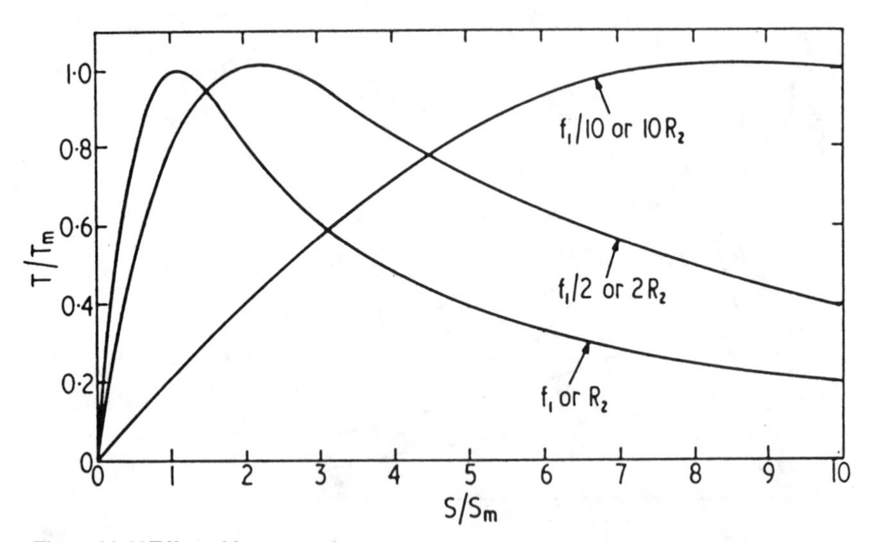

Figure 14.46 Effect of frequency change, or rotor resistance variation, on the normalised curve

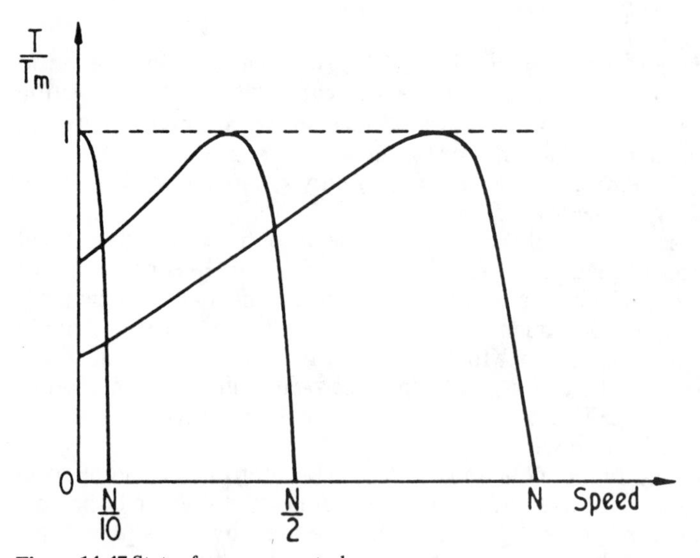

Figure 14.47 Stator frequency control

Equation (14.10) is used to calculate the torque for supply frequencies $f = f_1$, $f_1/2$ and $f_1/10$ where f_1 is the frequency used in Figure 14.44 and the results are given in Figure 14.46. From this it is evident that the available motor torque can be readily controlled to match the load and so give any desired operating speed. Furthermore, since the stator field speed need only be slightly greater than rotor speed, the resultant rotor current being just sufficient to overcome load torque, slip over the whole range can be very small giving high efficiency. In Figure 14.46 slip is plotted along the horizontal axis so that all the curves originate from the same point, but if torque is plotted against speed, as in Figure 14.47, the graphs terminate at their respective synchronous speeds.

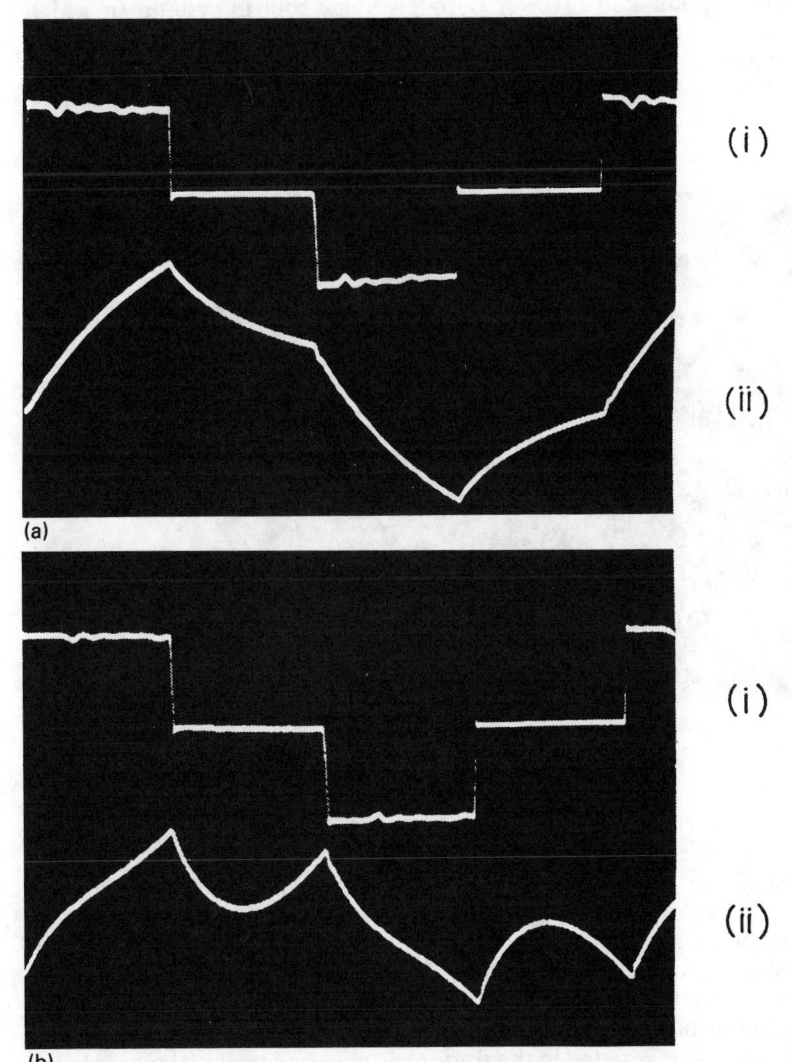

Figure 14.48 Motor waveforms for a bridge inverter using quasi-square wave voltage control: (a) motor stalled; (b) motor on no-load; (i) voltage; (ii) current

Due to the high efficiency at all speeds, variable-frequency control is suitable for motors of all sizes requiring infinite speed variation coupled with high performance. Frequency control can be accomplished statically by inverters or cycloconverters, as described in Chapters 10 and 13.

In Chapter 13 various forms of inverter voltage-control systems were discussed, and Figure 14.48 shows the motor voltage and current for quasi-square wave control. The stalled current corresponds to that discussed in Chapter 13, but when the motor is running, the back e.m.f. distorts this current waveform so that it is now difficult to determine equations for the ratings of the inverter design.

Figure 14.49 shows the output voltage and current waveforms for a pulse-width modulated inverter output voltage-control system, in which the current is seen to be relatively sinusoidal. Since the motor inductance is low a higher chopping frequency should be used to reduce further the high-frequency component in the current. This is evident from Figure 14.50, which also shows the current distortion caused by motor back e.m.f.

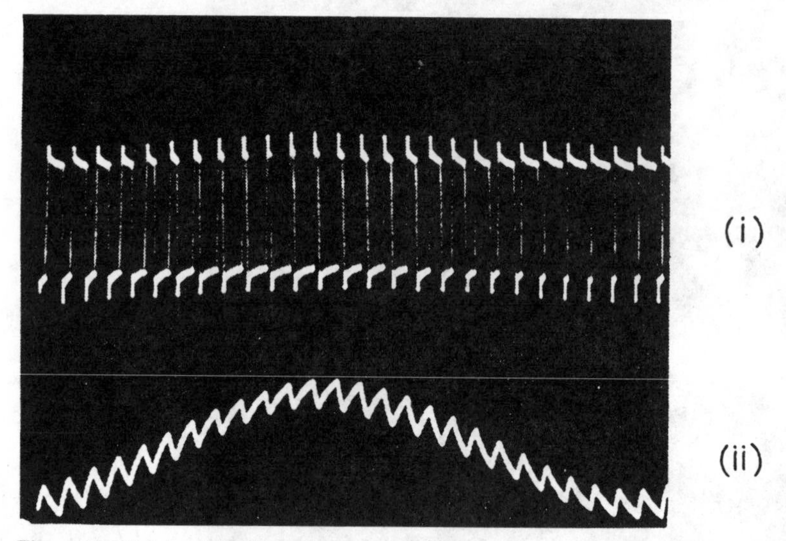

Figure 14.49 Motor-stalled waveforms for a bridge inverter using pulse-width modulation with a sine wave. $f_T = 555$ Hz, $f_S = 20$ Hz; (i) voltage; (ii) current

(3) Induced current control

In the previous sections the value of S_m was controlled by varying f_1. From equation (14.8) it is seen that the same effect can be accomplished by changing the rotor resistance R_2. Note, however, that S_m varies directly as R_2 but inversely as f_1 so that Figure 14.46 also represents the characteristics for variable rotor resistance control. There is one important difference between this method of speed control and that using variable frequency, since in this instance the stator field speed is constant so that at low speeds the slip energy is dissipated in the rotor resistance and the efficiency is low.

It is necessary to consider again the principle involved in the transfer of energy between stator and rotor. If the rotating field gives power P to the

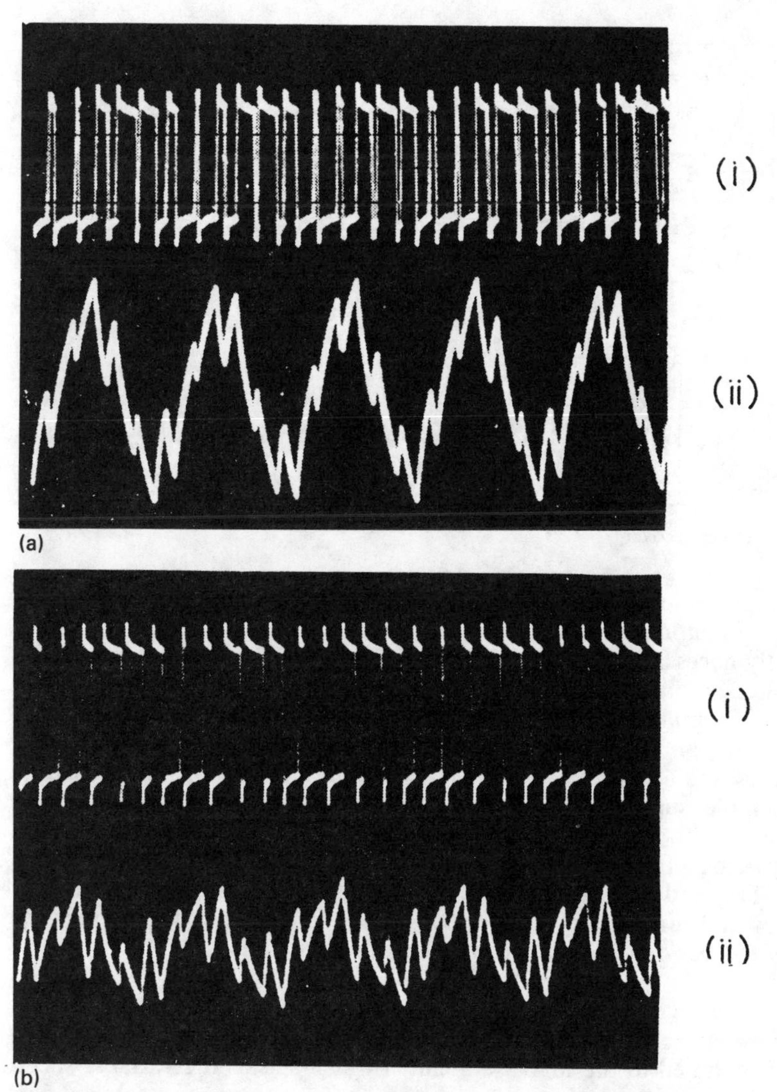

Figure 14.50 Motor waveforms for a bridge inverter using pulse-width modulation with a sine wave. $f_T = 555$ Hz, $f_S = 100$ Hz; (a) motor stalled; (b) motor on no-load; (i) voltage; (ii) current

rotor, then the quantity $(1-S)P$ is converted into mechanical energy at the shaft, the remainder (SP) being in the form of electrical energy in the rotor. If the rotor is closed on itself, or closed through a resistance, this energy is dissipated as heat, although if it is possible to extract the energy from the rotor, then the effect on the speed will be the same as if this energy were dissipated. The extracted energy can now be fed back to the supply to give a high overall efficiency. It is also possible to feed energy from an external source into the rotor, the slip now being made negative (rotor speed exceeding that of the stator field) and energy will flow from rotor to stator across the gap.

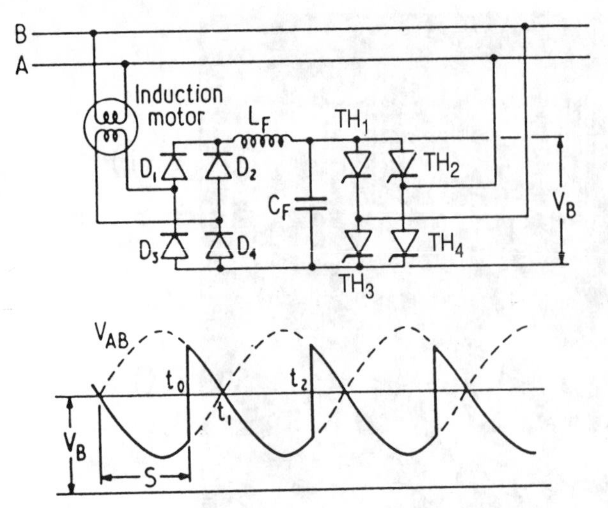

Figure 14.51 Slip recovery

A simple slip-recovery scheme is shown in Figure 14.51. Rotor voltage is rectified by diodes D_1 to D_4 and smoothed, although capacitor C_F is not normally necessary. Thyristor inverter TH_1 to TH_4 can be used to feed a controlled amount of the rotor energy back to the supply. As described in Chapter 9, unlike inverters used in frequency changers, this inverter connects to an a.c. reference line AB which determines its operating frequency and also provides the energy required to turn off the thyristors. For example, suppose TH_1 and TH_4 are fired at time t_0. The thyristors turn on and remain conducting even though the supply AB reverses at t_1, energy being fed from V_B to the a.c. lines over this negative voltage period. When TH_2 and TH_3 are fired at t_2 the voltage across the previous pair reverses and turns them off. The longer the firing delay, the greater the energy fed back from the rotor and the lower its speed.

(4) External clutch control

In this method of control the induction motor itself runs at a fixed speed, but it is coupled to the load via a clutch which acts as the variable-speed system.

Eddy current clutches are frequently used, in which an electromagnet is connected to the shaft of the induction motor and is rotated at the fixed speed. This produces a rotating field whose strength can be controlled by adjusting the coil current of the electromagnet. A disc is connected to the load shaft and is placed in the rotating field. Due to normal induction motor action the disc will rotate, the slip being adjusted by varying the magnetic field strength.

In effect there are now two induction motors in series. The first runs at a fixed (low slip) speed, whereas the second runs at adjustable slip. Clearly, the efficiency (1−slip) will be poor at low speeds and will be less than that of a single motor with slip control.

The advantage of using a clutch is that, since it dissipates most of the slip energy, it can be specially constructed to lose this heat more easily than if it were a motor although, because of the low efficiencies involved, this method of control is again limited to small motors.

Summarising the above, it can be seen that basically there are only two systems of speed control, (1) where the stator field speed is fixed, the slip being variable, and (2) where the stator speed is variable, the slip being small and relatively constant. The first method is inefficient. Voltage-control schemes are used in small systems with fan-type loads, or to provide constant torque over a limited range. To reduce dissipation in larger motors external eddy current clutches may be used. The efficiency of variable-slip systems may be increased by incorporating a form of slip-recovery scheme, but this necessitates the use of a wound rotor induction motor with associated slip rings and brushgear, which can make it unsuitable for certain applications, for example operation in hazardous atmospheres.

The most promising control system for a.c. motors is a variable-frequency drive. A cage rotor can now be used and efficiency over the whole speed range is also high. Such systems are normally the most expensive, but if a drive is required to provide a constant torque output for wide speed variations and the motor is located in an inaccessible position, variable-frequency control of a cage rotor machine is often the only practicable system.

14.3.4.3 Synchronous motor excitation and control

As stated earlier, the speed of a synchronous motor can only be controlled by means of frequency variation. Changing the rotor field does not affect the motor speed but does change the power factor that it presents to the supply. Since the field of the motor is mounted on the rotor, it is necessary to provided sliprings and brushes if the field excitation is to be supplied externally. This can be avoided by using a.c. exciters and rotating field diodes, as in brushless synchronous motors, one arrangement being shown in Figure 14.52. The armature of an a.c. exciter rotates with the

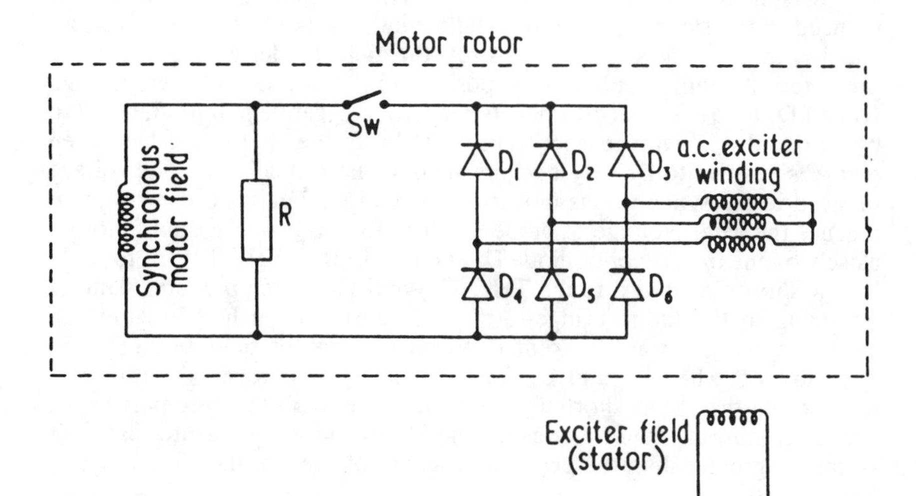

Figure 14.52 Brushless synchronous motor arrangement

synchronous motor field winding, the a.c. output from the exciter being rectified by rotating diodes D_1 to D_6 before being fed to the motor field. During run-up switch S_W is open and the field windings are shorted through resistor R, the a.c. exciter field being open circuit. When the machine nears synchronism centrifuge switch S_W closes, the exciter field now becoming energised so that d.c. is applied to the synchronous motor field and it locks onto the rotating stator field.

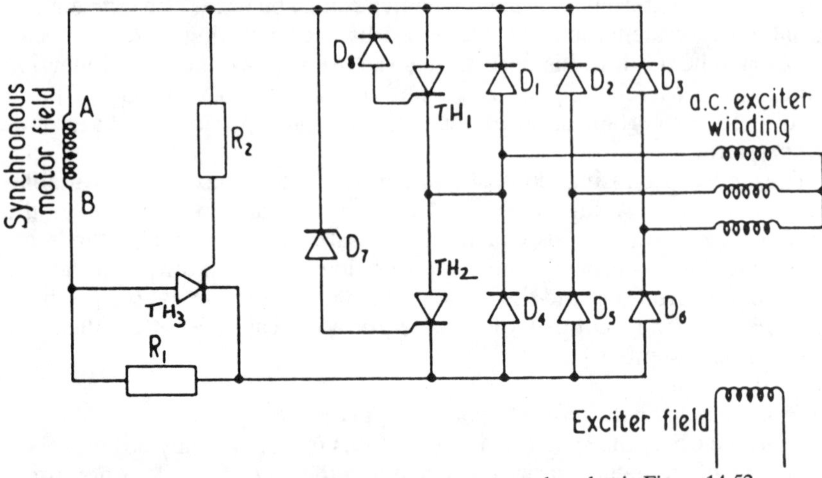

Figure 14.53 A more efficient brushless synchronous motor than that in Figure 14.52

The above system suffers from two disadvantages: (1) it is not fully solid state since switch S_W is used; (2) resistor R is connected in circuit throughout run-up and synchronous operation, so that the motor efficiency is reduced. Figure 14.53 shows a better arrangement. During run-up the a.c. exciter is de-energised as before. An alternating voltage is now induced in the synchronous motor field windings due to induction action, and it can be very large and destroy the field diodes unless these are protected. Assuming end B to be positive to A, current will flow through R_1 and D_1 to D_6. TH_3 is off since its gate is at a negative potential, equal to two diode drops, with respect to its cathode. In the next half cycle, when end A is positive to B, TH_3 is reverse biased and still held off. The voltage induced in the synchronous motor coil now rapidly builds up and when it reaches the zener voltage of diode D_7 thyristor TH_2 is turned on followed closely by the breakover of diode D_8 and the firing of TH_1. The motor field is now short-circuited through TH_1, TH_2 and R_1, which prevents damage occurring to the diode bridge. At speeds close to synchronism the a.c. exciter is energised and current flows into the motor field and TH_1 and TH_2 turn off when D_1 and D_4 first conduct. TH_3 is forward biased and turns on so that R_1 is short-circuited and no longer plays any part in the exciter operation. Zener diodes D_7 and D_8 are chosen to ensure that their voltage is greater than the peak a.c. exciter voltage, so that TH_1 and TH_2 are held off.

With the above scheme the starting and synchronous performance of the machine is identical to a conventional motor with slip rings, although since the efficiency of the rectifier is higher the overall motor efficiency is also greater. However, there is now random synchronism so that pull-in characteristics are inferior to those obtained with slip ring motors and slip frequency relay synchronisation. In the majority of applications this loss of pull-in torque is not critical, and this scheme finds use in systems such as reciprocating motor drives, cement mill motors which start unclutched, and synchronous condensers.

14.3.4.4 Braking

As with d.c. motors, there are three methods for electrically braking a.c. machines: plugging, dynamic braking and regeneration. During plugging the direction of rotation of the motor is reversed, which is done by reversing the direction of the rotating field by interchanging two of the stator winding connections. An induction and reluctance motor can be plugged in this manner, but a synchronous motor can only be plugged if it is fitted with induction windings so that during plugging it behaves as an induction motor. Since the rotor and stator fields are moving in opposite directions, large currents are induced in the motor, resulting in heat loss. The rate of plugging can be varied by controlling the supply voltage or frequency, so as to change its torque, which can be done by using thyristor phase-controlled circuits or frequency changers.

In dynamic braking the motor must act as a generator. For an induction motor to behave as a generator it must be driven above the speed of the stator field, which is not possible during braking if the normal a.c. supply is maintained to the stator. If the stator is supplied with d.c. the excitation necessary to induce currents in the rotor is present and at the same time the rotor exceeds the stator field speed. The operation is as in Figure 14.16, where since the stator poles are now stationary the rotor flux is also stationary. The system no longer behaves as a transformer and load current is not present in the stator. Therefore all the mechanical input is used in generating heat in the rotor, no power being fed back to the d.c. supply. Dynamic braking in a synchronous motor is similar to that used for d.c. machines. Since the rotor field is independent of the stator, all that is required is for the stator supply to be disconnected and braking resistors to be connected across the motor terminals, the rotor field still being maintained.

To regenerate in a synchronous motor the frequency of the supply must be reduced gradually so that the motor, now acting as a generator, is continually in step, the loss in rotor kinetic energy being fed back to the supply. The motor can also be disconnected from the source and its generated output, which will be a.c. whose frequency changes with the motor speed, can be rectified and fed into a d.c. source. An induction motor will generate so long as the rotor speed exceeds that of the stator field, provided this is not d.c. The motor inertia is converted partly into heat in the rotor and is partly fed back to the supply. The larger the difference in rotor and stator field speeds, the smaller the ratio of generated output to input power, and the greater the heat loss in the rotor. Clearly, regeneration into an a.c. source for all types of a.c. motors requires a controlled-frequency source.

14.3.5 A.C. generators

The principal objective in generator control is regulation of its output voltage. Referring to the elementary machine of Figure 14.11, the voltage induced in coil a-b is directly proportional to the rate with which it cuts the stator flux, i.e. to the relative speed between rotor and stator field and to stator field strength. In d.c. and synchronous generators, changing the rotor to stator field speed involves control of the mechanical drive, which is inconvenient. In a synchronous machine a change in speed also produces a change in output frequency, therefore field control is normally used, the power electronic drive system operating in d.c. line control or controlled rectification modes, depending on whether the supply to the controller is d.c. or a.c.

Since the excitation field in an induction generator is rotating, the speed of this field can be controlled by the use of thyristor frequency changers, so as to vary the generated output. It must be remembered that, due to the complex nature of the transformer action involved, only part of the induced rotor current is supplied to the stator output, the rest being converted into rotor heat. There is an optimum relative speed at which the output is a maximum. Excitation control may be used to control the magnitude of the output, although it is normally kept proportional to frequency, to maintain constant flux.

Synchronous generators, called alternators, are popularly used for power generation and these are treated here in further detail. Figure 14.54

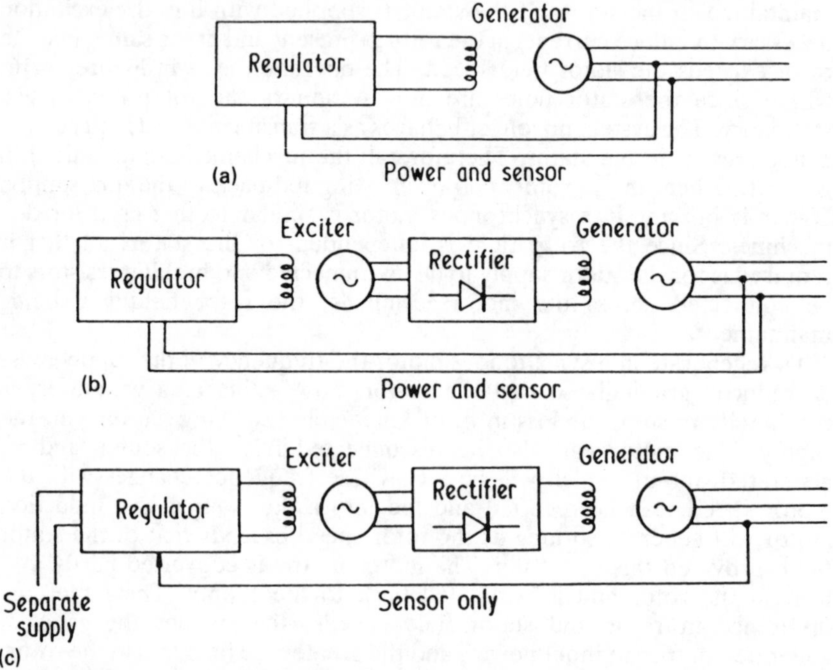

Figure 14.54 Different arrangements for alternator excitation: (a) direct self-excitation; (b) indirect self-excitation; (c) indirect separate excitation

shows several arrangements for alternator excitation, in all cases the regulator converting the input supply, if it is a.c., into d.c., and regulating the amount of the field excitation so as to maintain the output voltage constant, at a predetermined value, irrespective of the load on the machine. Figure 14.54(a) illustrates the simplest arrangement, which is used for smaller machines, the input to the regulator being derived from the a.c. output of the generator itself, these lines therefore performing the dual role of power supply and sensing for feedback voltage control. The regulator is normally a half-controlled single-phase thyristor bridge although half-wave circuits, using a single thyristor and a free-wheeling diode across the field, may be used. For larger machines the arrangement of Figure 14.54(b) is preferred, the regulator now controlling the field of an auxiliary exciter, its armature being mounted on the rotor and supplying the generator field through a rotating rectifier bridge. Both the above systems can be operated from an external supply if it is available, Figure 14.54(c) showing such a system for a large turbo-generator. The feedback line from the generator output is now required only for voltage-sensing purposes and carries no power.

The action of a thyristor regulator, when used in an alternator excitation system, is very similar to usual phase-control systems employed in controlled rectification. The greater the delay angle α in the firing of the thyristors during a half cycle, the lower its d.c. output and therefore there is a fall in alternator voltage, and vice versa. There are several other considerations which modify the simple control loop shown in Figure 14.54. For instance, it is often desirable to introduce a droop in the output voltage characteristic such that the terminal voltage falls with load, to enable better load sharing between parallel-connected machines. Similarly, alternators must be designed so as to give a large output current under fault conditions, which is necessary to operate various circuit trip systems. With a separately excited system this occurs naturally, as in Figure 14.55(a) where the short-circuit current is only limited by the alternator reactance. For self-excitation a fall in terminal voltage results in a reduction of the excitation current and a further fall in the generator output. The characteristic therefore folds back as in Figure 14.55(b) and sufficient current is not available to operate trip circuits. This can be overcome by feeding the alternator field in parallel with the regulator output, with current derived from the alternator lines through a current transformer.

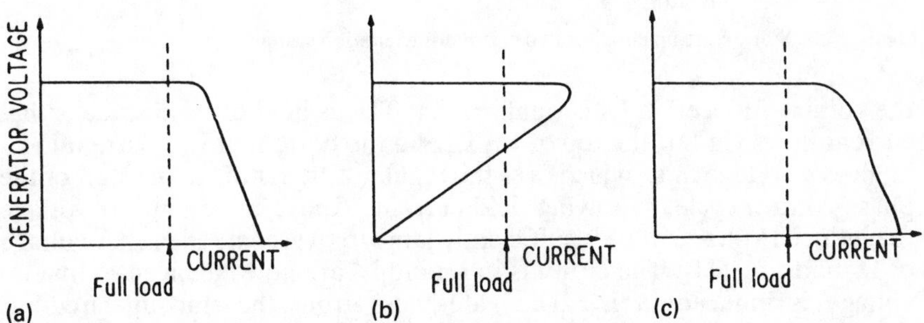

Figure 14.55 Generator characteristics up to short-circuit conditions: (a) separate excitation; (b) self-excitation; (c) self-excitation with current boost

An overload condition will now feed back field excitation and will be self-sustaining, the characteristics of the alternator being modified as shown in Figure 14.55(c). The similarity of Figure 14.55 to Figure 14.2 should be noted.

Another problem encountered with self-excited machines, which does not occur with separate excitation, is that of starting. When the alternator first runs up to speed, its output voltage is very low, being caused primarily by residual field flux, and it will be too small to operate the thyristor gate drive system, so that the thyristors will be held off and the machine will not be excited. To overcome this it is usual to apply continuous gate drive to the thyristors during the period that the alternator output is below a certain critical value. The easiest way of doing this is by a small normally closed relay connected between the anode and gate of the thyristors. Alternatively, a static control system may be used, one type being shown in Figure 14.56. When the a.c. line voltage from the alternator output is low,

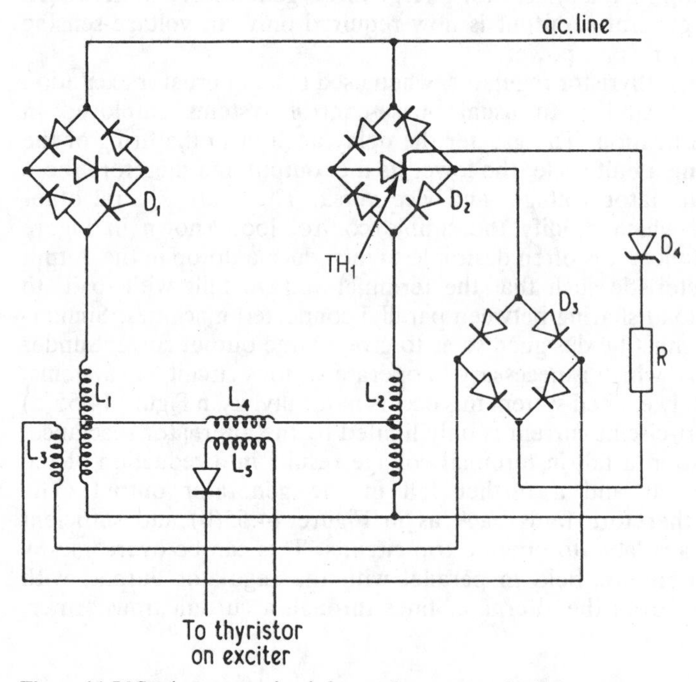

Figure 14.56 Static start-up circuit for an alternator excitation system

the voltage induced in L_3 is small so that TH_1 is held off. This means that current flows via D_1, the top of L_1, L_4 and the bottom of L_2. Current in L_4 induces a voltage in L_5 which fires the regulator thyristors at the start of the phase-control cycle, supplying field current. Above a certain line voltage the induced voltage in L_3 is sufficiently large to overcome the zener voltage of D_4 and fire TH_1. The currents in L_1 and L_2 are now balanced so that no voltage is induced in L_5. The gate pulse from the start-up circuit is therefore no longer applied to the regulator thyristors which now operate under normal phase control.

Electronic excitation circuits are small, robust and have fast response due to the low system lag. If there is a fall in the output voltage of an alternator, say due to the loss across the machine sub-transient reactance when an induction motor is switched on, it is necessary to increase the field current rapidly and therefore raise the output voltage again. This calls for a field current above the normal excitation level. Power electronic control circuits are well suited to provide this field-forcing action and the ratio of maximum to full-load excitation field voltage, called the forcing margin, is usually three.

14.4 Heating and lighting

Heating and lighting applications are frequently considered together since they both involve predominantly resistive loads. Several power electronic circuits can be used, Figure 8.1 already having introduced three variants of single-phase thyristor controllers, with waveforms for phase angle control. If the a.c. input supply has an r.m.s. value of V_{rms} and a peak of V_{pk} then the power P_L supplied to the load for a delay angle of α is given by

$$P_L = \frac{V_{rms}^2}{R}$$

$$= \frac{V_{pk}^2}{2\pi R}(\pi + \tfrac{1}{2}\sin 2\alpha - \alpha) \qquad (14.12)$$

Phase angle control results in a lagging power factor and harmonic generation, so for high current loads its use is usually not allowed by supply authorities. If the thermal time constant of the load is relatively high compared to the input a.c. period, then integral half cycle control can be used to vary the power to the load, as illustrated in Figure 8.18, although this technique cannot be used with incandescent lamps due their low inertia. In this method, if the control switch is on for n half cycles in a total period of m half cycles, then the power supplied to the load is given by

$$P_L = \frac{V_{pk}^2}{2R} \cdot \frac{n}{m} \qquad (14.13)$$

Figure 14.57(a) shows a simple circuit which can be used for phase angle control of a heating load. The capacitor begins to charge through the variable resistor once the input supply passes through its zero voltage point during each half cycle, and when this voltage exceeds the breakover voltage of diac D_1 it conducts and triggers triac TH_1. The delay is controlled by the charging of C and therefore by the value of resistor R.

An alternative circuit for a resistive single-phase load is shown in Figure 14.57(b) where a single thyristor is used within a bridge rectifier. The capacitor charges through resistor R as before, but since the voltage across it is d.c. only a silicon unilateral switch is required, the thyristor turning on when this breaks over. The range of control obtained with both the circuits given in Figure 14.57 is limited, since the supply voltage has to rise to a sufficiently high value before the trigger diodes break over. Zener diodes can be used in place of the trigger devices but then the magnitude of the

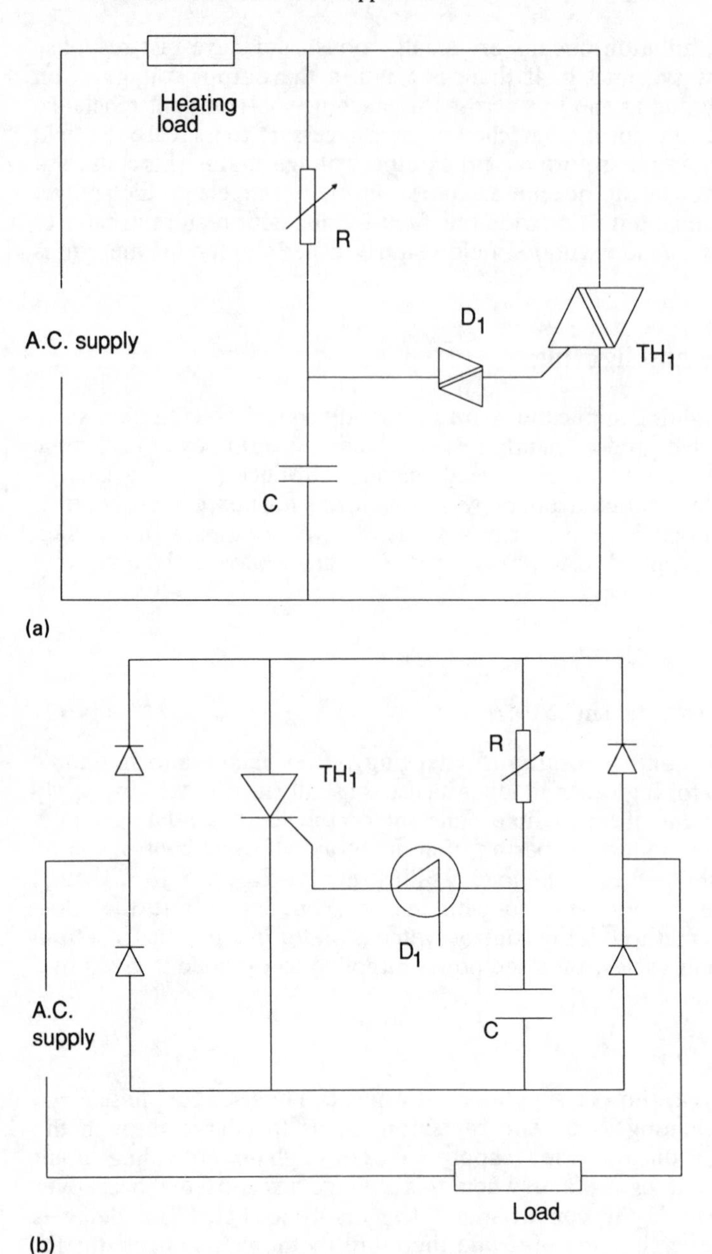

Figure 14.57 Simple phase-control circuits for resistive loads: (a) triac arrangement; (b) single-thyristor and diode bridge arrangement

trigger pulse is reduced, once the diodes start conducting, and the range of control is also less. Power loss in the controlling potentiometers of both these circuits is low, when once the power device turns on the voltage across the potentiometer falls to a very low value, the loss being minimum at full load, the delay angle being small.

Three-phase resistive loads can be controlled by using three-phase bridge circuits, fully controlled and half-controlled systems being shown in Figure 8.6. The power into the load will vary with the firing angle delay, and as this delay changes the number of phases to which power is simultaneously supplied also varies between two and three. For a star-connected load and a fully controlled bridge, the power is given by equations (14.14) to (14.16) for various delay angles, V_{pk} being the peak line voltage, as before:

$$P_L = \frac{3\,V_{pk}^2}{2\pi R}\left[\frac{\pi}{3} + \frac{\sin 2\alpha}{4} - \frac{\alpha}{2}\right] \tag{14.14}$$

for $0 < \alpha < 60°$

$$P_L = \frac{3\,V_{pk}^2}{4\pi R}\left[\frac{\pi}{3} + \frac{3}{4}\sin 2\alpha + \frac{\sqrt{3}}{4}\cos 2\alpha\right] \tag{14.15}$$

for $60° < \alpha < 90°$

$$P_L = \frac{3\,V_{pk}^2}{4\pi R}\left[\frac{5\pi}{6} + \frac{\sqrt{3}}{4}\cos 2\alpha + \frac{1}{4}\sin 2\alpha - \alpha\right] \tag{14.16}$$

for $90° < \alpha < 150°$

Similar equations can be derived for a half-controlled bridge, controlling a star-connected load, and these are given by equations (14.17) to (14.19), the delay angles at which the various equations applying now being different:

$$P_L = \frac{3\,V_{pk}^2}{2\pi R}\left[\frac{\pi}{3} + \frac{\sin 2\alpha}{8} - \frac{\alpha}{4}\right] \tag{14.17}$$

for $0 < \alpha < 90°$

$$P_L = \frac{3\,V_{pk}^2}{4\pi R}\left[\frac{11\pi}{12} - \alpha\right] \tag{14.18}$$

for $90° < \alpha < 120°$

$$P_L = \frac{3\,V_{pk}^2}{8\pi R}\left[\frac{7\pi}{6} + \frac{\sin 2\alpha}{4} - \frac{\sqrt{(3)}\cos 2\alpha}{4} - \alpha\right] \tag{14.19}$$

for $120° < \alpha < 210°$

Induction heaters are usually operated at a high frequency, in the region of 1–10 kHz, and they present a predominantly inductive load. They are usually driven from d.c. link inverters, of the type described in Chapter 13. Because of their high inductance a capacitor can be chosen which, when connected in series with the heater load, will cause resonance at the inverter frequency, as shown in Figure 13.18. Not only does this ensure that the inverter thyristors are commutated by series capacitor techniques, but a sine wave is also applied to the load.

Operating the induction heater from a d.c. link inverter requires that the a.c. input supply is first rectified and smoothed, to give the d.c. link voltage. Alternatively, cycloinverter arrangements can be used to provide the high-frequency output without the need for a d.c. link, as described in section 10.6 and shown in Figure 10.20, a centre-tapped heating coil being used. Generally, d.c. link inverters provide a more efficient and cheaper drive system and are more frequently used than cycloinverters.

14.5 Electrochemical

There are many electrochemical applications for power semiconductors although only electroplating, electrolysis and electrochemical forming are considered here. In electroplating a layer of metal is deposited from a metal anode onto the target, both of these being immersed in a bath of a suitable electrolyte. The current densities required for the process are low, the amount of metal deposited being directly proportional to the magnitude of this current and the time for which it flows. The supply must be d.c. although ripples in the current are acceptable, so a three-phase rectified source can be used. Generally, the magnitude of the voltage is also low, varying from 5 V to 100 V, and this is usually provided by the use of a thyristor bridge.

Electrolysis consists of extracting metals such as aluminium from a solution. Usually a low d.c. voltage is required, in the region of 5 V, but as the process proceeds this needs to rise to about 50 V. Several such electrolysis baths can be connect in series, so that the overall voltage requirements can reach in the order of 1 kV. High currents are necessary, often between 50 kA and 100 kA, and these are obtained by three-phase 12-pulse or 24-pulse rectifier systems, using parallel-connected thyristors. Current feedback is used to control the voltage and current supplied to the load, by adjusting the power thyristor conduction periods.

Some electrolysis processes require periodic reversal of the current flow, for a few seconds every few minutes, to prevent film formation at the electrodes, and this can be done by a reversing bridge arrangement.

Electrochemical forming is the reverse process to electroplating, where material is selectively removed from areas of the target. Low voltages are again required, in the region of 10 V to 30 V, and the current drawn by the load is high, in the region of several kilo-amperes. Therefore the drive circuits for these loads are similar to the d.c. power supplies used for electrolysis.

Appendix 1

List of symbols

Symbol	Meaning
a	cross-sectional area
A	amperes
A_l	absorption loss
B	lux density
B_l	multiple-reflection loss factor
c	specific heat
C	capacitance
C_d	depletion capacitance
C_{in}	input capacitance
c_{th}	thermal capacity
d	thickness or depth
D	diameter
D_n	electron diffusion constant
D_p	hole diffusion constant
dT	incremental temperature difference
di/dt	rate of change of current
dv/dt	rate of change of voltage
E	electric field or motor back e.m.f.
E_s	shielded electric field
E_u	unshielded electric field
E_{ST}	stored energy
f	frequency
f_{max}	maximum frequency
f_r	ripple frequency
f_S	supply frequency
F_C	commutation factor
h_{FE}	transistor common emitter gain
H	magnetic field
H_f	harmonic factor
H_s	shielded magnetic field
H_u	unshielded magnetic field

I current
I_a motor armature current
I_B transistor base current
I_{BR} transistor reverse base current
I_C transistor collector current
$I_{C(pk)}$ peak collector current
I_{CBO} collector base current with no emitter current
I_{CEO} collector emitter current with no base bias
I_{CO} transistor leakage current
I_D drain current of FET
$I_{D(av)}$ average diode current
$I_{D(rms)}$ r.m.s. diode current
I_{DRM} forward-blocking current
I_{DSM} forward-blocking transient current
I_f motor field current
I_F forward current
I_G gate current
I_{GD} gate non-trigger current
I_{GQM} gate turn-off current gain
I_{GT} gate trigger current
I_H holding current
I_l leakage current
I_{ld} leakage current (diffusion of minority carriers)
I_{lg} leakage current (charge generation)
I_L load current or latching current
$I_{L(av)}$ mean load current
$I_{L(min)}$ minimum load current
$I_{L(pk)}$ peak load current
$I_{L(rms)}$ r.m.s. load current
I_{MAG} magnetising current
I_p peak point current of UJT
I_{pk} peak line current
I_{rms} r.m.s. line current
I_R reverse current
I_{RRM} reverse-blocking current
I_{RSM} reverse-blocking transient current
I_S supply current
$I_{S(av)}$ mean supply current
$I_{S(min)}$ minimum supply current
$I_{S(pk)}$ peak supply current
$I_{T(av)}$ conducting average current
$I_{T(rms)}$ conducting r.m.s. current
$I_{TH(av)}$ average thyristor current
$I_{TH(pk)}$ peak thyristor current
$I_{TH(rms)}$ r.m.s. thyristor current
I_{TM} maximum forward-conducting current
I_{TSM} surge current
I_V valley point current of UJT
I_Z zener current

k constant
K constant
K_f flux constant
K_n speed constant
K_t torque constant

l length
L inductance
L_c critical inductance
L_e effective inductance
L_{PRI} primary inductance
L_{SEC} secondary inductance

m mass

n_e intrinsic electron concentration
$\overline{n}_n$ concentration of minority carriers in *n* region
$\overline{n}_p$ concentration of minority carriers in *p* region
N number of cycles or machine speed

p number of pulses per cycle
P power
P_c power (heat) loss due to conduction
P_{fl} power (heat) loss due to forced cooling, laminar
P_{ft} power (heat) loss due to forced cooling, turbulent
P_n power (heat) loss due to natural convection
P_r power (heat) loss due to radiation
P_{COM} power loss per commutation
$P_{G(av)}$ average gate power
P_{GM} peak gate power
$P_{GR(av)}$ average reverse gate power
P_{GRM} peak reverse gate power

q charge on an electron
Q_M maximum heat transport

r_{B1} emitter-base 1 resistance
r_{B2} emitter-base 2 resistance
r_{BB} interbase resistance
$r_{DS(on)}$ drain-source on resistance
R resistance
R_a motor armature resistance
R_{bk} bulk resistance
R_{ch} channel resistance
R_{ex} external resistance
R_f motor field resistance
R_l reflection loss
R_{pk} peak resistor value
R_{th} thermal resistance
R_{CE} collector emitter resistance of a transistor

R_E shielding effectiveness (electrical)
R_H shielding effectiveness (magnetic)
R_θ thermal resistance
$R_{\theta CS}$ thermal resistance case to sink
$R_{\theta jA}$ thermal resistance junction to ambient
$R_{\theta jC}$ thermal resistance junction to case

S shielding effectiveness or motor slip
S_m slip at maximum motor torque

t time
t_c switch closed time
t_d delay time
t_f fall time
t_{gr} gate recovery time
t_n reactor negative saturation time
t_o switch-open time
t_p reactor positive saturation time
t_q reverse bias time
t_r rise time
t_{rf} forward turn-on time
t_{rr} reverse recovery time
t_s storage time
t_{OFF} total turn-off time
t_{ON} total turn-on time
T temperature, or periodic time, or motor torque
T_c case temperature
T_j operating junction temperature
T_m maximum motor torque
T_{stg} storage temperature
T_A ambient temperature

v_a air velocity
v_c carrier velocity
V voltage
V_{av} average or d.c. voltage
$V_{av(\alpha)}$ average or d.c. voltage with α phase delay
V_c capacitor voltage
V_{fp} field plate voltage
V_g forbidden energy layer potential
V_j voltage across junction
V_n nth harmonic voltage
V_p peak point voltage of a UJT
V_{pk} peak voltage
V_{rms} r.m.s. voltage
$V_{rms(n)}$ r.m.s. voltage of the nth harmonic
$V_{rms(T)}$ total r.m.s. voltage
V_v valley point voltage of a UJT
V_B junction breakover voltage, or battery voltage
$V_{BE(sat)}$ base-emitter saturation voltage

$V_{(BO)}$ breakover voltage
V_{GD} gate non-trigger voltage
$V_{CE(sat)}$ collector-emitter saturation voltage
V_{DRM} repetitive forward blocking voltage
V_{DS} gate-drain voltage of a FET
V_{DSM} non-repetitive peak on-state voltage
$V_{DS(P)}$ pinch-off drain-source voltage
V_{GQM} gate turn-off voltage
V_{GS} gate-source voltage of a FET
V_L load voltage
$V_{L(rms)}$ r.m.s. load voltage
V_{PRIM} transformer primary voltage
V_{RRM} repetitive peak reverse voltage
V_{RSM} non-repetitive peak reverse voltage
V_S supply voltage
V_T threshold voltage
$V_{T(av)}$ average forward voltage drop
V_{TM} forward voltage drop
V_Z zener voltage

W width
W_{bd} width of depletion layer at breakdown
W_c width of channel
W_d width of depletion layer
W_n width of *n* region
W_p width of *p* region
W_i width of *i* region

X reactance

Z impedance
Z_c conductor impedance
Z_{ca} capacitor impedance
Z_d dynamic impedance
Z_{in} inductor impedance
Z_w impedance of electromagnetic wave
Z_T transfer impedance
$Z_{\theta(t)}$ transient thermal impedance
$Z_{\theta jc(t)}$ transient thermal impedance junction to case
$Z_{\theta jA(t)}$ transient thermal impedance junction to ambient

α common base gain, or delay angle
α_a average ionisation coefficient
α_e ionisation coefficient for electrons
α_m minimum delay angle
α_n delay angle (negative group)
α_p ionisation coefficient for holes, or delay angle (positive group)

β gain
β_{OFF} gate turn-off gain

Δ	skin depth of conductor
ε	emissivity
η	efficiency, or intrinsic stand-off ratio of a UJT
θ	angle
λ	wavelength
μ	permeability, or waveform overlap angle
γ	conductivity
τ	filter attenuation factor
τ_s	space charge generation lifetime
ϕ	power factor angle, or flux
ω	angular frequency ($2\pi f$)

Appendix 2

Glossary of terms

Anti-parallel: The connection of two unidirectional power devices in parallel, such that they conduct in opposite half cycles. This is also known as an inverse parallel or a back-to-back connection.

Avalanche breakdown: The conduction mechanism within a semiconductor when minority carriers attain sufficient energy to cause a chain reaction, resulting in very rapid build-up of current.

Bi-directional device: A power semiconductor capable of conducting in both directions.

Blocking voltage: A measure of the voltage which a power semiconductor can withstand, in the forward or reverse direction, without breaking down.

Breakdown voltage: The voltage at which a power semiconductor loses its voltage-blocking capability.

Bridge converter: An arrangement of power rectification devices which provides a d.c. output from an a.c. input, without requiring a centre-tapped load. (See also **push–pull converter**.)

Burst firing: See **Integral cycle control**.

Bypass: A component or a circuit used to carry the load current, so bypassing the main switching power device.

Chopper: A term commonly used for a circuit which regulates the amount of power flowing from a d.c. source to a d.c. load.

Commutation angle: The angle of overlap between phases of the a.c. supply, caused by the reactance in the supply, when conduction is changing over between the phases. This is also known as the overlap angle.

Commutation diode: See **Free-wheeling diode**.

Commutation process: The process of transferring current from one element to another, for example from one coil to another in a d.c. machine. Usually refers to the process used to turn off a power semiconductor by transferring the load current to a commutation diode.

Controlled rectification: The process of regulating the amount of power which flows from an a.c. supply to a d.c. load, this usually being achieved by bridge or push–pull connected converters.

Crowbar circuit: A circuit used to protect the load from overvoltages resulting from a d.c. supply by shorting the output of the supply when overvoltages are detected.

Current-fed inverter: A d.c. to a.c. converter which uses a high-valued reactor in the supply line, such that the current from the d.c. supply is substantially constant.

Cycloconverter: A frequency changer which converts a.c. at one frequency to a second, lower, frequency, without first going through a rectification stage. This is also referred to as a direct a.c. frequency converter.

Cycloinverter: A frequency changer which converts a.c. at one frequency to a second, higher, frequency, without first going through a rectification stage.

D.C. link frequency changer: See **Inverter**.

Delay angle: The delay between the start in conduction of a power semiconductor and the instance when the supply voltage across it begins to go positive. This is also referred to a phase-control angle or firing angle.

Depletion layer: The layer between a *p*-type and *n*-type region in a semiconductor which has very few free charge carriers.

d*i*/d*t*: The rate of change of current through a power semiconductor which, if it exceeds a certain maximum value, could destroy the device.

Doping: The process of adding impurities to a semiconductor to produce a *p*-type or *n*-type material.

dv/dt: The rate of rise of voltage across a power semiconductor which, if exceeded, could cause it to turn-on.

Feedback: The process of sampling the load voltage or current and feeding this back to the control device to enable exact regulation. Also the process of feeding back energy from the load to the supply.

Firing angle: See **Delay angle**.

Firing circuit: The circuit which provides current and voltage to turn on a power semiconductor, usually a thyristor or a triac. This is also called a gating circuit or trigger circuit.

Forced commutation: The use of external circuitry, such as a charged capacitor, to turn off a power semiconductor switch. This is primarily used in thyristor choppers and inverters. (See also **Natural commutation**.)

Free-wheeling diode: The diode which is placed across the load to carry the load current, when the power semiconductor switch turns off. This is also referred to as a commutation diode since it assists in commutation, or turn-off, of the power switch.

Full wave: An a.c. to d.c. converter in which the a.c. input current flows in both directions.

Fully controlled: An a.c. to d.c. converter in which the power can flow in either direction, this usually being achieved by the use of controllable devices, such as thyristors, in all the converter arms.

Gating circuit: See **Firing circuit**.

Half controlled: An a.c. to d.c. converter in which the power can only flow from the a.c. supply to the d.c. load. This is usually caused by converters in which only half the components are controllable (thyristors), or where a free-wheeling diode is place across the load.

Half wave: An a.c. to d.c. converter in which the current in the a.c. line flows in one direction only.

Holding current: The value of the current flowing through a semiconductor switch below which it will return to its off state.

Integral cycle control: A method for regulating the a.c. power to the load by controlling the number of whole or half cycles of the supply. This is also known as burst firing, and uses zero crossing control techniques.

Inverse parallel: See **Anti-parallel**.

Inverter: A converter which changes d.c. input to a.c. output. The d.c. may be derived from a battery, in which case forced commutation components are required for the power switches if they are thyristors, or it may be the d.c. energy from the load being fed back into an a.c. supply, as when a.c. to d.c. converters are operating in an inversion mode. The operating mode is now natural commutation.

Latching current: The value of current needed to ensure that a thyristor remains on once its gate drive has been removed.

Leakage current: The current which flows through a power semiconductor when it is in the off state. This can flow either in the forward or reverse direction, depending on the polarity of the voltage across the device.

Line commutation: See **Natural commutation**.

Mark-to-space control: Control of the amount of power delivered from a d.c. source to a d.c. load by varying the ratio of the power switch open to closed time in any cycle.

Natural commutation: The use of the energy available from the a.c. supply to turn off the power semiconductors, usually in an a.c. to d.c. thyristor converter operating in an inversion mode, or in a cycloconverter. This is also known as line commutation.

Overlap angle: See **Commutation angle**.

Peak inverse voltage: The maximum value of the reverse voltage which can be applied across a power semiconductor.

Phase control: A method for controlling the amount of power delivered to the load by varying the delay angle. (See **Delay angle**.)

Pulse number: The ratio of the number of output cycles to the input supply frequency for an a.c to d.c. converter.

Pulse-width modulation: A method for varying the mark-to-space ratio of the output voltage waveform during a cycle so as to minimise the magnitude of the harmonics in the output.

Push–pull converter: An arrangement for a power converter which uses a centre-tapped load. (See also **Bridge converter**.)

Regenerative gate: A thyristor structure which enables it to handle high rates of current rise.

Reverse recovery time: The time needed for a forward-conducting device to regain its blocking capability after it has been reverse biased. During this period its impedance is low and it can pass a large amount of reverse recovery current.

Ripple current: The a.c. component of current present on a d.c. supply, which has not been removed by filtering.

Safe operating area: The current and voltage area of a power transistor characteristic in which it must remain if it is to avoid thermal runaway. (See **Thermal runaway**.)

Shorted emitter: A thyristor construction technique which enables it to withstand rapid rate of rise of voltage across it.

Snubber circuit: A resistor–capacitor circuit connected so as to prevent the voltage from rising too rapidly across the power semiconductor.

Staggered phase carrier cancellation: A method of voltage control for an inverter in which the a.c. harmonics in the output are reduced.

Surge suppressor: A device which limits the value of the maximum voltage across a power semiconductor.

Thermal resistance: A measure of the increase in temperature of the semiconductor junction as the power dissipation within it increases.

Thermal runaway: A mechanism within a semiconductor in which the power increase results in a build-up of temperature which further increases the power, so that the device is rapidly destroyed.

Trigger circuit: See **Firing circuit**.

Voltage multiplier: A circuit which produces an output voltage which has a magnitude several times that of the input.

Waveform synthesis: A method for controlling the output voltage from an inverter, to approximate it to a sine wave and so reduce the amount of harmonics.

Zero crossing control: A technique for detecting the occurrence of the zero voltage points in the input a.c. voltage and of turning on the power switches at these instances, so as to minimise the peak voltage at which the devices switch and therefore reduce the generated radio frequency interference.

Bibliography

This section provides a list of some of the many books which are available, covering the field of power electronics. The comments made under each title are the author's personal views and because requirements on styles and depth of coverage can vary, readers may find that their views differ from his. A browse in a library is therefore recommended before any books are purchased.

B. R. Pelly (1971). *Thyristor phase-controlled converters and cycloconverters: operation, control and performance,* John Wiley, New York.

A very good reference book for those who wish to obtain a better understanding of cycloconverters, but not recommended as a general text on power electronics.

William F. Waller (ed.) (1972). *Rectifier circuits,* Macmillan, London.

A useful overview of the different elements involved in power electronics.

Raymond Ramshaw (1973). *Power electronics: thyristor controlled power for electric motors,* Chapman and Hall, London.

Describes the characteristics of thyristors and their application to a.c. and d.c. motor control. Useful for engineers working in the field of motor control.

S. B. Dewan and A. Straughen (1975). *Power semiconductor circuits,* John Wiley, Chichester.

Provides a fairly detailed mathematical analysis of the various thyristor circuits, but only a superficial introduction to thyristor characteristics and applications.

Sorab K. Ghandhi (1977). *Semiconductor power devices: physics of operation and fabrication technology,* John Wiley, New York.

Contains a detailed mathematical analysis of the characteristics and manufacture of diodes, transistors and thyristors. Does not cover circuits and therefore of use to those who are primarily interested in power component technology.

David Finney (1980). *The power thyristor and its applications,* McGraw-Hill, Maidenhead.

Deals with power thyristors only. Device characteristics are briefly introduced although it provides a useful introductory text to circuit applications. The book contains a large section on thyristor applications.

Graham J. Scoles (1980). *Handbook of rectifier circuits,* Ellis Horwood, Chichester.

Contains a useful collection of rectifier applications, especially multiplier circuits. No other power semiconductors are covered.

Richard A. Pearman (1980). *Power electronics: solid state motor control.* Reston Publishing Company Inc. (287 pages).

Provides an introduction to power electronic applications, but primarily with regard to thyristor circuits. Useful supplementary reading.

Cyril W. Lander (1981). *Power electronics,* McGraw-Hill, Maidenhead.

Covers the whole field of power electronics, although the emphasis is on power thyristors. Provides some useful worked examples and is recommended for supplementary reading.

Edwin S. Oxner (1982). *Power FETs and their applications,* Prentice-Hall, Englewood Cliffs, NJ.

Contains a detailed analysis of power FET characteristics and is of value to engineers who are primarily interested in these components.

B. M. Bird and K. G. King (1983). *An introduction to power electronics,* John Wiley, Chichester.

A good all-round text on power electronics and recommended for supplementary reading. It is weak on practical applications and on general power semiconductor characteristics, since it covers mainly thyristors and triacs.

M. Kubat (1984). *Power semiconductors,* Springer-Verlag. Berlin.

The majority of the book deals with the characteristics of power semiconductors, rather than its circuits or applications. The treatment is highly mathematical, and at times it is difficult to understand.

Albert Kloss (1984). *A basic guide to power electronics,* John Wiley, Chichester.

Primarily covers power thyristors used in bridge rectifier circuits. The treatment is analytical and generally this is not a beginner's book, although it contains some useful information.

S. K. Datta (1985). *Power electronics and control,* Reston Publishing Company, New York.

Covers the whole field of power electronics although in many instances this is at a fairly superficial level. Useful as an introductory text, but the emphasis is on thyristors and their applications, other power semiconductors being omitted.

Richard G. Hoft (1986). *Semiconductor power electronics,* Van Nostrand Reinhold, Wokingham.

Provides a mathematical treatment of thyristor circuits, and some beginners may find it a difficult book to follow. Weak on practical applications.

Index